Clinical Electroencephalography and Topographic Brain Mapping

F.H. Duffy V.G. Iyer W.W. Surwillo

Clinical Electroencephalography and Topographic Brain Mapping

Technology and Practice

With 192 Illustrations, 14 in Color

Springer-Verlag New York Berlin Heidelberg
London Paris Tokyo

FRANK H. DUFFY, M.D.
Associate Professor of Neurology, Harvard Medical School, Director of Developmental Neurophysiology, Director of the BEAM Laboratory, The Children's Hospital, Boston, MA 02115, USA

VASUDEVA G. IYER, M.D., D.M.
Chief of Division of Clinical Neurophysiology and Professor of Neurology, University of Louisville School of Medicine, Louisville, KY 40292, USA

WALTER W. SURWILLO, PH.D.
Professor and Director of Psychophysiological Research, Department of Psychiatry and Behavioral Sciences, University of Louisville School of Medicine, Louisville, KY 40292, USA

Library of Congress Cataloging-in-Publication Data
Duffy, Frank H.
Clinical electroencephalography and topographic brain mapping.
Includes bibliographies.
1. Electroencephalography. 2. Brain mapping.
I. Iyer, Vasudeva, G. II. Surwillo, Walter W.
III. Title. [DNLM: 1. Brain Mapping—methods.
2. Electroencephalography—methods. 3. Evoked Potentials. WL 150 D858c]
RC386.6.E43D84 1989 616.8'047547 88-32699

Printed on acid-free paper.

Typeset by Publishers Service, Bozeman, Montana.
Printed and bound by Edwards Brothers, Inc., Ann Arbor, Michigan.

9 8 7 6 5 4 3 2 1

ISBN-13: 978-1-4613-8828-9 e-ISBN-13: 978-1-4613-8826-5
DOI: 10.1007/978-1-4613-8826-5

Preface

Electroencephalography is truly an interdisciplinary endeavor, involving concepts and techniques from a variety of different disciplines. Included are basic physics, neurophysiology, electrophysiology, electrochemistry, electronics, and electrical engineering, as well as neurology. Given this interesting and diverse mixture of areas, the training of an EEG technician, a neurology resident, or an EEG researcher in the basics of clinical electroencephalography presents an uncommon challenge.

In the realm of technology, it is relatively easy to obtain a technically adequate EEG simply by learning to follow a protocol and by correctly setting the various switches on the EEG machine at the right time. But experience has shown that the ability to obtain high-quality EEGs on a routine, day-to-day basis from a wide variety of patients requires understanding and knowledge beyond what is learned by rote. Likewise, knowledge above and beyond what is gained by simple participation in an EEG reading is necessary to correctly and comprehensively interpret the record. Such knowledge comes from an understanding of the basic principles upon which the practice of clinical EEG is founded—principles that derive from the various disciplines cited.

While it is clear that some understanding of each of these disciplines plays an important role in the successful training of an EEG technologist, neurology resident, or EEG researcher, the depth and extent of the understanding pose a dilemma. How much and what kind of material should go into an introductory text?

The authors have attempted to meet the challenge of this dilemma in three ways. *First* of all, the text emphasizes concepts. For example, in discussions of Ohm's law the reader is not burdened with irrelevant computations of circuit parameters. Instead, the text shows how Ohm's law provides the basis for understanding EEG-based problems such as why low impedance leads are essential and how a modern impedance meter is able to measure the impedance of a single electrode by hooking it up in series with all the other EEG electrodes connected in parallel. *Secondly*, the text leaves out all but essential detail. Aside from material that is of historical interest, the criterion for inclusion is whether the material is essential for understanding the concepts. At the same time, special care was taken to avoid the risk of becoming simplistic. *Thirdly*, the text focuses on topics that are directly relevant to the recording of EEGs and to the understanding of the fundamental principles upon which EEG interpretation is based. A case in point is the Chapter 12 discussion of the principles of localization.

The text is intended for a wide and varied audience interested in electroencephalography. It requires no special knowledge beyond a familiarity with simple algebra and the elements of biology. To simplify presentation of the material, references to relevant literature, for the most part, have been left out. Chapter 19 and Appendix 7 are notable exceptions, mainly because they deal with newly-emerging areas of interest for which reference sources may not be readily available.

Chapters 1–9 comprise a useful primer for the beginner learning to record clinical EEGs for the first time. Those with some previous experience in EEG should also find these chapters useful, primarily as a review of essentials or a fresh way of thinking about the topics concerned. EEG technologists preparing for Board Examinations will find Chapters 10, 11, and 12 and Appendix 2 particularly helpful.

Chapters 10–16 and 21 were written with neurology residents who are just starting their EEG training in mind. The normal features of the EEG and some of the better-known, most-frequently encountered abnormalities are discussed and illustrated in Chapters 14 and 15. Chapter 21 briefly discusses the use of EEG in clinical diagnosis and its relationship to other neurologic tests; but detailed questions of interpretation are not taken up. Such matters are considered in more advanced, specialized texts. However, taken together with Chapters 1 and 2—which deal with background material and basic electrical concepts—this material provides the essentials for training of neurology residents as well as the elementary basics for training of EEG fellows.

The text includes a chapter each on the topics of seizure monitoring and average evoked potentials. In recent years, these have become important tools for the neurodiagnostician. A unique feature of the text is the inclusion of two chapters dealing with the topographic mapping of brain electrical activity. These chapters, which take up the basics of this exciting new method, will be of interest to anyone doing research in EEG and cortical evoked potentials as well as to clinical neurologists seeking new ways of interpreting the electrical activity of the brain.

Frank H. Duffy
Vasudeva G. Iyer
Walter W. Surwillo

Contents

Chapter 1

Brain Electrical Activity: An Introduction to EEG Recording

The ability to generate electrical activity is a common if not ubiquitous property of living tissue. The electrical activity produced by the heart and recorded in the electrocardiogram (ECG) and the electrical activity generated by muscles and recorded in the electromyogram (EMG) are familiar examples. Less well known is the fact that skin, stomach, and gut also produce electrical activity. Electrical activity has even been observed in plants. In view of this knowledge, it is not surprising that the brain should also generate electrical activity.

Historical Perspective

The existence of electrical activity and its association with life processes has intrigued and mystified scholars for more than two centuries. In the late 18th century an Italian physiologist, Luigi Galvani, made some remarkable observations while working with frogs' legs. Galvani noticed that the leg of a frog would contract violently when the nerve going to the muscle was touched by a brass or copper wire that was connected to an iron wire that made contact with the muscle. Galvani believed the muscle contracted because it was stimulated by "animal electricity" present in the tissue and conducted over the metal wires and nerve. But a contemporary scholar, the physicist Allesandro Volta, proved that Galvani's explanation of the phenomenon was wrong.

Volta demonstrated that the electrical current in Galvani's experiment was produced not by animal tissue but by the contact of the two dissimilar metals placed in a moist or liquid environment. Out of this work the electric battery or Volta's pile was born. The great significance of Volta's finding for the scientific community of the times can be appreciated by the fact that in the early 1800s, Volta was called to Paris to demonstrate his discovery to Napoleon Bonaparte. The events of this dramatic occasion were recorded in a painting by the French artist Alexandre Fragonard.

Despite the apparent setback, scientific interest in animal electricity or bioelectricity, as we call it today, continued. The realization that electrical activity of living tissue could be used as a sign of its function came in the mid 19th century. In 1848 Du Bois-Reymond, a German physiologist, demonstrated that an electrical signal occurred concomitantly with the passage of a nerve impulse. Of his remarkable discovery, Du Bois-Reymond said, "If I do not greatly deceive myself, I have succeeded in realizing in full actuality the hundred years' dream of physicians and physiologists, to wit the identification of the nervous principle with electricity."

Du Bois-Reymond's research served as an incentive to Richard Caton, a British physiologist. Caton reasoned that if activity in a peripheral nerve was accompanied by electrical activity, then a similar phenomenon might also occur in the brain. In 1875 Caton reported on some experiments done with monkeys and rabbits. These experiments demonstrated the existence of feeble electric currents in the brains of the animals. Nearly 50 years passed, however, before similar attempts were made to find out whether the human brain also generated electric currents.

Hans Berger, a German clinical neuropsychiatrist at the University of Jena, was the first to record the electrical activity of the human brain. Berger coined the term "elektrenkephalogramm," which is the German equivalent of electroencephalogram or EEG. Publication of his findings in 1929 followed years of careful, painstaking research in which he attempted to prove the cerebral origin of the phenomenon. However, his modest conclusion, "I therefore, indeed, believe that I have discovered the electroencephalogram of man and that I have published it here for the first time," was met with disbelief and mistrust. It was not until 1934, when Adrian and Matthews in England repeated Berger's experiments and confirmed his observations, that his work became accepted by the scientific community.

Berger's 14 published articles on the electrical activity of the human brain contain a large number of original observations on the EEG. Indeed, Berger was the first to observe and accurately describe many of the features of the EEG that we know today. He demonstrated that brain electrical activity consists more or less of a mixture of rhythmic, sinusoidal-like fluctuations in voltage having a frequency of about 1 to 60 oscillations per second. The waves most easily recognized had a frequency of about 10 oscillations per second, and these he called *alpha*. Alpha waves, he reported, tended to disappear with attention. Waves of frequencies greater than 15 oscillations per second were designated *beta*. Berger observed that the EEG had different features in neurological disorders such as epilepsy, trauma, and tumors. He was the first to record an epileptic seizure.

Following Berger's pioneering work, interest in electroencephalography and brain electrical activity—"brain waves"—became widespread in the late 1930s and 1940s. Frederick and Erna Gibbs, Hallowell and Pauline Davis, Donald Lindsley in the United States, Herbert Jasper in Canada, and W. Grey Walter in England, to name a few of the leaders, showed the importance of the EEG and its application in neurology and neurosurgery. Since that time, the EEG has become a routine clinical procedure of considerable diagnostic value as well as a powerful research tool in the neurosciences.

Recording Bioelectric Activity

There is hardly anyone who has not watched an ECG being recorded or who has not seen the short strip of paper upon which the ECG tracing appears. Figure 1.1 shows a sample of the ECG. The horizontal axis in the tracing represents time whereas the vertical axis represents amplitude or intensity of the electrical activity from the heart; the latter is measured in millivolts (thousandths of a volt, abbreviated mV). Like the pulse itself, the ECG is a transient phenomenon; a spike-like burst of electrical energy that occurs periodically, about once every second in adults. Note in the illustration that the ECG consists of some rapid upward and downward deflections followed by a single, more prolonged deflection.

The ECG tracing, of course, is taken by making an electrical connection between the patient's body and the ECG machine. This is accomplished by attaching *electrodes*—wires with some kind of conducting medium on one end—to portions of the body in the vicinity of the heart. The other ends of the wires are plugged into the ECG machine. With all electrodes connected in this manner, the ECG machine is turned on and the millivolt signals generated by the heart are amplified some 1,000 times until they are sufficiently large to drive the writer units that trace out the spikes on a moving strip of paper.

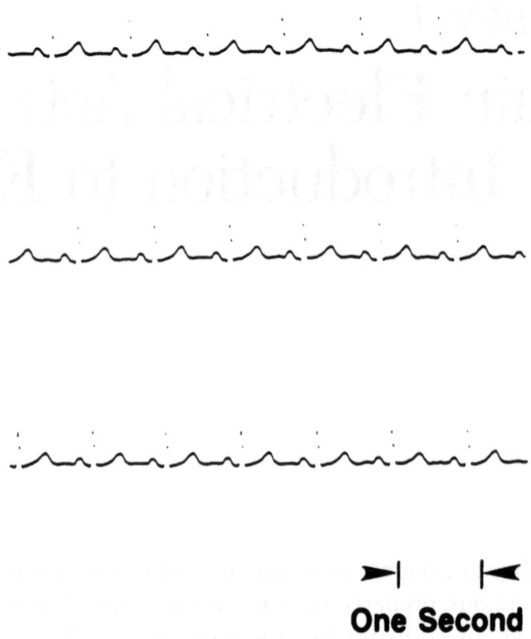

Figure 1.1. Sample recording of an ECG.

The recording of most kinds of bioelectric activity follows the same general plan. Specific methods employed and the particular instrumentation needed in a particular application are determined by two considerations: (1) the magnitude (voltage) and frequency composition of the bioelectric activity we wish to record, and (2) the nature and sources of the unwanted signals or artifacts that may be encountered. For reasons that will later become apparent, the recording of the EEG presents a formidable challenge to both the instrumentation and the technologist using it.

Some Characteristics of the EEG

Earlier we mentioned that Hans Berger characterized the EEG as a mixture of rhythmic, sinusoidal-like fluctuations in voltage. Although this description is somewhat of an oversimplification and does not admit of all the possible varieties of fluctuations in voltage that can be encountered in an EEG, it suffices as a starting point.

Sinusoidal fluctuations in voltage are described in terms of two characteristics or parameters. First, we specify how many oscillations or cycles occur in a standard time interval. The standard time interval universally used is one second; so we say that the sinusoidal fluctuations in voltage have a frequency of, or occur at the rate of, so many cycles per second. The term cycles per second is designated hertz or Hz, in honor of Heinrich Rudolf Hertz, the 19th century

German physicist. If we include nonsinusoidal activity as well, EEGs consist of voltage fluctuations in the range of 1 to 70 Hz.

Second, fluctuations in voltage are described by specifying their magnitude or amplitude, and this is measured in microvolts (millionths of a volt, abbreviated μV). A microvolt is an exceedingly small electrical signal. From the standpoint of instrumentation, it approaches the smallest voltage that may be detected by conventional methods of amplification. The amplitude of the fluctuations in the EEG may range anywhere from two to several hundred microvolts. For this reason, the EEG machine or instrument used to record brain electrical activity is a sophisticated device indeed. Scrupulous techniques need to be employed by the EEG technician in order to record these tiny signals.

Briefly defined, therefore, the EEG is a fluctuating electrical signal produced by the brain. The fluctuations can range in amplitude from two to several hundred microvolts when recorded from electrodes placed on the scalp. As we will see later, the waves recorded in this way originate mainly from the surface of the cerebral cortex. Some of the fluctuations in voltage are sinusoidal in character, some are not; these fluctuations cover a frequency range of approximately 1 to 70 Hz. The latter is referred to as the bandwidth or *frequency spectrum* of the fluctuations.

The EEG Frequency Spectrum

The frequency spectrum in EEG work is broken down into four subcategories. The bands of frequencies designated by these subcategories are identified by the Greek letters alpha, beta, theta, and delta.

The *alpha* band defines electrical activity in the range of 8 to 13 Hz. This includes the "alpha rhythm" or posterior-dominant rhythm, which is rhythmic activity normally recorded in the awake individual. Amplitude is variable, ranging from 5 to 100 μV, but is mostly below 50 μV. It is best seen when the subject's eyes are closed and under conditions of physical relaxation and relative mental inactivity. The amplitude of the alpha rhythm is attenuated by eye opening, attention, and mental effort. Other varieties of brain electrical activity are included in the alpha band. One of these is the mu rhythm, which is 7 to 11-Hz rhythmic activity occurring over central and centro-parietal regions during wakefulness. The mu rhythm is attenuated not by eye opening but by contraction of muscles on the contralateral side of the body.

The *beta* band includes frequencies over 13 Hz. The most common component of this band is the beta rhythm — rhythmic activity consisting of a variety of frequencies greater than 13 Hz and sometimes as high as 35 Hz. Amplitude is variable but is mostly below 30 μV. Beta rhythms are seen under a wide range of conditions. Very sharp transients, like spikes, usually fall into that part of the beta band greater than 35 Hz. Spikes, sharp waves, and other kinds of nonsinusoidal activity will be taken up later.

The *theta* band consists of electrical activity with a frequency of 4 Hz to under 8 Hz. Theta activity is normally seen in drowsiness and during the lighter stages of sleep, but may also be present in wakefulness. It may be strictly rhythmic as is the case with the alpha rhythm, or highly irregular in character. Irregular theta activity is sometimes referred to as being arrhythmic or polymorphic.

The *delta* band contains frequencies under 4 Hz. Delta activity is normally seen in the deeper stages of sleep and is a commonly observed abnormality in the waking state in adults. Like theta activity, it may be either rhythmic or irregular, in which case it is sometimes termed polymorphic delta. Delta activity has the highest amplitude of any activity recorded in the EEG (amplitudes as high as several hundred microvolts are sometimes recorded).

EEG technicians and neurology residents should become thoroughly familiar with the different bands in the EEG frequency spectrum. They should be able to quickly and accurately recognize the different kinds of activity in an EEG recording. Figure 1.2 shows some samples of the type of activity discussed. It is important to recognize that these sample recordings illustrate only a small proportion of the various different features present in the EEG. This will become apparent when we take up in greater detail the features of the normal EEG and describe and illustrate the more common EEG abnormalities in later chapters.

Fourier Series and Power Spectral Analysis

We have referred to the EEG as a mixture of rhythmic, sinusoidal-like fluctuations in voltage, and in the last section we considered the frequencies of the principal components recorded. At times, as when recording from a patient who shows a persistent alpha rhythm, the waveform looks a good deal like a pure sine wave. At other times, the waveform can be quite complex and bears little resemblance to a sine wave. Interestingly enough, such complex patterns found in the EEG can be simulated by adding together a number of sine waves having different frequencies, amplitudes, and phase relationships. In other words, a complex waveform can be *synthesized* from a number of simpler, sine wave components.

The reverse of this procedure of synthesis of complex patterns is known as *frequency* or *spectral analysis*. The method of spectral analysis separates a waveform into its different frequency components; it tells us the amplitudes of the different frequency sine waves of which the wave-

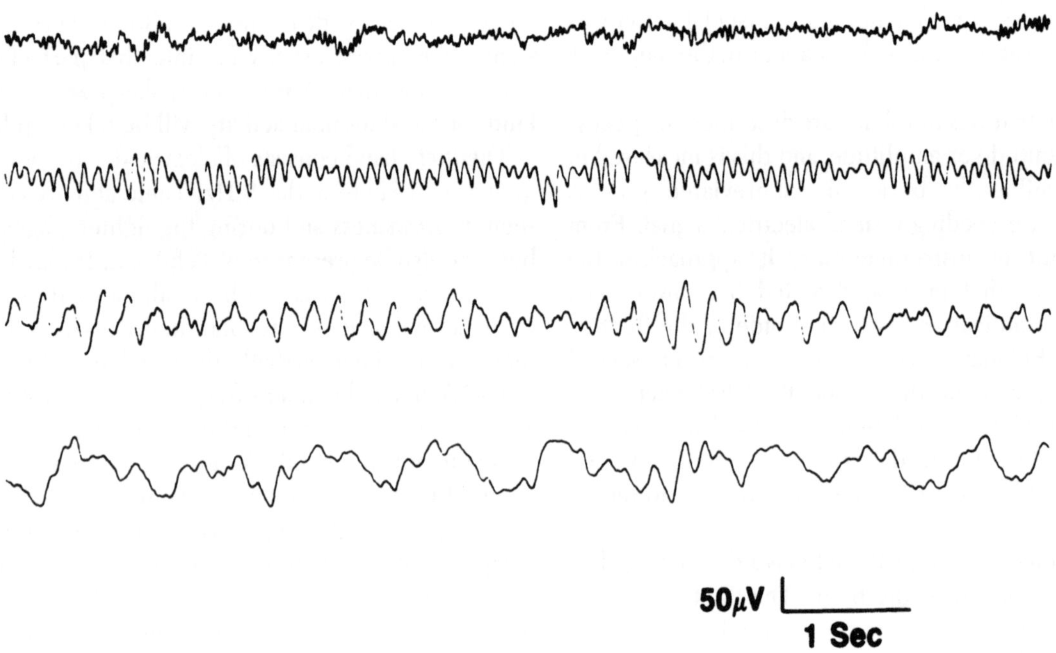

Figure 1.2. The four subcategories of the EEG frequency spectrum. The sample tracings show, from top to bottom, typical activity in the beta, alpha, theta, and delta bands.

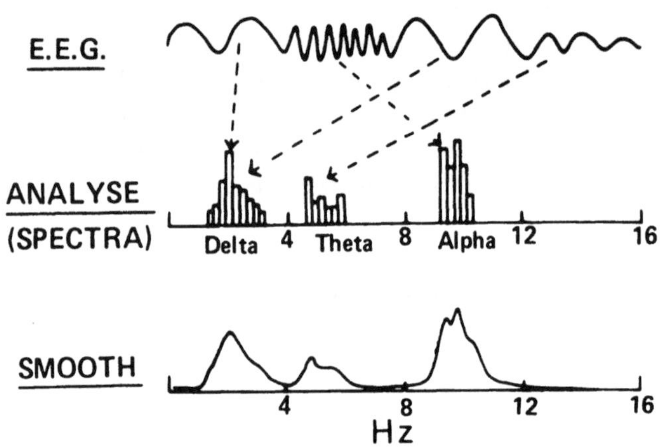

Figure 1.3. Diagram illustrating the method of power spectral analysis. A schematic representation of an EEG tracing of 4 seconds' duration is shown at the top, and the bar spectra resulting from Fourier analysis by computer is shown directly below. Smoothing of the bar spectra yields the power spectrum seen at the bottom. (From Fig. 10, p. 461, of Bickford RG: Newer methods of recording and analyzing EEGs, in Klass DW, Daly DD (eds): *Current Practice of Clinical Electroencephalography*. New York, Raven Press, 1979, by permission of the author and publisher.)

form is composed. This method of analysis is known as Fourier series analysis and was devised by the 18th to 19th century French mathematician Jean Baptiste Fourier. A Fourier series is just one of a variety of mathematical functions that are referred to as infinite series.[1] Theoretically, an infinite number of frequency components is needed in order to represent a complex waveform. In actual practice, however, an acceptable representation of the waveform frequently may be obtained by combining just the first eight or ten components in the series. Each component tells us the amplitude of the sine wave of specified frequency that goes into the composite. These data are then plotted, with frequency on the horizontal axis and amplitude on the vertical axis.

The component amplitudes of a Fourier series analysis are often expressed as mean square values. When presented in this way, the resulting plot of the data is called a *power spectrum*. By expressing the amplitude of each component in terms of its mean square value, the proportion of the analyzed waveform that is attributable to each particular frequency in the series can be determined. Figure 1.3 illustrates how an EEG tracing containing frequencies in the delta, theta, and alpha bands becomes transformed into a power spectrum.

It should be recognized that the power spectrum of a waveform represents a synopsis of the frequency compo-

[1] Another infinite series that has been used in EEG analysis is the Gram-Charlier series. This series differs from the Fourier series mainly in the assumptions that are made about the waveform analyzed. Thus, while the Fourier series assumes that the waveform to be analyzed is periodic—that is to say, it repeats itself exactly at regular intervals—the Gram-Charlier series does not and can deal equally well with aperiodic or periodic waveforms.

nents of but a short segment—10 seconds or less—of an EEG recording. To obtain a coherent picture of what is going on over a longer time period, the method of *compressed spectral arrays* was developed. With this method, a large number of comparable power spectra are plotted in close proximity to each other on the same graph. When presented in this fashion, a synopsis of the frequency components of 10 or more minutes of recording may be displayed on a single page. Figure 1.4 illustrates the method; in this case, a total of 48 power spectra are plotted together.

Despite its potential value, spectral analysis is not employed in routine clinical EEG work. Spectral analysis, however, is used in topographical brain mapping, a topic that is taken up in later chapters. The interested reader can consult more advanced texts or the EEG periodical literature for additional information.

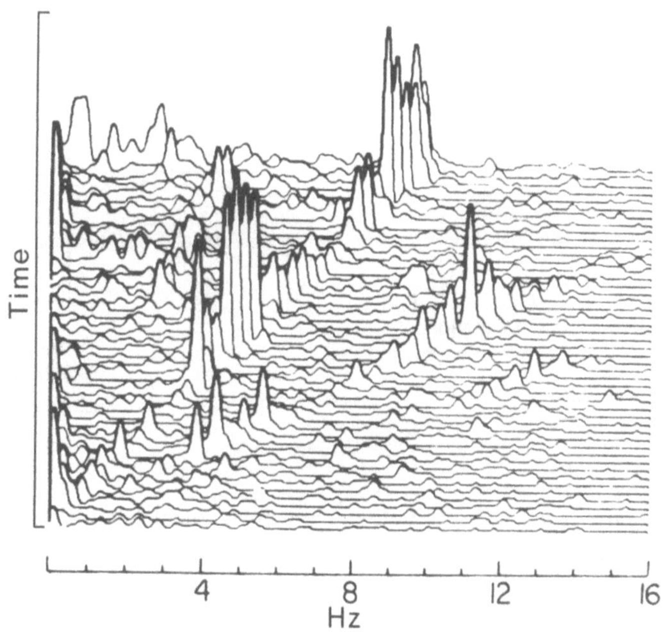

Figure 1.4. Compressed spectral array of a patient's EEG response to photic stimulation over the range of 1–16 flashes per second. Starting from the bottom of the figure, the flash rate was progressively increased, sweeping smoothly across the entire range. Note the peaks occurring at 4–6 Hz and at 9–10 Hz. The peaks signify the presence of high-amplitude activity at these frequencies and represent instances of photic driving. A separate series of smaller peaks can also be seen opposite the main peaks; they represent the second harmonic response. (Taken from Bickford RG, Brimm J, Berger L, et al: Application of compressed spectral array in clinical EEG, in Kellaway P, Petersen I: *Automation in Clinical Electroencephalography*. New York, Raven Press, 1973, p. 59, by permission of authors and publisher.)

Recording EEGs

Having thus defined the EEG and having considered briefly some of its major characteristics, let us turn next to one of the topics for which this text is primarily intended, namely, the recording of clinical EEGs. Up-to-date EEG technique involves obtaining a 20 to 30-minute sample of electrical activity from different combinations of 21 electrodes placed on the head according to the so-called 10–20 International System. Suffice it to say for the present that the 10–20 International System is simply a plan for placing electrodes on the scalp over specific strategic areas of the cerebral cortex. For now let us also assume that the 21 electrodes have already been attached to a patient and tested and are ready to be connected to the EEG machine. After becoming familiar with the EEG machine and with the details of recording EEGs, we will return to the matter of electrodes, electrode placements, and recording systems. The reason for this backwards approach is methodological. Experience has shown that the rationale of the electrode procedures used in taking an EEG is better understood after acquiring some knowledge of the workings of the EEG machine.

The EEG Machine: An Overview

Modern EEG machines are multichannel instruments. They consist of from 8 to 24 identical channels capable of recording simultaneously the electrical activity from as many different pairs of electrodes. Regardless of the manufacturer, every EEG machine consists of a number of particular structural units. Each of these units has a clearly defined function. In some machines, they are of modular construction to facilitate troubleshooting and to simplify servicing. From input to output end, the EEG machine consists of an electrode board, electrode selectors, amplifiers, filters, penmotors, and chart drive. In addition, a power supply is included for providing electrical power to run the different units, and an internal calibrator is furnished for testing and standardization.

Figure 1.5 is a block diagram of one channel of a typical EEG machine. Let us now briefly consider the function of each of these units in turn. In later, separate chapters we will discuss the design and operation of these units in some detail.

Electrode Board

Referred to also as the lead plug-in box or input box, this unit is a rectangular or square metal box frequently no larger than a small paperback book. Coming out of this box is a thick cable with a connector on the end that plugs into the EEG console. The electrode board is the means whereby the electrodes attached to the head are connected to the EEG machine. The connection is made by

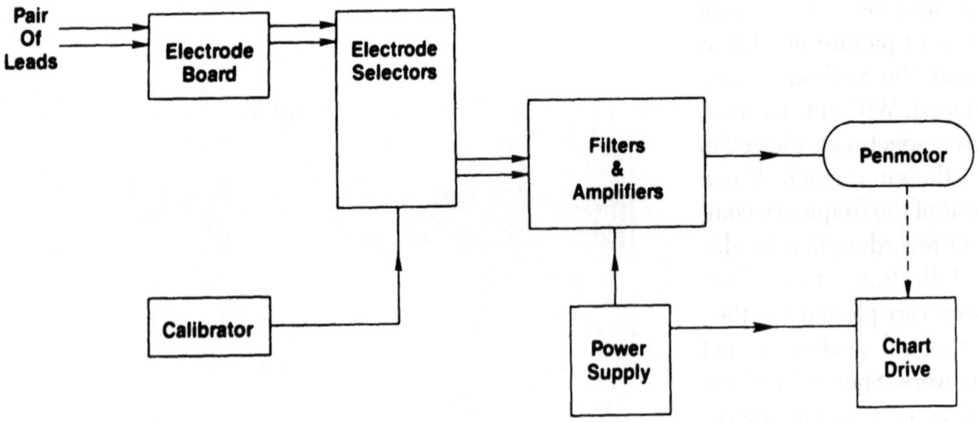

Figure 1.5. Block diagram of one channel of a typical EEG machine.

plugging the 1/16-inch diameter pin plugs on the ends of the electrode wires into mating pin jacks found on the front panel of the electrode board. These jacks are identified by the symbols used in the 10–20 International System, or are simply numbered. Several spare jacks are also included as well as jacks for connecting nasopharyngeal leads. The spare jacks have a variety of uses, among which are connecting electrodes for recording the patient's ECG or hooking up leads for recording his/her eye movements.

Electrode Selectors

We have said that to record the clinical EEG, a total of 21 leads is attached to the scalp over various strategic areas of the cerebral cortex. As we will see in the next chapter, electric currents flow in a circuit, not in a single conductor. This means that a *pair* of electrodes has to be connected to each channel of the EEG machine in order to record brain electrical activity. The process of recording from a pair of electrodes in an EEG channel and the recording obtained thereby is referred to as a *derivation*. With 21 electrodes to work with, it is obvious that many different derivations are possible.

To obtain an adequate sample of the electrical activity of all of the brain, it is necessary to record from many different derivations—more than can be displayed at one time even on a 24-channel EEG machine. For this reason, several different runs are required when taking an EEG, and a switching system is necessary to connect the various pairs of electrodes to the different channels of the machine. This function is performed by the electrode selectors or electrode-selector switches. These switches are designed for maximum flexibility so that the input connections to each channel of the machine may be connected to any combination of two of the 21 electrodes attached to the patient.

A particular arrangement by which a number of different derivations is recorded simultaneously on an EEG record is referred to as a *montage*. A variety of different montages are used in EEG recording. Derivations and montages are taken up in detail in a later chapter on recording systems. For now, we would like to mention that some EEG machines have a *montage switch* or master electrode selector. Such a device permits the EEG technician to change from one montage to another simply by indexing a single switch from one position to another. This greatly reduces the possibility of technician error in setting up the machine. It also saves considerable time since in the case of a 16-channel machine, the technician may have to reposition a total of 32 switches to change from one montage to another.

Amplifiers

Amplifiers are the heart of the EEG machine. For present purposes, let us say that an amplifier is simply a device for increasing the magnitude or amplitude of a voltage without introducing distortion.

In an earlier section of this chapter, we stated that the EEG contains fluctuations in voltage that are exceedingly small. This is especially true in the case of brain death recordings in which voltages of the order of 2 μV are encountered and need to be detected. To deflect the pens on an EEG machine, signals of the order of volts are required. This means that amplifications in the range of 100,000 to 1 million times are commonly found in an EEG machine. To achieve this degree of amplification, more than one amplifier is needed in each channel of the EEG machine. And so we find that each channel contains several "stages" of amplification.

The final stage of amplification is referred to as the *power amplifier* to distinguish it from the others, which are *voltage amplifiers*. The purpose of the power amplifier is to increase the current level of the amplified signal. This is necessary because it takes *power* as well as voltage to drive the penmotor, and power is equal to the product of voltage times current.

Although some of the fluctuations in voltage seen in the EEG can be very small, indeed, others may be quite large—several hundred microvolts in amplitude. In other words, the EEG is an electrical signal that has a wide *dynamic range.* Now what happens if an amplifier designed to handle signals in the 5 to 50-μV range is confronted with voltages that are five times as large? That some kind of untoward event takes place is intuitively obvious. In instrumental language, the amplifier "blocks," and the signal becomes distorted so that the output is no longer a faithful, amplified copy of the input.

To avoid distortion and to accommodate the wide dynamic range of the voltage fluctuations in the EEG, the amplifiers are provided with a means of changing amplification or "gain." So you find that each channel of an EEG machine has a conveniently placed switch for doing this. By increasing the gain of a channel, you increase the total amplification available, with the result that the amplitude of the voltage fluctuations traced out on the chart appears larger. The reverse happens when the gain is decreased. In most EEG machines, gain is referred to as *sensitivity* and is expressed in microvolts per millimeter deflection. This means that as gain *increases,* a millimeter corresponds to *fewer* microvolts and vice versa, so that gain and sensitivity defined in this way are inversely related. Machines having a large number of channels incorporate a master gain switch or all-channel-control switch so that the amplifications of all the channels may be changed simultaneously. Aside from being a convenience, this feature allows the EEG technician to minimize distortion by changing channel sensitivities quickly.

The amplifiers in the EEG machine are designed so they have the same distortionless output over the entire EEG frequency spectrum. This means that if the amplitude of theta activity present at a patient's scalp is four times greater than the amplitude of the patient's beta activity, then the amplitudes of theta and beta activity in the EEG tracing will stand in the same relationship to each other. To say the same thing but in more general terms, the gain of the amplifiers is independent of frequency over the EEG frequency range.

Filters

By attaching EEG electrodes to the head of a patient and connecting them to an EEG machine, the EEG technician expects to record the patient's brain electrical activity. In point of fact, however, brain electrical activity is not all that he or she will record. As we noted earlier, a variety of different kinds of living tissue generates electrical activity. Because the EEG electrodes placed on skin and scalp may be either over or in the vicinity of some muscles of the head and face, skin and muscle become sources of *contaminating electrical activity* in the EEG. The eyes are also an intrinsic source of contaminating electrical activity. The same is true of the heart, which generates a voltage large enough to be detected almost anywhere on the surface of the body. Such kinds of activity present in an EEG recording are collectively referred to as *artifacts.*

All sources of artifacts in the EEG are not intrinsic. The EEG can be contaminated by artifacts from extrinsic sources—electrical activity originating outside the body—as well. Such artifacts come from a wide range of different devices: x-ray equipment, relays and solenoids in a variety of different instruments, cardiac pacemakers, Ivac pumps, and winking fluorescent lamps, to name some of the more familiar sources. The most common extrinsic source of artifact, however, is the 120- and 240-V electric power lines that are present almost everywhere. These lines radiate 60 Hz electrical activity that is readily picked up and amplified by an EEG machine.

Since artifacts represent contaminating and hence unwanted electrical activity in the EEG recording, they need to be eliminated. The most effective and straightforward way of accomplishing this is to remove the source. In some instances, however, this is not practical and in others not even possible. Thus, for example, while it is possible to paralyze the muscles of the body with curare-like drugs, such methods of eliminating muscle activity are feasible only under the most unusual circumstances when the patient is on a respirator. Similarly, a 60-Hz artifact is readily eliminated by removing the 60-Hz power lines in the vicinity and operating the EEG machine on batteries. But, then, how would the electric lights be operated?

So we find that in EEG technology, the most direct method of getting rid of artifacts in recordings is not always feasible. Other techniques need to be applied to the problem. One such method uses a device referred to as a *filter.* Speaking in broad, general terms, filters are devices that selectively remove some components or ingredients of a mixture from other components or ingredients. Familiar examples are a sieve for separating large peas from small ones or a piece of filter paper for separating particles from a suspension.

The filters on an EEG machine are discussed in detail in a later chapter. For now it is sufficient to say that an EEG filter is an electrical or electronic circuit that is able to pass or transmit some frequency components in the electrical signal picked up by the EEG electrodes while rejecting or attenuating others. In other words, an EEG filter is a *frequency-selective device* that permits some frequency components to pass on and be amplified while other components are removed or diminished in amplitude. There are three different types of filters on modern EEG machines. There is the *low-frequency filter* that attenuates frequencies at the low end of the frequency spectrum but allows frequencies at the high end to be amplified. This is referred to also as a *high-pass* filter. The *high-frequency*

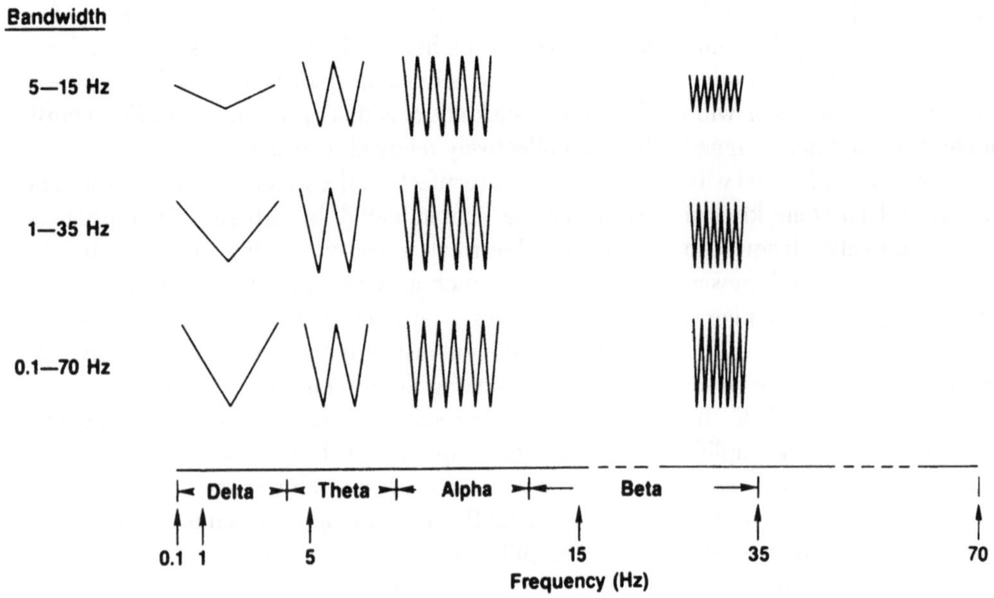

Figure 1.6. Basic function of the high- and low-frequency filters of an EEG machine in relation to the EEG frequency spectrum.

filter does just the opposite; it attentuates frequencies at the high end of the spectrum but allows frequencies at the low end to be amplified. Filters that perform this function are also called *low-pass* filters. Finally, many up-to-date EEG machines have a 60 Hz "notch" filter for reducing the artifact due to the presence of power lines in the vicinity of the patient. Figure 1.6 shows the effect of the high- and low-frequency filters on the EEG frequency spectrum.

The high- and low-frequency filters on an EEG machine are variable filters. This means that you can select the particular point in the frequency spectrum at which they begin operating—the so-called "cutoff point" or "roll-off point." In Fig. 1.6 the vertical arrows show the positions of some cutoff points in relation to the EEG frequency spectrum. Details are taken up later in the chapter on filters. Adjustment of the filters is accomplished by changing the position of a multiposition switch on each of the amplifiers; one switch is provided for the low-frequency filter and one for the high. In this way the EEG technician is able to adjust the high- and low-frequency cutoff points independently, and in so doing determine the *bandwidth* of the amplifier. As with the gain controls, some EEG machines have all-channel controls for the low- and high-frequency filters as well. The 60-Hz filter is controlled by a two-position filter-in, filter-out switch. In some machines the notch filters may be subjected to all-channel control just like the other filters.

It should be apparent that the filters are effective in eliminating unwanted signals from the EEG tracing if and only if the frequency characteristics of the unwanted signal are different from those of the desired signal. This is because filters are frequency-selective devices. The result, of course, is that artifacts in the delta, theta, alpha, and beta frequency bands cannot be reduced by filtering without also reducing the brain electrical activity as well. Fortunately, the electrical activity of skin and of muscle falls, respectively, in the low and high ends of the EEG frequency spectrum so that filtering becomes a practical way of eliminating these unwanted signals from the EEG record.

Aside from eliminating artifacts, the filters on the EEG machine may also be used to accentuate certain features of the EEG, which is discussed in the separate chapter on filters.

Penmotors

The penmotors are at the output or business end of the EEG machine. There is one for each channel, and they "drive" the pens that trace the voltage fluctuations in the EEG on the chart. The penmotor is quite properly named, for it is indeed a motor; a special kind of motor in which the armature rotates back and forth but never turns more than a fraction of a full circle. When the sensitivity of the amplifier driving the penmotor is correctly set, the deflections of the penmotor accurately describe the voltage fluctuations in the EEG.

As we will see later, the penmotors employ very strong permanent magnets in their design. Ordinary timepieces are easily magnetized by the strong field from these magnets. For this reason, you should avoid wearing your watch or avoid keeping any timepiece close to the penmotors unless they are solid-state digital devices or are specifically designated as being antimagnetic. Failing to do this will cause the watch or clock either to stop running or to run erratically until it is demagnetized.

Penmotors are sometimes referred to as *electrical-mechanical transducers*. A transducer is any kind of device that can change energy from one form to another, and the penmotor changes electrical energy to mechanical energy. It will be apparent that an electrical generator, which produces electricity, does the exact opposite and is referred to as a mechanical-electrical transducer. The same is true also of a microphone. It picks up the mechanical energy (vibrations) in a person's voice and transforms them into an electrical current.

Chart Drive

The chart drive pulls the paper chart (upon which the EEG is recorded) under the pens and through the machine, depositing it finally in the take-up tray. It does this at a highly accurate speed and with a minimum amount of weaving back and forth. One of the first things that the EEG technician-in-training learns is how to load paper into the chart drive. Chart paper comes in fan-folded or "accordion-folded" packs. Note that the chart paper is perforated at the folds; this is so that the pages fold easily and can be separated readily. A full pack usually consists of 1,000 pages, and they are numbered consecutively like the pages in a book. These numbers are a convenience; knowing the chart speed, the EEG technician can estimate the time duration of the recording simply by determining how many pages have gone through the machine. The page numbers may also be used to identify the location of interesting or particularly noteworthy events in a recording.

The chart drive has a control lever or a row of push buttons that allow the technician to change paper speed easily. Standard EEG chart paper speed is 30 mm/s. The reasons for settling on this particular speed will become apparent later. All EEG machines have, in addition, a slower and faster paper speed: usually one-half standard speed or 15 mm/s, and twice standard speed or 60 mm/s, respectively. Some machines provide for still other chart speeds as well, but these are not essential for routine clinical EEG recording.

It is important to recognize that the accuracy of the chart paper speed is of the utmost importance. Chart speed provides the *time base* in the EEG tracing. In practical terms, this means that the accuracy of our estimates of frequency of the brain electrical activity recorded on the chart is dependent upon the accuracy and reliability of the chart speed. Methods whereby this important machine parameter may be checked are taken up later.

Power Supply

The EEG machine contains a variety of electrical and electronic devices that require electrical energy to operate. But the circuits cannot operate on the 60-Hz, 120-V AC that is available from the power lines. They require low voltage DC instead. The power supply in the EEG machine provides this.

At the input end of the power supply is the thick line-cord with a three-prong connector on the end that plugs into a 120-V AC outlet. You can think of the power supply simply as a device that takes in 120-V AC house current at one end and puts out 12-V DC, accurate to very close tolerances, at the other. The power supply does this by transforming the 120-V AC to a lower voltage, converting the AC to DC by a process called rectification and smoothing, and finally regulating the 12-V DC output so that it is not only accurate but also stable. Details of this process need not concern us. Operation of the power supply is controlled by the main "power on" switch that is located in a prominent place on the machine console.

Calibrator

As we already know, the electrical activity of the brain recorded in a clinical EEG is measured in microvolts. Earlier we stated that a microvolt was a very small electrical signal and that the instruments used to detect low-level voltages were quite sophisticated devices. Because such sophisticated instruments are subject to failure, the EEG technician needs to know on a day-to-day basis whether his or her EEG machine is operating properly—whether it is doing exactly what it was designed to do. This is accomplished by the calibration procedure.

Calibration involves feeding a signal of known voltage and well-defined frequency characteristics into each channel of the EEG machine and observing the expected output on the chart. This signal, of course, needs to fall within the dynamic range of the EEG. A calibration is done just before an EEG recording is started on every patient and is repeated after the recording has been completed. These procedures are referred to as the precalibration or "pre-cal" and postcalibration or "post-cal," respectively. If the pre- and postcalibrations are satisfactory and yield identical tracings, we conclude that the EEG machine was functioning properly during the actual recording of the EEG. This assumes, of course, that nothing changed and then changed back again in the interval between the pre- and postcalibrations. Such intermittent faults or failures can occur and will be missed if they fail to show up during calibration. But in practice, it is usually only a matter of time before they are discovered.

All EEG machines have an internal voltage calibrator situated on the console. To perform a calibration, a number of simple steps are carried out. First, you need to connect the input of each channel of the EEG machine to the calibrator circuit by correctly positioning the electrode-selector switches or the montage switch if the machine has one. Second, you adjust the voltage of the calibration signal by indexing the *calibration level switch* to a selected value.

Table 1.1. Relationship between sensitivity of an EEG channel and pen deflection for a standard input signal of 50μV

Sensitivity[a] (μV/mm)	Pen deflection (mm)
2	25
5	10
7	7
10	5
20	2.5

[a]Note that sensitivity and gain are inversely related; thus, gain or amplification decreases as the numbers in this column increase.

The standard reference calibration signal in EEG work is 50 μV. Finally, you press and release the CAL button. In doing so, the pens are deflected first upward and then downward each time the CAL button is pressed and released. Note that the pens take some little time before returning to their original positions, so you need to wait before releasing the CAL button. The reason for this phenomenon and its significance are taken up when filters and calibration methods are discussed in detail.

Certain calibration standards are followed in EEG work. The gain setting most frequently used in recording clincial EEGs corresponds to a sensitivity of 7 μV/mm. This means, of course, that a 7 μV signal will deflect the pens exactly 1 mm. Using this sensitivity setting, the standard reference calibration of 50 μV should produce a 7 mm deflection of the pens. In the same way it will be seen that the standard reference calibration of 50 μV will produce a 5-mm deflection at a sensitivity of 10 μV/mm and a 10-mm deflection at a sensitivity of 5 μV/mm. Note that at the setting of 10 μV/mm the machine has *less* amplification or gain than at a setting of 5 μV/mm. Table 1.1 shows these relationships in tabular form.

Some newer machines provide for AC or sine wave calibration as well. Details of this method are taken up in Chapter 6, "Calibration and Calibration Methods."

In examining the calibration record it is important to verify that the tracings from all the channels of the machine are alike. This is because much of the information derived from an EEG is based on a comparison of the recordings from left and right sides of the brain, and for these comparisons to be valid, the channels recording activity from both sides have to be identical. To assist the EEG technician in recognizing any differences that may be present between channels, it is universal practice to do a so-called "biological calibration." This entails recording brain electrical activity from the *same pair* of electrodes, simultaneously, on every channel of the machine. These tracings are then examined carefully for evidences of differences between them. More will be said about this later.

Computerized EEG Machines

Although this term has appeared in the descriptive literature of some EEG machines, it is partly a misnomer. Computerized in this context refers to the computer control and display of EEG recording parameters, not to having the EEG technician replaced by a computer. Modern technology and computers notwithstanding, a well-trained EEG technician is essential for obtaining high-quality EEGs. Nevertheless, recent developments in microprocessor and microcomputer technology have had a significant impact on the design of modern EEG machines.

The most obvious change is in the physical appearance of some of the new models. There are virtually no more knobs to turn or dial plates to look at. Knobs have been replaced by touchpads, and dial plates by the screen of a cathode-ray tube (CRT). Thus, for example, the filters are set by pressing the appropriate touchpads and observing the result on the display of the CRT screen. Such machines have the convenience of being "menu-driven"; they provide for automated sequencing of montages and operations and afford the flexibility of user-programmable montages.

While state-of-the-art changes of this kind may be of considerable convenience to some users, they should be recognized as conveniences and nothing else. For example, the characteristics and operation of the low-frequency filter and its effects on the EEG tracing are the same regardless of whether the filter is switched in by turning a knob or by pressing a touchpad. Moreover, the ability to design and easily program one's own special montages may be of value, especially in research, but it is unlikely to have much impact on routine clinical electroencephalography where standardization is essential. For such reasons, these modern developments are not considered in the chapters dealing with the various basic structural units of the EEG machine.

By this time it will be apparent to the reader that the EEG machine is certainly not a simple instrument. Our brief survey merely serves as an introduction to what the machine does and how it is operated. To use the machine effectively on a routine basis, it is important to know something about how it actually functions and why it functions as it does. But this requires some background information. So you will find that it is first necessary to review some basic electrical concepts. These are taken up in the next chapter.

Chapter 2
Basic Electrical Concepts

In the brief survey of the EEG machine in the last chapter, many terms relating to electrical phenomena were used. We encountered terms like electric current, voltage, AC, DC, and electric circuit, to name a few. Thus far, these terms have gone undefined. To understand how the EEG machine does its job, some knowledge of these and other electrical concepts is essential. This is not to say that the EEG technician or neurology resident need to become adept in the area of physics, electronics, and electrical engineering. On the other hand, without some understanding of these concepts the EEG machine becomes a strange and mysterious device rather than a practical tool for recording a patient's EEG.

But this is not the sole reason for the present chapter. The EEG is itself an electrical phenomenon; an electrical sign of cortical activity. Much of today's knowledge of the EEG and of brain function in general came about as the result of important advances in physics, electronics, and electrical engineering. Indeed, it has been said that significant advances in neurophysiology have gone hand in hand with significant developments in these related areas. This being the case, it is appropriate for those concerned with the technology and practice of clinical electroencephalography to know something about electrical phemonena per se.

Electrical Currents

When we use the term "electrical activity" in speaking of the EEG, what we really are talking about is an *electric current*. What exactly is an electric current? For most of us, an electric current is what makes our lamps light and our motors run. Is this current the same kind of thing that is generated by the brain?

The physicist says that an electric current is a flow of electrons—a flow of negatively-charged particles in a conducting medium. To understand what this means and how it comes about, we need to review briefly some of the basics of atomic structure.

Atomic Structure

We know that all matter is made up of *atoms*. These minute particles, which are arranged in an organized structure called a molecule, consist of various kinds of *elementary particles*. Only the two major kinds of elementary particles, the *protons* and *electrons*, need concern us here. They are largely responsible for the electrical properties of substances.

The atom has an interesting structure. At its center is a relatively small nucleus that contains practically all of the weight of the atom. The particles called protons reside within the nucleus, and they carry a *positive* electrical charge. Surrounding the nucleus are tiny particles bearing a *negative* electrical charge; these are the electrons. Because unlike electrical charges attract each other, the electrons are attracted by and bound to the nucleus. The total negative charge of the electrons in an atom is equal in magnitude to the positive charge of the nucleus. Since equal charges of opposite sign exactly cancel each other, an atom is normally in electrical balance.

The protons and electrons of which all matter is constituted are identical. The fact that different substances display different properties arises in large measure from the circumstance that the number and arrangement of these elementary particles differ from one element to another. It is primarily the different electrical properties of substances that concern us here.

Conductors and Insulators

While the electrons in the atoms of various substances are all alike, the *strength* with which the electrons are bound to the nucleus of an atom differs in different substances. In some substances, the electrons are very tightly bound; in others, they are only loosely bound and free to move about. This difference is responsible for one of the important electrical properties of a substance, namely, its *conductivity*. Substances or materials that have electrons that are loosely bound to the nucleus are electrical *conductors*, while substances whose electrons are all tightly bound and not free to move from their natural positions in an atom are *insulators*. From this it follows that the movement of the electrons in a conductor is what we refer to as an *electric current*.

Most metals are examples of good conductors. Copper and silver are particularly good conductors, and this is the reason why they are used in electrical wiring. Gold is also an excellent conductor; because gold does not readily corrode and is not toxic, it frequently is employed in EEG electrodes. Glass, mica, and porcelain, as well as most plastics, are examples of poor conductors or good insulators. Most persons have seen the white porcelain insulators to which the power lines are attached as they enter a house or building. The insulators prevent the current in the wires from leaking in unwanted directions.

Electric currents also can flow in liquid media such as solutions. In such cases, the particles carrying the electrical charge are *ions*. Ions and the important topic of conduction in liquid media are taken up when we discuss electrodes in a later chapter.

Potential Difference and Voltage

Although electrons in conducting substances are only loosely bound and hence free to move about, they normally do not do so. To get the electrons in a conductor moving, a force has to be applied to them. In other words, we need to apply a *potential difference* between the two ends of the conductor.

An analogy is useful in helping to understand the concept of a potential difference. The electrons in a conductor that are free to move are analogous to water in a long, straight, horizontally positioned pipe. The water has the capability of flowing through the pipe; but it will only dribble out the ends as long as the pipe is exactly level. Only when one end of the pipe is raised above the other end will a flow occur. In the same way, electrons in a conductor will flow only when the electrical charge at the two ends of the conductor differs, that is, when there is a potential difference between the two ends. The potential difference is measured or expressed in volts, after the Italian physicist Volta, whose name was mentioned in the last chapter. Current flow is measured in amperes (A), milliamperes (mA), or microamperes (μA) after the 18th to 19th century French mathematician-physicist André Ampère. Without the presence of a voltage between the two ends of a conductor, there can be no flow of electrons; under such conditions, current flow is equal to zero.

Resistance

How much current flows in a conductor depends upon the voltage applied to it and upon the conductivity of the substance involved. We said earlier that there are good conductors and poor conductors. A good conductor is said to have high conductivity or to have a very low *resistance* to the flow of a current. A poor conductor, on the other hand, has low conductivity or a high resistance to current flow. Resistance, therefore, is inversely related to conductivity, or conductance as it is called. It is a parameter derived from the relation between the voltage applied to the conductor and the current flowing in it. We measure resistance in *ohms*, in honor of the 19th century German physicist George Simon Ohm.

Electrical Circuits

Connecting a voltage between the two ends of a conductor constitutes an electrical circuit. This, of course, represents the simplest kind of electrical circuit. In actual practice, electrical circuits are somewhat more complex; indeed, they frequently can become quite complicated. Nevertheless, the simple circuit does illustrate the important point that an electrical circuit describes a continuous pathway between the two points of a voltage source.

Figure 2.1A shows this circuit with a current-measuring device or ammeter in *series* with the conductor. In this instance a good conductor is connected up, and this is evidenced by the meter showing a relatively large current flowing in the circuit. Figure 2.1B shows the same circuit but with a poor conductor connected; note that considerably less current is flowing than in the previous case. Finally, in Fig. 2.1C, the conductor has been replaced by a good insulator whereupon the meter indicates that no current at all is flowing in the circuit.

We already mentioned in the last section that the current-carrying properties of a conductor are expressed in terms of its resistance and that resistance is measured in ohms. For this reason, a conductor is referred to as a *resistor* and is represented by a zig-zag line in a circuit diagram. This is illustrated in Fig. 2.2 where the circuits shown in Fig. 2.1 are drawn in the conventional manner with the conductors represented by resistors (abbreviated R).

It is obvious from Fig. 2.2 that the current flowing in a circuit decreases as the resistance increases. In other words, with the applied voltage kept constant, the current flowing in a circuit containing a resistor varies *inversely*

Figure 2.1. Simple electrical circuit containing a good conductor (A), a poor conductor (B), and an insulator (C). The arrows show the direction of current flow.

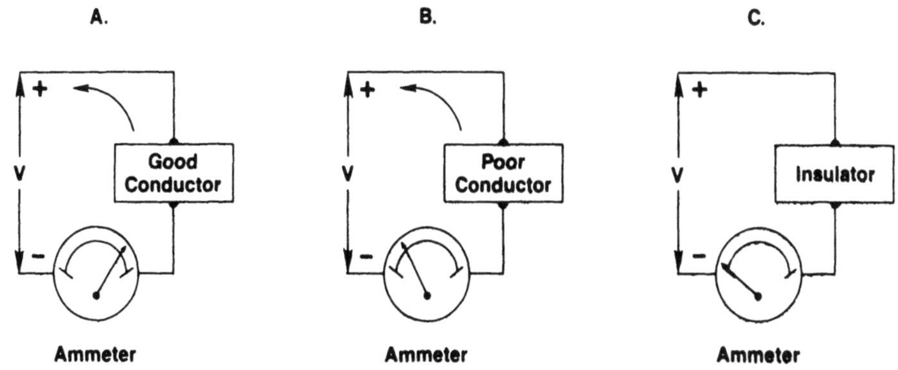

Figure 2.2. Simple electrical circuits shown in Fig. 2.1 drawn in conventional format used in circuit diagrams.

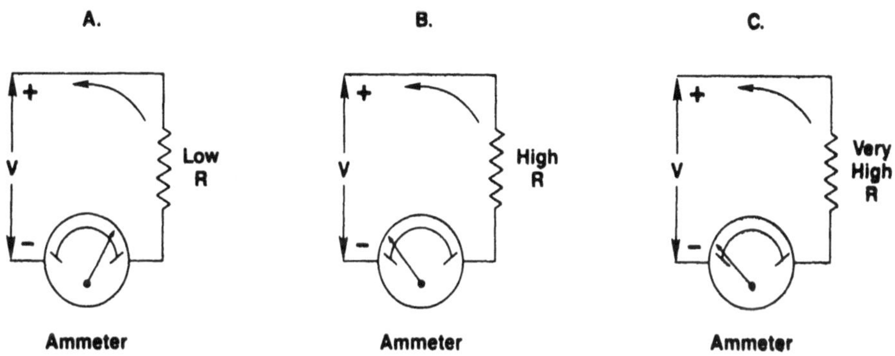

with the magnitude of the resistance. If, on the other hand, we kept the resistance in the circuit constant and allowed the applied voltage to change, we would discover that the current flowing varies *directly* with the magnitude of the voltage. The relationship between current, voltage, and resistance in an electrical circuit is defined by a famous formula referred to as Ohm's law.

Ohm's Law

Ohm's law states that in any electrical circuit,

$$\text{current} = \frac{\text{voltage}}{\text{resistance}}$$

or symbolically,

$$I = \frac{V}{R},$$

where I is current in amperes, V is the potential difference in volts, and R is the resistance in ohms. As with any algebraic formula, if you know the values of any two variables for a particular circuit, you can compute the third or unknown variable from the formula. Thus, for example, if in the circuit shown in Fig. 2.2B, V = 12 volts and R = 50,000 ohms,

$$I = \frac{V}{R},$$

$$I = \frac{12}{50{,}000} = 0.00024 \text{ A}$$

Note that if you doubled V, the current, I, would also be doubled; but if you doubled R instead, the current would be halved.

Ohm's law is a simple yet powerful formal rule that provides a means of analyzing simple circuits as well as many complex circuits. Figure 2.3 shows how the simple circuit in Fig. 2.2 may readily be made more complex by the addition of more resistors and branches with additional such elements. The analysis of these circuits requires two additional rules relating to the way in which separate resistors *in series* and *in parallel* are combined.

Series and Parallel Circuits

When two or more resistors are connected in series—that is, when one is connected to another in a single chain—the total resistance is simply the sum of the resistances of the individual elements. This rule is expressed by the formula

$$R_T = R_1 + R_2 + \cdots R_n,$$

where R_T is the total resistance and R_1, R_2, . . . R_n are the resistances of the individual elements. The second rule applies to resistances in parallel. In this case to calculate total resistance you add together the *reciprocals* of the branch elements. This yields the reciprocal of the total

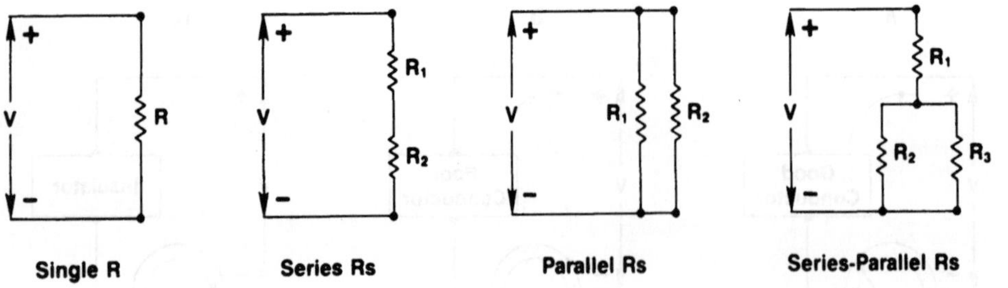

Figure 2.3. Basic forms of series circuits and parallel circuits.

resistance from which total resistance is readily computed. This rule is embodied in the formula

$$\frac{1}{R_T} = \frac{1}{R_1} + \frac{1}{R_2} + \cdots \frac{1}{R_n},$$

where R_T is the total resistance of the circuit, and R_1, R_2, . . . R_n are the resistances of the individual elements in parallel with each other.

It should readily be apparent that when resistors are in series, the total resistance will always be greater than the resistance of any of the individual *R*s. With resistors in parallel, on the other hand, the total resistance will always be less than the resistance of any of the individual *R*s. Why this is so may not immediately be obvious from the formula, although a simple example makes it quite clear. Thus, if R_1 = 10,000 ohms and R_2 = 10,000 ohms, then,

$$\frac{1}{R_T} = \frac{1}{10{,}000} + \frac{1}{10{,}000}$$

$$\frac{1}{R_T} = \frac{2}{10{,}000}$$

$$R_T = 5{,}000 \text{ ohms}$$

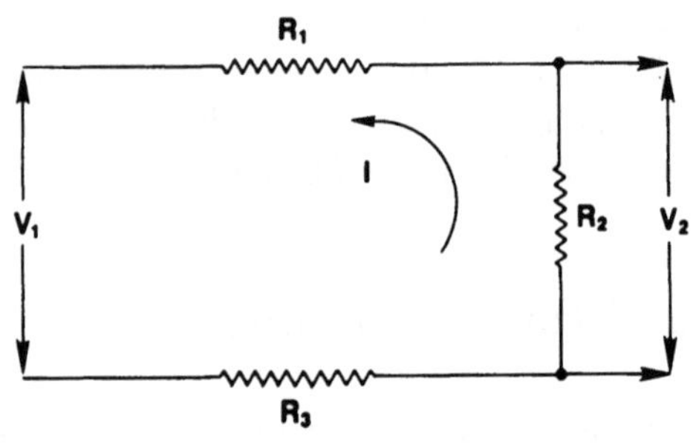

By Ohm's Law,

$$I = \frac{V_1}{R_T} = \frac{V_1}{R_1 + R_2 + R_3}$$

Again By Ohm's Law,

$$V_2 = R_2 I$$

$$V_2 = R_2 \left(\frac{V_1}{R_1 + R_2 + R_3} \right)$$

Figure 2.4. Application of Ohm's law in analysis of a series circuit.

In this case, R_T is one half the magnitude of either R_1 or R_2. The simplest way of understanding what happens with resistors in parallel is to recognize that by adding a resistor in parallel with another, you provide an additional pathway for current to flow in the circuit. By Ohm's law, you know that the current flowing in a circuit can be increased only by (1) increasing the applied voltage or (2) decreasing total resistance of the circuit. With voltage remaining unchanged, we have to conclude that adding the resistor in parallel results in a decrease in total resistance of the circuit.

The value of Ohm's law and of the rules for combining resistances in series and in parallel will become more apparent when we discuss the topics of electrodes and electrode impedance. In the meantime, the reader will find it helpful for future purposes to analyze the circuit shown in Fig. 2.4. The variable of interest is V_2, the voltage, across R_2. Figure 2.4 shows the general solution for V_2; as may be seen, this involves the use of Ohm's law and a little simple algebra. Now, if R_1 and R_3 are *very small* with respect to R_2, the formula for V_2 is approximated by the expression

$$V_2 = \frac{R_2 V_1}{R_2},$$

and V_2 will very nearly be equal to V_1. As we shall see in later chapters, this simple circuit analysis explains why in EEG work it is necessary for the impedance of the recording electrodes to be low and the input impedance of the EEG amplifiers to be very high. But more about this later.

Circuit Parameters

Earlier, we referred to resistance as a parameter of an electrical circuit. Our discussion turns now to the other parameters that are contained in electrical circuits.

There are but two other circuit parameters. They are called *inductance* and *capacitance*, and the associated elements are termed *inductors* and *capacitors* or *condensers*, respectively. It is surprising and indeed remarkable that electrical circuits consist only of these three parameters, no matter how complex they may become. Of course a circuit can contain other components like diodes, transistors, and vacuum tubes as well; but these are active elements and are not referred to as parameters.

Inductance is a circuit parameter of only minor interest in the area of EEG technology. Aside from the transformers and choke coils of the power supply, and the armature coils of the penmotors, there are no other inductors in an EEG machine. Moreover, inductance does not figure as a parameter in the electrical properties of living tissue. Therefore, no more will be said about inductance in this text.

Capacitance, on the other hand, is a parameter of considerable interest to both the technology and the practice of clinical electroencephalography. Not only are condensers essential components in the power supply and amplifiers of an EEG machine, but they also are the frequency-selective elements that make a filter work. We will hear more about this later in Chapter 4. Finally, living tissue displays electrical characteristics that resemble the electrical properties of a capacitor. For these reasons it is necessary to examine and to understand the electrical properties of this circuit parameter.

Capacitance

The physicist defines a capacitor as two conductors that are separated by an insulator. This, of course, is simply a structural definition that is reflected in the fact that a symbol used for capacitors consists of two short, parallel lines of equal length separated by a narrow space. Of considerably greater interest to the present topic, however, is the functional definition, and here the use of an analogy will be helpful. A capacitor or condenser has the same relationship to electrons that a pail has to water. From this it may be inferred that a condenser is capable of storing electrons. This is indeed the case. A condenser's storage capacity is measured in units called farads (in honor of the 19th century English chemist-physicist, Michael Faraday). Because a farad is an enormous quantity, the practical unit used in the circuits we deal with is the microfarad or millionth of a farad (abbreviated μF or MFD).

To understand the way in which a capacitor affects the functioning of an electrical circuit, it will be useful to return for a moment to the other parameter of electrical circuits that we have already discussed, namely, resistance. For reasons that will later become apparent, we need to address the topic of the *transient response* of electrical circuits.

Transient Response

The transient response of a circuit refers to the behavior of the circuit during the interval of time that a *change* is applied to it and the circuit is still adjusting to the change. This is the opposite of the circuit's *steady-state* response, which refers to the condition of the circuit after it has once again settled down. You can think of transient response and steady-state response in terms of what happens to a person's pulse rate as there are shifts in his or her level of physical activity. While at rest, your pulse rate is, say, 60 to 70 beats per minute. Suppose that at time zero, you begin running. What happens to pulse rate? You would find that pulse rate undergoes a rapid, transient change, shooting up to perhaps 100 to 110 beats per minute. As you continue to run, pulse rate begins to drop, and after a time levels off to perhaps 80 to 90 beats per minute once you attain your normal pace. From then on, until fatigue sets in, it shows only small fluctuations if you maintain a regular pace. This is the steady-state response.

With electrical circuits, the same principles are applicable. The transient response of an electrical circuit is commonly observed by applying an instantaneous change in voltage to the circuit and then measuring the change in current over the interval of time it takes the circuit to adjust to the change. The instantaneous change in voltage is referred to as a *step function*.

Let us begin by examining the transient response of an electrical circuit containing only a resistor. Figure 2.5A illustrates what happens when this is done. With no voltage applied to the circuit, the current flowing, of course, is zero. Now, let us instantaneously change the voltage from zero to some steady, finite value. The plot of voltage versus time in Fig. 2.5A shows this as a step increase in voltage. If the pointer on the current-measuring meter connected in series with the resistor had no inertia, you would see an instantaneous change in its position from zero to some value; there it would remain as long as the step in voltage continued to be applied to the circuit. The plot of current versus time shows this graphically. Note that the transient response of the resistance to a step function is itself a step function. In other words, the changes in current flowing in the circuit follow the changes in applied voltage perfectly. To put it another way, the circuit attains steady-state instantaneously so that, practically speaking, there is no transient response.

What happens when the circuit contains a condenser instead of a resistor is quite different indeed. The outcome is illustrated in Fig. 2.5B. Observe that as with the resistor, the current changes instantaneously from zero to a finite value when the step function is applied. It does not remain there, however, but begins immediately to fall, first rapidly and then slowly until it once again is zero. Note that in doing so the condenser is displaying the characteristics of both a conductor and an insulator. At the instant the voltage is connected to its terminals, the condenser behaves like a good conductor; but then as this voltage remains connected, it becomes a poorer and poorer conductor until current finally ceases to flow, and the condenser displays the properties of an insulator. This outcome is readily

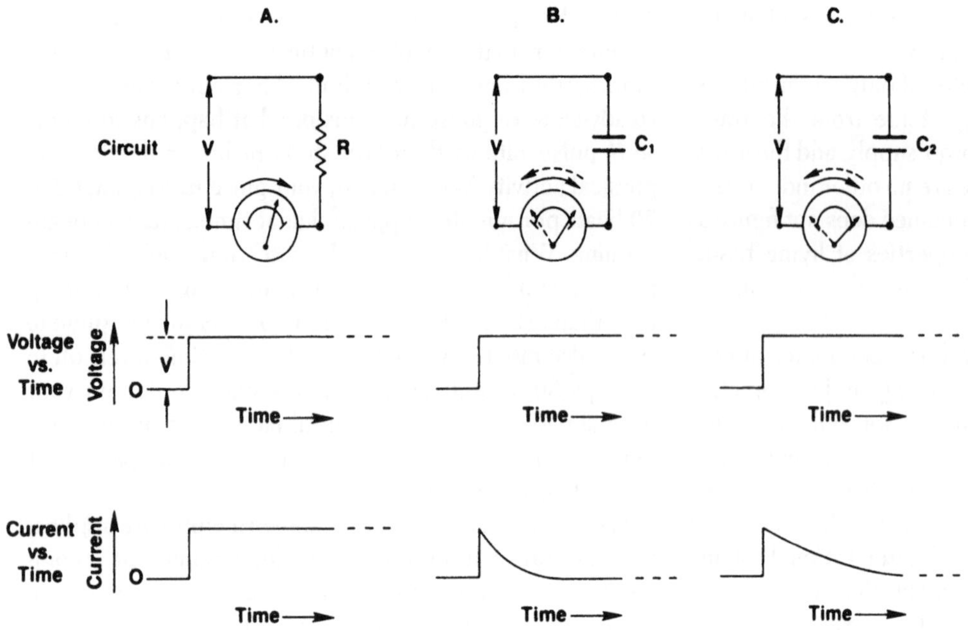

Figure 2.5. Transient response of electrical circuits containing (A) resistance only, (B) capacitance only, (C) capacitance only with C_2 greater than C_1.

understood by referring to our analogy of the water and pail. Assuming that the pail is an enclosed container that cannot overflow, the water will cease to flow in as it becomes filled. In much the same way, the flow of current in the condenser ceases when the condenser becomes filled with electrons or "charged."

The transient response of a capacitor, therefore, is quite different from the transient response of a resistor. While the resistor allows current to flow exactly in phase with the applied voltage, the capacitor exerts a counteracting force upon the flow of current set up by the change in voltage. For this reason, capacitance is considered to be a *reactive* element, and condensers are referred to as *capacitive reactance* in a circuit. We will go into the question of how capacitive reactance is measured later in this chapter in the section on impedance.

After the transient response is over and steady-state has been achieved, note that current through the condenser is zero. In other words, a condenser behaves like an insulator or an "open circuit" to a steady or unchanging voltage. This characteristic is of considerable practical value as it means that a condenser can be used in a circuit to block a steady voltage. As we will see later in a chapter on the differential amplifier, the various stages of amplification in an EEG machine are commonly coupled together by means of condensers. When used in such an application, the condensers are referred to as *blocking capacitors*.

What happens to the transient response as we increase the size of the capacitance? This is shown in Fig. 2.5C. Note that although the current level attained is the same as was the case with C_1 in the circuit, the current falls at a much slower rate. This happens because the larger capacitance has a greater capacity for storing electrons and, hence, takes longer to fill up or become charged. The fact that the condenser is actually charged may be demonstrated by connecting it to a voltmeter and observing that a voltage is present between the two terminals. Since very large condensers can store large numbers of electrons, they are capable of generating large currents when discharged. Charged condensers, therefore, constitute a shock hazard and can be potentially dangerous if you happen to touch their two terminals simultaneously; for this reason, they should be handled very cautiously.

Series R–C Circuit

In the last section we saw that the transient response of a capacitor to a step function was described by a current rising instantaneously to a maximum and then falling off to zero, first rapidly and then more slowly. We now consider the transient response of an R–C circuit—a circuit containing both resistance (R) and capacitance (C) in series. This is an extremely important circuit for the EEG specialist to know about since such circuits are incorporated in the frequency filters on the EEG machine.

Figure 2.6 gives the circuit diagram of a series R–C circuit and shows the response of this circuit to a step function. As in the case of a circuit containing only capacitance, the current rises instantaneously to a maximum value and then falls off, eventually returning to zero. Note in the plot of current versus time that the maximum value of current is equal to V/R, a value that looks like the right side of the equation for Ohm's law. The mathematical function describing the way in which the current varies with time is

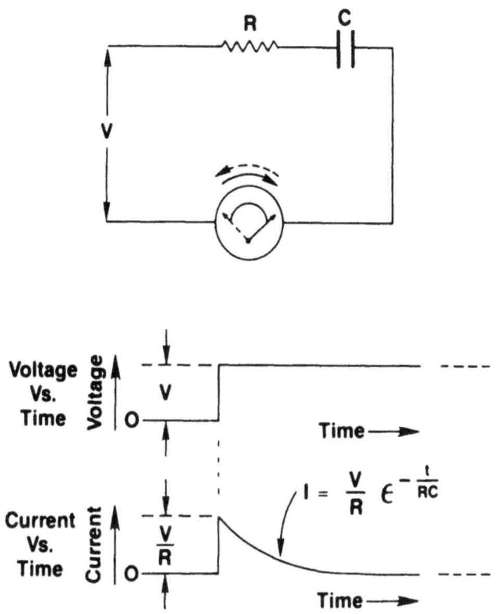

Figure 2.6. Transient response of a series R–C circuit.

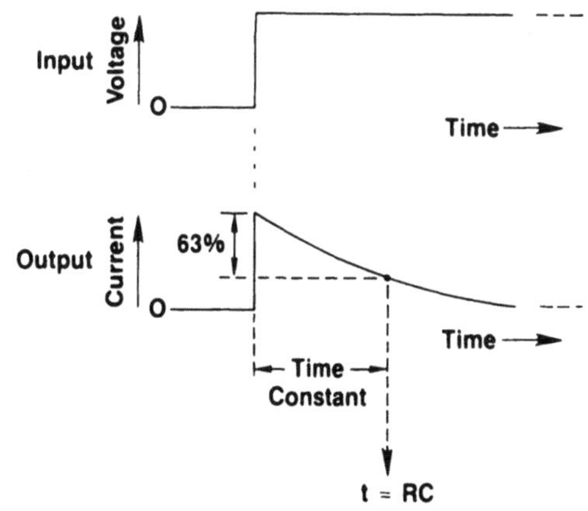

Figure 2.7. Time constant of a series R–C circuit.

referred to as a decaying exponential. It is given by the formula:

$$I = \frac{V}{R}\varepsilon^{-(t/RC)}$$

where V = applied voltage, R = resistance, C = capacitance, t = time in seconds, and ε is the Napierian constant, the number 2.718. . .[1]

While seemingly complex at first glance, this formula is relatively simple when you consider it by parts. Thus, the first part, $I = V/R$, is the now familiar Ohm's law. The rest of the formula corresponds to the quantity ε taken to an exponent, or $2.718^{-(t/RC)}$. Since the exponent is negative, this quantity will always be equal to 1 or <1. Observe that when the time t is equal to zero, the whole exponent also is zero and the quantity $2.718^{-(t/RC)}$ is equal to exactly 1 (remember that any number taken to an exponent that is zero is always equal to 1). Therefore at the instant the voltage is applied to the circuit, $I = (V/R) \times 1$ or $I = V/R$. For values of t greater than zero, the current is equal to some fraction of V/R. Note that the current becomes smaller and smaller as t becomes larger and larger. Theoretically, current approaches but never reaches zero; for practical purposes, however, it is equivalent to zero within the span of several *time constants.* This term is taken up in the next section.

Time Constant

It will be clear from an inspection of the equation for current in the series R–C circuit that the rate at which the decaying exponential falls depends on the value of the exponent, which, in turn, depends on the values of R and C in the circuit. When RC (the product of R and C) is large, the negative exponent is small and the current will fall to zero slowly. On the other hand, when RC is small, the negative exponent is large and the current falls more rapidly.

The rate at which current falls to zero during the course of the transient response is conveniently expressed by the term *time constant.* The time constant of a circuit is the length of time it takes the current to make 63% of its total transition from initial state to steady-state. This is illustrated for the series R–C circuit in Fig. 2.7. Notice that another way of looking at the time constant derives from the fact that it corresponds to the time at which the current is only 37% of its initial value. In terms of the circuit

---

[1] The mathematically sophisticated reader familiar with the calculus may be interested to learn the origin of this equation. We already know that for a circuit containing resistance alone, the current flowing in the circuit varies directly with the applied voltage and inversely with the resistance. In other words, $I = V/R$. In the case of a circuit containing only capacitance, the current varies directly with the capacitance and with the *rate of change* of the applied voltage. This means that time is a significant variable and current has to be expressed as a *derivative* of voltage. The equation for the current, therefore, is:

$$I = C\,dV/dt$$

Putting the two parameters together, the basic voltage equation for a circuit with R and C in series is:

$$RI + (1/C)\int_0^t I\,dt = V$$

Differentiating this equation with respect to t yields:

$$R\,dI/dt + I/C = 0$$

From which

$$I = (V/R)\,\varepsilon^{-(t/RC)}$$

parameters, the time constant is equal to the product of *R* and *C*.[2]

AC and DC

Most everyone has at one time or another noticed the identifying plaques that are affixed to electric irons, hair dryers, and a variety of other electrical appliances. Many of these plaques carry the warning "for use on 120-V AC only." The AC, of course, stands for alternating current, while the 120 V refers to the voltage of the electric power that is almost universally available from the wall, floor, and ceiling outlets in homes, offices, and hospitals. But what exactly is alternating current?

One way of approaching this question is to consider first the opposite of AC, namely, direct current or DC. Direct current is current in which the flow of electrons is in one direction only. It results when you apply a steady voltage to a circuit. Batteries are the most common source of DC; they produce DC electrochemically. Various different kinds of batteries capable of generating a wide range of voltages are available. They are found in automobiles, radios, flashlights, and other appliances. Despite the differences in their physical appearance, all have one characteristic in common: they provide a constant voltage so that the current flowing in the circuit to which they are connected moves in only one direction.

Alternating current or AC is a pulsating or fluctuating electric current that alternately flows in one direction and then in another. It results when an alternating voltage is applied to a circuit. Alternating current is usually produced by electric generators—huge rotating machines—at electrical power stations. As is the case with DC, we also speak of the voltage of an AC source. But because the voltage fluctuates or alternates, it is also necessary to specify the frequency of the alternation. The electricity used every day in homes, hospitals, and industry is 120-V 60 cycle or 60 Hz AC. Because AC is cheaper to produce than DC and also because it can be transmitted by power lines over long distances more efficiently than DC, AC is used in preference to DC as an energy source.

A 60 Hz AC completes one full cycle in 1/60 second (i.e., 0.0167 second). It flows in a positive direction for half the time and a negative direction for the remaining half cycle. The changes in voltage with time are sinusoidal, that is, they are described by a sine curve.

It is important to recognize that AC does not refer only to 60 Hz AC—the electrical energy that lights our homes and powers our appliances. Alternating current has a much broader meaning. The EEG is AC. It is not strictly sinusoidal, but it is AC nonetheless. The electric current generated by a microphone, an audio disc or tape, and a videotape is also AC. As a matter of fact, most sources of electric current are classified as AC sources.

AC Circuits

Thus far in this chapter, we have dealt only with the behavior of electrical circuits when a steady voltage is applied and with the transient response to a single, instantaneous change in voltage. All this falls under the heading of DC circuit analysis. We now turn to the topic of AC circuit analysis to examine the behavior of the same circuits when alternating instead of steady voltages are applied to them. We will deal only with the steady-state response—the response after any transients associated with connecting up the circuit to the source of voltage have subsided.

If all the circuits that we needed to deal with contained only the parameter resistance, our discussion could quickly be terminated. For circuits containing resistance only, Ohm's law applies regardless of whether AC or DC is involved. A resistor behaves in the same way regardless of the frequency of the current flowing. In other words, resistance is independent of frequency of the current flowing in a circuit. If, however, a capacitor is added to the circuit so that C is in series with R as in Fig. 2.6, Ohm's law needs to be modified. The formula in this case is changed to

$$I = \frac{V}{Z}$$

where Z is equal to the *impedance* of the circuit.

Impedance

The concept of impedance is important for the EEG specialist to understand. He or she deals with it on a daily basis whenever EEG leads are attached to a patient

[2] The reader may demonstrate this fact for him or herself by some simple algebra. From the last section we know that current in a series R–C circuit is:

$$I = (V/R)\,\varepsilon^{-(t/RC)}$$

Setting $t = RC$, we have

$$I = (V/R)\,\varepsilon^{-(RC/RC)}$$
$$= (V/R)\,\varepsilon^{-1}$$
$$= (V/R) \times (1/\varepsilon)$$

Since $\varepsilon = 2.718$,

$$I = (V/R) \times (1/2.718)$$
$$I = 0.37\,(V/R),$$

which means that when $t = RC$, the current is 37% of its initial value.

or when an EEG recording is interpreted. Thus, the EEG technician measures the impedance of each lead or each pair of leads before starting to take a recording. She/he knows that the impedance has to be low but not too low in order to obtain a satisfactory record. Similarly, the person interpreting the record needs to know that lead impedances were comparable whenever significant amplitude asymmetries show up in the tracings. With all this in mind, let us proceed with a discussion of impedance.

Impedance of an R–C circuit is the combined effect that the two parameters of resistance and capacitance have on the flow of current produced when an alternating voltage is applied to the circuit. Mathematically, impedance is equal to

$$Z = \sqrt{R^2 + \left(\frac{10^6}{2\pi fC}\right)^2}$$

where R is the value of resistance in ohms, C the value of capacitance in μF, f the frequency of the alternating voltage in Hz, and π the familiar constant that is equal to 3.14 . . . The term $10^6/2\pi fC$ is referred to as the capacitive reactance.

As is the case with resistance, Z the impedance is also measured in ohms. Let us consider the formula for Z carefully and list what it tells us about the characteristics of Z:

1. The value of Z depends on the values of the three quantities, namely, R, C, and f.
2. If C and f are held constant, Z varies directly with R.
3. If R and f are held constant, the term $10^6/2\pi fC$ or the capacitive reactance increases as C decreases and vice versa, so that Z varies inversely with C.
4. If R and C are held constant, the capacitive reactance increases as f decreases and vice versa, so that Z varies inversely with f. Note particularly that as f approaches zero, Z becomes very large indeed. In the limit when $f = 0$, the applied voltage is no longer alternating but becomes steady; under these conditions we are dealing with DC not AC, and the rules of DC circuit analysis would apply.

Let us summarize. Impedance is a *frequency-sensitive* quantity. Z varies with changes in frequency of the applied voltage as well as with changes in C and R. For this reason it is necessary to specify the particular frequency of the applied voltage whenever we talk about impedance. Thus, for example, we say that a particular circuit has an impedance of 10K ohms at 30 Hz. This property of impedance is uniquely due in the series R–C circuit to the capacitive reactance—to the presence of the capacitor. As we will see in a later chapter, this property of capacitance is the basic principle upon which the operation of the filters on an EEG machine is based.

Computational Example

The fact that differences in frequency have a profound effect on impedance is readily apparent from a simple example. Suppose in series R–C circuit (Fig. 2.6) that R = 10K ohms and C = 1 μF. What is the impedance at 1 Hz? The answer is obtained by substitution of these values in the formula for Z. Thus,

$$Z = \sqrt{R^2 + \left(\frac{10^6}{2\pi fC}\right)^2}$$

$$= \sqrt{(10^4)^2 + \left(\frac{10^6}{2 \times 3.14 \times 1 \times 1}\right)^2}$$

$$= \sqrt{(10^4)^2 + \left(\frac{10^6}{6.28}\right)^2}$$

$$= \sqrt{10^8 + 159{,}236^2}$$

$$= 159{,}550 \text{ ohms}$$

The reader should verify, by similar computation, that impedance at 35 Hz is equal to 10,986 ohms. Note that there is nearly a 15-fold difference in impedance at frequencies of 1 Hz and 35 Hz.

Frequency Response

The fact that impedance of a circuit varies with frequency finds expression in an important measure used to characterize the behavior of electrical circuits. This measure is the *frequency response* of a circuit. Whereas the transient response of a circuit is its response to an instantaneous step change in applied voltage, the frequency response is the response of the same circuit to an alternating applied voltage of constant amplitude that is allowed to vary in frequency. The frequency response of a particular circuit is reported as a *frequency-response curve*, the points for which are obtained by measuring the *amplitude* of the output voltage when voltages of different frequency but the same amplitude are applied to the input of the circuit.

The concept of frequency response will be familiar to readers who own or have used high-fidelity audio reproduction equipment. We know that sounds correspond to mechanical vibrations and that pitch is related to the frequency of these vibrations. Audible sounds have a frequency range of 20 to 20,000 Hz. Audio-reproduction systems simply pick up the mechanical vibrations via a microphone or phono pick-up, convert the vibrations to an alternating electrical voltage, amplify them, and then convert them back to mechanical vibrations in a loudspeaker. The fidelity of such systems is expressed by the frequency-response curve. Systems that reproduce music at very high fidelity have a frequency-response curve that is essentially

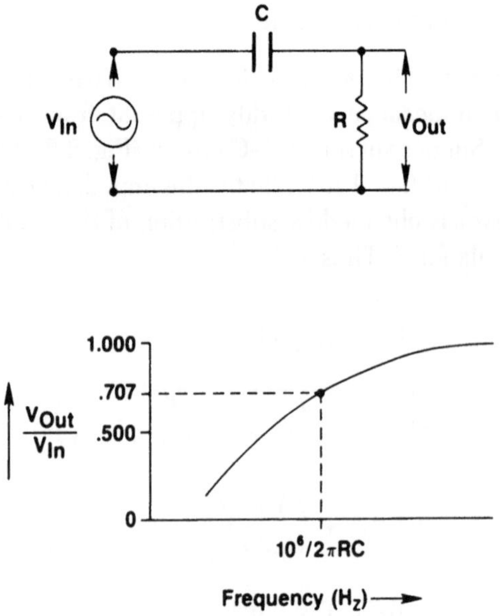

Figure 2.8. Frequency response of a series R–C circuit.

flat from 20 to 20,000 Hz. Low-fidelity systems, on the other hand, severely attenuate frequencies at both ends of the frequency spectrum so that the range of audible frequencies actually transmitted may be limited to only 100 to 8,000 Hz.

Although the frequency response of a circuit may be obtained empirically by applying an alternating voltage of variable frequency but constant amplitude to the input and then measuring the amplitude of the output, we may also calculate the frequency response from the circuit parameters. Figure 2.8 shows the frequency-response curve for the case of the now familiar series R–C circuit. This curve is derived in the following manner: taking the output of the circuit as the voltage across the resistor, we have from Ohm's law:

$$I = \frac{V_{in}}{Z} = \frac{V_{in}}{\sqrt{R^2 + (10^6/2\pi fC)^2}}$$

and again by Ohm's law

$$V_{out} = RI = \frac{R}{\sqrt{R^2 + (10^6/2\pi fC)^2}} V_{in}$$

Observe in the above formula that, when f is allowed to increase so that it becomes large, the quantity $10^6/2\pi fC$ will approach zero. This makes the fraction

$$\frac{R}{\sqrt{R^2 + (10^6/2\pi fC)^2}}$$

very nearly equal to

$$\frac{R}{\sqrt{R^2}} = 1,$$

so that V_{out} is almost equal to V_{in}. If we go in the opposite direction, i.e., allow f to decrease and become small, the denominator of the fraction becomes much larger than R the numerator, and V_{out} is a progressively smaller fraction of V_{in}. This is shown graphically by the frequency-response curve in the lower part of Fig. 2.8. Amplitude is plotted as the ratio of V_{out}/V_{in}. Since this ratio is always less than 1, we say that this circuit attenuates the input, with lower frequencies being attenuated more than higher frequencies. The dividing point between low and high frequency is arbitrarily taken to be the point at which frequency is equal to $10^6/2\pi fC$ Hz. This point is designated the *cutoff frequency* and corresponds to a 30% attenuation of V_{in}; in other words, $V_{out}/V_{in} = 0.707$ (see Fig. 2.8).[3]

What is the cutoff frequency for the series R–C circuit considered earlier in the computational example when R = 10K ohms and C = 1 μF?

$$\text{Cutoff frequency} = \frac{10^6}{2\pi RC}\text{ Hz}$$

$$= \frac{10^6}{2 \times 3.14 \times 10{,}000 \times 1}$$

$$= \frac{10^6}{6.28 \times 10^4} = \frac{100}{6.28}$$

$$= 15.9 \text{ Hz}$$

This means that a 15.9 voltage applied to the input will be attenuated by 30% when it appears at the output of the circuit. The practical significance of this finding will become apparent in Chapter 4.

[3] This is readily confirmed by some simple algebra. We have shown that

$$V_{out} = \frac{R}{\sqrt{R^2 + (10^6/2\pi fC)^2}} V_{in}$$

At the cutoff point,

$$f = 10^6/2\pi RC$$

Substituting this in the equation for V_{out}, we have

$$V_{out} = \frac{R}{\sqrt{R^2 + \left(\frac{10^6}{2\pi \times 10^6/2\pi RC \times C}\right)^2}} V_{in}$$

$$= \frac{R}{\sqrt{R^2 + R^2}} V_{in}$$

$$= \frac{R}{\sqrt{2R^2}} V_{in} = \frac{R}{\sqrt{2}\,R} V_{in}$$

$$= \frac{R}{1.414\,R} V_{in}$$

so that

$$\frac{V_{out}}{V_{in}} = 0.707$$

Chapter 3
The Differential Amplifier

In Chapter 1, we referred to the amplifier as the heart of the EEG machine. This chapter returns to the topic to fill in some important details. We begin with a consideration of electronic amplifiers in general, and then continue with a discussion and analysis of the differential amplifier. As we will see, the differential amplifier is a specially designed device for recording bioelectric activity, and it is employed universally in EEG machines. Our objective in this chapter is to find out precisely what this amplifier does and how it accomplishes its purpose.

Historical Background

Electronic amplifiers have an interesting developmental history, particularly so since their history illustrates the importance of serendipity as well as careful technological development in scientific research. A very brief look at the high points in this history illustrates nicely the principles upon which electronic amplifiers operate.

Everyone knows that Thomas Edison invented the electric light bulb. Few are aware, however, that he also discovered a phenomenon called the "Edison effect," which provided the basis for the development of the electronic amplifier. The year was 1884 and Edison was experimenting with his electric light bulb. Recall that a light bulb is nothing more than a glass envelope from which the air has been evacuated and which contains a loop of wire called the filament. Figure 3.1A shows a light bulb with a voltage source (a battery) connected to the filament. With the battery connected, the filament is made incandescent by the passage of an electric current, thereupon emitting light as well as heat.

In the course of his experiments, the purpose of which was to discover a method to prevent the bulb from darkening with use, Edison placed a second element—a metallic plate—inside the glass envelope. For reasons that are not entirely clear, he also connected a current-measuring meter between the plate and the heated filament. This circuit is shown in Fig. 3.1B. Upon connecting it up, Edison discovered that the meter showed a current was flowing in the plate circuit. When the battery was disconnected and the filament allowed to cool, the current flow ceased. Edison did not know the explanation for the current; nevertheless, he patented the device and the phenomenon became known as the Edison effect.

We now know that this remarkable discovery has a relatively simple explanation. The filament inside the bulb is a conductor of electricity. As we learned in Chapter 2, conductors have electrons that are only loosely bound to the nucleus. Connecting a voltage to the ends of the filament causes the electrons to move, producing an electric current. With large currents flowing, the filament heats up, causing the electrons to become more active until some of them are "boiled off" the filament. Electrons, of course, are negatively charged particles. The plate inside the glass envelope is positive, as it is connected through the meter to the positive terminal of the battery. Since unlike electrical charges attract, the electrons coming off the filament are attracted to the plate and flow in that direction. A flow of electrons, of course, is an electric current, which accounts for the deflection observed on the current-measuring meter.

Edison can hardly be blamed for not knowing the explanation of this phenomenon at the time of its discovery. Indeed, he had no way of knowing precisely what was happening in his circuit. The correct explanation had to await the discovery of the electron by Joseph J. Thompson some 13 years later.

The next event in our historical survey concerns the work of John A. Fleming, a British electrical engineer. In 1904, while experimenting with the Edison effect, Fleming discovered that Edison's device conducted in only one direction. What Fleming did was to hook up a battery in

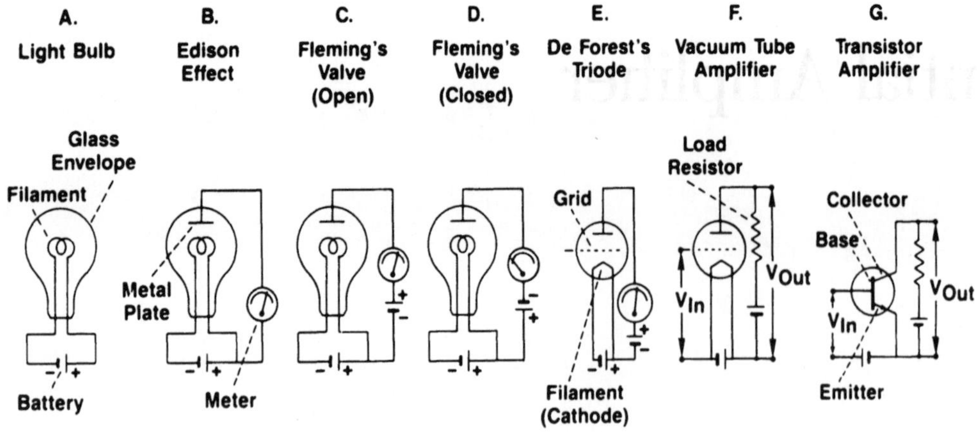

Figure 3.1. Development of the electronic amplifier.

series with the current-measuring meter; by this stroke of insight, he created the first *diode*. When the positive pole of the battery was connected to the plate, as shown in Fig. 3.1C, the meter registered a flow of current. Moreover, increasing the voltage of the battery resulted in a proportional increase in the amount of current flowing. When the polarity of the battery was reversed, however, as in Fig. 3.1D, no current flowed in the plate circuit. The reason for this is obvious. With the plate negative instead of positive, the electrons coming off the filament are repelled. The upshot of this was that you could control current flow by the voltage on the plate of the device. Fleming likened the action of this device to that of a "valve," a term used in the United Kingdom for a vacuum tube.

Valves are devices capable of controlling the flow of energy, usually with very small changes in energy level at the control end. In other words, by exerting a relatively small force at the control end of the valve, you can generate a very large flow through the device. A common water faucet is a familiar example; by using only a small force on the faucet, a huge stream of water may be turned on and off. Fleming's valve had the capability of doing the same for an electric current.

Lee De Forest, an American inventor, was impressed with Fleming's valve and began studying methods for improving the device. Through a stroke of inventive genius, he added a third element to Fleming's valve, whereupon the *triode* or three-element vacuum tube was born. The third element was called the *grid*, a plate-like object consisting of a mesh of fine wire that was positioned between the filament and plate. De Forest's triode is shown in Fig. 3.1E. The glass envelope in the illustration is drawn in the traditional manner of a vacuum tube, i.e., as a closed circle.

De Forest discovered that the current (I) flowing in the plate circuit of the triode could be controlled by the voltage applied between the grid and the filament. With the grid at zero volts, the plate current was at maximum. By making the grid negative with respect to the filament, the plate current could be reduced in proportion to the degree of negativity. As the grid became more and more negative, a point was reached where the current stopped flowing. The explanation of this phenomenon is simple. There is a competitive action on the electrode stream. The *anode*, or positive plate, attracts the electrons coming from the filament or *cathode* of the tube, whereas the negatively charged grid repels the electrons. The number of electrons flowing to the plate is determined, therefore, by the net effect of these two opposing forces. De Forest found that very small changes in grid voltage could produce a considerable change in the flow of current in the plate circuit of the tube.

From De Forest's grid-controlled triode, it was but a simple step to the vacuum-tube amplifier. This was accomplished by the addition of a resistor to the plate circuit—the so-called load resistor. The circuit is shown in Fig. 3.1F. Since Ohm's law states that $V = IR$, it should be clear that by allowing the changes in plate current to flow through a resistance, changes in voltage can be developed. The changes in plate voltage are an exact copy of the changes in grid voltage, with the exception that they have been amplified and are considerably larger. We see, therefore, that an electronic amplifier does not, in reality, magnify or amplify a particular voltage. Instead, it uses this voltage to control or modulate the current flowing through a resistor in a separate circuit. The voltage developed across the resistance in this circuit (Fig. 3.1F) then becomes the "amplified" voltage.

The final step in the development of the electronic amplifier is the substitution of the transistor for the vacuum tube in the circuit. Although it was an eminently useful device in its time, the vacuum tube had some serious limitations. To obtain the electrons that make it work, it is necessary to heat a filament to incandescence by passing an electric current through it. To do this, large amounts of electrical energy are expended, which, in turn, generate considerable heat. When a circuit requires a large number of vacuum tubes, the cost of heating the filaments and then cooling the circuits to avoid damage from excessive heat

Figure 3.2. Development of the differential amplifier.

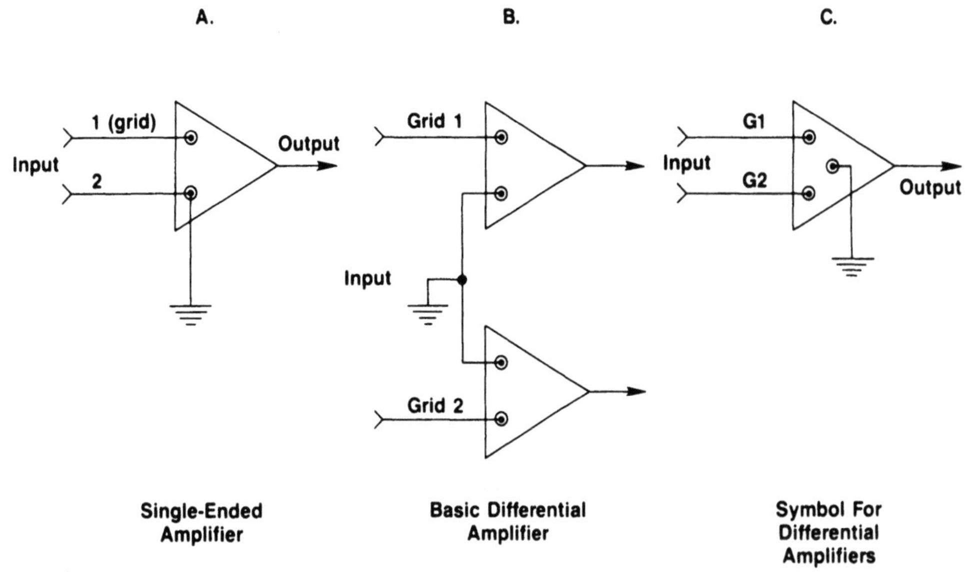

can be considerable. These problems were readily overcome by the development of the transistor that consumes no energy in heating filaments. In addition, the transistor has other advantages over the vacuum tube, namely, its smaller size, lighter weight, greater efficiency, and ability to operate with low voltage and current.

Transistors are devices composed of solid-state materials —substances called semiconductors. In terms of their electrical conductivity, semiconductors fall between conductors and insulators. The structure of these substances is such that not all of their electrons are bound tightly to the nucleus; in fact, they can be shaken loose at room temperature. This is accomplished by applying voltages to the semiconductors; when this is done, the device conducts an electric current in a manner analogous to what happens in a vacuum tube. There are important differences in the way a transistor functions when compared with a vacuum tube. These differences need not concern us, as the principles of amplification are basically alike in both devices.

Figure 3.1G gives a circuit for a simple transistor amplifier. Note the similarities to the vacuum-tube amplifier shown in Fig. 3.1F. Like De Forest's triode, the transistor has three elements: the *base*, which corresponds to the grid of the vacuum tube; the *collector*, which corresponds to the plate; and the *emitter*, which corresponds to the filament or cathode.

Single-Ended Amplifier

The earliest practical electronic amplifier was the *single-ended* or *single-sided* amplifier. It, of course, employed a vacuum tube as its active element and was operated on batteries. Figure 3.1F is a circuit for such an amplifier. Gain or amplification of the device is equal to the ratio of the change in output voltage to change in input voltage. Although this amplifier has two inputs, it is referred to as a single-ended amplifier because one of the inputs is a reference point that is connected to the earth or "ground."[1]

Note in Fig. 3.1F that the active input of the amplifier is the grid of the vacuum tube. In the case of a transistor amplifier, the active input would be connected to the base of the transistor. While most amplifiers have abandoned vacuum tubes in favor of transistors, the term "grid input" is still used in referring to the active input. Figure 3.2A shows the symbol used in a circuit diagram to denote a single-ended amplifier. Grid input is input 1; input 2 is connected to earth, as shown by the symbol for ground, which consists of a series of short horizontal lines, one below the other, with the lower lines being progressively shorter than the lines above.

Amplifying Bioelectric Activity

Historically, the first amplifiers used to amplify bioelectric activity were single-ended amplifiers. Although these amplifiers were adequate for dealing with high-level electrical signals like the ECG and the action potentials from some peripheral nerves, they proved to be of little if any value in the case of low-level signals like the EEG. There were three major reasons for this.

First, the single-ended amplifier is sensitive to outside interference like the 60-Hz activity from the power lines. This happens because one side of the 60-Hz power line—the neutral or "cold" side—is connected to ground. As input 2 of a single-ended amplifier is also connected to

[1] Ground is used universally as an electrical reference because the earth is the most stable source of zero voltage available to us.

ground, the grid (input 1) serves as an antenna and "picks up" the 60-Hz activity, which is then amplified. There may be more than 0.1 V of 60-Hz activity present at the input, in which case brain electrical activity would be obliterated. The 60-Hz artifact is all that would be recorded. It is alleged that during the 1930s, some electrophysiologists working in hospitals powered by DC actively resisted the conversion to AC because of the introduction of 60-Hz artifact into their recordings.

Second, the single-ended amplifier is sensitive to artifacts, like the ECG, that are generated within the body. The ECG artifacts can be picked up anywhere on the surface of the body and, in some instances, may be several millivolts in amplitude on the head. With such a large voltage present at input 1 of the amplifier, and with input 2 at ground potential, all but the very highest amplitude brain electrical activity would be obscured by the ECG. This may readily be demonstrated by converting one channel of the EEG machine to a single-ended amplifier as follows: Using the appropriate switch on the lead selector panel, connect input 2 of a channel on the EEG machine to ground. Connect input 1 of this channel to any EEG lead already on the patient. Now with all the other leads connected in the normal way, turn on the machine—first being careful to switch in the 60-Hz filter and to turn down the gain of this channel. What you see on the chart is an ECG whose voltage is so large that the gain of the amplifier cannot be raised sufficiently to bring the channel within the dynamic range of any EEG voltages. This dramatic effect is simply the result of grounding the differential amplifier, thereby converting it into a single-ended amplifier.

Third, the single-ended amplifier does not permit simultaneous, *independent* recording of EEGs from multiple placements on the scalp. This results from the fact that input 2 of a single-ended amplifier is connected to ground; if several amplifiers are involved, all will be connected to the same common point. Removing the ground connection has little effect since input 2 of the amplifiers is also connected to the power supply of the machine, which is a point common to all the channels. Thus, an event that occurs at *any* of the electrodes connected to input 2 of the amplifiers will appear in each and every channel. This means that it is all but impossible to obtain multichannel, bipolar recordings using single-ended amplifiers.

It should be clear from the foregoing discussion that EEG recording as we know it today would not be feasible if only the single-ended amplifier were available. To overcome the serious shortcomings of the single-ended amplifier, the push-pull amplifier and the *differential* amplifier were developed. We will consider only the differential amplifier as the push-pull amplifier was merely a step in the development of the differential amplifier.

The Differential Amplifier—Basic Concept

The differential amplifier was uniquely the result of a collaborative effort on the part of electrical engineers, electronic engineers, and neurophysiologists. It was developed in the 1930s primarily to meet the needs of multichannel EEG recording. Numerous well-known persons were involved; among these were E.D. Adrian, Alexander Forbes, B.H.C. Matthews, Franklin Offner, J.F. Toennies, and W. Grey Walter.

In principle, the differential amplifier is nothing more than two identical single-ended amplifiers connected back-to-back. This is illustrated in Fig. 3.2B. Note that the grids of the two amplifiers become the two inputs of the differential amplifier, while the other inputs are joined together and then connected to ground. By convention, the two inputs are designated grid 1 (G1) and grid 2 (G2). Because the inputs each go to a grid (base in the case of transistors), they are isolated from ground and from the power supply. This means that with differential amplifiers, many channels can be connected to the patient simultaneously, without any of the inputs being joined together by the machine. Figure 3.2C shows the symbol used to denote a differential amplifier.

A second important feature of the differential amplifier derives from the fact that the two halves of the amplifier are *balanced*, with one half being the "mirror image" of the other half.[2] The result is that the differential amplifier amplifies the *difference* between the voltages simultaneously present at the two inputs. This means, of course, that the output of this amplifier will be zero whenever identical voltages are present at both inputs and either negative or positive when the voltages at the inputs are different. Electroencephalographic convention dictates that all channels on the EEG machine deflect upward when the voltage at Gl is *negative* with respect to the voltage at G2 and *downward* when the voltage at Gl is *positive* with respect to the voltage at G2. Another way of saying the same thing is that a channel deflects upward when G2 is positive with respect to Gl and downward when it is negative with respect to Gl. Note that we say positive or negative "with respect to." This means that the polarity at any point is not an absolute quantity but is the polarity relative to the voltage present at another point.

The unique ability of the differential amplifier to amplify the difference between the voltages simultane-

[2] Because of this, differential amplifiers are sometimes referred to as "balanced" amplifiers, or amplifiers with a balanced input. This is in direct contrast to single-ended amplifiers, which are also referred to as amplifiers having an unbalanced input.

ously present at the two inputs gives rise to two important concepts in EEG technology, namely, the phenomena of *cancellation* and *summation*. The importance of these concepts can hardly be overemphasized as they determine the size, shape, waveform, and polarity of the tracing seen in the EEG record. They are discussed in detail in a later chapter that deals with the topics of polarity and localization. For the present, we will be concerned only with the phenomenon of cancellation. Moreover, we deal here specifically with cancellation as it relates to the problem of 60 Hz and ECG artifacts in an EEG recording. In this context, cancellation comes under the heading of an important descriptive term applied to differential amplifiers. This is the term *common-mode rejection*.

Common-Mode Rejection

We said earlier that a differential amplifier consists of two balanced, single-ended amplifiers appropriately joined together and that the output of this device is proportional to the difference between the voltages simultaneously present at the two inputs. These inputs can be either *out-of-phase* signals or *in-phase* signals; in-phase signals are also referred to as *common-mode* signals. Out-of-phase signals are signals in which the voltages simultaneously present at both inputs of the differential amplifier are different, whereas in-phase signals are signals in which the voltages at both inputs are the same. Brain electrical activity is a good example of the former, as there may be gross differences in the voltages present even at two closely spaced points on the scalp. The ECG and 60 Hz artifacts are examples of in-phase or common-mode signals.

The result is that brain electrical activity, being primarily an out-of-phase signal, gets amplified by the differential amplifier. On the other hand, 60-Hz activity and the ECG are cancelled out. The degree to which the in-phase voltages are cancelled out is determined largely by the extent to which the two halves of the amplifier are balanced. How well common-mode voltages are cancelled out or *rejected* by a particular differential amplifier is expressed in terms of the amplifier's *common-mode rejection ratio* or CMRR. The CMRR of an amplifier is estimated from the equation:

$$\text{CMRR} = \frac{\text{voltage out for out-of-phase voltage in}}{\text{voltage out for in-phase voltage in}}$$

or, alternatively,

$$\text{CMRR} = \frac{\text{Out-of-phase gain}}{\text{In-phase gain}}$$

where the in-phase and out-of-phase input voltages are of equivalent amplitude. The EEG amplifiers may have CMRRs ranging from 1,000:1 to 20,000:1.

A simple example will help to clarify what CMRR means and how it is used. Suppose that your EEG machine has a CMRR of 1,500:1. This means that common-mode voltages are not amplified—or multiplied by 1, which is the same thing—while out-of-phase signals are amplified 1,500 times. Now consider an EEG voltage that is 7 µV. Amplifying this voltage 1,500 times yields 7 × 1,500 = 10,500 µV or 10.5 mV. So we see that a 7-µV EEG voltage produces the same deflection on the chart as a 10.5-mV ECG voltage. If the EEG machine is set to the standard deflection sensitivity of 7 µV/mm, it is apparent that an ECG voltage as large as 10.5 mV would have to be present between each of a pair of EEG leads and ground before being detected as a 1-mm deflection in the tracing.

The CMRR of the amplifiers in an EEG machine has the same effect on external sources of voltage as well, that is, if they are common-mode signals. Of these, the most troublesome for the EEG technician is the 60 Hz artifact from the AC power lines. The simple example used to explain what happens in the case of an ECG artifact applies equally well to 60 Hz artifact. It is important to recognize in this context that the degree of rejection that may be realized applies to the common-mode voltage only. If the source of the common-mode voltage is from within the patient's body, as in the case of the ECG, it can happen that the voltages at the two inputs of the amplifier will not be exactly identical. This comes about mainly because the electrodes on the patient are not the same distance from the heart, which is the source of the voltage, or because there is a wide gap between the electrodes. The important thing to recognize here is that voltage, at any point, varies inversely with the distance from the source. This is one reason why ECG artifacts in the EEG are always larger with widely spaced electrodes and vice versa.

An additional point that will be taken up in detail later when we discuss the topic of electrodes and electrode impedances deserves brief mention here. In order to realize the CMRR quoted in the specifications of a particular amplifier, it is essential for the input circuit of the amplifier to be balanced. This means that the impedances of both leads in the pair need to be very nearly the same. If the impedance of one lead happened to be substantially higher than the impedance of the other, say 4 or 5 times higher, the input circuit would become unbalanced and the common-mode rejection capabilities of the amplifier would be significantly degraded. We see, therefore, that when correctly utilized, the differential amplifier's ability to reject common-mode voltages constitutes a powerful method of eliminating artifacts from an EEG tracing.

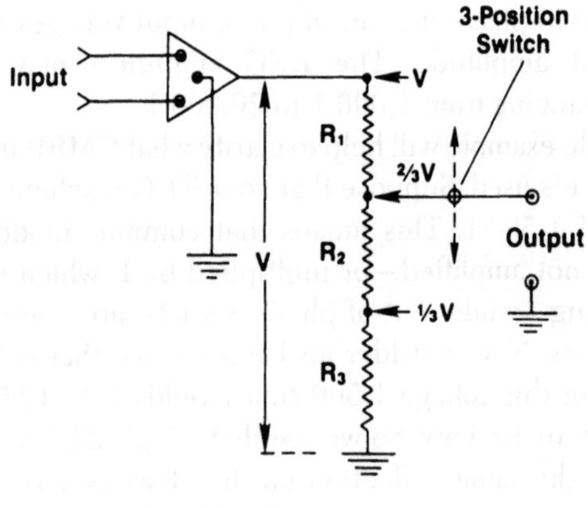

Figure 3.3. Output of a differential amplifier connected to a three-position voltage divider.

Sensitivity or Gain

As was mentioned in Chapter 1, the voltages commonly recorded in the EEG cover a wide dynamic range. At the low-amplitude end of the scale, voltages as small as 2 μV need to be detected in brain death recordings, while at the high end voltages as large as 2,000 μV or 2 mV are encountered in hypsarrhythmia. To provide for this wide dynamic range, all EEG amplifiers have a switch for changing the sensitivity or gain to accommodate these different voltages. A frequently used design employs a 12-position rotary switch that permits the EEG technician to select deflection sensitivities from 1 μV/mm to 70 μV/mm. Some EEG machines also employ a two-position *gain multiplier switch*. When this switch is in the "μV/mm" position, deflection sensitivity is read from the 12-position rotary switch in microvolts per millimeter. When, on the other hand, the gain multiplier switch is in the "mV/cm" position, deflection sensitivity is in millivolts per centimeter. The latter range of sensitivities gives flexibility to the machine. When hooked up with the appropriate transducer, these settings are used to record other physiological phenomena like pulse rate, respiration rate, body temperature, blood pressure, etc.

The gain changes are accomplished by a simple application of Ohm's law. Figure 3.3 shows how this is done. The output of the differential amplifier is connected to one end of a chain of three resistors (R_1, R_2, and R_3) in series, the other end of which is connected to ground. With an out-of-phase voltage applied to the input of the amplifier, a current will flow through the resistors. By Ohm's law,

$$I = \frac{V}{R_1 + R_2 + R_3}$$

Solving the equation for V, we have

$$V = (R_1 + R_2 + R_3)\, I$$

or

$$V = R_1 I + R_2 I + R_3 I$$

which means that V, the output voltage of the amplifier, is equal to the sum of the voltages across each of the resistors. In other words, the total voltage divides itself across the three resistors in proportion to their values. If R_1, R_2, and R_3 were all equal to each other, the voltage at the top of R_3 would be equal to 1/3V. With the three-position switch at the top of R_2, as shown in Fig. 3.3, the output of the switch is 2/3V. The chain of resistors wired to a switch in this fashion is referred to as a *voltage divider*. The selector switch may employ any number of different positions. For each additional position, an additional resistor is connected in series with the rest.

Amplifier Noise

Ideally, the tracings on the chart corresponding to each channel of the EEG machine should each be a completely straight, horizontal line when the inputs of the amplifiers are connected to zero volts.[3] Although this should be the case for a properly operating EEG machine switched to a standard gain of 7 μV/mm, it is rarely true when the machine is adjusted for the maximum gain of 1 μV/mm. Note that at such high gains, the pens do not describe an even trace but appear to wander about randomly. This is the so-called internally generated *noise* and is normal for all machines at maximum gain. For routine clinical work, noise level should be less than 2 μV peak-to-peak referred to the input.

Where does this noise come from? Noise or *random voltage fluctuation* is an inherent characteristic of all resistors, vacuum tubes, and transistors of which an amplifier is constructed. Some of these devices produce less noise than others of their kind by reason of their construction. Thus, for example, so-called low-noise resistors are used in certain sensitive portions of the circuit of an EEG amplifier. The connections in a rotary switch, or for that matter any kind of switch, can also generate noise. This is the reason why EEG technicians are frequently seen to be repeatedly indexing the switches on a noisy channel of the EEG machine. By this maneuver they hope to clean the switch contacts and hence reduce the noise level. A leaky capacitor can also produce noise, and even the solder joints used

[3] How do you go about connecting the inputs of a differential amplifier to zero volts? A little thought will show that the simplest way of doing this is to connect G_1 and G_2 together. This maneuver is referred to as "short-circuiting" the input of the amplifier; it brings both inputs to the same voltage so that the difference between the two is zero volts.

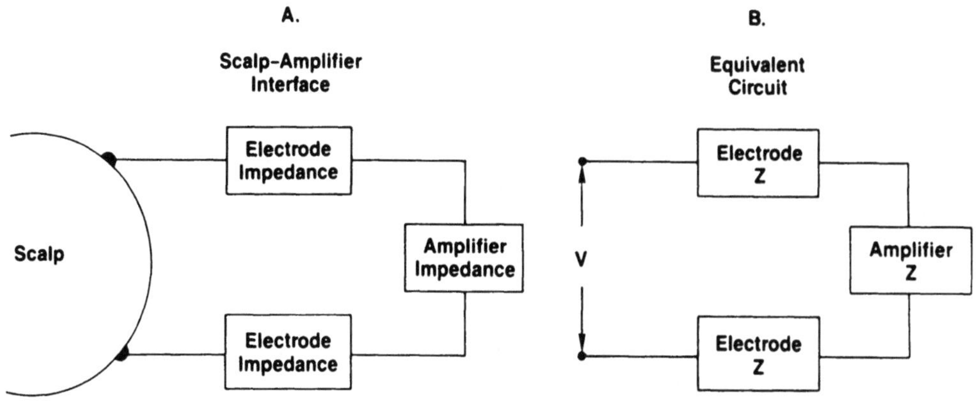

Figure 3.4. Schematic of the connections from scalp to electrodes to amplifier in EEG recording (A), and equivalent circuit of the same (B).

to connect the various components found in an amplifier may be noisy. Modern technology notwithstanding, the maximum noise-free gain required by an EEG amplifier is close to the theoretical limit available by ordinary methods.

Input Impedance

We have discussed a number of important characteristics of a differential amplifier—common-mode rejection, gain, and noise. One very important characteristic remains to be considered and that is *input impedance.* We have already discussed impedance in general. Recall that impedance is designated by the symbol Z; in the circuits that the EEG specialist deals with, Z consists of two components, resistance and capacitive reactance.

It will be helpful to begin our discussion of this topic by considering an important property that is shared by all *measuring instruments.* To yield valid measurements, a measuring instrument must not have any backward effect on the phenomenon being measured. By way of example, suppose that you wished to measure a patient's respiration rate by tying a loop of wire containing a strain gage transducer around the patient's chest. Each inspiration and expiration would be picked up by the strain gage. By appropriately displaying these on a channel of the EEG machine, you could count the number of waves occurring in a minute's time and, in so doing, estimate the respiration rate. But the respiration rate estimated in this way would only be the true rate if the loop of wire had no effect upon the patient's breathing. If, for example, the wire were too tight, it could interfere with normal breathing and yield a spurious result. This is an example of the backward effect of a measuring instrument upon the phenomenon being measured.

Each channel of the EEG machine may be considered an instrument for measuring various features of the EEG. Among these features is the voltage of the waves recorded. To validly measure voltage, the voltage appearing at the scalp must not be modified by connecting the electrodes and amplifier to the source. To satisfy this requirement, the input impedance or *impedance looking into the input* of the amplifier must be very high—many, many times higher than the impedance of the electrodes. Input impedance of the differential amplifiers in an EEG machine is commonly 20 million ohms (abbreviated 20 megohms, or 20M ohms) or higher.

Why is a high-input impedance essential? The reason may be understood from a perusal of Figs. 3.4A and B. The circuit in Fig. 3.4B will be recognized as a voltage divider to which V, the true EEG voltage is connected. This voltage divides itself across the electrode impedances and the amplifier impedance *in proportion to* their magnitudes. If the amplifier impedance is not very high with respect to electrode impedance, then a substantial fraction of V will appear across the electrodes and the voltage appearing across the amplifier will be significantly less than V. We speak of this as a condition in which the signal being observed is "loaded" by the input circuit of the measuring device. It is obvious that as amplifier Z increases and becomes very large, the voltage appearing across the amplifier will very closely approximate the voltage at V, the true EEG voltage. Refer to Fig. 2.4, Chapter 2, for computational formulas.

To obtain the high input impedances required for an EEG amplifier, a specially designed transistor is used in the input stage. This device is referred to as a field-effect transistor or FET. Recent advances in transistor technology have resulted in the development of new and even better FETs. Among them are the junction field-effect transistor (JFET) and the insulated gate FET (IGFET). But these are highly technical subjects that are unlikely to be of concern to the EEG specialist.

Special-Purpose Connections

Most modern EEG amplifiers have an output jack that is used for connecting the amplified EEG voltage to some peripheral device such as, for example, a cathode-ray oscilloscope, an instrumentation tape recorder, a computer, or the like. This jack can be found somewhere on the front panel of the amplifier or in a central location on the machine console. The voltage available at this jack is referred to as the IRIG output, which is nothing more than a 2.8-V (peak-to-peak) signal that corresponds to a 25-mm pen deflection on the EEG chart. The letters IRIG stand for Inter-Range Instrumentation Group.[4]

Some EEG machines have an IRIG input jack as well. This jack is used to play back signals from a tape recorder onto the EEG chart. A 2.8-V (peak-to-peak) signal fed into this jack will result in a 25-mm pen deflection.

These special-purpose connectors are not used in most clinical EEG laboratories. They are mentioned here, however, because a few laboratories do store EEG records on magnetic tape. Moreover, their use is essential whenever video display techniques are used, as in seizure monitoring (See Chapter 18).

The EEG Amplifier as a Whole

The differential amplifier is employed only at the first stage of amplification in an EEG machine. Once the advantages of the differential amplifier have been realized, the differential signal is converted to a single-ended signal and the rest of the necessary gain is provided by single-ended amplifiers. Details of the conversion process need not concern us here. The various stages of amplification are coupled together by means of capacitors. These so-called blocking capacitors are used to prevent any DC voltages from being transmitted from one stage to another and thereby being amplified. More will be said about the operation of blocking capacitors in the chapter on filters.

[4] The Inter-Range Instrumentation Group was a group of engineers from various guided-missile-testing ranges around the country. This group felt the need for standardizing the telemetry systems that were being used in guided-missile testing so that a system used on one range would be interchangeable with a system used on another. One device used extensively in telemetry systems is the instrumentation tape recorder, or "mag" tape recorder as it is called. For this reason, standardization procedures included specifications for mag tape recorders. Among the specifications called out is the standard input voltage operating level. Therefore, when we say that an EEG amplifier has an IRIG output available or is IRIG compatible, this means that the amplifier may be directly connected to (interfaced with) any modern mag tape recorder.

Chapter 4
Filters

If the amplifier is the heart of the EEG machine, then the filters could rightly be referred to as the brain. The filters are the selective devices that screen out unwanted signals and determine which features of the EEG will appear in the tracing. As we already mentioned in Chapter 1, there are three different types of filters on modern EEG machines: low-frequency filters, high-frequency filters, and 60-Hz "notch" filters. All three types operate according to the same basic principles. We will consider the principles of operation first before going into the details of how they actually function.

The Need for Filtering

In our treatment of amplifiers in the last chapter, little was said about the frequency characteristics of the voltages that were amplified. Indeed, aside from the fact that the 60-Hz power line voltage is discriminated against, the differential amplifier we discussed amplifies—within certain practical limitations—voltages over the entire spectrum of frequencies as well as any DC voltages. This means that we would see all varieties of electrical activity in the output in addition to the EEG voltages. For example, there could be static (DC) potentials from the electrodes, sweat artifacts, galvanic skin potentials, and eye movement artifacts in the low end of the frequency spectrum. In the intermediate range of frequencies, there could be muscle action potentials present, while at the high-frequency end we could observe transmissions from local radio stations that send out especially strong signals. In other words, aside from the common-mode rejection feature, the amplifiers in an EEG machine have little or no selectivity as far as frequency of the input voltage is concerned. The filters provide the necessary selectivity.

Basic Concept and Function

Ideally, a filter should admit, without modification, voltages of all frequencies that are of interest; at the same time, it should completely reject voltages of unwanted frequencies. Thus, for example, if we were interested in voltages within the frequency range of 1 to 35 Hz, the "ideal" filter for this purpose would function as shown in Fig. 4.1. But as we will see, there is no such thing as an ideal filter. Indeed, the filters commonly found on an EEG machine are far from ideal. For this reason the individual interpreting EEGs, as well as the EEG technician, needs to be thoroughly familiar with the way in which filters function.

In Chapter 2 we showed that the impedance of an electrical circuit varies with changes in frequency of the voltage that is applied to it. A simple computational example revealed that the impedance of a series R–C circuit, where $R = 10\text{K}$ ohms and $C = 1\ \mu\text{F}$, changed from a value of 159,550 ohms at 1 Hz to a value of 10,986 ohms at 35 Hz. The reason for this difference in impedance is that the capacitive reactance, which is a component of the impedance, is a frequency-sensitive quantity. The fact that capacitive reactance changes with frequency is the operating principle upon which the filters in an EEG machine work. This makes the capacitor the most vital part of the filter circuit.

The EEG filters are just one of a class of so-called "tuned" circuits. A more common type of tuned circuit is the frequency-selective circuit in a radio or television receiver—the circuit that is adjusted when you tune in a particular station or channel. Such circuits, which in reality are sharply tuned filters, admit only a very narrow band of frequencies while sharply attenuating all the rest. These circuits, along with all other filter circuits, follow the same

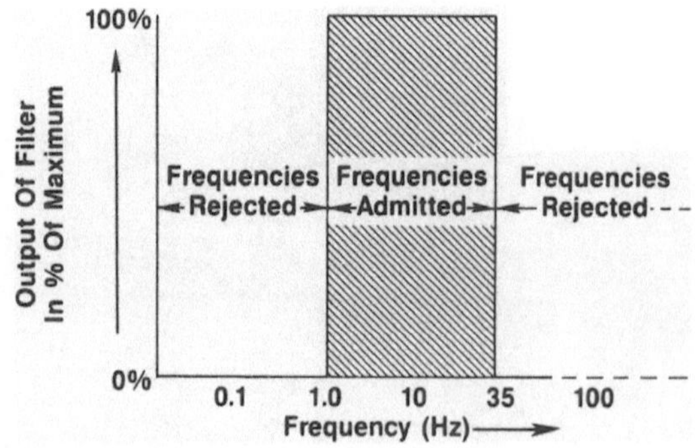

Figure 4.1. Response of an "ideal" filter.

basic design concept, namely, that impedance of an electrical circuit varies with changes in frequency.

Low-Frequency Filter

Although it was not pointed out at the time, the circuit (Fig. 2.8) that was analyzed in our discussion of frequency response is, in reality, a low-frequency filter. Let us return now to this circuit and discover how the circuit is used to prevent voltages at the low end of the frequency spectrum from getting into the EEG tracing.

Figure 4.2 shows a low-frequency filter with its input end connected to a pair of electrodes attached to the scalp of a patient and the output connected to the differential amplifier of one channel of an EEG machine. The filter circuit is balanced, that is, $R_1 = R_2$ and $C_1 = C_2$. As we pointed out in an earlier section of this chapter, the signal led off from the scalp by the electrodes consists of a variety of electrical phenomena in addition to the EEG voltages. This signal can include a wide range of frequencies and may include a DC voltage from the electrodes as well, as we will discover in a later chapter (Chapter 7). One purpose of the low-frequency filter is to prevent any such DC voltage from getting into the amplifier and being amplified. This is accomplished by an interesting property of capacitors that was discussed in Chapter 2. This property is simply that a capacitor has the unique ability to block the transmission of steady or unchanging voltages.

The impedance of each half of the circuit is given by the expression

$$Z = \sqrt{R^2 + \left(\frac{10^6}{2\pi fC}\right)^2}.$$

To simplify the analysis, we consider only one half of the filter; the other half being identical, the same analysis applies. Taking the voltage between electrode A and electrode C (ground) as V_{in}, the input to the upper half of the filter, and the voltage across R_1 as V_{out}, the output, we have, by Ohm's law,

$$I = \frac{V_{in}}{Z} = \frac{V_{in}}{\sqrt{R_1{}^2 + (10^6/2\pi fC_1)^2}}$$

and again by Ohm's law,

$$V_{out} = R_1 I = \frac{R_1}{\sqrt{R_1{}^2 + (10^6/2\pi fC_1)^2}} V_{in}.$$

The ratio of V_{out} to V_{in}, which is referred to as the *transfer characteristic* of the filter, is equal, by simple algebra, to

$$\frac{V_{out}}{V_{in}} = \frac{R_1}{\sqrt{R_1{}^2 + (10^6/2\pi fC_1)^2}}$$

Note in this formula that for large values of f, the capacitive reactance or $10^6/2\pi fC_1$ can become quite small so that V_{out}/V_{in} approaches its maximum value of 1. The opposite effect takes place at the other end of the frequency spectrum. Thus, as f decreases and attains smaller and smaller values, capacitive reactance becomes larger, the overall impedance increases, and V_{out}/V_{in} becomes smaller and smaller.

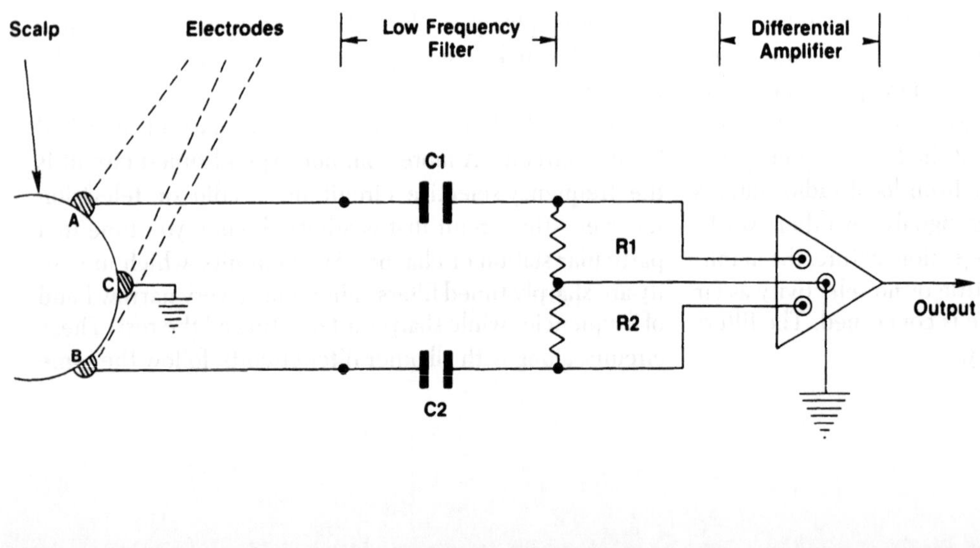

Figure 4.2. Circuit for a low-frequency filter in an EEG machine.

Figure 4.3. Asymptote plots showing the frequency response of a low-frequency filter. Solid line, cutoff frequency = 1 Hz; dashed line, cutoff frequency = 0.3 Hz.

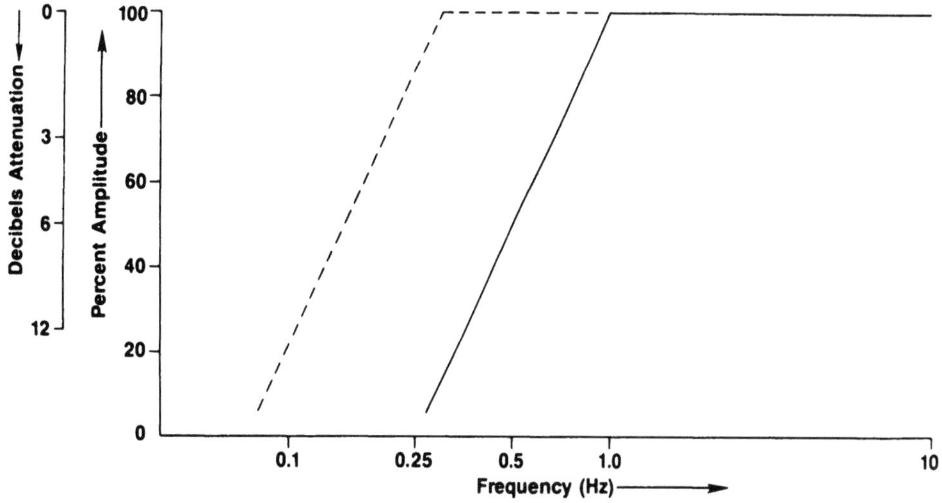

As was mentioned earlier in Chapter 2, the ratio of V_{out}/V_{in} when frequency is allowed to vary is termed the frequency response of the circuit. Frequency-response curves are plotted with frequency in hertz units on the horizontal axis and amplitude on the vertical axis. Frequency is always plotted on a logarithmic scale, while amplitude is plotted on a linear scale as a ratio or percentage of the maximum voltage, or on a logarithmic scale in decibels (abbreviated dB). In the case of a filter, V_{in} is the maximum voltage so that the ratio V_{out}/V_{in} comes out to be less than 1.

Low-Frequency Response—Asymptote Plot

Although the frequency response of a circuit is described by a curve and not by a straight line, for the sake of simplicity it frequently is approximated by straight lines. Figure 4.3 shows the straight-line approximation for the low-frequency filter shown in Fig. 4.2. The two solid lines in the graph define the response of the filter when it is set for a cutoff frequency of 1 Hz, the standard EEG setting. Observe that these lines meet at the cutoff point, with the line to the right of the cutoff point being parallel to the horizontal axis. The straight line to the left of it falls off at the rate of 6 dB/octave; which means that when frequency is halved, amplitude is also halved or attenuated by 50%. Thus, the amplitude at 0.5 Hz is one half the value at 1 Hz.

The straight-line approximation of the frequency-response curve is called the *asymptote plot*. This name derives from the fact that the straight lines define the asymptotes of the actual or true frequency-response curve. Characteristics of the true curve are discussed later in this chapter.

Figure 4.3 also shows that changing the cutoff point of the low-frequency filter from 1 to 0.3 Hz simply shifts the asymptote to the left by 0.7 Hz at the 100% amplitude point. This is shown by the dashed lines that meet at the 0.3 Hz point. Note that in this case as well, the line defining the cutoff falls at the rate of 6 dB/octave, which means that when frequency is halved or equal to 0.15 Hz, amplitude is also halved or attenuated by 50%. The same principle applies to the other settings on the low-frequency filter. Thus, for example, in the case of a cutoff frequency of 5 Hz, the asymptote plot is shifted to the right.

In Chapter 2 we learned that in terms of the circuit parameters, the cutoff frequency of a series R–C circuit is equal to $10^6/2\pi RC$ Hz. In practical terms this means that when you switch from one low-frequency filter setting to another, what you are doing is changing the values of C or R, or the values of both in the circuit. Larger values of C or R go along with lower cutoff frequencies, whereas smaller values go along with higher cutoff frequencies.

High-Frequency Filter

Surprising as it may seem, the high-frequency filters on an EEG machine employ the same kind of circuit as the low-frequency filters. Both use a series R–C circuit, the only morphological difference between the two being the position of the circuit parameters with respect to the output. As we have seen in Fig. 4.2, the output of the low-frequency filter is taken across the resistor. In the case of the high-frequency filter on the one hand, the output is taken across the capacitor in the circuit. Another difference between the two kinds of filters is their location within the EEG amplifier. Thus, while the low-frequency filter is commonly placed at the input to the differential amplifier or in the very early stages of amplification, the high-frequency filter is located down the line from the input.

Figure 4.4 gives the basic circuit for a high-frequency filter. Note that the circuit is placed after the differential amplifier and after the voltage of interest has been converted to a single-ended signal. This being the case, only a single R–C branch is required. As in the instance of the low-frequency filter, the current in the circuit is given by the expression

$$I = \frac{V_{in}}{\sqrt{R^2 + (10^6/2\pi fC)^2}}$$

and V_R, the voltage across the resistor, is

$$V_R = RI = \frac{R}{\sqrt{R^2 + (10^6/2\pi fC)^2}} V_{in}.$$

Since V_{in} divides between the two elements in the circuit,

$$V_{in} = V_R + V_{out}$$

or, rearranging terms,

$$V_{out} = V_{in} - V_R$$

Substituting for V_R in this equation, we have

$$V_{out} = V_{in} - \frac{R}{\sqrt{R^2 + (10^6/2\pi fC)^2}} V_{in}$$

$$V_{out} = V_{in}\left(1 - \frac{R}{\sqrt{R^2 + (10^6/2\pi fC)^2}}\right)$$

and

$$\frac{V_{out}}{V_{in}} = 1 - \frac{R}{\sqrt{R^2 + (10^6/2\pi fC)^2}}$$

In the formula for V_{out}/V_{in} it is apparent that when f is very small, the quantity

$$\sqrt{R^2 + (10^6/2\pi fC)^2}$$

can become quite large. This causes the fraction

$$\frac{R}{\sqrt{R^2 + (10^6/2\pi fC)^2}}$$

to become very small, with the result that V_{out}/V_{in} approaches the maximum value of 1. On the other hand, as f becomes larger and larger, the fraction

$$\frac{R}{\sqrt{R^2 + (10^6/2\pi fC)^2}}$$

approaches a value of 1 and V_{out}/V_{in} approaches zero. In other words, higher frequencies are increasingly attenuated by the circuit. Note that this is the exact inverse of what happens in the case of the low-frequency filter.

Another way of interpreting the manner in which the high-frequency filter works is to focus on the position of the capacitance in the amplifier circuit. An examination of Fig. 4.4 shows that C is in parallel with the input to the next stage of amplification. This being the case, the current flowing through R must divide between C and the amplifier of the next stage. Since impedance due to the capacitor decreases with increasing frequency, the C branch of the circuit is a low-impedance pathway for the high frequencies. Therefore, increasingly higher frequencies may be thought of as being *shunted away* from the amplifier input by the C.

High-Frequency Response—Asymptote Plot

The frequency response of a high-frequency filter can be approximated by straight lines, as with the frequency response of the low-frequency filter. Figure 4.5 shows the straight-line approximation or asymptote plot for the high-frequency filter in Fig. 4.4. The two solid lines in the graph define the response of the filter when it is set to a cutoff frequency of 35 Hz. These lines meet at the cutoff point, with the line to the left of the cutoff point being parallel to the horizontal axis. The straight line to the right falls off at the rate of 6 dB/octave, which means that the amplitude is attenuated by 50% at a frequency of 70 Hz.

The dashed line in Fig. 4.5 shows the corresponding asymptote plot when the cutoff frequency of the filter is set instead to 70 Hz. Observe that the result is to shift the entire graph to the right. The dashed-line graph is interpreted in the same way as the solid-line graph.

High- and Low-Frequency Response Combined—The True Curve

If the reader will visually combine the asymptote plots of Figs. 4.3 and 4.5, it will be apparent that the combined frequency-response curve of the amplifier-filter combination is described by a flat-topped pyramid. This pyramid becomes flatter and flatter as the bandwidth of the amplifier—which is to say, the range of frequencies that are amplified—becomes wider. Referring back to Fig. 4.1, it is obvious that there is a vast difference between the frequency response of a real filter and the ideal filter. While real filters—the filters on the EEG machine— attenuate the unwanted frequencies at a rate of 6 dB/octave, the ideal filter achieves 100% attenuation instantaneously. The ideal filter is analogous to an ideal racing automobile—a car that would be capable of accelerating from zero to say 150 miles/hour *instantaneously*. For obvious reasons, such a racing car is only a dream. So also is the ideal filter.

Figure 4.4. Circuit for a high-frequency filter in an EEG machine.

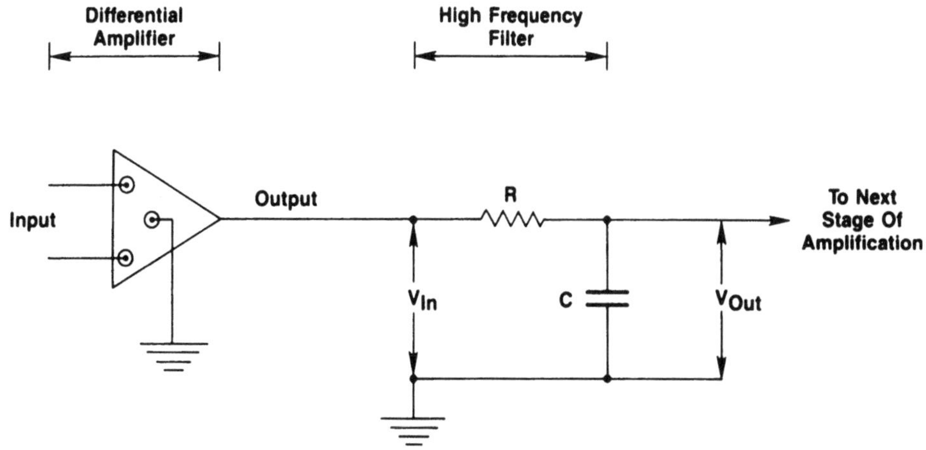

Figure 4.5. Asymptote plots showing the frequency response of a high-frequency filter. Solid line, cutoff frequency = 35 Hz; dashed line, cutoff frequency = 70 Hz.

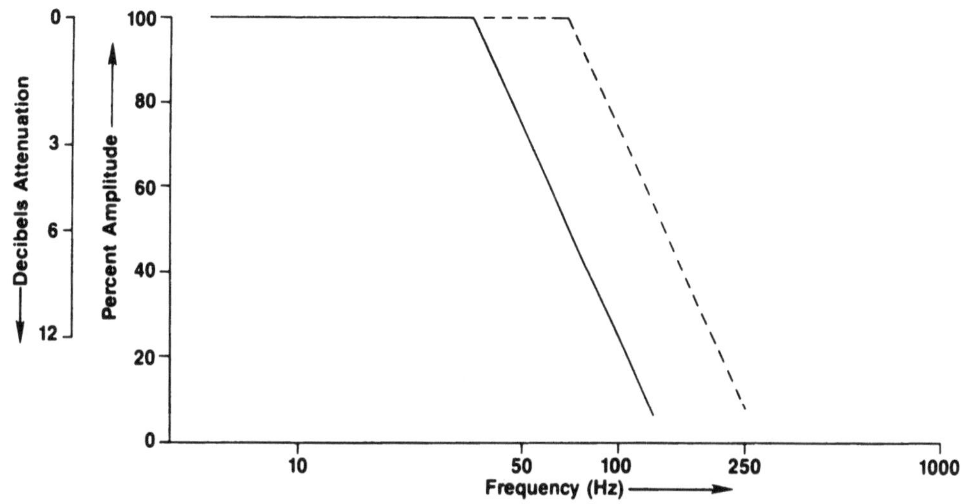

But we have been speaking only of the asymptote plot of the combined low- and high-frequency filters. The true plot—the true frequency-response curve—represents yet an additional step removed from the ideal. Figure 4.6 shows the difference between the asymptote plot and the true plot of the combined curve for a low-frequency setting of 1.0 Hz and a high-frequency setting of 35 Hz. By convention, the standard points of reference in dealing with the true curve are the amplitudes at the high and low cutoff points. We see in Fig. 4.6 that the amplitudes at 1.0 Hz and at 35 Hz in the true curve are only 70% of the amplitudes in the asymptote plots at the corresponding points. In other words, amplitude is down 30% at the cutoff frequencies. Note that amplitude is also reduced, but to a progressively lesser and lesser degree, on either side of the cutoff frequencies.

Figure 4.6 defines the true frequency-response curve (low-frequency cutoff of 1 Hz, high-frequency cutoff of 35 Hz) for the amplifiers plus filters of most EEG machines. The single exception to this occurs in the EEG machines made by the Grass Instrument Company. By virtue of a somewhat different design, the true curve of their filters is down only 20% from the asymptote plot at the cutoff frequencies. Also, when the cutoff frequency is halved, the amplitude is down by exactly 50%. This imparts an "S" shape to the curves of the Grass machines.

It is important to point out that the frequency-response curves as discussed in this chapter are for the amplifiers plus filters with the appropriate high- and low-frequency settings. In other words, the response curves do not reflect the effects due to the writer unit and pen. The pens on the EEG machine, of course, have inertia, and for this reason they have a profound effect on the overall high-frequency response. This topic is discussed in detail in the next chapter.

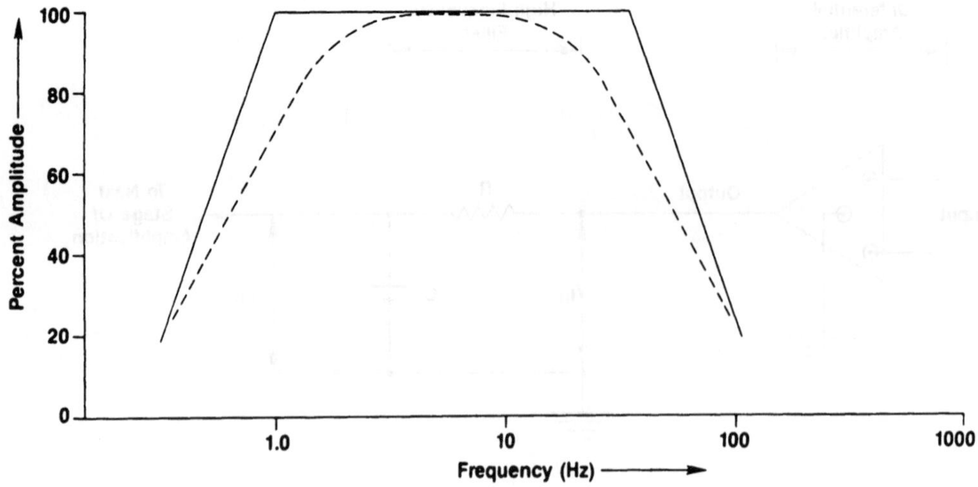

Figure 4.6. High- and low-frequency response curves combined. Solid line, asymptote plot; dashed line, true frequency-response curve.

60-Hz Notch Filter[1]

Despite the common-mode rejection feature of the differential amplifiers used in an EEG machine, 60 Hz artifact can be a formidable problem when EEGs are recorded. There are three major reasons for this. In the first place, some environments in which the EEG machine is operated have such high 60-Hz levels present that the common-mode rejection ratio cannot eliminate them. Second, although 60-Hz activity is a common-mode or in-phase signal, there are times when the amplitudes at both inputs of the differential amplifier are not identical. When this happens, we end up with some out-of-phase 60-Hz activity that gets to be amplified along with the EEG voltages. Finally, the circuit connected to the inputs of the differential amplifier may not be balanced, as when the impedances of the two electrodes are different. This degrades the common-mode rejection ratio of the amplifier and results in an elevated 60-Hz level in the EEG recordings.

The answer to all these problems is the 60-Hz notch filter. This type of filter is referred to as a band-elimination filter or trap since it attenuates a narrow band of frequencies that are unwanted. Here, again, we can think of an "ideal" case. The ideal 60-Hz filter is a device that would remove only the 60-Hz activity without affecting any adjacent frequencies. Although the ideal is not attainable, the 60-Hz filters on present-day EEG machines are more sharply tuned than the low- and high-frequency filters that we have been discussing.

[1] Some countries, including all of Europe, use 50-Hz instead of 60-Hz AC. In such cases, the problem is 50 Hz artifact and a 50-Hz notch filter is employed.

Figure 4.7 shows the frequency-response curve of such a filter. This is a true curve. Note in the figure that although the cutoff is considerably sharper than the 6 dB/octave cutoff of the other filters, a substantial proportion of the adjacent frequencies is attenuated by the 60-Hz filter. Thus, for example, we see that amplitude at 40 Hz is attenuated by about 30%. If the high-frequency filters on the EEG machine happened to be set to a cutoff frequency of 70 Hz, it is clear that the overall high-frequency response would be determined primarily by the 60-Hz notch filter and not by the former. For this reason, routine use of the 60-Hz filter is discouraged.

The 60-Hz notch filters are a fairly recent addition to the EEG machines used in routine clinical electroencephalography. Their presence is attributable to some technological advances that were not readily available in the 1960s. For this reason, some machines in use may not have the notch filter available. The technological advances and the methods whereby the sharper cutoff of the notch filter is accomplished need not concern us here.

Interpreting the Frequency-Response Curve and the Use of Filters

Suppose that the filter settings on an EEG machine are as shown in the frequency-response curve of Fig. 4.6. In other words, the low-frequency filter switch is set to 1 Hz, and the high-frequency filter switch is set to 35 Hz. Suppose, also, that the machine is calibrated and set for a deflection sensitivity of 7 μV/mm. This means that a 10-Hz signal at 50 μV will yield a pen deflection of 7 mm on the chart. What pen deflection will a 1-Hz signal at the same voltage yield?

The answer to this question comes directly from Fig.

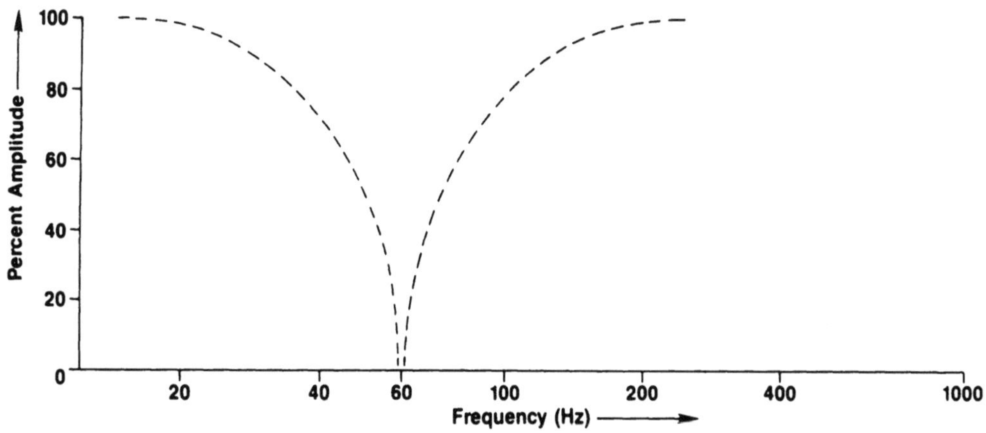

Figure 4.7. A 60-Hz notch filter as used in an EEG machine.

4.6—from the frequency-response curve. Observe in Fig. 4.6 that amplitude at 1 Hz is only 70% of the amplitude at 10 Hz. Therefore, the deflection for a 1-Hz signal will be 0.7 times the deflection of a 10-Hz signal or 0.7×7 mm = 4.9 mm. This means that when we observe delta activity at 1 Hz in a tracing with the low-frequency filter of the machine set to 1 Hz, this activity is in reality somewhat larger than it appears in the tracing.

The same logic may be applied to electrical activity of any frequency. For example, Fig. 4.6 shows that a 35-Hz signal also will yield a 4.9 mm deflection for an amplitude of 50 μV. The frequency-response curves for all filter settings will be found in the instruction manual that goes with an EEG machine. The EEG technician and interpreter alike should become thoroughly familiar with these curves.

Although we have spoken of the filters primarily as a means of eliminating artifacts or unwanted voltages from the EEG voltages, the filters have another important use. Some features of the EEG may be accentuated while others may be reduced by careful use of the filters. For example, suppose that some portions of a patient's sleep record suggested that there may be an asymmetry in the sleep-spindle activity, but that the recording was partially obscured by the high-amplitude, slow-wave activity of sleep. Here is where the low-frequency filters come in. Switching the low-frequency filter from 1 to 5 Hz will eliminate a greater portion of the slow-wave activity without significantly attenuating the sleep-spindle activity. This simple device permits a closer appraisal of the possibly asymmetry.

Similar use may be made of the high-frequency filter at the opposite end of the EEG frequency spectrum. Thus, it may happen that a patient's alpha rhythm becomes obscured by excessively high-amplitude beta activity. By switching the high-frequency filter from a cutoff frequency of 70 Hz to a cutoff frequency of 35 Hz, the beta activity in the tracing will be reduced without appreciably changing the amplitude of the alpha rhythm. Other examples of a similar nature could be mentioned. We leave these for the reader to discover.

The standard settings for the filters in routine clinical electroencephalography are 1-Hz low-frequency cutoff and 70-Hz high-frequency cutoff. However, as we will see in the next chapter, the upper limit of the high-frequency response with the high-frequency filter set to 70 Hz is limited by the inertia of the writer unit and pen. Standard practice requires that the 60-Hz notch filter be used only when it is absolutely necessary.

The recommended high-frequency setting requires some explanation. Knowing that the upper limit of beta activity is about 35 Hz, why is a 70-Hz cutoff frequency employed? The answer is that spikes, which are a feature of seizure activity, are equivalent to high-frequency activity and cannot be recognized in a tracing unless the maximum high-frequency response is employed. The reader can verify this for him or herself by recording spike activity on adjacent channels of the same machine, with the high-frequency filter of one channel set to 70 Hz and the high-frequency filter of the other to 15 Hz. If this is done, it will be apparent that the character of the spikes is profoundly changed at 15 Hz. By comparison with the tracing at 70 Hz, the spikes are rounded and have lost an important feature of their identity. For this reason, the standard filter settings should be strictly adhered to. There should always be a good reason for using settings that are not standard. Filters should never be used to indiscriminately "clean up" a record.

On some of the older vintage EEG machines, the high-frequency filter occasionally was referred to as a "muscle filter." Should this filter be used to remove muscle-activity artifacts from the EEG? The answer to this question is a qualified yes. The high-frequency filter is used for this purpose only when all other methods of eliminating muscle-activity artifacts have failed. The reason for this is that the tips of any remaining muscle spikes become significantly rounded when the high-frequency filter is set below 70 Hz. This being the case, the muscle activity—especially that appearing in the anterior regions—can easily be mistaken for beta activity. By providing the patient a comfortable cot or lounge chair to recline on and helping him or her to relax, it is frequently possible to eliminate muscle artifacts at their source, thereby avoiding use of the high-frequency filters. More will be said about this in Chapter 13.

Summary

At this point the reader may find it useful to refer back to Fig. 1.5, which summarizes the basic function of the high- and low-frequency filters in relation to the EEG frequency spectrum. This figure shows the way in which the EEG tracings differ when different bandwidths are used. How the frequency response of an EEG machine is assessed and the role played by the time constant in assessing frequency response are taken up in Chapter 6, "Calibration and Calibration Methods."

Chapter 5
The Writer Unit

The writer unit is the business end of the EEG machine. Although the workings of the rest of the EEG machine are easily forgotten or taken for granted, the writer unit and its problems are a part of an EEG technician's day-to-day routine. To make the best use of the writer unit and to aid in the interpretation of the tracing, some knowledge and understanding of the way in which it functions is valuable.

Regardless of its design or manufacturer, the writer unit consists of four major units: (1) the penmotors or direct-writing oscillographs, sometimes referred to as galvanometers; (2) the pens and pen mounts; (3) the inking system; and (4) the chart drive. We consider each of these in turn.

Penmotors

The penmotors are electromechanical transducers that convert electrical energy into mechanical movement. Movements of the pen—the deflections from the baseline—are proportional to the voltage of the signal applied to the input of the machine. This being the case, the penmotor functions basically as a recording voltmeter. As we already mentioned in Chapter 1, a penmotor has the attributes of an electric motor. Like an electric motor, it consists of an armature mounted between the poles of a magnet in such a way that it is free to rotate. Wound around the armature is a coil of wire that turns with the armature when it rotates. This arrangement is shown in schematic form in Fig. 5.1. One major difference between a penmotor and the common variety of electric motor is the range of rotation that is possible. Whereas the armature in an electric motor makes a complete, full-circle rotation, the armature in a penmotor is limited to a rotation of less than 45 degrees of full circle.

The operation of a penmotor depends on the principles of electromagnetic induction. Most important of these is the principle discovered in the 1830s by Michael Faraday, whom we had occasion to mention in Chapter 2. Faraday showed that changing the magnetic field surrounding a conducting circuit caused a current to flow—to be *induced*—in the circuit. By way of example, consider the illustration in Fig. 5.1. If you rotate the armature through an arc of, say, 10 degrees, a change is produced in the magnetic field surrounding the coil of wire wound around the armature. Because the magnetic field is "cut" by the armature, a current is induced in the armature circuit. In other words, mechanical movement produces an electric current.

Now it happens that the phenomenon just described works in the opposite direction as well. If you connect the armature circuit to an external voltage, so that current flows in the coil of wire, the armature will move (rotate). Briefly, this happens because the current flowing in the coil of wire sets up a magnetic field that interacts with the field of the magnet. This interaction of forces produces a rotation of the armature. The amount and direction of the rotation depend on the strength and direction of the current flow.

Penmotor Frequency Response

It was mentioned in the chapter on filters that the high-frequency response at a cutoff frequency of 70 Hz is determined largely by the characteristics of the penmotor. The reason for this will be apparent from a careful perusal of Fig. 5.1. Note that the armature terminates in a shaft to which the pen is attached. Not shown in the schematic diagram is the structure—the rod and bearing—that supports the armature and pen assembly. Although this assembly is designed and fabricated in such a way so as to be as light as possible, its weight nevertheless is not negligible. This being the case, the armature and pen assembly cannot be driven without significant loss of amplitude at the higher

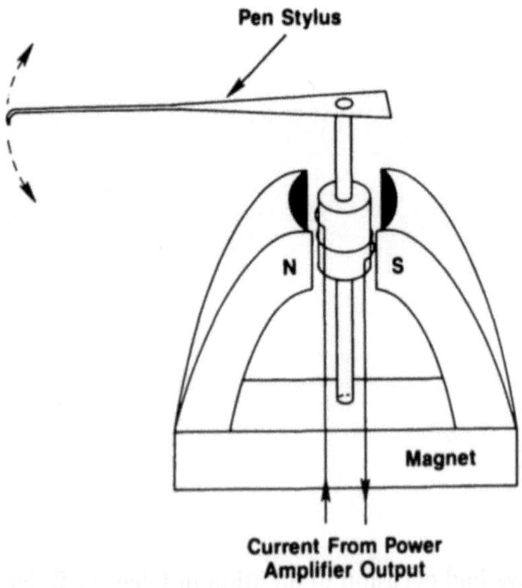

Figure 5.1. Schematic diagram of a typical penmotor.

frequencies, or frequencies above 70 Hz; thus, at 95 Hz the amplitude on most machines is already down by 50%. Although this is not objectionable in EEG work, it is important to keep these facts clearly in mind when the high-frequency filter is set to a cutoff frequency of 70 Hz.

Pens

A variety of different kinds of pens and pen mounts are found on different EEG machines. The pens may provide either *curvilinear* or *rectilinear* tracings, although curvilinear tracings are standard on most machines used for routine clinical work. Pens are narrow-gage metal tubes, one end of which is curved and drawn out to a fine tip—the writing tip. The metal tube is glued or soldered to a thin metal strip that serves as a stiffener. The end of the stiffener opposite the writing tip is fitted with a shaft, rigidly mounted at right angles to the tube, which snaps into place in the pen mount. Sometimes the pen has a sapphire tip inserted into the writing end of the tube. A sapphire tip provides for considerably longer wear but is more fragile than the metal writing tip.

Error of the Arc

Curvilinear tracings are produced by straight pens that are rigidly attached to the shaft of the armature. Such a pen is shown in the diagram of Fig. 5.1. When this pen is deflected, it does not trace out a line perpendicular to the baseline but describes, instead, an arc of a circle having a radius equal to the length of the pen. The difference between this arc and a line perpendicular to the baseline is referred to as the *error of the arc*. The magnitude of this error depends on the length of the pen and is best appreciated by deflecting a pen on one channel of the machine when the paper is not moving. Note that the error is the distance between the position of the pen and the location of a corresponding spot on the perpendicular line drawn from the pen's baseline position. Error of the arc increases in magnitude as the amplitude of the pen deflection becomes larger; it can readily be perceived when there are large pen deflections.

It is essential to keep the error of the arc in mind when making comparisons between tracings on different channels, as when a record is examined with the purpose of determining the origin of multifocal spike activity. If the amplitude of the activity displayed on the channels that are compared is not the same, some misleading conclusions may be drawn. Consider, for example, a spike that appears as a 5-mm deflection from the baseline position on channel 1 and that also appears on channel 4 as a 5-mm deflection from the peak or trough of a slow wave. Although the spike appears to occur earlier in time on channel 1 than on channel 4, the appearance is clearly an artifact resulting from the error of the arc.

Problems associated with the error of the arc are avoided by the use of rectilinear recording. Indeed, a rectilinear recording is the only tracing that shows the true waveform of the EEG. Rectilinear recording is accomplished with the use of pens that are attached to the armature shaft of the penmotor via a coupling mechanism that corrects for the error of the arc mechanically. Unfortunately, this mechanism is quite delicate and easily damaged by heavy routine clinical recording with restless patients. For this reason, rectilinear recording is not in common use.

Pen Mounts

The pen mounts, or cradles as they are sometimes called, serve two purposes. First, they hold the pens firmly to the armature shaft while at the same time providing for quick removal and replacement. Second, they provide a means of adjusting the amount of pressure that the pens exert on the chart paper while the recording is going on. Sufficient pressure is needed for the ink to flow properly. At the same time, too much pressure can reduce the high-frequency response of the channel and, in some cases, modify the time constant.

The manual that comes with the EEG machine should be consulted for information about pen-pressure adjustment. Some machines provide a gage—a scale-like device—for measuring pen pressure.

Inking System

Virtually all machines used in routine clinical electroencephalography make use of ink-writing pens. The chief reason for this is cost. By utilizing ink as the recording medium, the chart on which the EEGs are traced can be made of low-cost paper. This is an important consideration as large quantities of charts are used up in recording EEGs. For example, at the standard chart speed of 30 mm/s, a 30-minute recording requires a strip of recording chart 54 m long—which is a lot of paper.

Some inking systems have a single, common ink supply for all the pens on the machine, whereas others have a separate inkwell for each pen. Systems using a common supply have the advantage of being easier and faster to fill. Only one inkwell needs to be filled rather than 16, 18, or 24 separate inkwells. On the other hand, the priming mechanism for systems having a separate inkwell for each pen is simpler in design and much less likely to malfunction. Technological improvements notwithstanding, the simple cup-type inkwell fashioned from metal or plastic, with a tight-fitting cover that fits the inkwell like the wall of a cylinder, is difficult to improve upon. To prime the pen, you simply lift up the cover, place your finger tightly over the small hole in the top, and press down briskly.

Although the inking systems on routine clinical EEG machines employ a capillary-feed system for delivering ink to the paper, in some instruments the ink is sprayed as a jet stream. It is not at all clear whether the advantages of a jet stream outweigh its higher cost and complexity.

Inking-System Maintenance

One of the most bothersome and frustrating experiences for the EEG technician is to have the inking of one or more channels fail during the course of taking an EEG. Unless there is some foreign material present or bacteria growing in the ink, this should not happen—assuming, of course, that the inkwells are kept filled, that the machine is used on a daily basis, and that a rubber or plastic "dam" is placed under the tips of the pens when the machine is not actually in use.[1] Some commercially available inks contain chemical additives that retard bacterial growth. Use of such inks reduces the possibility of unforeseen clogging. When inking problems do arise, they usually are the result of the machine being used on an irregular schedule or only on a casual basis. Under such conditions, drying and caking of ink in the pens is inevitable. To a degree, this can be avoided if the machine is turned on daily and a calibration is run so that there is some flow of ink through the pens. By carefully observing the density of the tracings at this time, the EEG technician can usually predict which pens may be in jeopardy of clogging. These pens will display a tracing that is slightly ragged and a bit darker than the others. To avoid eventual clogging in such cases, most experienced EEG technicians use the priming mechanism to force a small quantity of ink through the pens. This simple expedient will usually restore the inking to an acceptable level.

While actual clogging of pens is rare if the procedures already outlined are followed, pens sometimes do get stopped up. When this happens, a number of steps should be followed. The first is to find out exactly where the blockage resides. Although the problem is usually in the pen itself, it is good practice to remove the plastic tube that feeds ink to the pen and then prime the inkwell to verify that ink is flowing freely through the tubing. If ink does flow freely, the problem is clearly in the pen.

At this point a simple tool is useful. Attach a short length of the plastic tubing that is used on the pens to the blunt needle of a 3-mL hypodermic syringe. Short lengths of this tubing are usually provided as spare parts with the EEG machine. Now, fill the syringe with water and attach the tubing over the writing tip of the blocked pen, which has been removed from the machine. With the tube firmly in place, try to force the water through the pen by exerting pressure on the syringe. In most cases minor blockages are cleared immediately by using this procedure.

If this method fails, try threading a fine steel wire through the pen to clear the blockage. Most troubleshooting or repair kits that come with an EEG machine contain a supply of such fine wires. Because the writing tip is a smaller bore than the rest of the pen, the wire should be threaded from the tip end. Care should be taken that the pen does not get bent out of shape or that the wire is not kinked or broken off inside the pen. Some practice is needed before the procedure can be carried out successfully. If this method also fails to clear the clogged pen, the only alternative is to soak the pen in a detergent solution overnight and then repeat the procedure using the fine wire.

The EEG technician should always make a determined effort to clear a blocked pen rather than replace it with a new one. There are a number of reasons for this. First, pens are expensive. Second, the pens supplied are not all exactly the same length. This means that the time-axis alignment needs to be checked whenever a pen is replaced and the penmotor realigned, if necessary, to bring the new pen into

[1] Some newer EEG machines have a device that automatically lifts the pens and moves the pen dam into place when a switch is closed. The dam is automatically retracted when recording is initiated or resumed.

alignment with the others.[2] Finally, the tip of the new pen may not be exactly parallel with the writing surface, in which event the pen may skip. To correct this condition, the tip may need to be lapped.

Obviously, the technician will have no choice but to replace a blocked pen if it cannot be cleared. When this happens, he/she should consult the instruction manual accompanying the machine for information concerning penmotor realignment and the technique for pen lapping.

Chart Drive

The purpose of the chart drive is to pull the chart paper through the machine at a constant rate and at the speed selected by the operator. Standard EEG chart speed is 30 mm/s, but all machines have a number of other speeds. For routine clinical EEG work, only one-half standard speed (15 mm/s) and twice standard speed (60 mm/s) are essential. The reader should recognize that the use of 30 mm/s as a standard is not a purely arbitrary choice. At 30 mm/s, activity in the theta, alpha, and beta bands (which includes frequencies in the range of 4–35 Hz) is readily appreciated and easily identified by eye.

The slower chart speed of 15 mm/s is used mainly to aid in identifying delta activity. Thus, while activity at frequencies less than 4 Hz is difficult to resolve visually at the standard chart speed, it stands out clearly at the slower speed. For example, if a delta wave focus appears to be present when the chart is run at standard speed, a short run at 15 mm/s may be helpful in confirming its existence. The slower chart speed may be thought of as functioning much like a high-frequency filter. By using a speed of 15 mm/s, the closely spaced waves at the high end of the EEG frequency spectrum merge together and become less obvious to the eye. At the same time, the widely spaced waves at the low end become more obvious by being pressed closer together.[3]

A chart speed of 60 mm/s has two major uses. In the first place, it provides an easy way of verifying the presence of 60 Hz artifact in the tracing. At this speed, the individual 60-Hz waves are readily identified and can be counted. The faster speed is also helpful in assessing whether one focus fires before another when a record shows evidence of multifocal spike activity. It should be recognized that the higher chart speeds place greater demands on the inking system. This means that inking problems are more apt to show up; thus, a channel functioning only marginally at 30 mm/s may fail completely at 60 mm/s.

To ensure proper tracking of the chart paper through the machine, paper should be loaded and tensioning devices set in exactly the manner spelled out in machine's instruction manual. Seemingly small matters such as the position of the pack of paper on the feed tray can affect performance. When properly functioning, the chart drive should operate smoothly and quietly, without excessive chattering or scraping of the paper. Two common problems encountered are *weave* of the chart paper as it goes through the machine and *breakage* of the paper under the drive roller or rollers. In severe cases of weave, the chart runs out of line so badly that it buckles and tears up at the drive roller. Although different EEG machines employ somewhat different designs for their chart drives, all are subject to these problems if the mechanisms are not precisely adjusted. Adjustments should be carried out carefully and only after reading the instruction manual and/or consulting with the manufacturer. It is not uncommon for a slight turn of an adjustment screw to make the difference between a malfunctioning or a perfectly functioning chart drive.

Marker Pens

Although they are not essential in routine EEG work, some EEG machines have one or two additional channels that serve as markers. These are the so-called marker pen channels. The pens for them are placed at the very top and very bottom of the chart, and as they deflect only a few millimeters, they take up little additional space. One of them is invaribly a *time marker* that produces a small, sharp deflection on the chart at regular intervals—commonly once every second. This marker provides an on-going, minute-by-minute check on the accuracy of the chart speed—assuming, of course, that the timer activating the pen is both accurate and reliable. The time-marker channel is also useful in identifying the particular chart speed being used. It is especially helpful if there is a shift back and forth between different chart speeds during the course of a recording.

Various other kinds of information can also be displayed on a time-marker channel at the same time. To avoid confusion, the pen deflections are in the opposite dirction to the time marks. For example, the particular montage currently being run can be identified by a coded deflection pattern appearing at regular intervals on the chart. This feature is a convenience for the person reading the record as he or she need not go back to the beginning of the run to find

[2] Because all pens are not exactly the same length, the EEG technician should also be especially careful not to mix up the pens if they are removed from the machine. In other words, the pen used on channel 1 should remain there and should not inadvertently be connected up and used on another channel.

[3] The slower chart speed is also useful in polysomnography. In addition to rendering the slow activity of the deeper stages of sleep more readily interpretable, the slower chart speed conserves chart paper—a consideration during all-night recording.

out what derivations are being displayed. Of considerably greater importance is the use of the time-marker channel to indicate the light flashes during photic stimulation (see Chapter 16). This information is essential in order to identify and document the presence of photic driving; unless a suitable marker channel is available for this purpose, one of the EEG channels must be sacrificed.

If an EEG machine has a second marker channel, it usually is under the control of the operator. This additional marker pen may be activated by manually pressing a button or moving a switch. The variety of functions that can be displayed on this marker channel is limited only by the imagination and ingenuity of the user. For example, a button for activating the pen can also be given to the patient. Such an arrangement is useful for having the patient signal his or her response to questions when recording during absence seizures.

The EEG technician should recognize that marker pens are more likely to present inking problems than the pens tracing out the EEGs. The reason for this is that with deflections of only a few millimeters, marker pens use up ink at a much slower rate than the pens on the EEG channels. When this happens, evaporation becomes a significant factor, and ink in the wells gradually becomes thicker and thicker—a process that ultimately leads to clogging. This problem is easily avoided by periodically drawing out (with a syringe) and discarding the ink from these inkwells and refilling with fresh ink.

Chapter 6
Calibration and Calibration Methods

Every EEG begins and ends with a calibration. Frequently, the calibration becomes so routine a procedure that it is easy to pass over it with little thought to its meaning and purpose. This is especially easy to do when the machine used contains from 18 to 24 channels, and one is eager to get on to the important part of the record, namely, the EEGs themselves. Nevertheless, the calibration is a vital part of every EEG and for this reason we devote a chapter to it.

Purpose and Basic Concept

The purpose of calibrating any instrument is to demonstrate that it is a valid and reliable device for measuring the phenomenon of interest. Calibrating an EEG machine involves connecting voltages having known characteristics to the inputs of the machine and verifying that the pen deflections traced on the chart conform to certain specific standards. The calibration voltages, of course, should be capable of testing the machine within the spectrum of frequencies for which it will be used. During the course of a routine calibration, the following points need to be verified:

1. A standard input voltage yields a standard pen deflection. For example, at a deflection sensitivity of 7 μV/mm, a 50-μV calibration signal should deflect the pens a total of 7 mm.
2. The deflection sensitivity is *linear*. This means that if a 50-μV calibration signal deflects the pens by 7 mm, a 100-μV signal should produce a 14-mm deflection.
3. The frequency response conforms to the conventions employed in clinical EEG work.
4. The noise level is within acceptable limits.
5. There are no perceptible differences in deflection sensitivity, frequency response, and noise level among the channels of the machine.
6. All channels are in accurate alignment—in other words, the deflections of the pens on all the channels fall on a vertical, straight line of the chart.

The foregoing address the question of the validity of the instrument. If all the requirements are satisfied, the machine is a valid instrument. The degree to which these standards are reproducible on a day-to-day basis refers to the reliability of the instrument. If the machine behaves differently from one day to the next—for example, if noise level of one channel was acceptable yesterday but is excessive today—it is not reliable. Validity, of course, is meaningless without reliability. An unreliable machine simply cannot be trusted.

Some of the six requirements listed above are self-explanatory; others require additional discussion. We take these up in turn.

Voltage Calibration—Deflection Sensitivity

This is accomplished by connecting a signal of known voltage to the inputs of all the channels. The calibration signal is accurate to $\pm 2\%$, and on most machines a range of calibration voltages is available—2, 5, 10, 20, 50, 100, 200, and 500 μV. Usually, the calibration voltage is produced by pressing a button on the machine console. Direct current and sometimes AC calibration signals as well are used, although most EEG machines employ only DC calibration. The reason for this will become apparent later in this chapter.

When AC calibration is used, the calibration signal is usually a 10-Hz sine wave derived from a signal generator (referred to as an oscillator) inside the EEG machine. The calibration signal as displayed on the chart is measured from the very top of the peaks of the waves to the very bottom of the troughs. If a 50-μV, 10-Hz signal is used, we

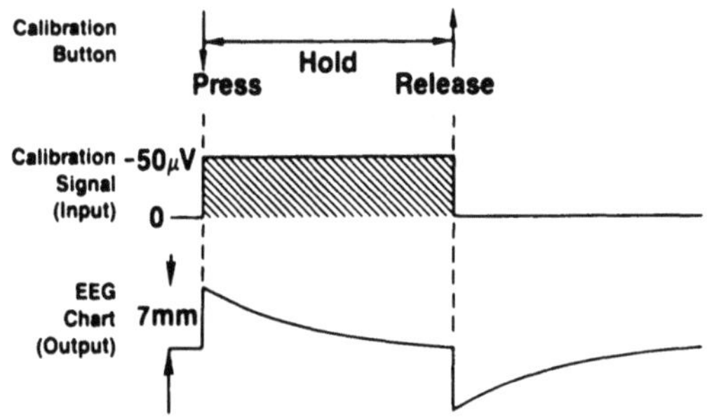

Figure 6.1. The process of DC calibration; 50-μV input at a deflection sensitivity of 7 μV/mm.

speak of it as measuring 50 μV *peak-to-peak*. For standard gain, this signal should yield a pen defection of 7 mm between the peaks and troughs of the waves. Although only the newer model machines provide AC as well as DC calibration, many do have a jack to which an accurately calibrated external oscillator may be connected for doing an AC calibration.

Direct current calibration uses rectangular-wave signals, sometimes called DC "pips." When the calibration button on the machine console is pressed, the voltage present at the inputs of the channels changes instantaneously from 0 μV to, say, 50 μV and remains at that level until the calibration button is released. The sequence of events is illustrated in Fig. 6.1 along with the output on the chart of the machine. The fact that the calibration signal as seen on the chart is quite different from the signal applied to the input should come as no surprise to readers who have already studied the material in Chapter 2.

The tracing on the chart, of course, describes a decaying exponential; it is the result of the capacitor in the circuit of the low-frequency filter. The reader will recall from Chapter 2 that a capacitor responds only to a change in voltage. Pressing the calibration button produces a change from 0 μV to 50 μV. This results in a corresponding deflection of the pen and a change in the position of the tracing; but as the voltage remains steady at 50 μV, the tracing decays back to zero. When the calibration button is released, the steady voltage of 50 μV is abruptly removed and the input is returned to 0 μV. Accompanying this change is a deflection of the pen and a change in the position of the tracing, which is *equal and opposite* to that observed when the calibration button was pressed. But, again, the tracing decays back to zero as the voltage at the input remains steady. As the pen deflects upward when the calibration button is pressed, we infer, in agreement with EEG convention, that the calibration signal is 50 μV negative at G1.

Needless to say, there should be no perceptible difference in the deflection sensitivities of all the channels on the machine when their gain settings are identical. Small differences that may be present are easily corrected by the EEG technician by turning a screw-driver-adjusted control that is usually located on the front panel of each amplifier. Details are given in the manual that comes with the machine.

Linearity

Linearity of the channels is assessed by repeating the calibration procedure using a variety of different calibration voltages at the same gain setting. In each case the maximum deflection on the chart is accurately measured. These deflections should be in proportion to the calibration voltages. This means that if 50 μV yields a deflection of 7 mm, 20 μV should give a 2.8-mm deflection; 10 μV, a 1.4-mm deflection; 100 μV, a 14-mm deflection; and so on. Normally, the linearity is very reliable and does not change unless some component in the amplifier circuit fails.

Frequency Response

At first glance the most obvious way of checking the frequency response of an EEG machine would seem to involve the use of a variable frequency oscillator. If the machine has a jack for connecting an external oscillator to the inputs, this can, indeed, be done.[1] The EEG technician would simply run the oscillator through a number of different frequencies from 0.1 Hz to say 90 Hz, making sure that the amplitudes of all the calibration signals of different frequency used were identical. She/he would then measure the peak-to-peak deflection on the chart for each calibration frequency. These values would be plotted against frequency to obtain the frequency-response curve, one curve for each channel.

It is readily apparent that this procedure would be quite time consuming and hardly possible to carry out prior to taking an EEG. As it happens, it is not even necessary. A short-cut method for obtaining information about the frequency response of the EEG machine is available. This short-cut method is based on the fact that a precise mathematical relationship exists between the frequency response of an electrical circuit and its transient response.[2]

[1] The internal AC calibrator on the machine will not serve this purpose as it provides only a sine wave of a single frequency, usually 10 Hz.

[2] This mathematical relationship is embodied in the Laplace transformation developed by Pascal Laplace, the late 18th century French mathematician. The Laplace transformation *(LT)* provides a way of relating a function of time $f(t)$, the response of a circuit to a step function, to a function of frequency, the frequency response of the same circuit. This relationship is given by the equation

$$LT\{f(t)\} = \int_0^{\infty} f(t)\,\varepsilon^{-st}\,dt$$

where s is a complex variable related to $2\pi f$, the reciprocal of the time constant of the circuit. Note the quantity ε^{-st}, which is called the kernel of the equation. This, of course, is the now familiar expression for a decaying exponential.

Briefly, all electrical circuits may be thought of as having a characteristic response in the *time domain* and a characteristic response in the *frequency domain*. In the time domain we have the circuit's transient response, while in the frequency domain we have the frequency response. For simple series R-C filter circuits, the salient variable in the time domain is the time constant, while in the frequency domain the corresponding variable is the cutoff frequency. These two variables are related according to the following equation:

$$\text{Cutoff frequency} = \frac{1}{2\pi TC}$$

where *TC* is the time constant in seconds. The formula shows that as *TC* increases, cutoff frequency decreases and vice versa.

This simple relationship tells you that if you know the value of the *TC* for a series R-C circuit, the cutoff frequency can be calculated from the formula. Knowing the cutoff frequency, you can readily plot the frequency-response curve—that is, the asymptote plot—for the circuit. In practical terms, this means that the same information about the circuit may be derived by obtaining the transient response and estimating the *TC* as by laboriously obtaining the circuit's response to a wide range of frequencies and plotting the frequency-response curve. If the reader will briefly turn back to and review the section on transient response in Chapter 2, he/she will discover that the DC calibration technique provides a way of obtaining the transient response of the circuits in the EEG machine to a step function.

In Table 6.1 we have listed the values of the time constants corresponding to the commonly used low-frequency filter cutoff frequencies. Figure 6.2 shows that there is a profound change in the DC calibration as we go from a cutoff frequency of 0.1 Hz to a cutoff frequency of 5 Hz. In fact, as Table 6.1 shows, there is a 50-fold increase in the time constant as the cutoff frequency is changed from 5 to 0.1 Hz. Although the time constant may readily be checked by careful measurement of the tracings obtained during the DC calibration, this is rarely necessary. It is simpler to keep a template available that was made at a time when the time constant was known to be correct. With this template in hand, it is an easy matter to match it up to the routine calibrations to see if there is any deviation from the standard.

Although not as obvious, the transient response or DC calibration also provides some information regarding the frequency response of the high-frequency filter circuit. This will become apparent from a careful perusal of Fig. 6.2. As shown in this figure, the three rows of calibrations are for high-frequency filter settings of 70, 35, and 15 Hz, respectively, from top to bottom. Examination of the upward deflections of the pens reveals that the tracing rises more rapidly in the case of the 70-Hz high-frequency filter setting than for the 15-Hz setting. Note also that the terminal point of the tracing is sharp for the 70-Hz setting but rounded in the case of the 15-Hz setting. The calibration with the high-frequency filter set to 35 Hz falls somewhere in between. This aspect of the calibration comes under the heading of *rise time* of the circuit.

Table 6.1. Relationship between time constant (TC) and cutoff frequency in a series R–C circuit used as a low-frequency filter

Cutoff frequency (Hz)	Time constant (seconds)
0.1	1.59
0.15	1.00
0.3	0.53
0.5	0.30
1.0	0.159
1.5	0.100
5.0	0.032

If the chart had been run through the machine at a much higher speed, the difference in rise times between the settings would have been more apparent. Indeed, at the higher chart speed, it would be seen that the tracings do not rise instantaneously but describe what appears to be an exponential function—even at the 70-Hz setting. This is indeed the case. These differences are of practical value. Thus, the experienced EEG technician can readily detect differences in the high-frequency response of the different channels by examining the tracings and visually measuring the rise times.

Biological Calibration

We have discussed AC and DC calibration of the EEG machine in considerable detail. A third calibration method is also employed in EEG work. This is the so-called *biological calibration* or *bio cal*. The bio cal is used extensively, and machines having a montage switch as described in Chapter 1 frequently have a position included in the switch for the bio cal.

Despite the formidable name, the bio cal is nothing more than the simultaneous recording on all channels of the electrical activity from the same pair of electrodes. Its primary purpose is to obtain an additional, quick check of the frequency response of all the channels. It is also useful for picking up gross differences in the time axis alignment of the pens. To perform its function effectively, the pair of electrodes used in the bio cal should tap voltages from the entire EEG frequency spectrum. This is the reason why the frontal pole and occipital electrodes—usually Fp_1 and O2—are selected for this purpose. In the awake and relatively alert patient, tracings from these leads normally

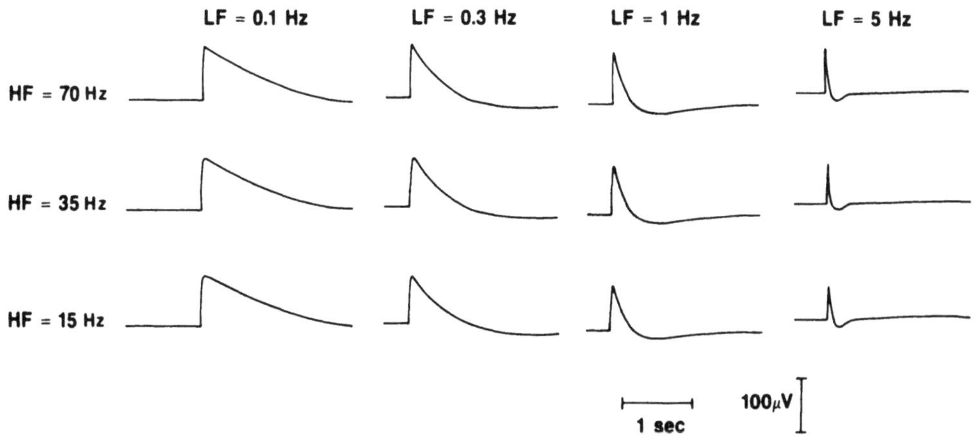

Figure 6.2. DC calibrations at different high-frequency filter settings (down), and different low-frequency filter settings (across).

include the alpha and beta rhythms, the central rhythms, as well as the lower-frequency eye movement potentials.

The biological calibration is used mainly to check the EEG machine for differences between the channels. By running his or her trained eye down the chart, the experienced EEG technician or the person interpreting the EEG record can pick up even small differences between the various channels. It is essential, of course, for all the channels on the EEG machine to be alike. This point cannot be overemphasized as much of the value of the clinical EEG rests upon the ability of the test to detect differences—sometimes of a subtle nature—between the two hemispheres of the brain. It is obvious that any such differences could not reliably be detected unless the recording channels themselves were identical.

Noise Level

At the standard deflection sensitivity of 7 μV/mm, no noise should be perceptible during routine calibration. This means that the tracings taken during the DC calibration should be smooth and regular. It is a good idea to record, in addition, a page or two with the controls in the calibration position but without generating the calibration signal. These tracings should show smooth, perfectly horizontal straight lines, with no evidence of any deviation above or below the horizontal. All machines used in routine clinical work should be capable of satisfying these standards.

When the EEG machine is used at higher gain—as, for example, in brain death recordings—the requirements are more stringent, but some noise will become perceptible. The noise at routine calibration appears as a random wavering of the tracings and should be less than 2 μV peak-to-peak with reference to the input. Noise in the context of the present discussion should be understood to be internally generated noise, or noise produced by the EEG machine itself. Noise can, of course, be externally generated as well. To rule out the possibility that any noise seen in the calibration is externally generated, the inputs of the machine should be short circuited and the tracing taken again. Details of the procedure are taken up in the chapter on troubleshooting.

Postcalibration

The calibration procedures that have been outlined and discussed refer to the precalibration, or calibration that is performed immediately prior to the time that a patient's EEG is taken. Immediately upon completion of the EEG, a postcalibration is done. The postcalibration is identical to the precalibration with the exception that it also includes DC calibrations at any deflection sensitivities that were used in addition to the standard of 7 μV/mm.

If the postcalibration is no different from the precalibration and both satisfy the standards set forth, we usually conclude that the machine is both a reliable and a valid instrument. The conclusion concerning reliability, however, is only an inference. The postcalibration only tells us that the machine is functioning exactly like it did some 30 minutes ago before the patient's EEG was begun. We cannot be certain that something did not change sometime during the 30 minutes and then changed back again before the postcalibration was started. This kind of thing has happened and is not uncommon when the machine has some kind of intermittent fault. On the other hand, if pre- and postcalibrations are no different from each other over weeks and months of routine use, the probability is high that the machine is indeed reliable.

Chapter 7
Recording Electrodes

The role of recording electrodes in EEG technology and intepretation is a relatively simple one in practice but a complex one in theory. In this chapter our attention is primarily directed to the practical issues involved. It is hardly possible to obtain a satisfactory EEG without having good quality electrodes that have been properly connected to the patient. But what is a good electrode? What constitutes proper electrode application technique? How do departures from these standards affect the EEG record? The present chapter addresses all these questions. At the same time, in order to understand what actually takes place when you attach a pair of electrodes to a person's scalp and connect the other ends of the wires to a channel of the EEG machine, it is necessary to consider some of the basic theory involved.

Basic Concepts

In practical terms, the electrodes—or leads as they are also called—are simply the means whereby the electrical activity of the brain is communicated to the input circuits of the amplifiers in the EEG machine. Although a remarkable variety of different types of electrodes have been used for this purpose, there is a fundamental component or element that is common to all of them. This component is the *metal-electrolyte interface.* The metal is the material of which the electrode is composed, while the electrolyte may be a conducting solution, gel or paste, or it may be the fluids of living tissue as when an electrode is inserted below the skin. It is at the metal-electrolyte interface that current flow within the brain becomes electron flow in the electrodes and electrode wires. By what mechanism is this activity able to pass across the interface and be recorded?

In Chapter 2 we discussed electric currents, electrons, the electrical properties of conductors and insulators, and the way in which electrical currents flowed in metals. To understand something about the way in which electrical currents pass across the metal-electrolyte interface, we need to have some basic knowledge concerning the electrical properties of electrolytes or conducting solutions. To do this, it is essential to become familiar with the concept of an ion.

Ions

Everyone is familiar with what happens when you pour ordinary table salt into some water and stir the mixture. Assuming that a large amount of salt is not used, the salt crystals eventually disappear and what remains is a clear liquid. We say that the salt has completely dissolved in the water or has gone into solution. In purely physical terms, the salt has changed state completely. For present purposes, however, some more important changes have taken place.

In chemical terms, what has happened is that the elements of which the table salt is composed have become dissociated. In going through this process of dissociation, *ions* are formed. Ions are particles in solution that bear an *electrical charge.* For our example of the table salt—which is sodium chloride or NaCl—going into solution, we end up with sodium ions designated by the symbol Na^+ and chloride ions designated by Cl^-. The sodium ions have a positive electrical charge while the chloride ions have a negative electrical charge.

The ions in a solution have a number of interesting properties. Of particular interest in the present context is the fact that ions are free to move about in the solution. If a voltage is applied between two points in the solution, an electric current can be made to flow in it. The current is carried by the ions in the solution in the same way that a current is carried by the loosely bound electrons in a metallic conductor. Thus, a simple analogy is appropriate, namely, that ions are to electrolytes and conducting solutions as the loosely bound electrons are to metals.

In the context of our discussion of recording electrodes, it will be apparent that the metal-electrolyte interface is the junction where a flow of ions is converted into a *flow of electrons*. In other words, it is the place where an electrochemical phenomenon is converted into a purely electrical phenomenon. This is why recording electrodes are sometimes referred to as transducers. The reader may find it useful at this point to turn back to Chapter 1 where transducers were defined and discussed briefly.

The Electrical Double Layer

Although almost any kind of metal may be used as an electrode, the electrolyte chosen for recording leads is usually some kind of salt solution, principally sodium chloride. Two major reasons dictate this choice. First, sodium chloride is very soluble in water. For this reason, a sodium chloride solution is able to contain a high concentration of ions, which means that the solution will be a really good conductor. Second, sodium and chloride ions are a major constituent of the body fluids and hence are compatible with them.

Despite the fact that any metal that happens to be a good conductor could serve as a recording electrode, some metals are, at best, only poor materials for this purpose. This state of affairs arises mainly from the fact that a metal electrode discharges positive ions into solution when it comes in contact with an electrolyte. Some of these discharged ions may be tightly bound to the surface of the electrode. Concomitantly, an adjacent layer of oppositely charged ions from the solution is formed, resulting in the creation of the so-called *electrical double layer* at the metal-electrolyte interface. These two processes occur at different *rates*, depending on the species of metal used for the electrode and the type of electrolyte. The difference in the rates of these two processes results in a voltage appearing at the electrode. This voltage is termed the *electrode potential* or half-cell potential. The latter term is appropriate since the potential of an electrode itself is always measured with respect to a reference electrode, it not being possible to measure the voltage of a single electrode with respect to a solution.

Polarization and the Double Layer

The characteristics of the double layer vary with different electrode materials. These characteristics determine whether an electrode will be *polarizable* or *nonpolarizable*. With some electrodes there is a free exchange of charges (ions) across the double layer. Such electrodes are termed nonpolarizable or reversible, and the silver-silver chloride electrode is a common example. With this electrode (designated by the symbol Ag-AgCl) in a solution of sodium chloride, there is a tendency for Ag to go into solution forming Ag^+, but also an opposite tendency for the Cl^- ions in solution to combine with the Ag^+ to form AgCl. The result is an electrical balance between the two opposing processes.

With other electrode materials, only a minimal transfer of charges occurs across the electrical double layer. Such electrodes are termed polarizable electrodes. Polarizable electrodes have an electrical charge on them. For this reason they display the characteristics of a capacitor, which means that they do not pass DC and act as a low-frequency filter. Most recording electrodes available are of the polarizable type. Nonpolarizable electrodes are expensive and technically more difficult to work with. Fortunately, a variety of polarizable electrodes are quite satisfactory for recording clinical EEGs. They are discussed later in this chapter. Nonpolarizable electrodes are not essential in EEG work since DC or very low-frequency voltages are not recorded in routine clinical EEGs. For this reason, nonpolarizable electrodes will not be dealt with any further in this text.

Electrode Potentials

We said earlier that when a metal electrode is placed in contact with an electrolyte, a voltage develops between the metal and the electrolyte. This voltage was referred to as the electrode potential or half-cell potential. The term half-cell potential implies that the single electrode is acting as if it were half a battery, and this is indeed the case. Because it is like any other voltage connected to the input of an amplifier, we can expect the electrode potential to be amplified as any other voltage would be. Such being the case, the electrode potential would appear as an artifact in the EEG tracing.

But this does not normally happen for two reasons. First, we know that two electrodes are required to record an EEG and that if the electrodes are identical, the same voltage will be present on each of them. Therefore, the electrode potential will appear as a common-mode signal at G_1 and G_2 of the amplifier and be rejected by the CMRR. Second, the electrode potential is a DC voltage. If this voltage were relatively stable, the capacitor in the low-frequency filter of the EEG machine would block it out before it had a chance to be amplified.

As it happens, different metals have different electrode potentials. For example, the base metal lead has an electrode potential of hardly more than 0.1 V, whereas aluminum has an electrode potential of about 1.7 V. The presence of a difference in voltage between dissimilar metals is the principle upon which the battery or voltaic pile is based and is of some interest to the EEG technician. Thus, if a pair of EEG leads attached to a patient happened to be made of different metals, there could be a substantial voltage between them. This voltage would not necessarily be objectionable if it were stable, since DC voltages are blocked by the capacitor in the low-frequency filter. In practice, however, such voltages are rarely stable and for this reason they represent a source of artifact in the EEG recording. Such artifacts are sometimes referred to as a "battery effect." The upshot of this is simply that the two recording electrodes of a pair should always be made of the same material.

Residual Potentials

Even though both electrodes are made of the same material, in practice some voltage frequently can be measured between them. In other words, the electrode potentials of the two leads may not be identical. A number of factors can be responsible for such *residual potentials.* Our discussion of them includes the following ways of alleviating their effects:

1. There may be impurities in the metal, or the surface of the electrodes may be contaminated by foreign metal ions. To avoid the former, only high-purity metals are used in recording electrodes. In the case of silver electrodes, for example, only silver designated as "high fine" is used. Care in cleaning and storing is necessary to prevent surface contamination.
2. There may be foreign metal ions present in the electrolyte. To avoid this possibility, electrode pastes and gels should be carefully selected and protected from contamination during use. Any tools used in lead application should be kept scrupulously clean.
3. There may be differences in the concentration of the electrolyte at the two electrode sites through lack of homogeneity in the electrode paste or gel used.
4. There may be a difference in temperature of the skin at the two electrode sites. This happens because some metals used in electrodes have electrode potentials with temperature coefficients in excess of 100 $\mu V/^{\circ}C$.

Taken together, these factors could result in significant differences in voltage between the two electrodes of a pair. But here, again, the presence of a voltage is not objectionable in clinical electroencephalography if the voltage is stable. Problems arise only when the residual potentials fluctuate. When this happens, the variations in voltage constitute a source of artifact in the EEG tracings.

Types of Electrodes

By and large, most clinical EEGs currently are done using *surface electrodes.* The advantage of surface electrodes over needle electrode is obvious, in terms of both convenience and comfort for the patient, as well as relative freedom from infection. Less obvious but also a significant factor is the lower electrode impedances that are frequently possible with surface electrodes (see later section entitled "Factors Affecting Electrode Impedance"). For these reasons, this text deals only with surface electrodes.

By far the most popular surface electrode used in clinical EEG work is the *metal-disk electrode.* Disk electrodes are circular pieces of thin metal that may be flat or slightly cup-shaped to hold the electrolyte that forms the metal-electrolyte interface. The diameter may vary from 4 to 10 mm, the smaller disks being used mainly for recording in infants. Some cupped-disk electrodes have a hole in the center through which the electrolyte can be introduced after the electrode has been attached to the scalp. Tin, lead, solder, silver, and gold have been used in the construction of disk electrodes. Gold electrodes, however, are not pure gold but are simply gold plated over high-fine silver. The noble metals like gold, being less reactive than the base metals, make the most stable, drift-free electrodes.

The disks are soldered to a flexible, insulated wire, and this junction is carefully covered by a plastic material to prevent moisture or any of the electrolyte from reaching the solder joint. Should a breakdown of the plastic material cause penetration by water and electrolyte, an active battery would be created at the junction of the dissimilar metals resulting in large, electrode-generated artifacts. For this reason, disk electrodes need to be handled carefully. The EEG technician should never scratch the surface of an electrode or bend or pull the electrode at the junction between wire and disk. The electrodes should be kept dry when not in use; avoid soaking them in water for long periods of time. Appendix 7 takes up methods of disinfecting metal-disk electrodes.

Application of Surface Electrodes

The hair and surface of the scalp should be clean and free of hair oils, pomades, or other hair dressings before electrodes are applied. For this purpose, it is desirable for the hair and scalp to be washed the night before the EEG is taken. As an alternative, topical cleaning at the measured locations using an alcohol prep swab may be tried. After the electrode sites have been measured off and marked,[1] an

[1] Details concerning measuring and marking electrode sites are found in Appendix 5.

electrolyte is rubbed into them. Some EEG technicians call this procedure "scrubbing," although, in practice, only gentle rubbing is used. Preparations that contain free chloride ions and a mild abrasive for reducing the resistance of the skin are available for this purpose.[2] These are best applied using a cotton-tipped wood applicator stick. To protect both patient and technician from possible infection, special care is necessary so as not to scratch or break the skin. Care also needs to be taken to avoid spreading the material over an area much wider than the diameter of the metal disk, as in doing so the effective area of the electrode is increased beyond the limits of the metal disk. When this happens, the localization capabilities of the electrode are significantly reduced.

An extreme case of spreading the electrolyte too far occurs when two adjacent electrode sites are involved. By carelessly spreading the electrolyte over too large an area, the two areas scrubbed may inadvertently overlap. This condition, known as a *"salt bridge,"* creates a short circuit between the two electrode sites. Leads attached to these sites would not function as two separate electrodes but as a very large, single electrode. More about salt bridges later in this chapter.

Many different techniques have been used to secure the metal disks to the scalp. The simplest and speediest method is to use a *conductive electrode cream* or *paste* that has adhesive properties sufficient to hold the disk in place. In this case, a cupped disk is chosen. With the cup filled with paste, the electrode is pressed firmly against the scalp after the hair at the site has been separated. To keep long hair out of the way, some EEG technicians braid the hair or tie sections of it together with rubber bands. A small square of gauze is placed over the electrode to retard drying of the paste.

Although paste is a satisfactory means of securing the electrodes when the patient is cooperative and relatively quiet, the leads are easily pulled away by the movements of a convulsing patient or a very restless child. In these situations, the use of collodion is the best alternative. When collodion is employed, a cupped disk with a hole in the center is the appropriate choice. The electrode is held in place over the measured location and collodion is spread around its edges. A stream of compressed air must be used to dry the collodion quickly. Special care should be taken to avoid getting any collodion into the patient's eyes. With the electrode firmly in place, electrolyte is introduced through the hole in the disk by means of a hypodermic syringe with a blunt, wide-gage needle.

Other materials and methods have been employed to secure the recording electrodes on the scalp. Thus, for example, some laboratories have used bentonite or low-melting-point wax. An entirely different approach makes use of a cap or helmet that is tied to the patient's head and holds the electrodes in place. These methods are not discussed in this text. Interested readers may consult the literature on EEG technology for further information.

[2] A preparation called "Omni-Prep®," which is commercially available, has been used with considerable success by the authors.

Electrode Impedance

Assuming that the electrodes have been applied to the measured-off locations on the patient's scalp following one of the procedures outlined above, the next step is to "check" the electrodes. This involves measuring their electrical impedance.

Impedance is measured by applying a small, external voltage to the electrodes and then measuring the amount of current flowing in the circuit formed by the leads. The reader will recall from Chapter 2 that Ohm's law states that

$$I = \frac{V}{Z}$$

or, rearranging the terms in the equation

$$Z = \frac{V}{I}$$

This formula tells us that the impedance may be calculated simply by dividing the applied voltage *(V)* by the current *(I)* flowing through the circuit. Impedance meters used to check electrode impedance allow the EEG technician to read the lead impedance directly so that no computation is necessary. Details relating to the impedance meter are taken up in a later section.

Because impedance in the present application consists of a combination of resistance and capacitive reactance (see Chapter 2), AC is used to measure impedance of the leads. Typically, the measurement is made by using a 10 to 30-Hz AC signal, which is well within the range of the EEG frequency spectrum. Since the patient forms part of the electrode circuit, the current also flows through the patient. However, as current levels are very low—only a few microamperes—the current poses no danger whatsoever and is rarely even perceived by the patient.

For optimal recording, it is generally agreed that the impedance of a surface electrode should be less than 5K ohms. Recording problems usually arise when the impedance is either too high or too low. To understand what happens when lead impedance is very high, it is helpful to look at the circuit that is formed with the patient by the two electrodes and the amplifier. Such a circuit is seen in Fig. 7.1A, where V_{EEG} is the voltage of the brain electrical activity picked up by the electrodes, Z_{E1} and Z_{E2} are the impedances of the electrodes, Z_A is the input impedance of the amplifier, and V_A is the voltage appearing at

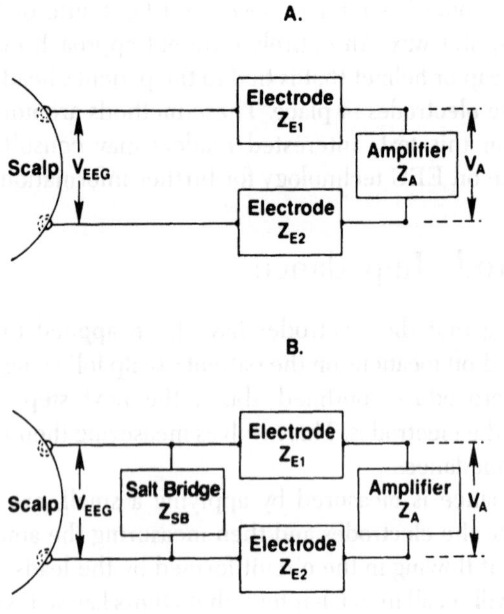

Figure 7.1. Equivalent circuits showing a pair of electrodes attached to a patient's scalp and connected to the input of an EEG amplifier. (A) normal circuit; (B) circuit showing a salt bridge.

the input of the amplifier. This basic circuit was already analyzed in Chapter 2. Although in that instance only resistance was present in the circuit, the same analysis is applicable when impedance is substituted for resistance.

Referring back to Fig. 2.4 in Chapter 2, we find that the EEG voltage divides across the three impedances so that the voltage at the amplifier is always some fraction of the EEG voltage. For the circuit in Fig. 7.1A, this voltage (after rearranging the terms somewhat) is given by the formula

$$V_A = \frac{Z_A}{Z_A + Z_{E1} + Z_{E2}} V_{EEG}$$

Let us examine what happens in this formula when Z_{E1} and Z_{E2}, the electrode impedances, are allowed to vary.

Suppose Z_A is equal to 2M ohms, and Z_{E1} and Z_{E2} are each 5K ohms. This means that

$$V_A = \frac{2{,}000{,}000}{2{,}000{,}000 + 10{,}000} V_{EEG}$$

$$V_A = 0.995\ V_{EEG}$$

or the voltage present at the input of the amplifier is very nearly identical to the EEG voltage. On the other hand, if the electrodes were poorly applied to an unwashed scalp, the impedance of each might be as high as 500K ohms. Under these conditions,

$$V_A = \frac{2{,}000{,}000}{2{,}000{,}000 + 1{,}000{,}000} V_{EEG}$$

$$V_A = 0.666\ V_{EEG}$$

or the voltage present at the input of the amplifier is only two thirds of the true EEG voltage. It should be clear from this example that, to avoid excessive reduction of the amplitude of the EEG voltages, electrode impedance must be low by comparison with the input impedance of the EEG amplifiers.

But the reduction in amplitude of the EEG voltages that results from using high-impedance electrodes is not the only reason why low-impedance leads are essential. A more serious consequence is concerned with the reception (pick up) of 60-Hz artifact by the EEG amplifiers. It happens that the amount of 60-Hz artifact that gets into the amplifiers from the external environment is directly proportional to the impedance of the circuits connected to their inputs. This means that higher 60-Hz levels in the tracings go along with leads of higher impedance and vice versa. In the limiting case when impedance is equal to zero, no externally generated 60-Hz activity would be present in the tracings. This is the reason why the inputs of the amplifiers are short-circuited to test for the presence of 60-Hz activity in the external environment. If a 60-Hz artifact disappears when the inputs are short-circuited, we conclude that the 60 Hz is external to the EEG machine since the short circuit is virtually the same as connecting zero ohms to the amplifier inputs.

Based on the formula for V_A that we have been dealing with, it would seem that the lower the electrode impedance, the better the recording. In practice, however, this is not the case. Recordings taken with leads of very low impedance—measuring less than 500 ohms—usually indicate that another phenomenon may be taking place. This phenomenon is the salt bridge that was mentioned earlier in the chapter. A salt bridge acts as a short circuit or pathway of very low impedance between the two electrodes. Figure 7.1B shows the equivalent circuit for a pair of recording electrodes that have been bridged by the application of too much electrolyte. With Z_{SB} less than 500 ohms, it is easy to see that the brain electric currents will be shunted away from the amplifier circuit. As a result, V_A will be very low amplitude and, in the limit, will be equal to zero. Whenever a single channel on the EEG machine shows activity that is of markedly lower amplitude than the activity on the other channels, the person reading the record as well as the EEG technician should suspect a salt bridge.

Factors Affecting Electrode Impedance

Because electrode impedance is such a critical variable, it is necessary to consider the factors that affect it and the ways in which it may be reduced to acceptable levels. First, we have the factor of *surface area*. Large-diameter disks have more surface area than smaller diameter disks and for this reason have a lower impedance. This is why needle electrodes, which have a small surface area by comparison, may show a relatively high impedance even though they penetrate the skin. Unfortunately, it is not practicable in EEG work to reduce electrode impedance by increasing the size of the electrode. When electrodes are larger than 10 mm in diameter, accurate localization becomes impossible. Moreover, with very large disks the interelectrode distances become so small that the danger of having a salt bridge is significantly increased.

Empirical studies have shown that the impedance of a pair of surface-recording electrodes on a patient is mainly due to the skin on which the electrodes are placed. This being the case, removal of the skin directly under the electrodes would greatly reduce the impedance, but obviously this is not possible in routine clinical electroencephalography. Nevertheless, the practical alternative of gently rubbing a preparation containing an electrolyte and a mild abrasive into the skin under the electrodes is usually quite successful. Additionally, it also helps to allow a little time for the skin to become hydrated and the electrolyte to soak in before checking the electrodes. While waiting, the technician can spend the time filling in the clinical details in the worksheet.

Electrode-Induced Artifacts

We already mentioned one major source of electrode-induced artifacts in our discussion of residual potentials. It was noted that for various reasons, a pair of electrodes may act as a battery. This phenomenon, which is known as the battery effect, may constitute an important source of artifact in the recording if the voltage generated by the electrodes is not stable.

Let us assume that a pair of electrodes of high inherent stability is correctly and carefully attached to a patient's scalp. Will these leads yield a satisfactory recording? The answer to this question is that it depends on the degree of *mechanical disturbance* to which the electrodes are subjected. In this case we are speaking of disturbance of the metal disks themselves, not disturbance of the wires to which they are attached. Artifacts produced by displacement of the wires are termed movement artifacts and are obviously avoided by preventing the lead wires from being disturbed.

Mechanical disturbance of the disks produces an artifact usually referred to as an electrode "pop." Such artifacts appear to result from a disturbance or instability of the electrical double layer. According to Geddes (Geddes LA, 1972), electrodes that are relatively free of such artifacts are those in which the electrode-electrolyte or metal-electrolyte interface is removed from direct contact with the patient. This would suggest that the metal disk itself should not touch the patient's scalp but should be "floated" on the surface of the electrolyte. The method of applying leads that employs electrode cream or paste would seem to satisfy this requirement admirably.

Detection of Electrode Artifacts

Electrode artifacts are best detected and identified by including the suspected electrode as a common electrode in a bipolar montage (see Chapter 11). What happens is best described by an example. Suppose that of five electrodes designated A, B, C, D, and E, electrode C is suspected of producing an artifact. To confirm this, the electrodes are connected to four EEG channels, as shown in Fig. 7.2.

As electrode C in Fig. 7.2 is common to two channels, an electrode artifact like a "pop" will be present in both, namely, channels 2 and 3. Note that the deflections are of equal size and of opposite phase in the two channels so that one appears as the mirror image of the other. This happens because electrode C goes to opposite grids of channels 2 and 3. Note also that despite their large size, the deflections are present only in channels 2 and 3, suggesting that the disturbance has no field. This confirms that the deflections observed are artifacts and do not originate in the patient. Indeed, an artifact should be suspected whenever a deflection pattern like that shown in Fig. 7.2 is observed. Had the deflections originated in the patient, there would be a field surrounding the focus at electrode C. Depending on the distances involved, this field could be picked up by the electrodes connected to channels 1 and 4, which would then show deflections similar to but of smaller amplitude than those in channels 2 and 3. This is taken up in Chapter 12 where the rules of localization are discussed.

Impedance-Measuring Devices

We have discussed electrode impedance and the theory of measuring impedance in some detail. It remains yet to consider the actual methods that are employed in practice.

Although it is easy to think of the impedance of a *single* electrode, it is no simple matter to measure it. A little thought will explain why. Impedance is measured by passing a current through the element to be measured. But how do you pass a current through a single electrode? The answer is that you cannot. An electric circuit always has two connections, whereas a single electrode attached to the patient has only one. What is needed to make a circuit is a pair of electrodes. But this means that the measuring current passes through two electrodes connected in series so that the impedance measured will be the impedance of

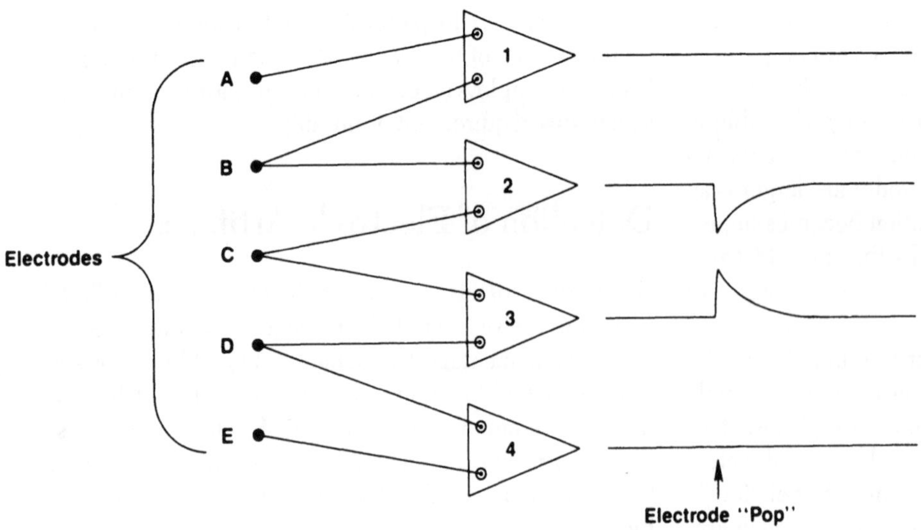

Figure 7.2. Appearance of an electrode artifact in bipolar recording. The tracings show the occurrence of a "pop" in electrode C. Note that the artifact appears only in the two channels that are common to the defective electrode and that the deflection in one channel is the mirror image of the deflection in the other. Electrode "pops" bear a striking resemblance to the deflections traced out by DC calibration signals.

both electrodes combined. In estimating the impedance of one of the electrodes, we have to assume that each electrode is one-half the total impedance.

Although the method just described for measuring the impedance of recording electrodes is used extensively, it has a serious problem. The problem is simply that we have no assurance that the impedances of both leads in the pair are the same. The same meter reading could be obtained as well if one lead were high impedance and the other were very low, and vice versa. A better method is clearly necessary.

By the ingenious application of Ohm's law and the rules governing the combining of impedances, it becomes possible to solve this dilemma. To understand how this is accomplished, consider the circuit shown in Fig. 7.3. This circuit, which is a series-parallel circuit, consists of a single impedance, Z_X, connected in series with a total of 20 separate impedances that are hooked up in parallel. In Chapter 2 we learned that the total impedance of a group of impedances connected in parallel is equal to

$$\frac{1}{Z_{1-20}} = \frac{1}{Z_1} + \frac{1}{Z_2} + \frac{1}{Z_3} + \cdots \frac{1}{Z_{20}}$$

or

$$Z_{1-20} = \frac{1}{\frac{1}{Z_1} + \frac{1}{Z_2} + \frac{1}{Z_3} + \cdots \frac{1}{Z_{20}}}$$

The total impedance in the circuit, Z_{total}, is equal to

$$Z_{\text{total}} = Z_X + \frac{1}{\frac{1}{Z_1} + \frac{1}{Z_2} + \frac{1}{Z_3} + \cdots \frac{1}{Z_{20}}}$$

Suppose, now, that all of the 20 Zs in this equation were approximately equal to each other in magnitude. Under such conditions,

$$Z_{\text{total}} = Z_X + \frac{1}{\frac{20}{Z_n}}$$

or

$$Z_{\text{total}} = Z_X + \frac{Z_n}{20}$$

If $Z_X = Z_n = 5$K ohms, total impedance is

$$Z_{\text{total}} = 5{,}000 + 250$$
$$= 5{,}250 \text{ ohms}$$

which means that the total impedance of the circuit differs but little from the impedance of Z_X.

The circuit just discussed embodies the principle upon which modern impedance meters used to measure impedance of a single recording electrode operate. Z_X is the electrode whose impedance we wish to measure, while Z_{1-20} are the impedances of the remaining 20 leads used in taking a routine clinical EEG. Inside the impedance meter box is a complex switching system. This switching system alternately connects each of the electrodes attached to the patient in series with all the other electrodes that themselves have been hooked together in parallel. Note that the impedance of any lead may be estimated to within 5% of its actual value by means of this circuit.

The method of measuring impedance just described is normally carried out before the patient is hooked up to the EEG machine. Many EEG machines provide a means of quickly measuring impedance of electrodes at any time while the EEG is being run. This is a convenience since an

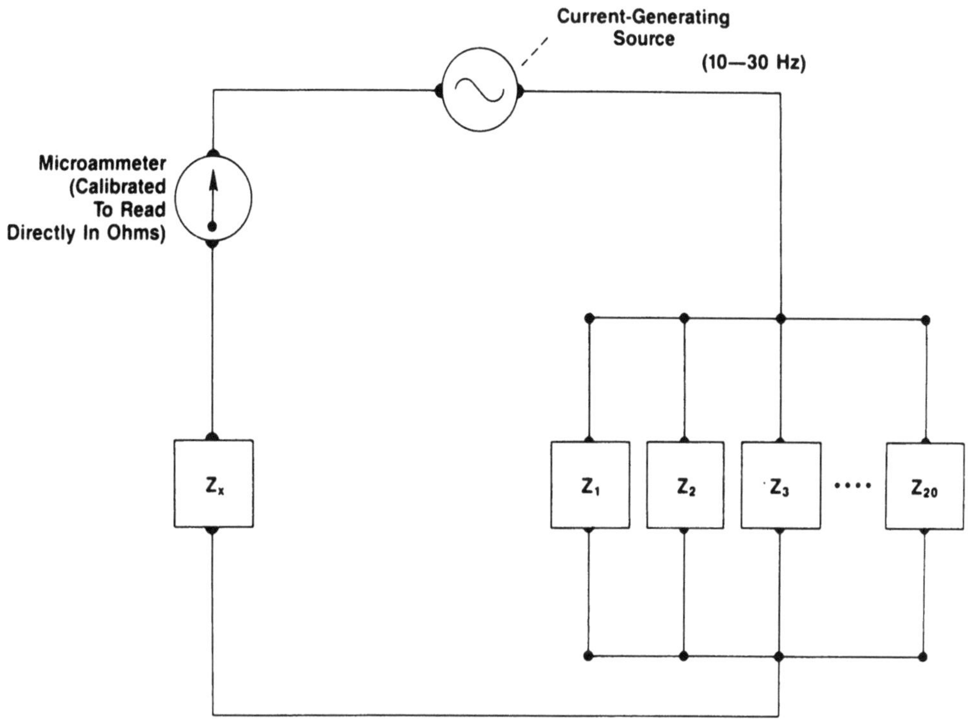

Figure 7.3. Simplified circuit of an up-to-date impedance meter showing the series-parallel circuit used for measuring the impedance of single electrodes.

electrode suspected of having a high impedance can be checked without interrupting the test.

This feature is enabled by closing the *electrode test switch* on the console of the machine and then pushing the *electrode test button*. The electrode test switch automatically adjusts the sensitivity of all channels to a standard value, say, 7 μV/mm, and turns on a 30-Hz test oscillator. Pushing the electrode test button feeds the 30-Hz signal to the electrode pair of each channel. This signal is recorded on each channel of the machine, and the recorded amplitude of each tracing is proportional to the impedance of the electrode pair connected to it; a peak-to-peak deflection of 1 mm = 2,000 ohms. By using a bipolar montage so that some electrodes are common to two channels, the EEG technician can figure out which electrode of a pair is high impedance.

Reference

Geddes LA: *Electrodes and the Measurement of Bioelectric Events.* New York, Wiley-Interscience, 1972.

Chapter 8
Electrical Safety

Everyone knows that it is potentially hazardous to be standing outdoors in the rain during an electrical storm. Most persons attribute the danger to the fact that the body provides a pathway for the flow of electric current to ground from a nearby discharge of lightning. In other words, there is a risk of being struck by lightning. Since the current levels in a typical lightning stroke can exceed 20,000 A, the danger is indeed a real one.

Common knowledge tells us that it can be dangerous for a person to touch an uninsulated live electric wire. Yet it is well known that birds often perch on such wires without danger of electric shock. The difference lies in the fact that in the former case, the body provides a pathway for the flow of current to ground if the person is directly or indirectly in contact with the earth. Thus, it is the flow of current and not simply the presence of a voltage that constitutes the danger, and it is the amount of current forced through the body that is the real measure of a shock's intensity. How much current constitutes a danger? To answer that question, it is essential to consider where a voltage is applied and the pathway the electric current takes through the body.

Macroshock and Microshock

The term macroshock applies to the type of hazardous situations we have just described. Macroshock involves currents that pass from one *external* surface area of the body to another and are perceptible to the person exposed to them. Although the effects will depend partly on the actual pathway of the current through the body, Fig. 8.1 shows, in general, the physiological effects of various current densities. Recalling that the current flowing in a circuit is directly proportional to the applied voltage and inversely proportional to the impedance, it is readily apparent that 120 V applied to a person whose impedance across the connections is 5K ohms will cause 24 mA to flow, which, by Fig. 8.1, is a dangerous shock. It should be obvious, therefore, that ordinary 120-V AC house current used to power the EEG machine can represent a potential danger to the EEG technician and the patient. Indeed, persons have been electrocuted by all kinds of appliances that use ordinary house current.

Microshock is another electrical hazard that EEG technologists and others dealing with EEGs need to be concerned about. Microshock is a term used to describe very small currents that nevertheless may be lethal because of where they are applied. Patients with indwelling electrodes, catheters, or implanted transducers in the heart are at considerable risk because these devices may provide a pathway for the flow of electric current. Such currents may be very small indeed—as small as 100 μA. However, although 100 μA is harmless and not ordinarily perceptible when applied to the external surface of the body, it may be lethal when introduced directly to the heart. More will be said about this problem and how it is resolved later in the chapter.

In taking EEGs, electrical safety is concerned chiefly with protecting the patient and the EEG technical from inadvertently becoming a pathway for the flow of electric currents to ground. To understand how this can come about and how it may be prevented from occurring, it is essential to understand the way in which "ground" is involved in electrical circuits. For this reason, we begin our discussion of electrical safety with a consideration of the topics of ground and grounding.

Ground and Grounding

In the domain of electrical phenomena, "ground" can take on a variety of meanings. The term itself refers to the ground of the earth; indeed, the British speak of "earthing

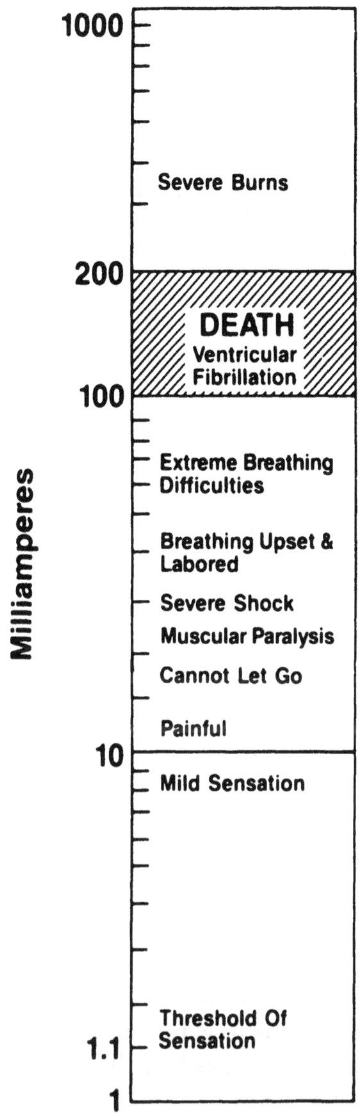

Figure 8.1. Physiological effects of electric currents. (From The fatal current, *Tektronix Service Scope*, Dec 1965, #35, 1-2, by permission of Tektronix, Inc., Beaverton, Oregon.)

a circuit" when they refer to "grounding." Thus, ground can refer to any point in the earth to which an electrical circuit may be connected.

Why does the earth play such an important role in electrical matters? There are a number of reasons. In the first place, the earth is an electrical conductor. This fact is of considerable practical importance. To save on the cost of insulated wire, many early electrical generating systems used the earth as a return path for the flow of electricity in their power distributing circuits.[1] Except in cases of high-voltage transmission, this practice has been largely abandoned in modern power distributing circuits where an insulated wire is used for the return path. Nevertheless, the earth continues to be of importance as this return wire itself is grounded. This connection to earth provides a stable reference of zero volts for the circuit as well as a safe pathway to earth for electrical currents produced by lightning in a storm.

Inside the home as well as in the EEG laboratory, the term ground takes on a more specific meaning. In this case, ground refers to the third contact on the 120-V AC electrical outlets found along the walls of any room. This is the contact that mates with the third contact (the long prong) on the plug of the power cord running to the EEG machine. If you trace where the long prong on the plug goes, you will find that it is connected to an insulated wire (the green wire) in the power cord.[2] Once inside the machine, this wire is ultimately connected to a number of different points. Among them are the ground jack on the electrode board of the machine, the ground connections of the differential amplifiers, and the console of the machine itself. Note that the console is made of metal, which is a good conductor of electricity, and that the metal chassis completely encloses the electrical and electronic circuits contained in the machine.

Going now in the opposite direction, that is to say, from the previously mentioned third contact in the wall socket *into* the wall structure, you will find that this contact is hooked up to a wire covered with green insulation. Note that this wire matches the color of the wire inside the power cord to which it corresponds; green always indicates a ground wire in modern electric power distributing systems. Following the wire to its source, you will discover that it is connected to the earth. Connection to ground is usually made by way of a large-diameter cold water pipe or by a metal rod(s) or plate buried in the earth. As the plumbing in most houses or structures is metal, and it makes contact in one way or another with the earth, the various pipes themselves become grounded. Electrically speaking, therefore, a person touches earth whenever he or she comes in physical contact with any metal part of a house's permanent structure.

Given these basic wiring facts, it can be deduced that (1) the console of an EEG machine is (or should be) grounded when the power cord is plugged into an appropriate 120-V AC socket, and (2) any metal fixtures in the EEG laboratory that are a permanent part of the room (electrical switch plates, heating pipes and ducts, metal sinks, water pipes, faucets, drain pipes) are (or should be) grounded.

[1] The reader will recall from Chapter 2 that two connections are always needed to make an electrical circuit. In this example, the earth serves as one of these connections.

[2] The other two wires in the cable—one with black insulation (the hot wire) and the other with white insulation (the return pathway)—provide the 120-V AC that runs the machine.

This is a well-thought-out plan for which there is good reason, which is embodied in the first rule of electrical safety for the EEG laboratory. Simply stated, this rule says that for *safe operation* all metal objects and exposed metal surfaces in the EEG laboratory must be connected to ground.

Why is grounding an essential part of safe operation? We can answer this question best by considering what could happen if a fault developed inside the EEG machine. Suppose one of the many current-carrying wires that are inside the machine came loose and touched the inside wall of the machine's chassis. What would happen? With the chassis of the machine properly grounded, the current would be carried away harmlessly to ground without offering any danger to the EEG technician operating the machine. On the other hand, if the EEG machine were ungrounded and the EEG technician happened to be touching it while at the same time making contact with a heating duct or pipe in the laboratory that was grounded, current would flow to ground *through the technician's body*. In other words, being in contact with ground is dangerous if there are sources of electric current nearby that can make connection with the body. This is analogous to what happens when a person stands on the ground outdoors during an electrical storm. The result in either case could be a serious injury, if not loss of life.

Because proper grounding of the EEG machine is so important, the EEG technician should periodically check the installation to verify that the ground connection is intact. How this is done is taken up in Appendix 3.

Leakage Currents

As we just mentioned, a potentially dangerous situation exists if the ground connection of the EEG machine is interrupted and a fault were at the same time to develop within the machine. But now suppose that the machine became ungrounded, as in the previous example, but *no fault* was present in the machine. Is there still a potential for harm?

To understand what happens in this case, it is important to recognize that the power cord and parts of the power supply of the EEG machine behave like a number of capacitors connected together in parallel. The green ground wire acts as one side of the capacitors while the current-carrying wires of the power cord act as the other. The effect, which is known as *stray capacitance*, is distributed along the entire length of the cord. With 60-Hz 120-V AC applied to the hot wire of the power cord, a current will flow through the capacitive reactance formed by the stray capacitance. This current is called the *leakage current*. Its magnitude depends on the length and characteristics of the power cord and on the design and construction of the power supply. Leakage currents are not negligible; for a 6-ft power cable they may range between 7 and 60 μA.

Normally, the leakage currents generated in the EEG machine pass harmless to ground. However, should the ground wire that connects the machine to earth be interrupted, these currents need to find alternative pathways. One pathway is via the chassis of the machine. Another is back through the input and the recording electrodes. This means that the EEG technician and the patient hooked up to the machine are both at risk from these currents—the patient because he or she is connected to the EEG electrodes and may also be connected to some other equipment that is grounded, and the EEG technician because he or she may be touching something connected to ground. Because of this risk, *leakage current limits* have been specified for EEG machines. The maximum allowable chassis leakage current with the ground wire of the power cord interrupted is 100 μA. For the inputs of the machine, a leakage current of up to 50 μA between electrodes and ground is allowed. It should be obvious from our discussion thus far that protection against chassis leakage currents is afforded by making sure that the EEG machine is connected to ground when the power cord is plugged in. How chassis leakage current is measured is taken up in Appendix 4.

Patient Grounding

Thus far we have been talking about the grounding of the EEG machine. In routine clinical EEG recording, a ground lead is usually attached to the patient. By means of this electrode, the patients gets connected to earth via the ground wire of the EEG machine. But if being in contact with ground is a potentially hazardous condition, does not the deliberate use of a ground connection pose a serious danger to the patient? And why is a ground lead attached to the patient in the first place? We take up the latter question first. But before doing this, some background information is in order.

We all know that radio and ordinary television are methods of communication that employ no wires between transmitter and receiver; they are wireless systems. In principle, their operation is relatively simple. A transmitter beams out an alternating electrical voltage that produces a widespread electrical field. At a distance from the transmitter is the receiver to which an antenna is connected. The antenna is nothing more than an electrical conductor—in its simplest form, a loop or coil of wire. In Chapter 5 we noted that an electric current could be induced in a conducting circuit that is placed within range of a changing magnetic (electrical) field. In the case of radio and television, such currents are induced in the antenna of the receiver. These currents are very feeble indeed. Nevertheless, when amplified and suitably processed (demodulated), they emerge as radio and television programs.

It happens that by attaching EEG leads to a patient's head, we create what turns out to be a pretty good antenna. In some cases this antenna has been known to pick up strong local radio and television stations. As might be expected, such signals are capable of introducing artifacts into the EEG tracing. In practice, however, interference from radio and television is generally not a problem in EEG work since the high-frequency filters largely eliminate these sources. What is a problem, however, is the 60-Hz power lines. These power lines, which act like a transmitting antenna, are almost everywhere. The result is that a 60-Hz electrical field is almost impossible to avoid under ordinary circumstances.[3]

When placed within range of a 60-Hz electrical field, a patient with EEG leads attached serves as an excellent receiving antenna. Because the body acts like a capacitor with respect to the earth, 60-Hz AC will flow in this antenna circuit. Although the current flowing in this antenna circuit is normally too small to pose any danger to the patient, it frequently generates voltages that are large enough to be recorded by the EEG machine. In some cases, 60 Hz artifact can obliterate the EEG tracings.[4] By placing a ground electrode on the patient, we effectively ground the antenna. As a result, these currents are significantly reduced, especially if the ground electrode is located somewhere on the patient's head. This means that the differential amplifiers in the EEG machine have less 60-Hz activity (less common mode voltage) to discriminate against.

So we see that the patient's ground connection is essential for reducing 60 Hz artifact in the EEG tracings. Indeed, the presence of 60 Hz in a recording is frequently indicative of a poorly attached ground lead. But the advantage gained by using a ground lead is not obtained without some cost. By deliberately grounding the patient, we subject the patient to the risk of electrical shock should he or she come into contact with a live voltage or with a current-carrying wire. When this happens, the patient serves as a direct pathway for the flow of current to ground. How might such a situation come about in practice?

There are a number of possible ways in which a patient's body might provide a pathway for the flow of current to ground in the EEG laboratory. In discussing the ways, the reader will argue that they are all highly unlikely events. This, of course, is true. Nevertheless, when talking about safety, unlikely events need to be considered because they are known to have occurred at one time or another. Indeed, patient safety may aptly be defined as protecting the patient against all unnecessary risks no matter how improbable their occurrence.

One possible source of risk to the patient occurs if the patient's body happens to be touching the chassis of the EEG machine or some other metal fixture in the room that should be grounded, but through some fault is not. When this happens, the patient serves as the pathway to ground for the leakage currents discussed in the last section. Moreover, the patient would be at additional risk if an electrified device or current-carrying wire happened to make electrical contact with the ungrounded chassis or metal fixture.

Another way in which the patient's body could provide a pathway for the flow of current to ground would occur if any of the 21 EEG electrodes attached to the patient's head were suddenly to acquire a voltage other than the voltages derived from the patient. This might come about if an internal failure occurred in a differential amplifier and the power supply voltages of the EEG machine somehow became connected to the amplifier's input circuit. Needless to say, the probability of this happening is very, very low indeed. Nevertheless, to minimize this potential hazard, some EEG machines employ optical coupling devices in the input circuits to isolate the patient from inadvertent contact with dangerous power-supply voltages.

The greatest possible risk to a patient who is having an EEG taken occurs when other electrical devices besides the EEG machine are connected up to the patient at the same time. This is not an uncommon occurrence, especially when EEGs are done in hospital rooms or in the intensive care unit (ICU). As a general rule, the more instruments and devices of various kinds one connects to a patient, the more chances there will be of encountering a fault of some kind. Devices such as ECG machines, electrically powered blood pressure monitors, instruments connected to indwelling catheters, and electrically powered implants may all have their own ground connection with the patient. If any of these ground connections (or the EEG ground) developed a fault, leakage current from the device could flow through the patient on its way to ground via the remaining intact ground connections. Depending on the precise pathway followed by the current through the body, the result could be destructive if not actually life threatening.

Effect of Patient Impedance

By now it should be clear to the reader that the danger from leakage currents and currents resulting from some fault in any electrical device connected to the patient comes from the possibility that these currents might flow to ground *via* the patient's own body. The actual amount of current that could flow through the patient depends on his

[3] So-called electrically shielded rooms have been employed to screen out 60-Hz fields. However, such rooms are rarely used, if at all, in routine clinical EEG work. For this reason, they are not discussed in this text.

[4] In some cases it is possible to measure as much as a volt or two of 60-Hz AC between a point on the body and the earth.

or her impedance. As we know from Chapter 2, this relationship is given by Ohm's law, which states that current flow in a circuit is equal to the voltage divided by the circuit's impedance. This means that the patient's impedance is an important factor in electrical safety: if the patient's impedance is high, less current will flow through his or her body than if impedance is low.

The impedance of an ordinary surface EEG electrode attached to the patient is attributable largely to the impedance of the skin under the electrode. For this reason, electrical contacts that bypass the skin and go directly inside the body may offer extremely low-impedance pathways for the flow of electrical currents. For example, a catheter that carries fluids to or from the body presents a current pathway having an impedance as low as 500 ohms. If such a low-impedance pathway were connected inadvertently to a current source, injury or death could result. This is one of the reasons why taking EEGs in the ICU presents a greater potential risk of electrical shock to the patient than taking them in the EEG laboratory.

The EEG Technician's Role in Patient Safety

Aside from ensuring that the EEG equipment used is designed, constructed, and maintained with a view to the operator's and patient's safety, what can the EEG technician do to minimize the risk of an electrical accident happening?

First of all, it is important to identify what electrical equipment, if any, is already hooked up to a patient before attaching EEG leads. Remember that the more wires you connect to a patient, the more chances there are of faults being present or developing and the greater the risk of electric shock. When an electrical device already connected has a ground of its own on the patient, do not attach an EEG ground to the patient but use the already existing ground instead. If this results in unacceptably high levels of 60 Hz artifact, remove the existing ground lead and replace with the standard EEG ground electrode placement—usually the center of the patient's forehead. A good general rule to follow is that there should be only one ground electrode attached to the patient at any one time.

Second, an electrode board having an *isolated ground* or *isoground* can be employed instead of the commonly used unit. The isolated ground limits current flow to ground via the EEG ground electrode to safe levels regardless of any faults that might develop in other connections to the patient. How this device works is taken up in the next section of this chapter. Finally, if other electrical devices are connected to the patient in addition to the EEG machine—as frequently happens in the ICU—the use of a so-called *bipotential isolator* electrode board is strongly recommended. The bipotential isolator ensures that the EEG lead connections are neither the source of hazardous currents flowing through the patient, nor the *means* whereby dangerous currents can find their way to ground.

Isolated Ground and Biopotential Isolator

Recognizing the potential hazards associated with connecting a patient to ground, there is obvious need for some kind of fail-safe method or device to protect the patient. The isolated ground, which is a product of modern technology, is just such a device. What exactly is the isolated ground, and how does it function to protect the patient?

Properly speaking, the isolated ground does not really isolate the patient from the ground at all. If we wanted to "isolate" the patient from any connection with ground, the ground electrode could simply be left off the patient. But as we learned earlier in this chapter, a ground electrode is important—it is needed to minimize the amount of 60 Hz artifact in the EEG tracings. In reality, the isolated ground is a sophisticated *current-limiting device*. One such device employs a solid-state component that is connected in series with the ground electrode connected to the patient. This component acts like a resistor in the circuit, like a *variable* resistor. With little or no current flowing in the ground circuit, the solid-state component has a relatively low resistance that ranges between 4K and 6K ohms. However, should the current flowing in the circuit increase and approach hazardous levels, the resistance of the component increases and quickly can become very high. Since by Ohm's law current is inversely proportional to resistance, this circuit prevents dangerous levels of electric current from flowing through the patient's body on into ground.

The isoground device can protect the patient only from electrical currents finding their way to ground via the ground lead. In routine clinical electroencephalography, there are 21 other electrodes attached to the patient. If a fault developed in any one of these input connections to the EEG machine, and if any other electrical device besides the EEG machine happened to be hooked up to the patient, the patient could serve as a pathway for the flow of hazardous currents between these points. To protect the patient against this potential danger, the *biopotential isolator* may be used. This device is a special electrode board in which a current-limiting solid-state component like that used in the isolated ground is connected in series with each of the 21 leads (as well as the ground lead) that are attached to the patient's head. By using the biopotential isolator, none of the electrodes connected to the patient during EEG recording can serve as a pathway for the flow of dangerous electrical currents through the patient.

Although the safety provided the patient by the isoground and the biopotential isolator is of prime concern, if should be recognized that the benefits are not obtained without some cost. The solid-state component that is the heart of these devices has a resistance that is considerably greater than zero even at very low current levels. As one of these components is connected in series with each EEG lead, the resistance of the component gets added to the electrode resistance. This being the case, the EEG technician should recognize the consequences of using these devices.

In the first place, when an electrode board having an isoground is used, the EEG technician will discover that impedance measurements of the ground lead are always higher than the impedance measurements of the other electrodes. This obviously happens because the impedance-measuring device displays impedance of the electrode *plus* impedance of the current-limiting device that is connected in series with the electrode. Secondly, EEGs taken when using the biopotential isolator will generally show higher levels of 60 Hz artifact than EEGs taken using the standard type of electrode board. The reason is that the impedance of the current-limiting device increases the total impedance of a pair of electrodes, and 60 Hz artifact is always greater with higher electrode impedance. The upshot is that the EEG technician needs to use special care in attaching EEG electrodes to ensure low impedances when using a biopotential isolator.

Ground Loops

A ground loop is a condition that occurs whenever more than one ground wire is attached to a patient. The loop, of course, is produced by the patient making an electrical connection between the two ground wires. Earlier in this chapter we noted that placing more than a single ground connection on the patient was potentially dangerous. This was to be avoided because the two ground connections might be at different voltages, thereby causing a current to flow through the patient's body. But although patient safety is the most important reason for avoiding a ground loop, it is not the only reason. Previously we mentioned that the wires running to the EEG electrodes attached to the patient create an antenna that picks up 60 Hz artifact. The same is true of a ground loop. As the loop is surrounded by a 60-Hz field, Faraday's principle of electromagnetic induction applies, and a 60-Hz current will be induced in the loop. Depending on the strength of the field and the area enclosed by the loop, a 60 Hz artifact may be recorded in the EEG tracings.

A ground loop does not necessarily have to involve a patient. If two separate pieces of electrical equipment each having a ground wire of its own happen to have their metal chassis touching, a ground loop will be formed. This can happen in the EEG laboratory when the chassis of a photic stimulator grounded by its power cord is allowed to come in contact with the chassis of the EEG machine. Being surrounded by a 60-Hz field, the loop will have a current induced in it and thereupon a voltage will be developed. By Ohm's law, this voltage is equal to IZ, the product of the current and the impedance; if large enough, it may be recorded as a 60 Hz artifact in the EEG tracings. Many seemingly mysterious 60 Hz artifacts observed in the EEG laboratory are explained in this way. As Ralph Morrison, author of a text on grounding techniques, rightly stated, "Basic physics, when properly applied, explains all known electrical phenomena . . . Most . . . grounding problems are just Ohm's law" (Morrison R, 1967).

The general grounding rule to follow is that all electrical devices, as well as the patient, should have their own single connection to ground, and the ground connection should be common to everything.

Reference

Morrison R: *Grounding and Shielding Techniques in Instrumentation.* New York, John Wiley & Sons, 1967, p. viii.

Chapter 9
Elementary Practical Troubleshooting Methods

In the early days of electroencephalography, a person wanting to take an EEG frequently had to play the role of an electrical engineer and electronic technician, as well as an all-around mechanic. The EEG machines in the 1930s through early 1950s were often makeshift, home-made rigs that were unreliable and sometimes temperamental—or so it seemed. It was not uncommon to interrupt an EEG recording to correct some malfunction caused by a noisy or microphonic vacuum tube or a discharged or leaky battery, or to eliminate an artifact produced by the corroded contacts on a switch or connector. Because 60-Hz notch filters and amplifiers with high common-mode rejection were not yet available, 60 Hz artifact was a constant source of harassment. Such problems, it can be imagined, made the outcome of a recording session uncertain.

The availability of today's more sophisticated EEG machines has ended all that. The EEG technician can now obtain a recording without having to minister to the needs of a malfunctioning machine. Nevertheless, modern machines are not entirely free of malfunction, and the EEG technician or the person interpreting the record is sometimes called upon to diagnose the cause of a fault or breakdown. Any but minor repairs or adjustments that may be necessary, however, typically are left to the manufacturer of the machine or to its representative. Additionally, the EEG technician or interpreter may sometimes have to discover the source of artifacts appearing in a record that originate outside of the EEG machine.

Basic Principles

Despite its complexity, an EEG machine is relatively easy to troubleshoot. This is occasioned by the fact that EEG machines are composed of a number of identical channels, each usually consisting of several separate, *interchangeable modules*. In many machines, moreover, components such as the montage switch, power supply, calibrator, and chart drive are of modular construction as well and may easily be removed and replaced. From these facts the two main methods of troubleshooting emerge. These are, first, the method of *substitution* for diagnosing malfunctions and breakdowns limited to a single channel of the machine, and second, the method of *isolation or elimination* in the case of malfunctions or breakdowns that are common to all channels of a machine.

Single-Channel Problems—The Principle of Substitution

The simplest kind of single-channel problem is when a channel fails completely. You turn on the machine and despite your double checking and triple checking the setup and connections, nothing happens. Although the other channels are all operational, this channel will record neither an EEG or a calibration; it is completely dead. At this point you could employ the method of substitution that will shortly be discussed. But the most likely cause of this kind of breakdown is a burned-out fuse if the circuit that provides the power needed to run the amplifier is fused. So it is worth checking the fuse(s) in the amplifier first.

The instruction manual that comes with the machine shows how to locate and replace amplifier fuses; spare fuses are usually provided in a repair kit that also comes with the machine. If a fuse needs to be replaced, it should be recognized that a burned-out fuse is frequently only a symptom—a symptom that there may be some malfunction present within the circuit protected by the fuse. If the fuse you replace burns out repeatedly, the amplifier will need to be returned to the manufacturer for repair. On the other hand, if a burned-out fuse is not the problem, it will be necessary to make use of the substitution method to locate the source of the problem.

Despite its simplicity, the method of subsitution is a powerful technique for discovering the source of a malfunction. Nevertheless, there are pitfalls associated with its use that need to recognized. These can best be appreciated by looking at a simple example.

Suppose that in calibrating your machine prior to taking a patient's EEG, you suddenly discover that channel 5 is behaving peculiarly. Although all the other channels of the machine display a horizontal, perfectly straight line, the pen of channel 5 wanders about erratically. A short strip of recording reveals that this channel is tracing out a voltage in the theta and delta frequency range. Since the artifact has a magnitude of about 50 μV, it can significantly contaminate the EEGs recorded on this channel.

Where does the artifact come from? Before even addressing this question, the experienced EEG technician will quickly double check the machine and verify that channel 5 is set up correctly and that all the switches are in their proper positions. Some technicians may even index all the switches on this channel back and forth several times to assure that the switch contacts are free of any foreign material. (The role of switch contacts in producing artifacts is discussed later in this chapter.) Having done this, however, you discover that the artifact is unchanged. Now the only recourse is to apply the method of substitution.

After selecting one of the good channels for the substitution, say channel 3, you need to decide which of the modules to substitute. You can, of course, routinely substitute all of them, one at a time. But experience has shown that the most likely source of a problem such as this will be the amplifier, or preamplifier if the preamplifiers in your machine happen to be separate, modular components. So you remove the amplifier in channel 5 (the bad channel) and replace it with the amplifier from channel 3 (the good channel). The malfunction disappears. Channel 5 is now O.K., and you conclude that the amplifier that was originally in channel 5 is faulty. But in doing so, you would be making an inference and leaving yourself open for a possible error.

The error we refer to derives from the fact that the malfunction could have been due to the connector that the amplifier was hooked up to as well as the amplifier itself. It is not unusual for the contacts in a connector to become noisy spontaneously and for the problem to be cleared simply by breaking and remaking the connection, as happens when an amplifier is removed and replaced. To rule out this possibility, it is essential to do a double substitution; in other words, you need to observe the result of installing the amplifier from channel 5 (the bad channel) in channel 3 (the good channel) as well. If after doing this the artifact moves with the amplifier, i.e., channel 3 now shows the same artifact that was seen on channel 5, you can be reasonably sure that the channel 5 amplifier is the culprit. As a double check, it is good practice to return the amplifiers to their original positions and verify that the artifact moves from channel 3 back again to channel 5. In the event that the malfunction was caused by a dirty contact in the connector, the problem may be corrected by cleaning the contacts. The procedure to follow is detailed later in this chapter under "Connector and Switch Contacts."

Before going through the double-substitution procedure, a few precautions should be observed. First, for safety's sake make sure that the main power switch on the machine is off while you make the substitutions. Second, consult the instruction manual that comes with the machine before removing an amplifier or, for that matter, any other modular component. Many manufacturers have special methods and some even have special tools that aid in removing the modules. Third, all connectors should be handled carefully and protected from dust, dirt, and especially from electrode paste or jelly. They should be plugged into their mating connectors firmly but never forcefully as the contacts are easily bent or damaged.

The method of substitution can be applied to check the power amplifiers, lead selector switches, filters or penmotors, assuming, of course, that they are all of modular construction. Once the fault has been localized to a particular module, it is a relatively easy matter to correct the problem. Thus, most manufacturers of EEG machines will ship you a "loaner" module to replace a malfunctioning one if you telephone and describe the symptoms of the malfunction, explaining what tests you have carried out to localize the fault to a particular module. Use the packing material from the loaner to package the faulty module and return it to the factory for repairs. To document the fault, it is helpful to send a strip of record from the machine showing the nature of the malfunction as well as a brief explanation along with the defective module. The reason for this is that rough handling that might occur during shipping could cause an artifact produced by a poorly soldered joint in a circuit to momentarily disappear. Should this happen and the factory had no record documenting the malfunction, the credibility of the person returning the module might be questioned.

Single-Channel Problems Observed During EEG Recording

In the just-completed discussion, the artifact present in channel 5 was observed during the calibration done prior to taking an EEG. Suppose, instead, that the same peculiar signal was observed on a single channel of the chart while an EEG was being taken. Now, thinking that this signal is an artifact of the kind discussed in the last section, you quickly switch the machine over to calibration to observe it more closely. In doing so, the peculiar signal disappears and the channel calibrates properly. Could this voltage, this peculiar signal, be an instrumental artifact?

If the recording happened to be from a referential montage, it would be impossible to judge whether the voltage in question was an instrumental artifact, an electrode artifact, or a voltage generated by the patient. To investigate, you switch to a bipolar montage. In so doing, you find that the peculiar signal is present now on *two* channels, both of which have a single electrode in common. This finding rules out the possibility that the voltage is due to an intermittent malfunction in the channel on which the artifact was first seen, or to a fault in the lead selector switch or montage switch. However, it still could be an instrumental artifact that is associated with the electrode board, an electrode artifact, or a voltage from the patient—the latter possibility depending on the presence and distribution of the voltage in adjacent leads. Artifacts associated with the electrode board are taken up in the next section. The other possibilities are considered in Chapters 7, 11, 12, and 13.

Electrode-Board Artifacts

While electrode-board artifacts are not common, they can occur if the EEG technician is careless about using electrode paste and prep materials like Omni-Prep®, allowing them to be smeared on the plug ends of the electrode wires. When this happens, the electrode paste or other material may be transferred to the jacks on the electrode board where it dries and cakes up. Sometimes, the salt present in these materials actually causes corrosion of the jacks. The caking and/or corrosion produces a high resistance contact between the plug of the electrode wire and the electrode-board jack. The result of this condition may be a variety of curious, frequently intermittent artifacts in the tracing, or tracings if the lead is connected to more than one channel of the machine.

Electrode-board artifacts are particularly troublesome because they are often mistaken for electrode artifacts. When first observed, the EEG technician will invariably reapply the electrode involved—sometimes several times—without any effect whatsoever. After doing this, most technicians would suspect the problem to be a faulty electrode and would replace it with a new one, but again to no avail. At this point, the evidence clearly suggests that the voltage in question either is generated by the patient or arises from a fault in the electrode board and/or connector of the interconnecting cable. Whether the patient or the machine is to blame may readily be decided by plugging the electrode under consideration into a spare jack on the electrode board and adjusting the appropriate lead selector switch to this new position. If the peculiar voltage disappears, it is obvious that the voltage is an instrumental artifact.

What can be done to correct this problem? Dried electrode paste or other material clogging a jack in the electrode board can sometimes be removed by dipping a wooden dowel the size of the pin plug on an electrode wire into some alcohol and working it back and forth in the jack. Great care must be taken that the dowel does not break off in the jack, or that the inside of the jack is not damaged. If the jack is already corroded, this procedure will probably be futile. In such a case, the only recourse is either to discontinue using the defective jack and use a spare one on the board instead—remembering that the lead connected to it will need to be switched in separately if you use a montage switch—or to ship the electrode board back to the factory for repair.

A better way of dealing with this problem is simply to prevent it from occurring. This is readily accomplished by the technician ensuring that his/her hands are free of electrode paste and other such materials before the plug ends of the lead wires are handled. Similarly, care in handling the electrodes after they have been removed from a patient also can help. Experienced technicians never let the plug ends of their electrodes touch the disk ends if the latter are covered with electrode paste.

Rarely, the hypothetical artifact we have been discussing may originate not from the electrode board itself but from a faulty contact in the connector of the interconnecting cable (the cable that connects the electrode board to the EEG console). What to do in this event is discussed later in this chapter under "Connector and Switch Contacts." Such artifacts are particularly bothersome, frequently coming and going without apparent rhyme or reason.

Problems Common to All Channels

These include both the simplest and the most difficult problems to diagnose. They may have their origin either inside or outside the EEG machine. Artifacts or breakdowns common to all channels that originate *inside* the EEG machine arise from a variety of causes. Artifacts originating *outside* the EEG machine may be caused by disturbances in the power line, noise in the earth or ground connection, and noise in the environment in which the EEG laboratory is situated. We will consider externally generated artifacts first, but only very briefly because these artifacts are not normally within the purview of the EEG technician. Dealing with them first does not imply that externally generated artifacts take priority over artifacts originating inside the machine in practical cases of troubleshooting.

Externally Generated Artifacts

These can be the most difficult to diagnose and eliminate. The suspicion that an artifact originates outside the EEG machine can be arrived at in a number of ways. You might already have searched for the source in the machine itself and failed to find it. Or you may have tried another EEG machine in the identical location and experienced much the same problem. Or the pattern and time of appearance of the artifact may suggest that its occurrence is coincident with some activity or event in the vicinity of the laboratory.

If you suspect that a particular artifact does originate outside of the EEG machine, you need first to isolate the problem to either the power line, the ground, or the environment. Artifacts from the environment invariably get into the EEG machine by way of the inputs. To identify the environment as the source of the artifact, or to rule it out, you need to find out what happens when you short-circuit the inputs of all the channels. This is done for each channel by switching both G1 and G2 to the same position. To avoid interaction between the channels, choose a different lead selector switch position for each channel. If after carrying out this test the artifact disappears, it can be concluded that the artifact originates somewhere in the environment.

Artifacts originating in the environment can come from a variety of sources. The power lines and transformers in the vicinity of an EEG machine, which operate at 60 Hz in the United States and 50 Hz in Europe, are the most common points of origin. Note that although their origin may be the same, the place of entry of power line artifacts that we discuss next is different. Thus, while environmental artifacts are coupled through the inputs, power line artifacts get into the machine directly via the power cable that runs to the machine. Another possible source of environmental artifacts is a winking fluorescent lamp in the vicinity of the EEG machine. A less likely but possible source is a radiotelegraphy station located close to the EEG laboratory.

The latter is of particular interest because one of the authors has had some experience with it. For months the EEG laboratory was plagued by the occurrence of medium-to-high amplitude spikes on the tracings of all channels of the machine. These spikes, which occurred in a variety of apparently random patterns, would appear suddenly at any time during the day; they continued for as much as a couple of hours, and then disappeared as abruptly as they had started. They clearly originated from the environment because they disappeared when the inputs were short-circuited. Where did the artifacts come from? Well, one day while recording and observing them for at least the hundredth time, the thought occurred that the spikes were not appearing entirely at random but followed some organized pattern. A quick reference to the Morse code reference in the dictionary revealed that this was indeed the case. We were able not only to decode the message being sent at the time, but also to get the transmitter's call letters. The transmitter, we found, was owned by a ham operator—a student at the university. When he learned about our problem, he obliged us by going off the air whenever we were taking EEGs.

If after short-circuiting the inputs of all the channels the artifact still remains, you need to consider whether the power line into the EEG machine is the source. When an EEG laboratory shares the same power line with electrical equipment having heavy start-up current loads like large motors and x-ray machines, power line artifacts are not uncommon. To identify the offending piece of equipment, it is necessary to check out the EEG machine when the equipment in question is not in operation. Sometimes, as in the case of elevator, pump, or fan motors, this may be only on nights and holidays. Should you find that a particular piece of equipment using the line is responsible for the artifact, it may help to plug the EEG machine into a different electrical outlet if one is available in the laboratory. It is not unusual for a single room to be served by two different branches of the power line and for an artifact to be worse on one branch than on the other. If this fails to alleviate the problem, the only choices are to eliminate the offending equipment, to move the EEG laboratory to another location, or to install a surge suppressor or a special isolation transformer on the power line feeding the EEG machine. But this decision is best left to a specialist —an electrical engineer or biomedical technologist.

Once you have eliminated the environment and the power line as sources of an artifact and you are sure that it does not originate within the EEG machine, the source remaining is the ground. Noise originating in the ground wire is most troublesome because it may be extremely difficult to diagnose, costly and sometimes impossible to eliminate. The first thing to do is to make sure that any auxiliary electrical equipment used in the EEG laboratory has a wire of its own going directly to ground, i.e., that other equipment *does not* get its connection to ground via the EEG machine. Be sure also that the chassis of the EEG machine and any auxiliary equipment used are not in actual contact with each other. This is important because with the two chassis touching, a ground loop is created. A ground loop provides an alternative pathway to ground for the auxiliary equipment via the EEG machine and vice versa. If the auxiliary equipment should happen to be generating some kind of electrical noise and leading it onto the ground wire, the noise could find its way into the EEG machine. The topic of ground loops is taken up in some detail in the chapter on electrical safety.

If the artifact is not generated locally, you will need to look outside the EEG laboratory for the source. This is done by disconnecting, one at a time, all other pieces of electrical equipment that are hooked up to the same ground wire used by the EEG machine and then observing the result. In some cases, this test is impossible to carry out as the exact pathway taken by the EEG laboratory's ground wire on its way to earth may be unknown. An alternative is to move the EEG machine temporarily to another location where the ground wire is closer to the earth connection and is free of branch connections to any electrical equipment. If the artifact remains, the entire ground system may be noisy, and you will need to consult a specialist for help.

Sometimes, problems with ground noise can be corrected by providing an entirely separate ground system for the EEG laboratory. This is accomplished by having several long, copper-clad steel rods driven into the earth nearby and connecting the ground wire of the EEG machine to these rods. Obviously, this is not a good solution if the building housing the EEG laboratory sits on rock or on sandy, dry soil. Another possible solution is to have a special circuit—a wave trap—designed and installed in series with the ground wire to the EEG machine. By providing a high impedance pathway for the noise, the artifact is kept from entering the EEG machine. But these are matters for the specialist and obviously outside the realm of the EEG technician's responsibilities.

Internally Generated Problems—Breakdowns

Breakdowns affecting all channels of an EEG machine are the easiest to diagnose. Following is an example that may be familiar to some EEG technicians.

Upon turning it on, you discover that the EEG machine is dead. Something has gone wrong with the machine. But wait! Before looking inside for the source of the problem, it is good practice to verify that the fuse or circuit breaker in the power line to the machine has not burned out or been tripped. The simplest way of doing this is to plug a lamp or other electrical device that is handy into the same outlet used by the EEG machine to see if it works. While this test is so obvious that it hardly seems worth mentioning, it is frequently overlooked. The result of the oversight can be a fruitless seach for a nonexistent problem inside the machine.

If the electrical outlet checks out all right, but the machine is dead—i.e., no pilot lights on, no pen deflections regardless of what you may do—the problem very likely is the power supply. A fuse may have burned out. Spare fuses for the power supply usually come with the machine and replacement instructions are found in the instruction manual. For safety's sake, disconnect the power cord to the machine before removing and replacing any fuses. Should the newly installed fuse or fuses burn out, locate and disconnect the cable that carries power from the power supply to the amplifiers and other components. If new fuses are again installed and these too burn out, a malfunction is present in the power supply. Contact the manufacturer for a loaner unit and instructions so that the defective power supply can be returned to the factory for repair. If, on the other hand, the fuses burn out only when the power supply is connected up to the amplifiers and other components, there may be a short circuit in the cables distributing power within the machine. A telephone call to the manufacturer or his/her representative will inform you how to proceed in this instance.

Another breakdown that affects all channels happens when there is a failure in the calibration circuit. The machine seems able to record EEGs, but the record taken during calibration no longer shows the proper deflection sensitivities—or the calibration waves may be totally absent—when you press the calibration switch. To aid in troubleshooting such problems, some machines have a test jack on the front panel for measuring the voltage feeding the calibration circuit. The instruction manual explains how this voltage may be checked. If the measurement carried out shows that no voltage is present, a fuse protecting the circuit may be burned out and should be replaced. If the voltage is too low, the calibration circuit may have developed a fault. In this event, the module containing it should be returned to the factory for repairs. Another possibility, if the machine is of older vintage, is that the battery providing the calibration voltage is run down and needs to be replaced. Here, again, a few moments spent with the instruction manual or a telephone call to the manufacturer will tell you how to proceed.

Machines that have malfunctioning all-channel controls will display breakdowns that affect all the channels. To determine whether a problem is attributable to the all-channel control circuit, set all the amplifiers for independent channel control. If in so doing normal operation is restored, the malfunction resides in the all-channel control unit. Check the fuses in this module to determine whether or not they are burned out. If fuses check out, the only choice is to return the module to the factory for repair.

Internally Generated Problems—Artifacts

Artifacts that are present on all channels and originate within the machine can arise in two different ways. They are caused either by malfunctions in components common to all channels, or by a malfunction present in a single channel that is *spread to the others* by connections or pathways and components that are common to all the channels. Assuming the power supply adequately isolates the EEG machine from the power line, the ground circuit is the only *pathway* that is common to all the channels. Machine *components* having a common relationship to all the channels are the power supply, the calibrator, and the all-channel control circuit if the machine has one.

Artifacts caused by a malfunction in a single channel and spread to the others by pathways and/or components common to all the channels are the easiest of the group to diagnose. For this reason, you should test for them first.[1] If the machine has an all-channel control feature, begin by setting the switches on all the channels for independent channel control. Should this cause the artifact to disappear, we would conclude that the all-channel control module was involved—either causing the artifact or spreading it from another source to all the channels. On the other hand, if the artifact remains after all the amplifiers are set to independent channel control, the all-channel control module is not involved. In the latter case, continue the troubleshooting by removing and replacing each amplifier, one at a time, and carefully observing, in each case, the effects on the remaining channels. If the artifact in question disappears after one of the amplifiers is removed, that amplifier is clearly suspect. Next, you need to find out whether the connector hooked up to the amplifier is involved. To do this, remove another amplifier from its bin and put the suspected amplifier in its place. If now the previously observed artifact returns, the suspected amplifier is clearly at fault and should not be used until replaced.

The situation in which the artifact disappears when the controls of all the amplifiers are set for independent channel control needs yet to be considered. This problem may be caused either by a malfunction in the all-channel control or by a malfunction in a single channel that is spread to the others by the all-channel control. To decide which it is, you need to discover whether one of the amplifiers is faulty. Amplifiers in particular are focussed on because of all the components in a channel, they are the most likely to malfunction. The necessary troubleshooting procedure is tedious but simple. With the controls for all the channels set to all-channel control, remove and replace each amplifier, one at a time, observing the effects of this procedure. If the artifact disappears after removal of any single amplifier, the amplifier in question is responsible for the artifact. On the other hand, if the artifact remains unchanged despite the removal of any of the amplifiers, the problem most probably lies with the all-channel control.

Artifacts originating with the power supply of the EEG machine are difficult to localize. If another EEG machine of the same type and using exactly the same model power supply is accessible, the two power supplies can be exchanged and the result observed. Otherwise, about all that the EEG technician can do if a power supply malfunction is suspected is to measure the voltages produced. These voltages ought to be correct to within about $\pm 2\%$ of their stated values (see the instruction manual) and should be *very* stable. An accurate digital voltmeter or voltmeter with a scale sensitive enough to detect a 1% shift in voltage should be used. These measurements need to be made while the machine is in actual operation or while a calibration is being run. Low or unstable voltages are indicative of power supply malfunction, and your findings should be communicated to the manufacturer, who will instruct you further.

It is important to keep in mind that the validity of such measurements is determined to a large degree by the characteristics of the particular voltmeter used. Thus, although the power supply voltages you measure appear stable, they may, in reality, be unstable. This could happen if any fluctuations occurring in the power supply voltage were too rapid for the voltmeter to follow. Of course, more sophisticated measurements may be carried out using a cathode-ray oscilloscope. But these methods of measurement are not ordinarily available to the EEG technician.

Artifacts due to the calibrator usually originate in the switch that produces the calibration pulses or in the potentiometer that adjusts their amplitude. When they occur, such artifacts show themselves as intermittent, randomly occurring spikes or sharp waves having the same pattern in all the channels. Sometimes they may disappear briefly or even for long periods of time if you repeatedly close and open the calibration switch many times. The artifacts are due mainly to a high-resistance contact in the switch. If the artifacts persist, the switch will need to be replaced.

Artifacts originating in the potentiometer that adjusts the calibration-pulse amplitude have similar characteristics, although the amplitude of the calibration pulse may also show some erratic, spontaneous variations. If an accurate digital voltmeter is available for resetting the calibration voltage, the EEG technician can attempt to correct the problem his or herself. First, locate the potentiometer and the test point for measuring the calibration-circuit voltage inside the calibrator circuit. Next, using a screwdriver that fits the slot in the control shaft of this potentiometer, rotate the control shaft back and forth several dozen times and return it to its original position. Connect the digital voltmeter to the test point and adjust the potentiometer carefully until the voltmeter reads the exact voltage specified in the instruction manual. Replace any cover that may have been removed from the unit containing the calibrator circuit. Having done all this, now run a calibration in the usual way. If the artifacts were due to a high-resistance contact in the potentiometer, the simple

[1] This test may be carried out with a practice patient connected up to the machine using a bipolar montage. A more convenient arrangement is to connect a *separate* 5K ohm resistor (a dummy patient) to Gl and G2 of each channel of the machine. This is most readily accomplished by soldering pin plugs to the lead ends of the resistors and then plugging these devices into the electrode board.

procedure of cleaning the contacts by rotating the control shaft back and forth will frequently correct the problem.[2]

Connector and Switch Contacts

The role of connector and switch contacts in the generation of artifacts has already been mentioned several times in this chapter. The extensive use of modular construction in many EEG machines means that a lot of connectors are needed to hook the various modules together. Compared with a hard-wired connection like a solder joint, connections made by connectors can become very noisy. So we find that the convenience afforded by modular construction is not obtained without some cost to the user of the machine. Nevertheless, given that connector noise can readily be identified and easily corrected, the convenience far outweighs the costs.

Artifacts produced by noisy connector connections arise mainly in two different ways. In the first place, some contacts in the mating connectors may not fit together properly because they are defective or have been damaged. Connector contacts are readily liable to mechanical damage—they are easily bent, scored, or pitted. For this reason, they need to be handled carefully. It is essential not to use excessive force when plugging a module or cable into its mating connector.

A second way in which connector connections can become noisy results from the inherent tendency of metal to corrode or to become coated with foreign material. Metals, of course, are very good conductors. They offer a low resistance to the flow of electric current, which is why they are used in connectors. But if the junction between the contacts in two mating connectors becomes coated with corrosion or foreign material, the contact resistance may increase sufficiently to significantly resist the flow of current across the junction. What is important as far as the production of artifacts is concerned is that contact resistance under such conditions not only is elevated but also may become unstable—it may change rapidly and spontaneously from a low to a high value within seconds. This is what produces the characteristic noisy-contact artifact; electronic technicians and engineers sometimes check for the presence of such artifacts by sharply tapping a suspected connector with the handle of a light screwdriver.

Artifacts of this kind are quite common in some of the older EEG machines. This is because the connector contacts in the vintage machines were usually silver plated. As anyone using silver or silver-plated flatware knows, silver tarnishes quite readily in the atmosphere of our industrial society. The silver-plated contacts in a connector are no exception. Any tarnish occurring at the junction between two contacts can raise the contact resistance appreciably. The use of gold-plated contacts has greatly reduced if not eliminated this problem. Nevertheless, even gold contacts can become noisy because of the presence of foreign material on their surfaces. For this reason, the EEG technician needs to have some ready method of cleaning connector contacts.

The plug ends of some connectors can be cleaned rather easily by using the rubber eraser on the tip of an ordinary lead pencil. Rub the eraser in an even stroke across each of the contacts in the direction of the tips. Make sure first that the tip of the eraser has been rounded off and that it is scrupulously clean. If the contacts are flat, one or two strokes with the eraser are sufficient; round contacts will require several strokes. Carefully remove all eraser crumbs from the cleaned contacts before the plug is inserted again into the socket. Contacts on multiple-pin connectors are frequently not accessible with a pencil eraser. In these cases a thin sheet of rubber can sometimes be used instead. Draw a narrow strip of the material across the pins at several different angles. If the contacts are inaccessible even by this method, about all the EEG technician can do is to work the plug end of the connector back and forth about a dozen times in its mating socket. Files, emery cloth, or other abrasive materials must never be used on connector contacts.

The contacts in *switches* can present many of the same problems. As with connectors, the artifacts generated by switches are the result of intermittent, high-resistance contacts. Unfortunately, the methods used for cleaning connector contacts cannot be applied to switch contacts. Switch contacts are not readily accessible to a pencil eraser; indeed, some switches are completely enclosed in a container so that the contacts are not even visible. As mentioned here earlier, switch-contact artifacts can sometimes be eliminated by repeatedly indexing the switch back and forth about the suspected contact. A variety of contact cleaners are available commercially. These can be sprayed directly on the contacts of the troublesome switch if the contacts are accessible. In the authors' experience, however, they have not proved particularly useful. If simply indexing the switch does not correct the problem, the switch will have to be replaced in all probability.

The types of switches considered thus far in this chapter are the conventional mechanical types—rotary, toggle, and

[2] The EEG technician should be aware that this simple procedure cannot be carried out on all machines. Thus, some manufacturers put special seals on the chassis of their modules. If these seals are broken by the user to get to a circuit for troubleshooting and repair, the manufacturer's warranty is voided. Other manufacturer's sometimes put sealing wax on the shafts of potentiometers. If any adjustments are attempted, the wax will be broken and the warranty may be voided. Therefore, before attempting any such troubleshooting or adjustments, consult the instruction manual that comes with the machine.

lever switches. As a result of recent technological developments, the switches on some of the very newest-model EEG machines have changed greatly. The use of so-called "soft-touch" switches, the display of machine parameters by means of a cathode-ray tube, and touch-screen switching have noticeably modified the physical appearance of the EEG console. Also, rotary switches that control multiple-switching operations are being replaced by solid-state, electrical switches. These changes, in turn, have made menu-driven machines possible. While these switching devices show promise of being more reliable than their conventional equivalents, only time will tell whether they and the other recent modifications in the EEG machine will enhance overall machine reliability, reduce downtime, and decrease the amount of time expended in troubleshooting.

Chart-Drive Malfunctions

Some of the problems that beset the chart drive have been taken up already in the chapter on the writer unit. A few yet remain to be considered. Although the troubleshooting that the EEG technician can do on the chart drive is limited, a few simple diagnostic tests are possible. The methods employed follow the same logic that is used for troubleshooting the rest of the machine.

The heart of the chart drive is, of course, the motor that powers it. In many machines the current needed to run the motor is derived from the power supply for the machine, and in some cases this current is separately fused. Thus, the first thing to do if the chart drive fails to operate is to locate and check the fuses. Remember, however, that before touching these fuses you should make sure that the power cord to the machine has been disconnected from the wall socket. If the fuses check out, the problem is either the power supply or the motor. To help you decide between the two alternatives, many power supplies have test points on their chassis that permit easy measurement of the power-supply voltages. Check the instruction manual. If the voltage that runs the chart drive checks out O.K., the malfunction is in the motor; otherwise, the power supply is at fault. In some machines, modular construction extends even to the chart drive. This, of course, greatly simplifies servicing once a malfunction has been localized to the chart drive.

The chart drive contains an important switch that deserves special mention. This switch, sometimes called the *master writer switch*, has three positions and controls the actual operation of the machine. In the first or *off position*, the electronic portions of the machine are on (assuming, of course, that the main power switch is turned on), but the chart drive is not operating. In the next or *chart position*, the chart paper runs through the machine as well. This position is used to check tracking of the chart through the machine. Finally, in the last or *run position*, the amplifier outputs are connected to the penmotors so that the EEGs are traced out on the moving chart.

Because the switch we are discussing gets a great deal of use, it occasionally breaks down. Usually, the breakdown is such that only one channel of the machine is affected. The symptoms of the breakdown are classic. The channel is completely dead and nothing you do will make the pen deflect from its baseline position. Yet, the amplifier, penmotor, lead selector, filter—all the modules in the channel—test out O.K. A check with a voltmeter, however, shows that the operating voltage never reaches the penmotor because it is blocked by a faulty master writer switch. Replacement of the switch, of course, is necessary. In some machines, this is facilitated by the fact that the switch is incorporated in a separate module—the writer control module.

In feeding paper through the machine, the chart drive provides the time base for the EEGs traced out on the chart. Measurements of frequency of the waveforms of the tracings, therefore, depend on the accuracy with which the chart paper moves through the machine. The actual drive mechanism used is quite simple; it consists usually of a rubber pressure roller in direct contact with a knurled metal roller. The chart paper rides between the two rollers. Through use, the rubber may deteriorate or develop a flat spot on its surface. For this reason, the rubber pressure roller should be inspected occasionally and replaced when necessary as a worn out roller can degrade the accuracy of the chart speed. See also "Chart Drive" in Chapter 5.

The standard chart speed of 30 mm/s should be checked occasionally. This is done most readily by recording 60 Hz artifact on all channels of the machine and then finding out how long a segment of chart in millimeters is occupied by exactly 60 of these waves. Separate measurements should be made at different positions on a page of the chart to determine whether the paper is feeding accurately over the folds as well as over the smooth portions of the chart. The chart speed should be accurate to $\pm 2\%$.

Chapter 10

Neurophysiology

Scientists have diligently pursued the goal of unraveling the mystery of brain electric activity from the days of Hans Berger, but like many other natural phenomena, less is known about the exact genesis of the EEG than about its physiological and pathological variations. A number of techniques, including surface, depth, and intracellular, as well as in vitro recording methods, have been used to study this phenomenon, and a wealth of data has been accumulated. Before considering these data and entering into a discussion of the current ideas concerning the origin of the EEG, it is essential that we look at the neurons—the cellular units of the nervous system. We will examine them from the points of view of anatomical structure and generation of bioelectric activity.

Structure of the Neuron

There are two distinct varieties of cells in the nervous system: the nerve cells or the *neurons* and the support cells or the *glia*. Although glial cells are important in providing the appropriate environment for neuronal function and for myelination of the axons, it is the neuron that forms the ultimate unit of brain function. A typical neuron has three easily identifiable parts, namely, the *cell body* (soma, perikaryon), the *dendrite*, and the *axon* (Fig. 10.1). The cell body, which contains organelles such as the nucleus, endoplasmic reticulum, and Golgi apparatus, is the metabolic center of the neuron. It is covered by the *neuronal membrane*, which separates the cell contents from the extracellular fluid.

The dendrites, which are small multiple arborizations of the cell body (like the branches of a tree), serve as the points of input of signals to the neuron. The axon is like a tube extending from the cell body for long distances (as much as 1 m in the human); its distal end divides into many fine branches forming the synaptic terminals. These terminals are found near the cell bodies or dendrites of other neurons. The axon is often covered by a layer of myelin, which shows points of interruption known as nodes of Ranvier. The myelin sheath, which acts as an insulation for the axon, helps in the rapid conduction of electrical signals. The primary function of the axon is the conduction of signals from a neuron to one or more other neurons. It also transports materials like neurotransmitters synthesized by the cell body through the axoplasm onto the synpatic terminals (axoplasmic transport).

Although all neurons are essentially similar in possessing cell body and processes, there are considerable differences in their morphology. They may be *unipolar*, in which case there is only one process emerging from the cell—and dividing into branches that serve as dendrites and axon; *bipolar* wherein one process acts as a dendrite and the other as the axon; or *multipolar*, in which case there is one axon and several dendrites. Most cells in the human nervous system are multipolar. The advantage of such a neuron is that it can make contact with a very large number of other neurons; for example, the Purkinje cell, which is a large cell in the cerebellar cortex is known to receive as many as 150,000 contacts from other neurons.

The length of an axon can show considerable variation. As already mentioned, an axon can, in some cases, extend for very long distances before making contact with another cell. Neurons having such long axons are often referred to as relay neurons. They serve to relay information from one area of the nervous system to another, as, for example, from the motor cortex to the anterior horn cells in the spinal cord. Neurons with shorter axons make contact with cells in different layers in the same vicinity and are called interneurons.

Membrane Potential

The neural membrane displays some interesting and important properties. Of particular importance is that the

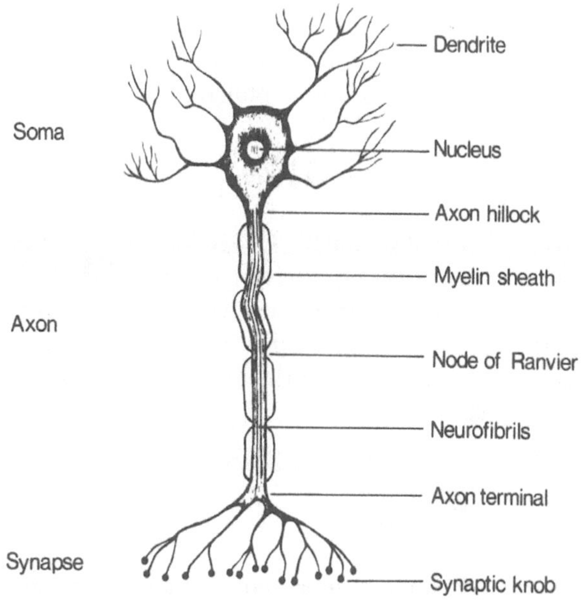

Figure 10.1. Schematic diagram of a neuron showing the major structural features.

resting membrane is electrically polarized; i.e., a difference in potential or voltage exists between the two sides. To understand the origin of this membrane potential, an understanding of the structure of the neuronal membrane and its ionic channels is essential. The following discussion takes up the bare essentials. Details are found in more advanced texts.

The neuronal membrane is made up of proteins and lipids and is about 10 nm (0.01 μ) thick. The lipid forms a double layer that, being immiscible with water, does not allow water-soluble ions to move in and out of the cell. However, the cell membrane contains areas in which there are protein "pores" through which ions can pass. These are called ionic channels and are of two varieties: passive and active. The passive channels are always open and allow diffusion of selected ions, depending on the concentration gradient across the membrane.

The active channels may be kept open or closed through gates strategically placed at one or both ends of the channels. When the membrane is at rest, most of the active channels remain closed. These channels open abruptly when there is a specific change in the membrane potential and hence are called voltage-regulated channels. The channels are selective, acting like filters, and often allow only one species of ion to pass through. Thus there are specific channels for sodium, potassium, and calcium ions.

The neuronal membrane separates the cytoplasm of the neuron (neuroplasm) from the extracellular fluid. If we analyze the composition of the intracellular and extracellular fluids, certain specific differences become evident at once. Thus, there is a higher concentration of potassium ions within the cell, whereas sodium and chloride ions show higher concentrations outside the cell (see Table 10.1). Such a difference in the ionic concentration is maintained by selective permeability of the membrane to specific ions and by active transport mechanisms that selectively pump certain types of ions in or out across the cell membrane. Let us examine each of these mechanisms.

The cell membrane at rest is highly permeable to potassium ions but poorly permeable to sodium ions. Since the concentration of potassium ions is 35 times greater inside the cell than outside, a concentration gradient exists across the membrane. This concentration gradient tends to drive potassium ions from the intracellular space into the extracellular fluid along the passive potassium channels. The outward diffusion of potassium ions, which are positively charged, leads to an excess of negatively charged ions inside the cell. This is because certain negative ions within the cell (the organic anions) are unable to pass through the cell membrane; instead, they tend to line up along the inner surface of the membrane. The excess of negative charge on the inside and positive charge outside the cell membrane soon prevents further escape of more potassium ions, although the actual concentration of potassium ions stays higher within the cell. Indeed, the loss of potassium ions from inside the cell is relatively insignificant because of the large number of potassium ions in the whole of the cytoplasm.

By the time the potassium ions stop diffusing out and there is an excess of negative ions on the inside and positive ions on the outside of the cell membrane, the positively charged ions—of which the most abundant are sodium ions—tend to line up along the outer surface of the cell membrane. But because the cell membrane is poorly permeable to sodium ions, these ions cannot enter the cell in sufficient numbers to neutralize the excess of negative charge inside the cell. Because of this selective permeability of the membrane and also because of an active pumping mechanism described later in this chapter, there is an accumulation of negative charges on the inner surface and positive charges on the outer surface of the neuronal membrane. As the membrane separates opposite charges on its two sides, it is said to be "polarized."

The cell membrane with its associated ions on either side resembles a charged capacitor since the lipid lining acts as a dielectric separating positive and negative charges. If we were to measure the charge across the membrane, we would find that in the resting state the inside of the cell is −40 to −90 mV compared with the outside. This is called the *resting membrane potential* (RMP). All signaling and transmission of information in the nervous system is based upon the induction of fluctuations in the RMP, which lead to the synaptic and the action potentials. For this reason, it is important to gain some insight into the ways in which such fluctuations take place.

Nernst Equation

Diffusion of potassium ions from the inside to the outside of a cell (efflux) occurs only for a brief period of time. Equilibrium is quickly established, and the potassium ions cease to move outward owing to the excess negative charge that accumulates inside and also because of the repelling force of the positive charge outside. The potential difference between the inside and the outside of the cell membrane due to potassium ions, when there is no net movement of potassium ions into or out of the cell, is called the potassium *equilibrium potential* or *Nernst potential*. The equilibrium potential for any ion may be defined as the electrical force required to balance the ionic movements caused by diffusion; it is proportional to the logarithm of the ratio of the concentrations of the ion to which the membrane is selectively permeable. Equilibrium potential can be calculated using an equation propounded by the German scientist Walther Nernst in 1888 on the basis of thermodynamic principles. In other words, the Nernst equation provides us with a method of calculating the potential at which the concentration gradient causing efflux of an ion is exactly balanced by the electrical gradient that opposes such efflux.

The Nernst potential for any ion is defined by the equation:

$$\xi = \frac{RT}{nF} \ln \frac{\{\Upsilon\}_o}{\{\Upsilon\}_i}$$

where R = the gas constant (8.316 joules per degree), T = temperature in degrees Kelvin, F = the Faraday (96,500 coulombs per mole), n = valence of the ion with its appropriate sign, and $\{\Upsilon\}_o$ and $\{\Upsilon\}_i$ are the extracellular (outside) and intracellular (inside) concentrations, respectively, of the particular ion. By substituting the constants and converting the natural logarithms to base 10 logarithms, the equation may be simplified to

$$\xi = \frac{1}{n} 61.5 \log_{10} \frac{\{\Upsilon\}_o}{\{\Upsilon\}_i}$$

which gives ξ in millivolts at 37°C (310° Kelvin). For a cell membrane that allows only one type of ion to diffuse, this formula can predict the membrane equilibrium potential accurately. Table 10.1 gives the Nernst potentials for sodium, potassium, and chloride ions on the basis of the above formula.

Table 10.1. Concentration of Major Ions in the Intracellular and Extracellular Fluids of Vertebrates and the Nernst Potentials

Ion	Intracellular Concentration (mmol/L)	Extracellular Concentration (mmol/L)	Nernst Potential[a] (mV at 37°C)
Sodium (Na^+)	14	142	+62
Potassium (K^+)	140	4	−95
Chloride (Cl^-)	4	115	−90

[a] These values are computed using the formula

$$\xi = (1/n)\, 61.5 \log_{10} (\{\Upsilon\}_o/\{\Upsilon\}_i).$$

For the chloride ion we have

$$\begin{aligned} \xi &= (1/-1)\, 61.5 \log_{10} (115/4) \\ &= -61.5 \log_{10} (28.75) \\ &= -61.5 \times 1.4587 \\ &= -90 \text{ mV}. \end{aligned}$$

Goldman Equation

The Nernst equation applies only if the membrane is permeable to a single ion and the concentration of this ion is greater on one side of the membrane than the other. The glial cells are known to be selectively permeable to potassium ions, and their membrane potential is close to the value estimated from the Nernst equation. However, the situation is more complex in the neuron as the neuronal membrane is permeable to several different ions. Thus, it is not only permeable to potassium but also to chloride and, to some extent, to sodium ions.

A number of factors seem to influence the membrane potential under these circumstances. The concentration gradient of each ion across the membrane, the polarity of each ion, and the degree of permeability of the membrane to each different ion are perhaps the most important factors. The Goldman equation, which is an extension of the Nernst equation, attempts to integrate the contributions of each different ion into a single equation by taking the permeabilities of the ions into account. When only sodium, potassium, and chloride ions are taken into consideration, the Goldman equation for the membrane potential is given by the expression

$$\xi_m = \frac{RT}{F} \ln \left(\frac{P_K \{K^+\}_o + P_{Na} \{Na^+\}_o + P_{Cl} \{Cl^-\}_i}{P_K \{K^+\}_i + P_{Na} \{Na^+\}_i + P_{Cl} \{Cl^-\}_o} \right)$$

where the constants R, T, and F are as previously defined, $\{K^+\}_o$, $\{Na^+\}_o$, $\{Cl^-\}_o$ are the extracellular concentrations of the ions, $\{K^+\}_i$, $\{Na^+\}_i$, $\{Cl^-\}_i$ the intracellular concentrations, and P_K, P_{Na}, and P_{Cl} represent the resting permeabilities of these ions when there is no net flow of ionic current.

It is clear from the above equation that the permeability of the membrane to each different ion is a crucial factor in the actual contribution of each ion to the membrane potential. Since the membrane is highly permeable to potassium in the resting state, the RMP is close to the potassium equilibrium potential.

The Sodium-Potassium Pump

The membrane is highly permeable to potassium, but it also allows some amount of sodium diffusion. Sodium ions tend to diffuse slowly from outside the cell to the inside, and if this were allowed to proceed unhampered, the charges eventually would become equal on either side of the membrane. The cell uses an ingenious active pumping mechanism (electrogenic ion pump) to prevent this from happening. Using active ion channels and energy from adenosine triphosphate (ATP), the cell pumps out sodium ions while taking in potassium ions. Three sodium ions are expelled for every two potassium ions taken in. This tends to reestablish the intracellular concentration of potassium and sodium continuously so that appropriate ionic concentrations are maintained to ensure a constant RMP. It will be seen later that during the generation of an action potential, the membrane becomes highly permeable to sodium for a brief period of time. This leads to sodium ion accumulation within the cell with a corresponding increase in positive charge. Thereupon, the electrogenic pump helps to reestablish the negative RMP by pumping out the sodium ions. The rate of pumping is increased when the sodium ion concentration increases within the cell.

As mentioned earlier, the RMP in nerve cells ranges from -40 to -90 mV, the inside being negative with respect to the outside. When there is an increase in the negative charge inside the cell membrane, the membrane is said to be *hyperpolarized*. When there is an increase in the positive charge inside the cell membrane, it is said to be *depolarized*.

Action Potential

Neurons communicate with each other through generation of action potentials, which are changes in membrane potential that are propagated along the axons. If we were to record an action potential from an axon, we would discover that the negative RMP suddenly becomes positive for a very brief period of time and then rapidly returns back to its original negative level. The initial stage of this rapid change, known as depolarization, is accompanied by an abrupt and massive change in the sodium permeability of the neuronal membrane. This occurs as a result of the opening of voltage-dependent sodium channels. These channels, which are protein-lined pores in the membrane, have activation gates situated toward the outer layer of the cell membrane and inactivation gates situated toward the inside of the membrane. A decrease in negativity of the membrane potential of 20 to 30 mV causes the abrupt opening of the activation gates, leading to an increase in sodium permeability by a factor of about 5,000 and resulting in a massive influx of sodium ions into the cell. The accompanying increase in the number of positive charges within the cell closes the inactivation gates of the sodium channels so that no more sodium ions can enter the cell.

At this point in the sequence of events, the gates of the potassium channels open and allow potassium ions to escape outside the cell; this leads to a decrease in the positive charge within the cell and reestablishes the original negative RMP. In other words, the membrane is repolarized. Once this happens, the active potassium channels close and in so doing prevent further escape of potassium ions. Although the concentration of sodium ions inside the cell membrane increases during the generation of an action potential, the increase is small when compared with the concentration of sodium ions in the extracellular fluid. Nevertheless, with repeated passage of action potentials, intracellular depletion of potassium and accumulation of sodium ions may occur. However, this is circumvented by the action of the sodium-potassium pump that was discussed earlier.

What initiates the action potential? Any condition that leads to a sufficient decrease in the negative charge inside the cell membrane can initate an action potential. The decrease in negativity may be produced chemically as, for example, by neurotransmitters, or mechanically or electrically. The most familiar circumstance is electrical stimulation of a peripheral nerve, as in the study of somatosensory-evoked potentials. In such studies the cathode of the stimulating electrode produces an excess of negative charge outside the nerve membrane. This is equivalent to reducing the negative RMP by an amount sufficient to result in the opening of the voltage-dependent sodium channels and the generation of the action potential. Similarly, potentials are induced at the synaptic terminals by chemicals like acetylcholine, which can open pores in the membrane and thereby result in sodium ion influx. Once an action potential is initiated, it is propagated along the axon by the opening of sodium channels in the adjacent areas of the axonal membrane, which leads to a wave of depolarization and subsequent repolarization.

Synaptic Potentials

As mentioned earlier, the most important function of neurons is communication. Neurons communicate with each other and also with effector organs such as muscles and glands. Such communication is accomplished by the process of synaptic transmission. A synapse is the junction between two neurons or between a neuron and an effector organ. In the former case, the junction is between a synaptic knob (see Fig. 10.1) of one neuron and a dendrite of another. Let us briefly look at the mechanism involved in synaptic transmission in the central nervous system.

The transmission of signals from one neuron to another across a synapse is achieved mainly through chemical transmitters. Chemical transmitters can either excite (depolarize) or inhibit (hyperpolarize) the neuronal membrane. For example, neurotransmitters like acetylcholine are excitatory whereas γ-aminobutyric acid (GABA) is inhibitory. Excitatory neurotransmitters evoke an electrical change in the postsynaptic membrane called the *excitatory postsynaptic potential* (EPSP). The transmitter first binds itself to a receptor in the postsynaptic membrane, which results in the opening of the sodium channels. This depolarizes the postsynaptic membrane, thereby inducing the EPSP.

The EPSPs are sometimes called miniature potentials. Unlike action potentials, they do not propagate but spread passively along the membrane, diminishing in amplitude as the distance increases. The usual size of the EPSP is around 5 mV. Normally, this value is not sufficient to trigger an action potential. However, when a number of synapses making contact with a dendrite or a cell body develop EPSPs, these potentials can summate and cause sufficient depolarization of the axon hillock (Fig. 10.1) through setting up of ionic currents in the extracellular and intracellular fluid to evoke an action potential. Summation may be of two different kinds, namely, spatial summation and temporal summation. *Spatial summation* is the summation of several EPSPs produced simultaneously or nearly simultaneously at different sites on the postsynaptic membrane. *Temporal summation* refers to the summation of successive potential changes at a single site on the postsynaptic membrane such that one EPSP is superimposed onto another.

Inhibitory neurotransmitters act in a different way. They seem to open the chloride channels, leading to an influx of chloride ions that, in turn, results in an increased negative charge inside the cell membrane. The potential produced in this way is known as an *inhibitory postsynaptic potential* or IPSP. It hyperpolarizes the cell membrane, thereby inhibiting depolarization and preventing the generation of an action potential.

Except for the fact that EPSPs are depolarizing and IPSPs are hyperpolarizing, both share the same properties. Thus, IPSPs show temporal and spatial summation, as do EPSPs. Inhibitory synapses are often strategically located so that they can prevent generation of action potentials from the axon hillock.

Membrane Equivalent Circuit

From what we know about the permeability of the neural membrane and the ionic composition of intracellular and extracellular fluids, it is possible to represent the membrane by an equivalent electrical circuit. Figure 10.2 is a simplified version that lumps together the membrane parameters (in actual practice, these parameters are distributed along the length of a membrane). The circuit diagram shows three separate conducting channels corresponding to the major ions. Each channel has a battery that represents the equilibrium potential produced by the concentration gradient for the particular ion and a leakage resistance that stands for the membrane's permeability for that ion. Note that the polarity of the sodium battery is opposite to that of the potassium and chloride batteries. The circuit also includes a capacitor that can store electric charge. As mentioned earlier in this chapter, this is because the lipid bilayer acts as a dielectric, separating positive and negative charges. The capacitor serves as the second main route whereby current may flow across the membrane. The equivalent circuit is useful in analyzing various details of neuronal functioning, but such particulars are beyond the scope of this text.

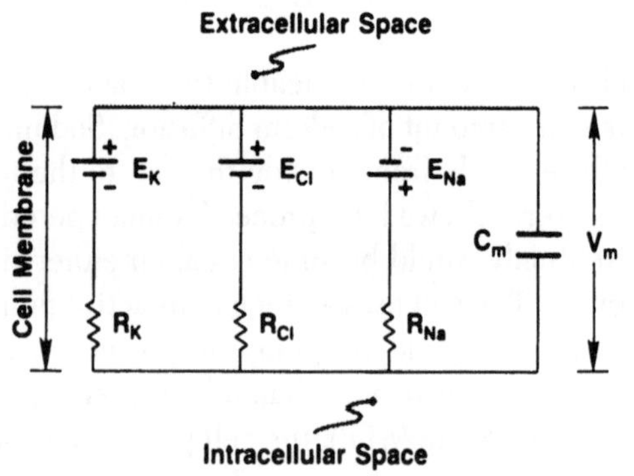

Figure 10.2. Equivalent circuit for a cell membrane. E_K, E_{Cl}, and E_{Na} are potassium, chloride, and sodium batteries respectively. The potentials of these batteries depend on the concentration gradients of the respective ions. R_K, R_{Cl}, and R_{Na} represent the leakage resistances, respectively, of the different ionic channels. C_m is the membrane capacitance; V_m is the membrane potential.

Membrane Potential Fluctuations and the EEG

It is not absolutely essential to know exactly how the EEG is generated for the limited purpose of its clinical interpretation. Nevertheless, an insight into the mechanisms of generation of the EEG is certainly helpful for a clearer understanding of the pathogenesis of the different EEG abnormalities seen in various disorders of the central nervous system.

Based on current knowledge, it is clear that the EEG as recorded over scalp represents fluctuations in the membrane potentials of a large number of neurons in the cerebral cortex. In earlier sections we saw that there are two major types of fluctuations in the membrane potentials, namely, the action potential and the synaptic potential. Since the EEG is an extracellular recording, the electrical potentials recorded would be expected to represent voltage changes in the extracellular space—field potentials secondary to changes in the membrane potentials of the neurons. Let us briefly explore the possible mechanisms involved.

When an EPSP is generated, there is a sudden influx of cations through the subsynaptic membrane. Such an influx would attract cations from the surrounding extracellular space and cause them to move toward that area. This phenomenon is similar to the flow pattern when the drain of a sink filled with water is suddenly opened; the extracellular space may be compared to the water in the sink and subsynaptic membrane to the outlet. The ionic movement sets up a field potential that would be negative in the region of the subsynaptic membrane and positive in the surrounding extracellular fluid. In the case of an inhibitory synapse, a different but related phenomenon occurs. When an IPSP is generated, there is an influx of anions like chloride through the subsynaptic membrane into the cell. Such an influx would result in an excess of cations in the extracellular space close to the synapse. This, in turn, leads to a flow of cations away from the synaptic area, thus setting up a field potential that would be positive in the region of the subsynaptic membrane and negative in the surrounding extracellular fluid.

It is true, of course, that action potentials in the axons could also set up ionic fluxes and field potentials similar to those just described. But because of its very brief duration (about 1 ms), the action potential is an unlikely source of EEG waveforms like the alpha rhythm. The longer duration (10 ms and greater) of the synaptic potentials and their graded nature make them a more likely source of the EEG waveforms. This is clearly brought out in Fig. 10.3, which compares the waveforms of the action potential and the EPSP.

The probable sequence of events just described has been verified by simultaneous recording from microelectrodes placed within the cell and in the extracellular space (Fig. 10.4). During an EPSP there is a positive deflection recorded from the intracellular electrode placed near the synaptic knob of the stimulating axon. At the same time, there is a large current flowing inward at the subsynaptic membrane and an equally large current flowing outward at numerous places some distance from the synaptic membrane, as shown in Fig. 10.4. The extracellular electrode situated near the synapse records a negative potential with respect to a distant reference point since it is near the sink that has been created. On the other hand, the extracellular electrode also placed near the membrane, but at a distance from the synapse, records a positive potential. Conversely, during an IPSP there is a negative deflection from the intracellular electrode, which generates an out-going current across the synaptic membrane. This site, therefore, is a source rather than a sink, and it produces a positive field potential in the region of the synapse and a negative field potential at the membrane sinks at a distance.

On the basis of large amounts of data accumulated from such microelectrode studies and from similar studies employing simultaneous surface recordings, it is now generally accepted that the EEG waveforms seen in scalp recordings or in recordings taken directly from the brain's surface represent summated field potentials set up by EPSPs and IPSPs from a large number of cortical neurons.

The Role of Different Types of Neurons in the Generation of the EEG

There are basically three types of neurons in the cerebral cortex. These are the pyramidal neurons, the stellate neurons, and the spindle neurons (see Appendix 2). To pick up electrical activity from such cells by means of electrodes placed on the surface of the scalp, sufficiently large field potentials need to be set up in the extracellular space surrounding the neurons. In other words, a relatively large area on the surface of the brain has to become either negative or positive for a measurable voltage to be recorded over the scalp. Taking into consideration the orientation of a scalp electrode in relation to the cortical surface, large numbers of vertically oriented dipoles are necessary. (Vertically oriented refers to the axis of the dipole being perpendicular to the outside surface of the cortex. See Chapter 12 for a discussion of dipoles.)

If one looks at the anatomical arrangement of the cortical neurons (see Appendix 2), it becomes obvious that the pyramidal cells are the most probable source of such electrical fields. They are oriented perpendicularly to the cortical surface, with apical dendrites ending superficially in layer I of the cortex, with cell bodies situated in layers II and III and to some extent in layer V. The current flow owing to the EPSP and IPSP could produce a field potential with negativity or positivity in the dendritic zone and opposite polarity at a distance from the surface, thus constituting a vertical dipole. Afferent input from the large number of neurons that make contact with the dendrites of the pyramidal cells in layer I might therefore result in the production of waveforms like those that characterize the EEG recordings.

Rhythmicity of the EEG Patterns

We have seen that cortical neurons can produce extracellular field potentials. But what leads to the rhythmic variations in voltage that are an intrinsic characteristic of the

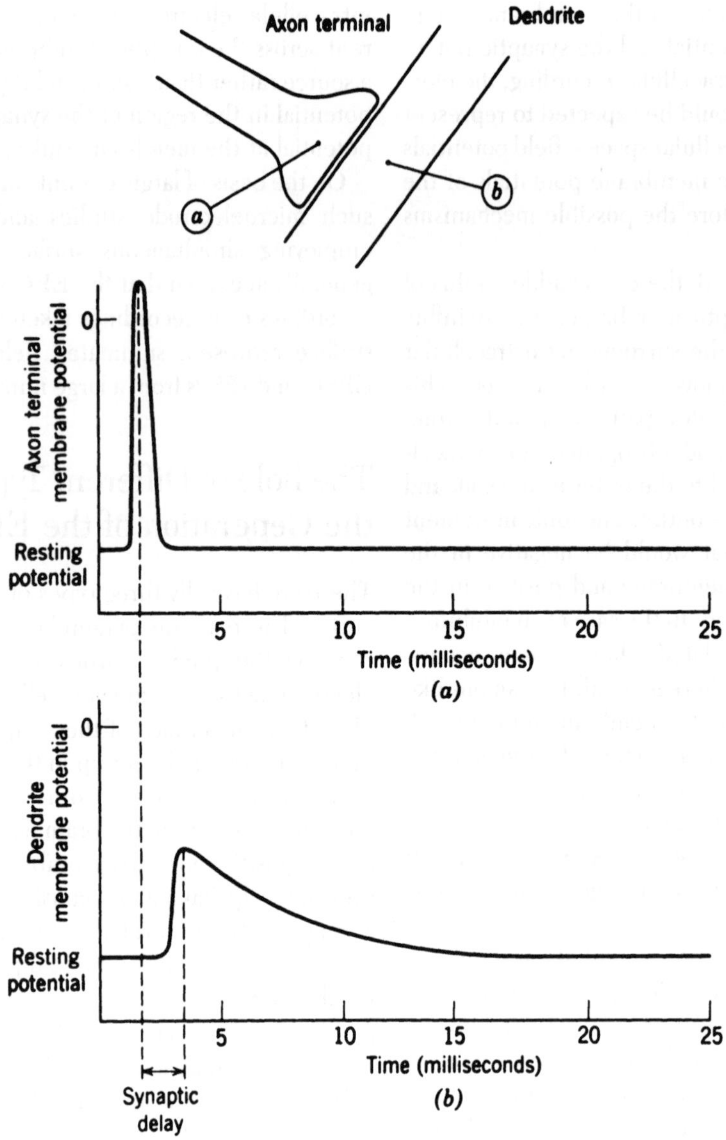

Figure 10.3. Schematic depicting (a) recording of membrane potential in an axon terminal showing an action potential and (b) recording of membrane potential in a dendrite showing an EPSP. Notice the difference in the amplitude of the two waveforms and the much longer duration of the EPSP. (From Stevens CF: *Neurophysiology: A Primer.* New York, John Wiley & Sons, 1968, p. 35, by permission of the author and holder of the copyright.)

EEG? This important question brings in the concept of a pacemaker that "drives" the cortical neurons. Various experiments have been conducted for the purpose of discovering the region of the brain and the mechanism responsible for producing such rhythmic activity of the cortical neurons. We know that if EEGs are recorded from animals subjected to brainstem sections at different levels such as the medulla, pons, or the midbrain, the rhythmicity seems to persist. However, when the thalamocortical connections are disrupted, the rhythmicity disappears. The thalamic nuclei have extensive connections with all parts of the cortex and, therefore, seem to be the most likely site for a cortical pacemaker.

Andersen and Andersson (1968) explained the role of the thalamus in the etiology of the EEG by their facultative pacemaker theory. Briefly, the theory assumes that rhythmic activity is an inherent property of groups of cells in the thalamic nuclei. The rhythmicity is produced by a simple mechanism in which the discharge of one thalamic neuron causes (via an inhibitory neuron) inhibition of many adjacent neurons. During the postinhibitory rebound that follows, many of the adjacent neurons discharge and the cycle thereupon repeats itself. This rhythmic activity is believed to be imposed upon the cortex by diffusely projecting thalamocortical fibers. In so doing, it could result in rhythmic activity in large numbers of cortical neurons. Figure

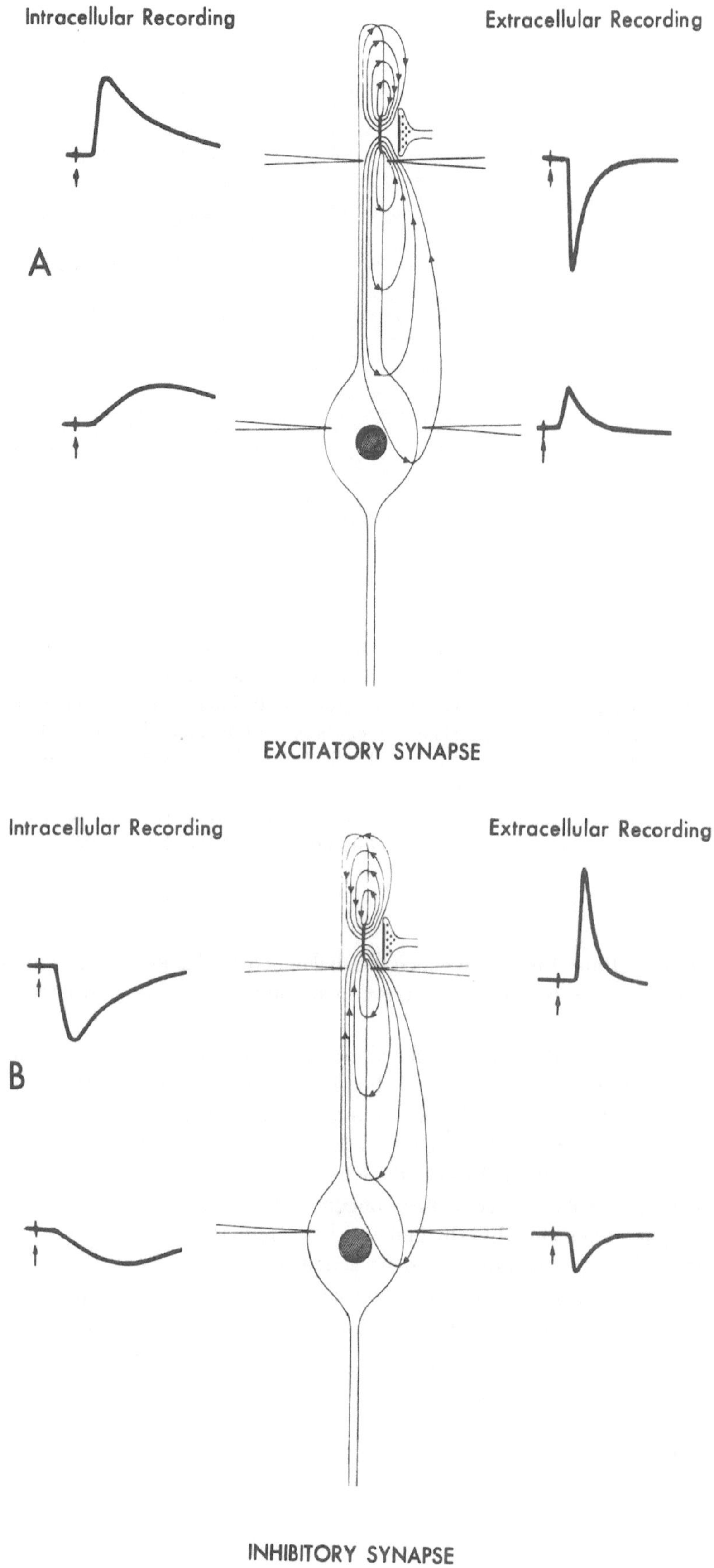

Figure 10.4. Diagram showing the fields generated by EPSPs and IPSPs in excitatory and inhibitory synapses of the central nervous system neurons. For simplicity's sake, the neurons are shown as having only a single dendrite; the synaptic knob from the connecting neuron is seen at the top and to the right of the dendrite. In all cases, positive is up. In A, an EPSP in the dendrite generates a negative field in its immediate vicinity, and a positive field at a distance along the dendrite and cell body of the neuron. In B, an IPSP in the dendrite generates a positive field in its immediate vicinity, and a negative field at a distance. (From Hubbard JI, LLiṇás R, Quastel DMS: *Electrophysiological Analysis of Synaptic Transmission*. London, Edward Arnold, 1969, p. 289; courtesy of the authors and holders of the copyright.)

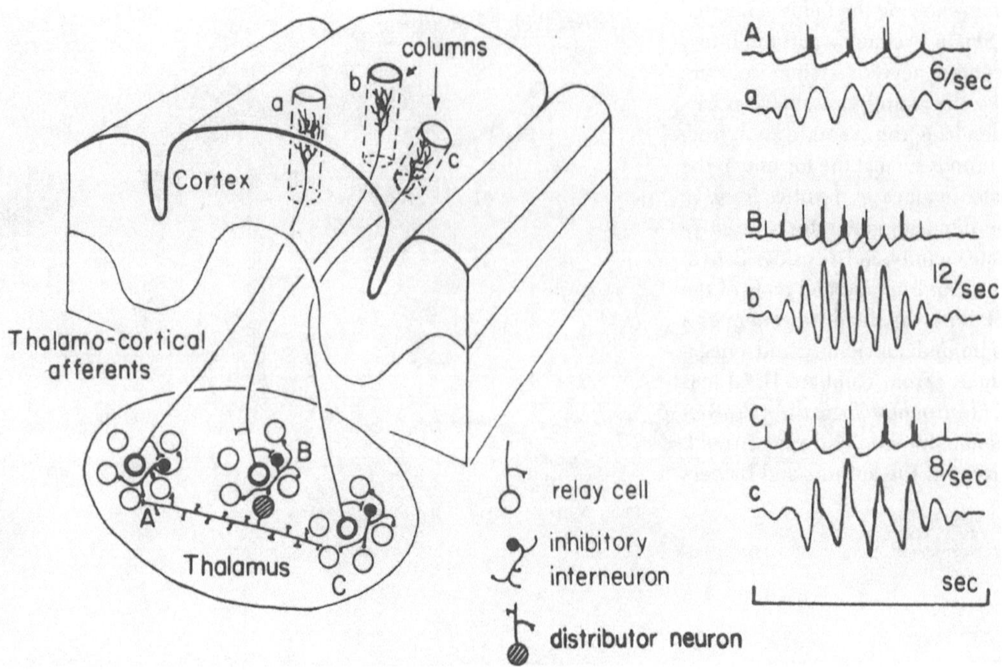

Figure 10.5. Diagram illustrating Andersen and Andersson's model for the generation of EEG rhythmicity by means of a thalamic pacemaker. A, B, and C are three groups of thalamic cells that send their axons via thalamocortical fibers to the cerebral cortex and activate columns of cortical cells designated a, b, and c. Collaterals of the axons of the thalamic cells excite inhibitory neurons (black) that can simultaneously inhibit a large number of thalamic neurons. During the postinhibitory rebound that follows, many of these neurons discharge and the cycle repeats itself. The different groups of thalamic cells have different frequencies of discharge, as illustrated in the right-hand column, rows A, B, and C. The corresponding cortical spindles, whose frequencies match the thalamic discharges, are shown in rows a, b, and c. (From Andersen P, and Andersson SA: *Physiological Basis of the Alpha Rhythm.* New York, Appleton-Century-Crofts, 1968, p. 59; courtesy of the authors and publisher.)

10.5 illustrates the way in which the thalamocortical afferent fibers are believed to impose rhythmicity upon the cerebral cortex.

Rhythmic activity such as the alpha rhythm and barbiturate spindles could conceivably be produced in the manner described. Because the ascending reticular activating system has a profound influence on the thalamus, it could synchronize or desynchronize the EEG pattern. However, while the thalamic pacemaker concept explains some of the phenomena in EEG rhythmicity, there are many unexplained areas. According to some scientists, rhythmicity may be associated with corticocortical connections. But this is a topic for more advanced texts.

To summarize, the EEG appears to be the result of synchronized variations in the membrane potentials (synaptic potentials) of large numbers of neurons in the cerebral cortex. Summated EPSPs and IPSPs in the dendrites of the vertically oriented cortical pyramidal neurons, which are caused by afferent activity coming from the thalamus, are the most likely causes of the rhythmic variations in voltage that characterize the scalp EEG.

Reference

Andersen P, Andersson SA: *Physiological Basis of the Alpha Rhythm.* New York, Appleton-Century-Crofts, 1968, pp. 3–83.

Chapter 11
Recording Systems

Previous chapters dealt with the instrumental aspects of the EEG recording system. Separate chapters were devoted to the topics of recording electrodes, the differential amplifier, filters, and the writer unit. This chapter will look at the principles and techniques of electrode arrangement used to obtain a maximally useful EEG recording.

Electrodes as Field Samplers

Brain electrical activity as seen in the scalp EEG is generated by the neurons of the cerebral cortex. There are billions of neurons in the cerebral cortex and any number of them may be active at a particular moment of time. Unlike the ECG where there are only five waves that normally repeat at regular intervals to reckon with, the EEG is a continuum of electrical discharges that may vary from moment to moment. The electrical activity is different in frequency and amplitude over different areas of the brain. Since the brain is suspended in the spinal fluid, which acts as a volume conductor, the electrical activity recorded on the scalp turns out to be extremely complex.

A simple analogy may be useful. Imagine dropping a stone in a pond of water and observing the ripples that are formed on the surface. These ripples travel outward in ever-widening, regular concentric circles. Now, drop several stones in different spots at the same time. Each stone will produce its own series of concentric ripples. In places where the ripples come together, a complex, seemingly chaotic pattern of ripples will appear on the surface of the water. The multiple generators of electrical activity in the brain present a similar picture. In this context, the EEG electrodes attached to the scalp become "samplers" of the electrical fields set up by the multiple generators of electrical activity.

To obtain a complete and valid picture of the brain's electrical activity, it is necessary to simultaneously record samples of activity from different areas of the scalp, on both sides, over a prolonged period of time. This is essential not only to adequately describe the features of the normal EEG, but also to localize abnormalities. To use another analogy, the localization of abnormalities may be compared with the practice of taking several samples from different areas of a polluted lake in order to find the source or sources of contamination. To accomplish this, it is apparent that some kind of standardized, logically based recording system is needed.

Historical Background

Historically speaking, the sophistication of EEG recording systems has closely followed technological developments in electronics. In the 1940s when technology was limited, EEG recordings were made on machines having only two or four channels. This limited the EEG to the simultaneous recording of samples from one or two pairs of electrodes on each side. Such limited sampling, of course, is totally inadequate for describing the electrical activity of the entire brain. Nevertheless, many of the features of the EEG that we know today were first recorded and described by researchers and clinicians using such primitive equipment.

When multichannel machines became available, the question of where to place the many electrodes that it was possible to use, and how to hook them up, became a major issue. Different EEG laboratories began to use totally different placements. To bring some harmony into the resulting chaos, the First International Congress of EEG held in London in 1947 recommended that an attempt be made to standardize the electrode systems used. Herbert Jasper studied the different systems used at the time and, in 1958, suggested adopting what is called the 10-20 International System of electrode placement. This systems is used by the vast majority of EEG laboratories around the world.

The 10-20 International System

The placement of electrodes in this system depends upon measurements made from standard landmarks on the skull. The system affords adequate coverage of all parts of the head, with electrode positions designated in terms of the underlying brain areas (i.e., frontal pole, frontal, central, parietal, occipital, and temporal) to which they correspond. These areas are abbreviated using capital letters, with F corresponding to frontal, C corresponding to central, and so on. A single-digit number goes along with the letter. Odd numbers designate left-sided and even numbers right-sided locations. Thus, for example, C3 corresponds to the central region on the left side.

The term "10-20" is used because the electrodes are placed either 10% or 20% of the total distance between a given pair of skull landmarks. The use of percentages to ascertain the distances between electrodes rather than absolute values allows for the normal differences in size and shape of the head between different persons. This means that the 10-20 International System is appropriate for use with small infants as well as adults having very large heads.

Measurements are carried out using a narrow measuring tape with markings preferably in centimeters and millimeters. For anteroposterior measurements, the distance between the nasion and inion over the vertex in the midline is taken. Five points are located along this line and designated frontal pole (Fp), frontal (F), central (C), parietal (P), and occipital (O). The point Fp is 10% of the nasion-to-inion distance above the nasion. F is located behind Fp at a distance of 20% of the nasion-to-inion distance; C is behind F at a distance of 20%, and so on. The points located are marked off directly on the scalp with a colored china-marker pencil.

The lateral measurements are made in the central coronal plane on the basis of the distance between the left and right preauricular points. Ten percent of the distance above the preauricular points marks the location of the T3 and T4 electrodes, C3 is at a distance of 20% above T3, and C4 is 20% above T4 (see Fig. 11.1). Details concerning the measuring techniques and of locating the positions of the various other electrodes are given in Appendix 5. Using these directions, a total of 19 electrode placements are marked off on the scalp. Together with the earlobe placements (designated A1 and A2), this comprises the 21 standard electrodes in the 10-20 International System. Figure 11.2 shows the approximate locations of the electrodes in relation to areas of the cerebral cortex. A recent study (Homan RW, Herman J, Purdy P, 1987) employing CT scanning to visualize brain structure suggests that the 10-20 System provides scalp locations that correlate well with the expected cerebral structures.

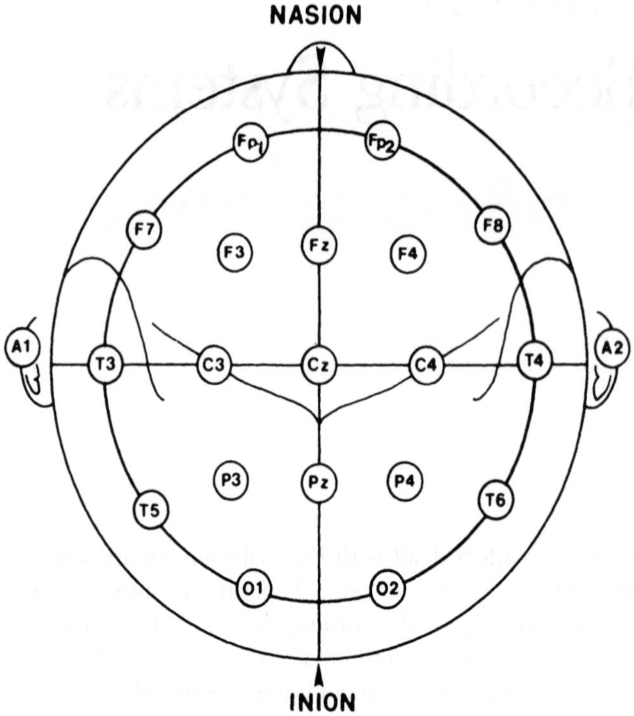

Figure 11.1. The 10-20 International System of electrode placement.

Derivations

With a total of 21 electrodes to work with, how should the electrodes be hooked up to best display the brain's electrical activity? In other words, what combinations of electrodes should be connected to the various channels of the EEG machine, and at what times?

Because two connections are needed to complete an electrical circuit (see Chapter 2), two electrodes have to be connected to each channel (amplifier) of the machine. As was mentioned earlier in Chapter 1, a particular pair of electrodes connected to a single amplifier is referred to as a *derivation*. Experience has shown that a machine that displays at least eight derivations simultaneously is necessary to adequately study the spatial characteristics of the brain's electrical activity. However, a 16-channel machine is preferable; an 18-channel machine is even better. The larger machines are capable of gathering more data in the same amount of time as well as providing better resolution of the spatial characteristics of the brain's electrical activity.

With the use of 21 electrodes, one can have a total of 210 different derivations.[1] But in actual practice, all possible

[1] The number of possible combinations of n things taken two at a time is equal to $\frac{1}{2}[n(n - 1)]$. For example, if $n = 3$, we have $\frac{1}{2}[3(3 - 1)] = 3$, which is readily verified by trial and error. When $n = 21$, the formula becomes $\frac{1}{2}[21(21 - 1)] = 210$.

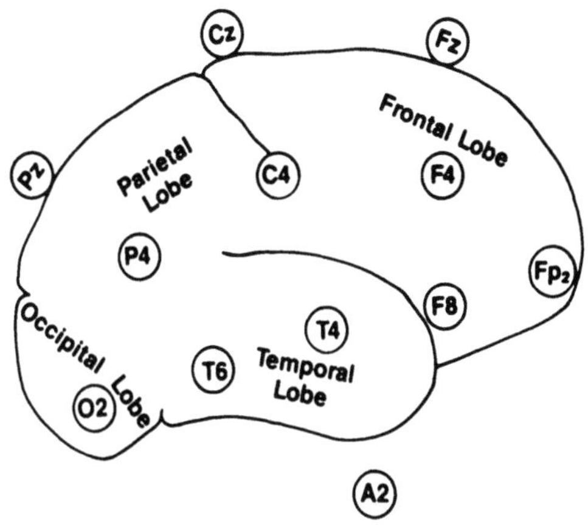

Figure 11.2. Topographic relationship between 10-20 System electrodes and areas of the cerebral cortex.

combinations of electrodes are seldom used. There is an important reason for this. Since interpretation of the EEG ultimately involves comparison of different derivations, it is essential to use derivations having comparable interelectrode distances. This requirement eliminates a large number of possible derivations. Indeed, most routine EEG work employs only 48 or even fewer derivations.

Montages – Rationale

The particular arrangement whereby a number of different derivations is displayed simultaneously in an EEG record is termed a *montage.* With even only 48 derivations, a large number of different montages can be designed. The main reason for using different montages is to make EEG interpretation as easy and accurate as possible. For this purpose, certain guidelines have to be followed, and the American EEG Society has given some recommendations in this regard.

First of all, a montage should be simple and easy to comprehend. Montages should follow some kind of anatomical order or pattern. For example, channels representing the more anterior electrodes should be arranged on the recording chart above those from the more posterior regions. Derivations from the left side should be located on the chart above derivations from the right side. This may be accomplished either by alternating the derivations, i.e., left, right, left, right, and so, or by placing derivations from the different sides in blocks, e.g., left, left, left, left; right, right, right, right. It is advantageous for a laboratory to use a few common or standard montages so that records from different laboratories can be compared with ease.

There are two basic types of EEG montage: referential and bipolar. It is advantageous in routine EEG work to use both. A brief discussion of each type follows, after which examples of some commonly used montages are given.

Referential Montages

The principle of this technique involves measuring the electrical activity at different electrodes simultaneously, in comparison with a common reference electrode. Ideally, the common reference electrode should be unaffected by cerebral electrical activity. Each electrode is connected to grid 1 of a different amplifier, and the single reference electrode is connected to the grid 2 inputs of all the amplifiers. The terms monopolar and unipolar have sometimes been used to refer to referential recording, but the use of these terms is discouraged.

The major advantage of referential montages is that the common reference allows valid comparisons to be made between amplitude measurements in different derivations. This is in contrast to the case of bipolar montages where amplitude measurements may be unreliable so that amplitude comparisons between derivations are invalid. As we will see in Chapter 12, localization of a discharge in referential montages is based on the presence of amplitude differences between channels. If we assume that the common reference electrode is unaffected by a particular discharge, then the discharge will appear with the same polarity in all the nearby electrodes and will show a higher amplitude in the electrode adjacent to the source in comparison with the surrounding electrodes.

The major disadvantage of referential montages lies in the fact that there is no ideal reference electrode. The commonly used sites of reference, namely the earlobes, are close to the temporal lobes and hence pick up a considerable amount of cerebral electrical activity from those areas. As will be detailed later in Chapter 12, this can lead to confusion in localization. Briefly, if there is an active spike focus at T3, the midtemporal electrode, then the left ear reference (A1) will also be "active." Under such conditions, the voltage between T3 and A1 may be less than the voltage between F7 and A1 or between T5 and A1. The result is a spike in the T3 derivation that is smaller in amplitude than the spikes in the F7 or T5 derivations; therefore, the spike focus is falsely localized to F7 or T5 instead of to T3 which is correct.

Another disadvantage of using the earlobes as a reference is that problems with ECG artifacts are more common. Why this happens is readily understood if the reader will recall our discussion of common mode rejection in the chapter on differential amplifiers. The ECG is a common-mode or in-phase signal, and common-mode signals are rejected by a differential amplifier. The degree to which the ECG will be rejected depends on the common-mode rejection ratio of the amplifier and on the extent to which the ECG voltages present at grids 1 and 2 of the amplifier are the same. Now, the ECG voltages appearing at the various EEG electrodes are rarely identical; their magnitudes depend on a number of factors, one of which is the distance

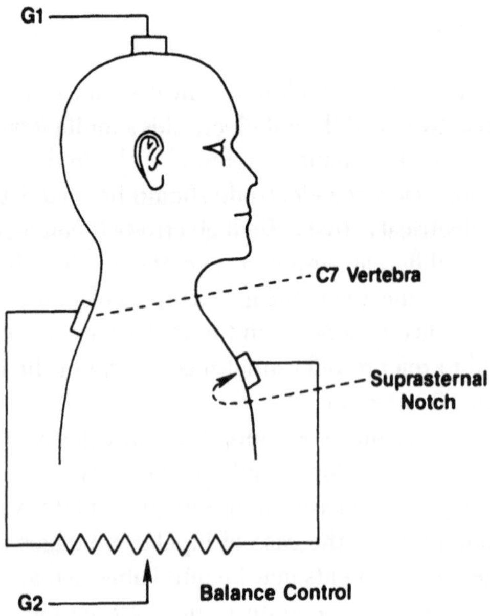

Figure 11.3. Noncephalic reference electrode. The balance control is a 50K-ohm potentiometer. To avoid unbalancing the amplifier inputs, the electrodes may be connected to the potentiometer via a buffer rather than directly as shown in the diagram.

of the electrodes from the heart. In the case of closely spaced electrodes, the differences in voltage may be quite small, but they can become large if the electrodes are widely spaced, as when one electrode of a pair is connected to an earlobe. These voltage differences are treated as out-of-phase signals by the amplifier; they are amplified along with the EEGs and appear as artifacts in the recording.

It is sometimes possible to reduce ECG artifacts in referential recording by connecting both earlobes together. This is referred to as a linked-ear or (A1 + A2) reference. Needless to say, it is essential for impedances of the A1 and A2 electrodes to be low (3K to 5K ohms) when the earlobes are employed in referential recording.

Extracerebral Reference Electrodes

In order to reduce the contamination of the common reference electrode by cerebral electrical activity, many other sites besides the earlobes have been proposed. Thus, for example, electrodes located on the angle of the jaw, on the chin, the tip of the nose, and the neck have been tried as references. These attempts have not been notably successful. The major problem is the contamination of all channels by ECG and electromyographic (EMG) artifacts.

Noncephalic electrodes have also been tried as common references. One noncephalic reference electrode described uses an electrode over the patient's C-7 vertebra, which is connected to another electrode over the suprasternal notch through a balancing potentiometer that is adjusted to minimize ECG artifacts (see Fig. 11.3). Although such an arrangement can reduce ECG artifacts, it is somewhat complex for routine EEG work, as rebalancing may become necessary when electrode impedances change or the position of the head vis à vis the heart is altered.

Average Potential Reference

Sometimes referred to as the Goldman-Offner reference, this system of obtaining a common reference was derived from a similar technique used in electrocardiography by N.F. Wilson and colleagues in 1934. The principle upon which the method is based is statistical. Assume that the voltages present at the EEG electrodes occur at random and are therefore independent of each other. In other words, when the voltage at some electrodes is positive, it is negative or zero at the others, and vice versa. Now, if all these voltages are summed and the mean is taken, it is easy to see that the mean will tend toward an absolute voltage of zero as long as none of the electrodes show any excessively large values that fall outside the distribution of the others. This average potential, thereupon, serves as a common reference and is connected to the grid 2 inputs of all the amplifiers. As is the case with the ear reference, each of the grid 1 inputs is connected to a different electrode. The practical circuits used to derive the average potential are somewhat complex and need not concern us here. The interested reader may consult a more advanced text for such detail.

The major disadvantage or shortcoming of the average potential reference is that the basic assumptions upon which the method is based are not entirely satisfied. In particular, the voltages present at the various EEG electrodes do not occur at random; nor do these voltages fit a normal distribution. Thus, for example, excessively large voltages associated with eye movements are commonly picked up by Fp_1 and Fp_2, and the high-amplitude vertex waves that accompany stage II sleep are of maximum amplitude in Cz. If these electrodes are included in the average potential reference, eye-movement potentials and vertex waves may appear as artifacts in the tracings from many of the electrodes. What happens is that these large, distinctive voltages contaminate the average and, in so doing, get into the amplifiers via the grid 2 inputs. To avoid such contamination, Fp_1, Fp_2, and Cz are never included in the average. F7, F8, Fz, and Pz are often excluded for the same reasons, leaving a total of 14 electrodes in the average.

Average potential reference systems have a provision that easily permits the technician to exclude any of the electrodes actually contained in the average. This is an essential feature of such systems because any electrode containing a high-amplitude discharge like a spike or sharp wave can contaminate the average in the same way that was already described. However, there is a limit to the number of electrodes that may be taken out of the average, as averages based on a small number of cases are unreliable. As a rule of thumb, averages containing fewer than 10 electrodes should be avoided. An example of average reference contamination by high-amplitude focal spike discharge is given in Chapter 12 (Fig. 12.13).

Despite these problems, recordings obtained with the average potential reference are often superior to those using an ear reference. Nevertheless, there are instances where the average potential reference can be grossly misleading. One interesting example is the case of a patient whose EEG shows an alpha rhythm of high amplitude, with the activity spreading into the midtemporal and central regions. When the average potential reference is used in such a case, a remarkable thing occurs. The alpha rhythm is seen in widespread distribution over the entire scalp, including the Fp_1 and Fp_2 electrodes. The finding is obviously an artifact; the alpha rhythm gets into the anterior channels via the grid 2 inputs from the contaminated average reference. This is readily verified by noting that the alpha waves appearing in the frontal derivations are opposite in phase to the same waves present in the occipital derivations (see Fig. 14.3).

Bipolar Montages

In bipolar recording, the potential difference between two electrodes placed on the scalp is displayed. Unlike the case of referential recording, both electrodes are considered to be active, and the varying difference in voltage between the two is recorded. Electrodes in a bipolar montage are connected in a sequential manner: the electrode going to grid 2 of the first derivation is also connected to grid 1 of the next derivation. These sequences can be arranged in chains, either in a longitudinal or a transverse array. With such montages, an electrical discharge originating in the common electrode of two adjacent channels will show the phenomenon of phase reversal in both longitudinal and transverse sequences.

As we will see in Chapter 12, the advantage of bipolar sequential montages is the ease with which localization can be made. Thus, phase reversals are very easily detected by eye in the tracings. On the negative side, bipolar montages do not provide an accurate or valid measure of the amplitude of the waveform of a particular event. Depending on the magnitude and phase of the voltages at the two electrodes, the signals are subject to cancellation or summation (see Chapter 12) in the amplifier. For this reason, bipolar recording merely provides a comparison of the voltage at one electrode of a pair with respect to the voltage at the other.

Table 11.1. Some Commonly Used EEG Montages

(A) Longitudinal Bipolar		(B) Transverse Bipolar		(C) Referential w/earlobes	
Fp_1 - F7	L	F7 - Fp_1	Fp	F7 - A1	L
F7 - T3		Fp_2 - F8		T3 - A1	
T3 - T5		F7 - F3	F	T5 - A1	
T5 - O1		F3 - Fz		F8 - A2	R
Fp_2 - F8	R	Fz - F4		T4 - A2	
F8 - T4		F4 - F8		T6 - A2	
T4 - T6		T3 - C3	C	Fp_1 - A1	L
T6 - O2		C3 - Cz		F3 - A1	
Fp_1 - F3	L	Cz - C4		C3 - A1	
F3 - C3		C4 - T4		P3 - A1	
C3 - P3		T5 - P3	P	O1 - A1	
P3 - O1		P3 - Pz		Fp_2 - A2	R
Fp_2 - F4	R	Pz - P4		F4 - A2	
F4 - C4		P4 - T6		C4 - A2	
C4 - P4		T5 - O1	O	P4 - A2	
P4 - O2		O2 - T6		O2 - A2	

Commonly Used Montages

As mentioned earlier, it is good practice to use routinely a few montages that are common to different laboratories so that EEG records from them can easily be compared. It is advantageous to use both bipolar and referential montages; in the case of bipolar montages, both longitudinal and transverse montages should be included.

Table 11.1 shows some commonly used montages. This table should be studied in conjunction with Fig. 11.1. In column A of Table 11.1 we have longitudinal bipolar arrays, namely, the temporal and the parasagittal bipolar chains on both sides. Note that the derivations from right and left sides are combined in blocks and that the temporal and parasagittal chains are joined together into a 16-channel montage. This montage is referred to colloquially as the "double banana."[2] If a machine with only eight channels is available, the temporal and parasagittal derivations can be recorded in separate runs.

Column B of Table 11.1 shows a standard transverse bipolar montage on a 16-channel machine. Note that the more anterior electrodes are located at the top of the page, while the more posterior electrodes are located at the bottom. Note, also, that the electrodes on the left side appear before those on the right. This configuration is commonly referred to as a coronal montage. As a careful perusal of column B will show, this montage is not readily convertible for use on an eight-channel machine.

[2] When viewed in the context of Fig. 11.1, the temporal and parasagittal chains combined take the shape of a banana. There is one of them on each side; hence the term "double banana."

Finally, column C in Table 11.1 shows a commonly used referential montage that employs the ipsilateral earlobes as common references. This same montage is also appropriate for the average potential reference, in which case the average reference is substituted for A1 and A2 in the table.

Reformatting of Montages

Recent advances in technology have afforded a means whereby the pattern of activity observed in one montage may be used to derive or predict the pattern of activity that would be seen in another. Although the method involves some extensive computational operations that are carried out by computer, it is simple in theory. An example will best serve to explain the process of reformatting.

A patient's EEG is recorded using a referential montage in which tracings from the 19 scalp electrodes are taken with respect to a common electrode. Each tracing, of course, shows the variation in voltage that occurs with time at a particular derivation. If, now, the voltages from two of these derivations—say T3 and T5—are combined algebraically over time, the result would be a time-varying voltage like that observed if T3 and T5 were connected in bipolar fashion to a differential amplifier. The same operation can be carried out for any two electrodes, and in this way a wide variety of montages can be created. In practical terms, the method permits the technician to take the EEG using but a single montage. Later, when the record is read, the electroencephalographer can reformat into any numbers of different montages, selecting those that provide the most accurate localization information.

Extension of the 10-20 System

Although sufficient for most routine EEG work, the standard 19 scalp and 2 ear electrodes do not adequately evaluate the electrical activity of the cerebral cortex in all circumstances. This is especially true in the case of topographic mapping of brain electrical activity, as the accuracy of information contained in a map is directly related to the total number of electrode locations from which the map is derived. To remedy this situation, a number of electrodes may be added to the standard 21 derivations.

The most commonly used additions to the 21 standard derivations are the T1 and T2 electrodes. These are located by first finding the point that is one third of the way from the external auditory meatus to the outer canthus of the eye, and then locating a point 1 cm directly above. Falling between and somewhat below F7 and T3 on the left, and F8 and T4 on the right, T1 and T2 are closer to the anterior part of the temporal lobes than F7 and F8, which are actually located over the inferior frontal area (see Fig. 11.2). T1 and T2 may be included as part of a bipolar chain or in a referential montage.

Additional electrodes may also be applied between the standard electrodes in the coronal rows. Thus, we have Fp_z between Fp_1 and Fp_2, F1 between Fz and F3, F5 between F3 and F7 on the left, F2 between Fz and F4, F6 between F4 and F8 on the right, and so on for central, parietal, and occipital rows. Together with the 21 standard derivations and the T1 and T2 electrodes, this yields a total of 37 derivations. Finally, additional electrodes may also be placed between the rows defined by the frontal pole and frontal electrodes, by the frontal and central electrodes, by the central and parietal electrodes, and by the parietal and occipital electrodes. In this way, more than 60 different derivations become available.[3] When electrodes are placed so closely to each other, smaller-diameter disks are advisable, and special care needs to be taken when applying them to avoid a salt bridge between electrodes.

Special Electrodes

The electrical activity of certain portions of the cerebral cortex, notably the basomedial parts of the temporal lobe and the orbital and medial parts of the frontal lobe, is not accessible to the electrodes taken up thus far. This sometimes leads to problems in accurately locating seizure foci, particularly in patients who are being considered for temporal lobectomy. To overcome such problems, a number of special electrodes may be used. These are described below. Refer to Fig. 11.4 for the approximate location of these electrodes.

Zygomatic Electrodes

Ordinary disk electrodes are used, and they are located over the easily palpated zygomatic arch, below and anterior to the T1 and T2 electrodes. Zygomatic electrodes are useful for picking up activity from the tips of the temporal lobes.

Nasopharyngeal (NPG) Electrodes

The tips of these electrodes are placed in contact with the roof of the nasopharynx, so that activity from the uncus, hippocampus, and orbitofrontal cortex may be picked up. An NPG electrode consists of a piece of insulated flexible

[3] These extensions of the 10-20 System have been organized into an expanded system of electrode placement—the 10% system—by a number of electroencephalographers. For details and suggested electrode-site nomenclature, refer to Nuwer MR: Recording electrode site nomenclature. *J Clin Neurophysiol* 1987; 4:121–133.

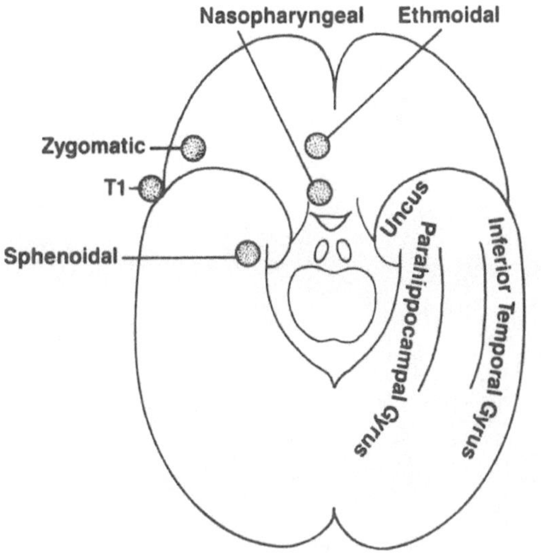

Figure 11.4. Basal view of the brain showing the approximate locations of special electrodes.

silver wire with a 3-mm silver or gold-plated silver ball at the tip. The silver wire varies in length from 5 to 15 cm (to accommodate pediatric and adult patients) and can be bent to suit the dimensions of the nasal cavity. A trained technician can place these electrodes easily and with only minimal discomfort to the patient, although in patients with deviated nasal septum insertion may be difficult and there may be some discomfort. Local anesthesia is seldom required.

The electrode should be autoclaved before use and may be lubricated with sterile conductive gel. Holding the electrode between the fingers, the ball at the tip is guided along the floor of the nasal cavity beneath the inferior turbinate until it has passed through the nasal cavity and is in contact with the pharyngeal mucosa (verify this placement by asking the patient whether the ball is touching the back of his/her throat). At this point, rotate the electrode so that it is pointing upward and outward. The electrode can be retained in that position by securing the connecting wire under mild tension to the chin or jaw with a piece of tape. To remove the electrode, retrace the same steps.

Nasopharyngeal electrodes are notorious for artifacts from respiration and pulse and may give false lateralization owing to the tips being close to the midline. If the activity observed in nasopharyngeal electrodes is totally confined to them without any reflection in other derivations, distinction from an artifact may be impossible.

Sphenoidal Electrodes

Thin, flexible insulated platinum wire is introduced along a spinal needle, under local anesthesia so that the tip of the wire lies in close proximity to the foramen ovale. After x-ray verification of the position of the electrode, the spinal needle is withdrawn. The wire can stay for several days for prolonged recording. Activity from the basal and mesial temporal cortex can be recorded without too much artifact. The procedure is usually well-tolerated but needs to be done by a physician familiar with the technique.

Ethmoidal Electrodes

The electrode is a flexible insulated silver wire with a bulbous tip. Under topical anesthesia it is introduced into the nostril and gently passed up so that the tip lies in contact with the cribriform plate of the ethmoid bone. Activity from the orbitofrontal cortex may be recorded using this technique.

Surgically Placed Electrodes

Patients undergoing evaluation for seizure surgery may need long-term recording by electrodes situated close to the cerebral cortex. The electrodes may be placed epidurally, subdurally, or within the brain substance (depth electrodes). For subdural recordings, a flexible plastic plate with up to 64 electrodes is used. During seizure surgery, specially designed electrodes may be placed over the exposed cortex to localize the sites of epileptiform activity accurately (electrocorticography). Details are found in more advanced, specialized texts.

Reference

Homan RW, Herman J, Purdy P: Cerebral location of International 10-20 System electrode placement. *Electroencephalogr Clin Neurophysiol* 1987; 66:376–382.

Chapter 12
Localization and Polarity

In plain and simple terms, the EEG is the recording of the difference in electrical potential or voltage between various pairs of sampling electrodes attached to the surface of the scalp. Seen in this limited context, there are three major questions that need to be answered in clinical electroencephalography, namely, (1) what is the magnitude of a particular voltage observed at a particular instant in time?, (2) what is its polarity, that is to say, is the particular voltage positive or negative?, and (3) from where does it originate? Whereas the answers to these questions can be relatively simple for a linear conductor, they are considerably more complex in the case of a *volume conductor* like the brain. The magnitude of the complexity already was recognized in 1853 by Hermann Helmholtz who noted that a given "electromotive surface" may reflect an infinite variety of internal electrical fields (Helmholtz H, 1853).

Volume Conductors

What is volume conductor? We can best define this by first reviewing what is meant by a linear conductor. All the electrical circuits we have heretofore dealt with that are part of an EEG machine are linear conductors. The electrical currents involved flow along limited pathways in wires and various elements like resistors and condensers. In a volume conductor, additional pathways have been added. The conducting medium occupies three-dimensional space so that current flow can take many pathways. These pathways may display characteristics that are quite complex, as when conducting media having different electrical characteristics are involved.

Such is the case with the brain and brain electrical activity where the recording electrodes are placed not directly on the brain but on the scalp. The brain itself is a volume conductor. But, in addition, the EEG signals must pass *through* volume conductors—cerebral spinal fluid, meninges, skull bone, and scalp (which in turn interfaces with the surrounding atmosphere)—before reaching an electrode. One result of this is a fourfold decrease in amplitude of EEG signals and a marked attenuation of the higher frequencies. The magnitude of the attenuation is rather dramatically illustrated by the phenomenon sometimes referred to as the "breach effect," or focal enhancement of the amplitude of beta activity resulting from a skull defect.

Concept of a Dipole: Fields and Equipotential Contours

How are voltages distributed in a volume conductor? And what is the relationship between the voltages within and those observed on the surface of a volume conductor? To facilitate discussion of this topic, the various voltage sources within a volume conductor are looked upon as simple *dipoles.* A dipole is an electrical unit composed of equal but opposite electrical poles or charges separated in space. It is analogous to a bar magnet where one end is north and the other end is south. Just as there is a magnetic field encircling a magnet, so there is an electrical field surrounding a dipole. Let us examine the nature of that field.

The electrical field around a vertically oriented dipole in a vessel containing a conducting medium is shown in Fig. 12.1. The horizontal line that is perpendicular to the axis of the dipole represents the interface between the conducting medium or volume conductor below and the air above. The curved lines represent equipotential contours; this means that the voltages (referred to a distant reference electrode) measured at all points along a given contour are the same. Note that the voltages are smaller as the contours move farther away from the electrical poles. Note also that the equipotential contours appear distorted and show a different configuration on the surface than they do within

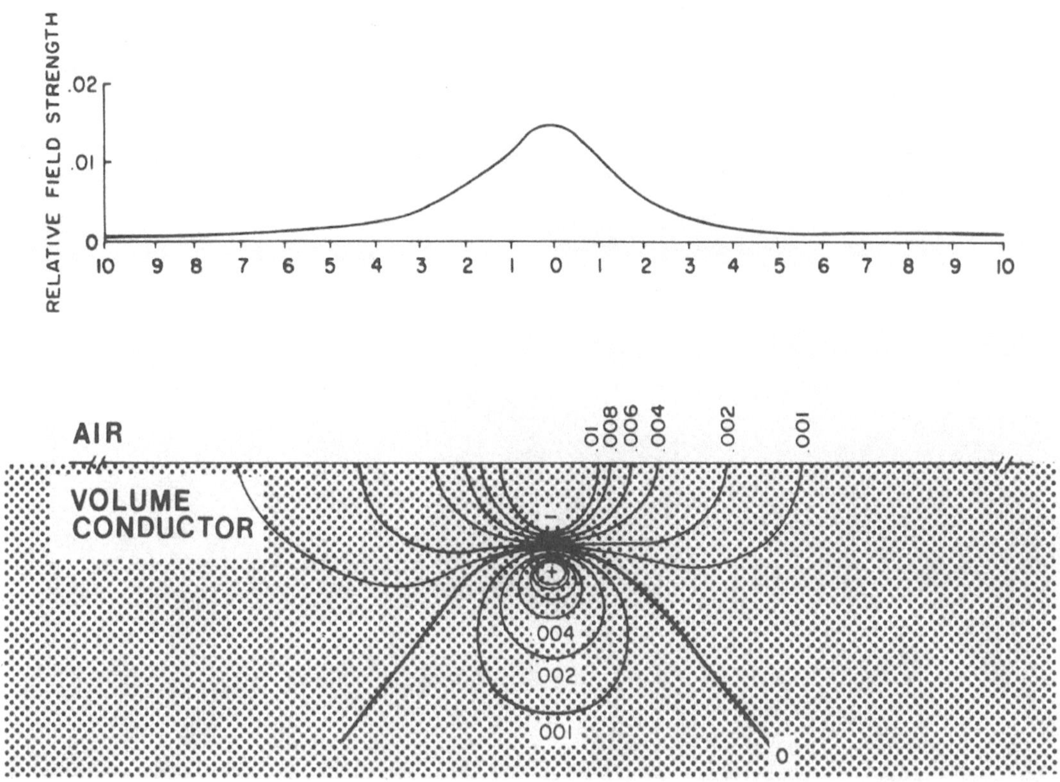

Figure 12.1. The electrical field around a vertically oriented dipole. The figure shows the surface and depth distributions of electrical potential for a dipole located 1 cm below the air-volume conductor interface. Field-strength values are relative to the strength of the dipole, which is equal to 1. Remote reference electrode.

the volume conductor. The absolute voltages are seen to be larger, and the field is more extended on the surface.

The upper portion of Fig. 12.1 shows a plot of strength of the field at various locations on the surface relative to the strength of the dipole, the voltage of which is taken to be 1.000. As will be seen, the field is strongest at the axis of the dipole, which is the line connecting the points of maximum negative and positive charge. In Fig. 12.2 we show the surface and depth distributions of potential when the dipole is oriented *horizontally* with respect to the interface between conducting medium and air. Note that when the dipole is oriented in this manner, a phase reversal is perceived in the voltage measured at the surface. The topic of phase reversals is taken up later in this chapter.

Theory of Localization

Localization in electroencephalography is based on the premise that brain electrical activity is generated by dipoles as described in the previous section. The location of the dipoles is inferred from the distribution and polarity of voltages observed on the surface of the scalp. This is accomplished by the use of multiple-channel recording in which voltages are recorded simultaneously from different derivations. For this purpose, the differential amplifier is essential. As mentioned in an earlier chapter, a differential amplifier records the difference in voltage between two electrodes or derivations. By convention, EEG machines are designed so that the pens deflect upward when grid 1 of the differential amplifier is negative with respect to grid 2. Although this may suggest to the reader that the derivation connected to grid 1 is electrically negative if the pen of that channel deflects upward, a little thought will show that this is not necessarily so.

Problems of Polarity

Consider some simple examples. Suppose that grid 1 is connected to -50 µV while grid 2 is connected to $+10$ µV. The voltage difference under such conditions is 60 µV, and since grid 1 is more negative than grid 2, the pen deflects upward. But now, suppose instead that grid 1 is connected to $+10$ µV while grid 2 is connected to $+70$ µV. In this case the voltage difference is again 60 µV, and since grid 1 is less positive (more negative) than grid 2, the pen again deflects upward. We find, therefore, that in both instances the pen deflects upward; but in the former case grid 1 is

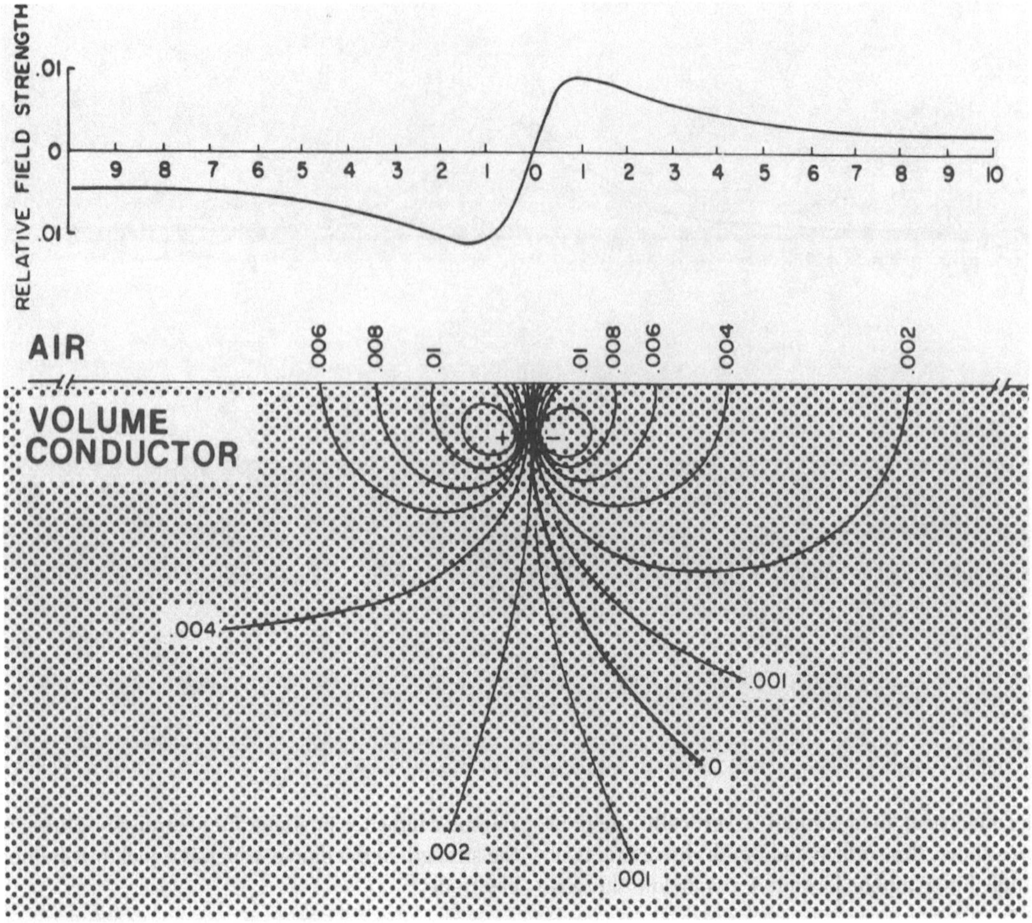

Figure 12.2. Surface and depth distributions of electrical potential as in Fig. 12.1, except that the dipole is horizontally oriented.

connected to a negative voltage, whereas in the latter it is connected to a positive voltage. The obvious conclusion is that we cannot infer the absolute polarity at a particular electrode simply by connecting it up to a differential amplifier and observing the output.

How then do we discover the polarity of an electrical event occurring within the brain from surface recordings of electrical activity? And how do we discover its location?

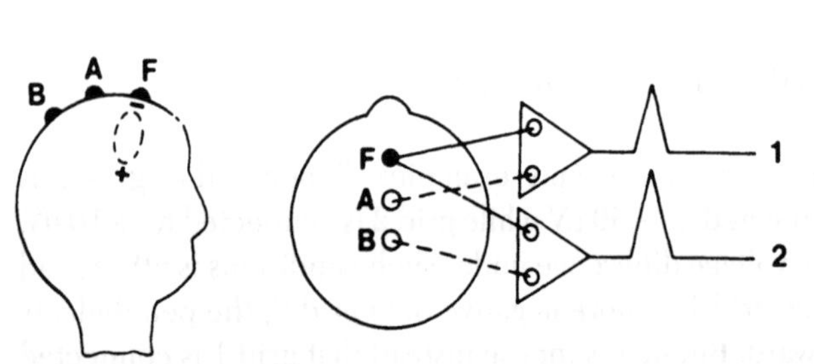

Figure 12.3. Principles of localization, rule 1. A, B, and F are scalp electrodes connected to the differential amplifiers of two EEG channels. The radial dipole is oriented so that the axis of the dipole is perpendicular to the scalp at F. Note that interelectrode distance FB is twice the distance in FA.

Concerning polarity, we are able only to deal with relative polarity—the polarity at a particular electrode *relative to* other regions on the scalp or body. The localization of the electrical event occurring within the brain from activity recorded by electrodes placed on the scalp is determined by five principles. These principles are discussed in the following section.

The Five Principles of Localization

These principles or rules are illustrated in the diagrams of Figs. 12.3 through 12.9. In these diagrams, F is the point of intersection with the scalp of the axis of a radially oriented dipole that is perpendicular to the scalp. F, then, represents a *focus* or the center of a limited region on the scalp displaying the electrical activity generated by the dipole. Following EEG convention, the differential amplifier is represented by a triangle with the apex pointing to the right. The solid line is the input to grid 1 and the broken line the input to grid 2. The voltage picked up from the dipole and amplified is represented as a triangular waveform. Polarity of the waveform is negative because the negative pole of the dipole is adjacent to the surface while the positive pole is buried deeper in the cerebral cortex.

Rule 1. If one of a pair of electrodes is at F, the focus, the amplitude of the recorded potential will increase as the distance between this electrode and the second electrode of the pair increases. Thus, in Fig. 12.3, the distance FB is greater than FA; therefore, the amplitude of the voltage recorded between F and B is greater than the amplitude recorded between F and A. In short, widely spaced electrodes record larger voltages than closely spaced electrodes.

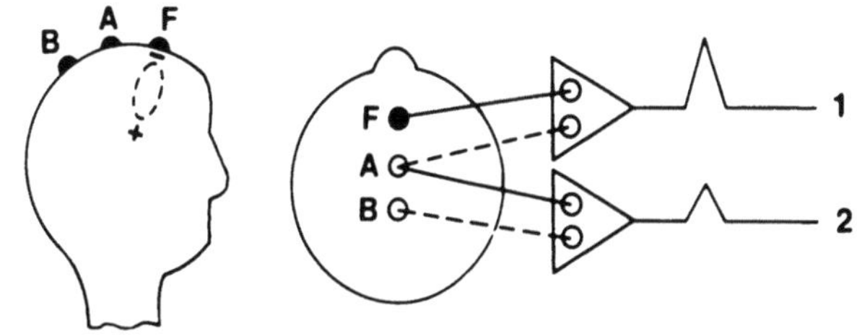

Figure 12.4. Principles of location, rule 2. Legend as in Fig. 12.3 except that interelectrode distances FA and AB are equal.

Rule 2. Given two pairs of electrodes having equal interelectrode distances, the potential recorded from the pair having one electrode at F will be greater than the potential recorded from the pair having neither electrode at F. In Fig. 12.4, the interelectrode distances FA and AB are equal. The voltage recorded is greater in channel 1 than channel 2 because one of the channel 1 inputs is connected to F, the focus of the activity.

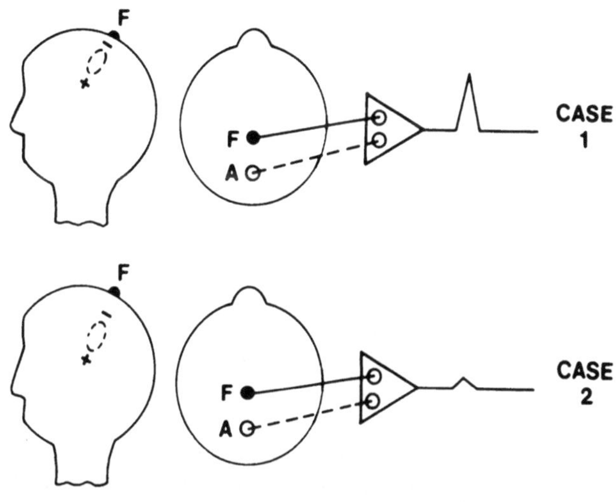

Figure 12.5. Principles of localization, rule 3. Cases 1 and 2 both have the axis of the dipole perpendicular to the scalp at F. But in case 1, the negative pole is close below the surface, whereas in case 2 it lies deep in the cerebral cortex.

Rule 3. The farther away the dipole is from the surface of the scalp, the smaller will be the potential observed at the surface and the smaller the voltage recorded between pairs of electrodes, interelectrode distances being constant. This rule is illustrated by two examples in Fig. 12.5. Note that the dipole in case 1 is nearer the surface than it is in case 2 so that the voltage recorded by electrodes at F and A is greater in case 1 than in case 2. Because the actual dipole generators from which we are able to record using surface electrodes are near the surface of the cerebral cortex, the rule has more theoretical than practical significance. It is possible, of course, for a dipole generator in a sulcus or on the mesial or inferior surface of the hemisphere to be a deep generator; but, in such a case, the dipole axis probably would not be perpendicular to the surface. This condition cannot be analyzed by current methods herein described as they assume that the dipole is perpendicular to the scalp. Generators located in the subcortical gray matter are of little practical interest because they are too far away from the scalp to be recorded by conventional EEG methods.

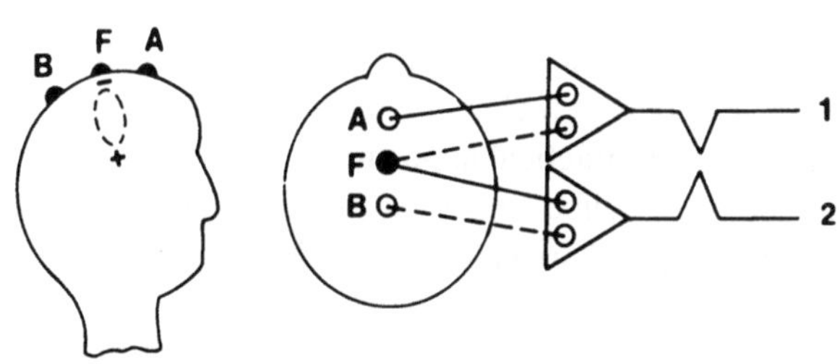

Figure 12.6. Principles of localization, rule 4. The radial dipole is oriented so that its axis is perpendicular to the scalp at F. The focus of activity, F, lies equidistant from the electrodes A and B. Channels 1 and 2 show an instrumental phase reversal.

Rule 4. If three electrodes are connected so that one is at F and is common to two recording channels, being the grid 2 input of the first channel and the grid 1 of the second, the pen deflections in the two channels will be in opposite directions. Figure 12.6 shows this set of conditions. The outputs of the two channels illustrate what is meant by the term "phase reversal," or what is really an *instrumental phase reversal* not a true phase reversal. Note that the reversal results from the fact that the shared electrode goes to opposing inputs and hence causes the opposing deflections

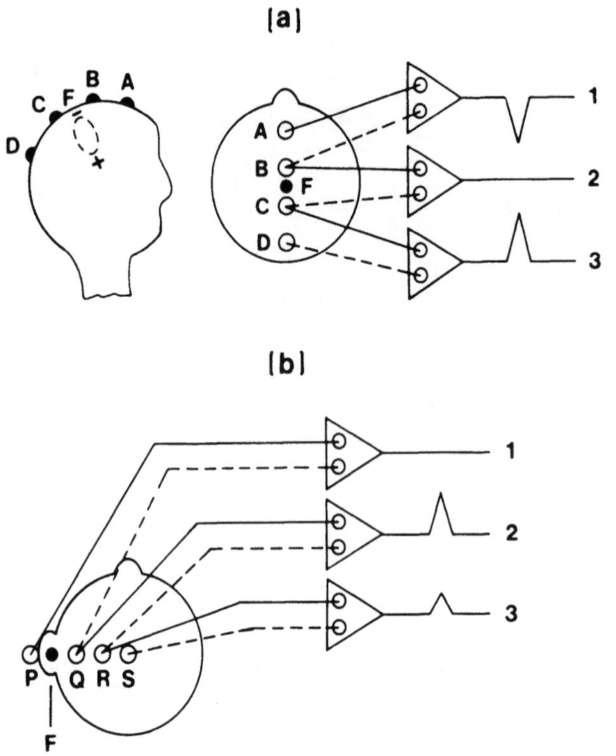

Figure 12.7. Principles of localization, rule 5. (**a**) The radial dipole is oriented so that its axis is perpendicular to the scalp at F. The focus, F, is equidistant from electrodes B and C, thereby forming an equipotential zone at these electrodes. Channel 2, therefore, records no difference in potential between them. Channels 1 and 3 show an instrumental phase reversal. (**b**) Rule 5 in the case of an "active ear." In this instance, the focus F is situated in the temporal area adjacent to the ear and is equidistant from electrodes P and Q. Channel 1 documents the presence of the equipotential zone at these electrodes.

to occur. A phase reversal identifies the electrode that is nearest to the point of maximum voltage, or the focus.

Rule 5. If two electrodes are equidistant from F the focus, no voltage will be recorded between them. In Fig. 12.7(a), B and C are equidistant from F and no voltage is recorded from the "equipotential zone" surrounding these two electrodes, which are the inputs to channel 2. This outcome is an example of *cancellation*, a phenomenon that will be taken up in a later section. Figure 12.7(a) also shows a phase reversal between channels 1 and 3.

An interesting, practical application of rule 5 occurs in the case of the so-called "active ear," in which a focus is situated in the temporal area adjacent to the ear. This is illustrated in Fig. 12.7(b), where electrodes are placed in a coronal chain across the top of the head, from left to right, starting with the electrode on the left earlobe. With the focus located midway between the earlobe and the midtemporal electrode, an equipotential zone is created about electrodes P and Q so that no voltage is recorded between the inputs of the channel 1 amplifier. The deflection observed in channel 2 is larger than that in channel 3 because Q is closer to the focus than R—a corollary of rule 2.

When an earlobe electrode is used as a common electrode in referential recording, an "active ear" can become a serious source of contamination in the derivations for which this electrode serves as a reference. Ear contamination is taken up in a later section of this chapter.

Figure 12.8(a) shows a special case of rule 5. In this instance the focus F is not on the line joining electrodes A, B, C, and D but instead is to one side. Nevertheless, F is equidistant from B and C so that the rule still holds. However, the exact position of F along the perpendicular from the midpoint of the line joining B and C cannot be determined using the configuration of electrodes shown in Fig. 12.8(a). To locate the focus in this dimension, a chain of electrodes perpendicular to the ABCD chain at the midpoint between B and C is applied. This configuration is shown in Fig. 12.8(b), where the electrode at R happens to be directly over the focus F. Note that there is an instrumental phase reversal at electrode R, which, by rule 4, localizes the focus to this electrode.

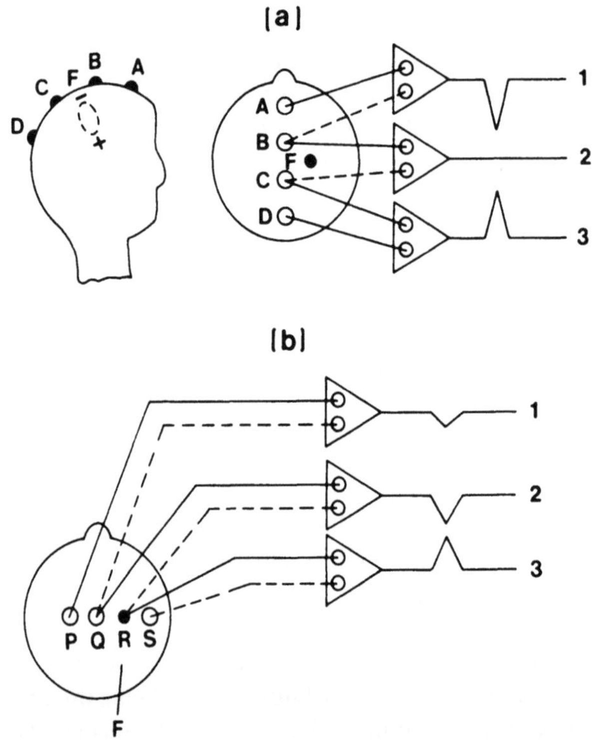

Figure 12.8. Special case of rule 5. (**a**) The focus, F, is equidistant from electrodes B and C, but is not on a line joining them. As in Fig. 12.7(a), channel 2 records no difference in potential between these electrodes. (**b**) Use of a horizontal line of electrodes to document the location of the focus in the horizontal dimension. The phase reversal in channels 2 and 3 indicates that electrode R is at the focus.

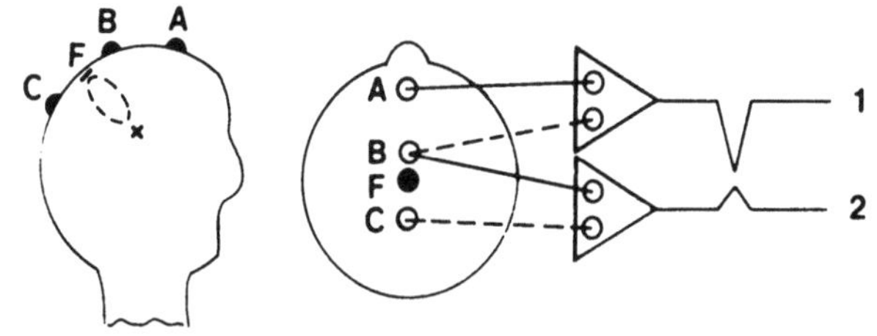

Figure 12.9. Another special case of rule 5. The focus, F, is between electrodes B and C, but is closer to one than the other. Channels 1 and 2 show an instrumental phase reversal of unequal amplitude, which indicates that the focus is between electrodes B and C, and nearer to B than C. What happens when F is closer to C than B?

Another special case of rule 5 occurs when F is between two electrodes, but nearer to one than the other. In such a case a voltage will be recorded between these two electrodes, but the voltage will be less than it is when one of the electrodes is directly over F. The rule is illustrated by the comparative amplitudes in channels 1 and 2 of Fig. 12.9. In this figure an instrumental phase reversal is observed between channels 1 and 2. This happens because grid 2 of channel 1 and grid 1 of channel 2 are connected together at B, a common lead. The phase reversal indicates that electrode B is nearest to the focus. Since the voltage between B and C is less than the voltage between A and B, F must be between B and C but closer to B than C.

All these rules are derived from the fact that around the point of maximum potential that is the focus, there are concentric isopotential lines (equipotential contours in three-dimensional space). Each successive line represents a uniform decrement in potential, and every point on a line has the same potential as every other point on the same line. The distance between successive concentric circles becomes greater as the distance from the focus increases, which indicates that the rate at which field strength is decreasing diminishes as one moves farther away from the focus. The magnitude of the potential recorded between any pair of electrodes in the field surrounding a focus will depend on the strength of the field at one electrode relative to the strength of the field at the other. The more isopotential lines between two electrodes, the greater the potential difference between them; two electrodes anywhere on the same isopotential line will have no potential difference between them.

Cancellation, Summation, and the Determination of Polarity

Cancellation and summation are phenomena that occur as the result of using differential amplifiers for recording EEGs. The importance of these phemonena cannot be overemphasized, as a thorough understanding of their effects is essential in order to correctly interpret an EEG record.

Cancellation occurs when input voltages of the same polarity are connected to grids 1 and 2 of a differential amplifier. Cancellation may be either complete or partial. We already encountered a case of complete cancellation in the last section. When a focus of electrical activity is equidistant from the electrodes connected to the inputs of a differential amplifier, the voltages at grid 1 and grid 2 are identical, and the channel records an output of zero volts. We explain this outcome by saying that the voltages at the two inputs, being identical, have cancelled each other out. The reader should recognize that any number of different positive and negative voltages as well as zero volts at both inputs can yield the same result, namely, an output equal to zero volts. This observation highlights the fact that a differential amplifier measures voltage differences not absolute voltages.

Partial cancellation occurs when the voltages connected to both grids have the same polarity but are of different magnitude. A few examples will show what happens when different combinations of voltages of the same polarity are connected to the two inputs of a differential amplifier. The effects are summarized in Table 12.1.

Table 12.1 illustrates three important points that readers should verify for themselves by reference to the various entries. In the first place, note that the magnitude and polarity of the output voltages are determined by algebraically subtracting the input voltage on grid 2 from the input voltage on grid 1. Secondly, the output voltage is always smaller than either input voltage, an outcome that is predicted by the term cancellation. Lastly, it is impossible to work backward and deduce either the magnitude or polarity of the input voltages from the magnitude and polarity of the output voltage. This will be apparent by noting in Table 12.1 that an output of $-5\ \mu V$ can be generated

Table 12.1. Some Examples of Cancellation in a Differential Amplifier[a]

Input Voltages (μV)		Output	
Grid 1	Grid 2	Voltage (μV)	Polarity
−15	−10	5	−
−20	−30	10	+
+10	+ 6	4	+
+50	+80	30	−
−25	−20	5	−
+20	+16	4	+
+25	+30	5	−

[a] To simplify the table, output values shown assume that the amplifier has a gain or amplification factor of exactly 1. This makes the amplifier a buffer amplifier.

Table 12.2. Some Examples of Summation in a Differential Amplifier[a]

Input Voltages (μV)		Output	
Grid 1	Grid 2	Voltage (μV)	Polarity
− 20	+ 30	50	−
+ 100	− 75	175	+
− 30	+ 20	50	−
+ 25	− 150	175	+
− 3	+ 2	5	−
− 2	+ 3	5	−

[a] To simplify the table, output values shown assume that the amplifier has a gain or amplification factor of exactly 1. This makes the amplifier a buffer amplifier.

by grid 1 and grid 2 voltages of − 15 and − 10 μV, − 25 and − 20 μV, and + 25 and + 30 μV. These, of course, are only three of an infinite number of different combinations of grid 1 and 2 voltages that can produce an output of − 5 μV.

Summation occurs when input voltages of opposite polarities are connected to grids 1 and 2 of a differential amplifier. The examples in Table 12.2 show what happens when different combinations of voltages having different polarities are connected to the two inputs of a differential amplifier. Note that, as was the case with cancellation, the magnitude and polarity of the output voltages are determined by algebraically subtracting the input voltage on grid 2 from the input voltage on grid 1. Also, observe that the output voltage is always larger than either input voltage—an effect suggested by and in harmony with the term summation. Table 12.2 also shows that it is impossible to infer magnitude of input voltages from magnitude of the output voltage. Thus, for example, note in Table 12.2 that a 50-μV output arises from grid 1 and 2 inputs of − 20 and + 30 μV or − 30 and + 20 μV, and an output of 175 μV from + 100 and − 75 μV or + 25 and − 150 μV.

Because the polarity of every output voltage shown in Table 12.2 is the same as polarity of the corresponding input voltage to grid 1, it appears that output polarity might serve as a clue to the absolute polarity of the input voltages. Unfortunately, this does not happen in practice, the reason being that we can never be absolutely certain that a particular event in an EEG tracing is the result of summation. For example, although in Table 12.2 an output voltage of − 5 μV is present only when there is a negative voltage on grid 1 and grids 1 and 2 have voltages that sum to 5 μV, this outcome is not unique to summation. Thus, as Table 12.2 shows, an output of − 5 μV can be produced by a variety of different positive as well as negative voltages connected to grid 1.

In concluding this section, the reader should be clearly aware of two essential facts concerning EEG recordings. We reiterate them here at the risk of seeming repetitious because they are so important. Firstly, it should be recognized that the voltage displayed at any point in time in an EEG tracing represents the *difference* between the voltages present at the two electrodes to which the differential amplifier is connected. Furthermore, it is important to understand that any polarities assigned to various electrical events observed in an EEG tracing are not absolute polarities but are only relative polarities. Thus, we cannot say that a particular deflection in the tracing results from the voltage at a particular electrode being negative; we can only say that it is negative *with respect to* the voltage at the other electrode in the circuit—that the voltage is *relatively* negative. Fortunately, it happens that from the neurophysiological standpoint, most focal activity such as spikes is likely to be negative. So when a spike is seen in a clinical EEG, the best guess is that it is electrically negative. But the EEG tracing does not tell us that, even though the tracing happens to be compatible with this conclusion. It only tells us that the focus is negative with respect to the area surrounding it. Information concerning absolute polarity can be gained only from other types of recordings using different kinds of methods.

Phase Reversal

A phase reversal can be either of two different types, *instrumental* phase reversal or *true* phase reversal. Instrumental phase reversal is by far the most common type encountered in clinical EEG. As mentioned earlier in this chapter, it occurs when grid 2 of one channel and grid 1 of a second channel are both connected to a single electrode situated over a focus of activity. As seen in Fig. 12.6, opposing deflections occur in these two channels because the identical voltage goes to opposing inputs. The term instrumental, therefore, simply reflects the fact that the phenomenon is the result of the particular way in which electrodes are connected to the EEG machine. Instrumental phase reversals are the key to localization when employing a bipolar recording system—a recording system in which adjacent channels are connected in a chain. Recording-system details have already been discussed in the previous chapter.

A true phase reversal occurs when the axis of a dipole that is the source of electrical activity recorded from electrodes on the scalp is not perpendicular to the scalp. This would happen if, for example, the dipole were located in a sulcus of the cerebral cortex, which could place its axis almost parallel to the scalp. The result of such a condition is analogous to the situation and outcome shown in Fig. 12.2. Note that a phase reversal occurs in the plot of the voltage measured at the surface. This is a true phase reversal since the voltages are all measured with respect to the same distant electrode.

Thus far in this chapter, we have discussed phase reversals as if they occurred only in response to the presence of a brief electrical event like a spike generated in a limited cortical region. This is clearly not the case. A phase reversal is not synonymous with the presence of a spike; nor is it necessarily indicative of an abnormality in the EEG. An instrumental phase reversal may occur whenever the same electrode is connected to opposing inputs of two adjacent channels. Thus, for example, the alpha rhythm recorded in T5-O1 can show some phase reversals with the alpha rhythm recorded in O1-O2, but this has no clinical significance.

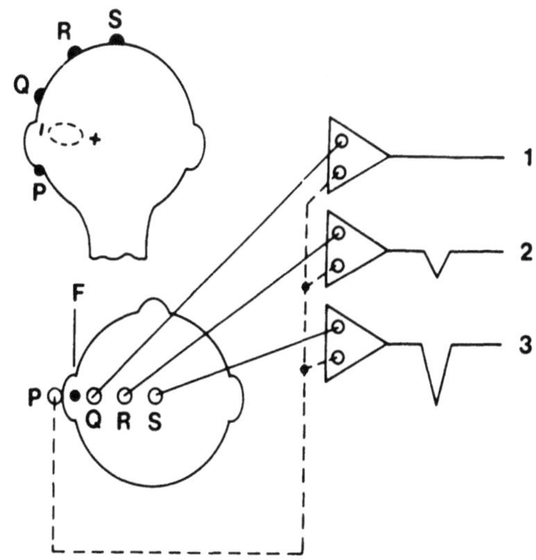

Figure 12.10. Contamination of a referential recording employing an ear reference electrode by an "active ear." F, the focus, is equidistant from electrodes P and Q. Channels 2 and 3 are contaminated with activity picked up from F by P, the reference electrode.

Localization in Referential Recording

In the last section we mentioned the case in which the voltages present at various points along a surface are all measured with respect or reference to a single electrode. This condition is referred to as referential recording, and the method has already been discussed in some detail in the previous chapter. Our purpose here is simply to show how referential recording is used in localization.

In referential recording, a different scalp electrode is connected to grid 1 of each differential amplifier, while a common electrode goes to grid 2 of each amplifier. This common or reference electrode may be either a single electrode, as in the case of an ear reference, or a composite of numerous electrodes, as in the case of an average reference. The basic convention discussed earlier still holds, namely, that a channel will deflect upward when grid 1 is negative with respect to grid 2, and downward when grid 1 is positive with respect to grid 2. This means that a phase reversal observed in a line of referentially connected electrodes is a true phase reversal. Indeed, a true phase reversal is best detected and localized using referential recording. The other clue to localization in referential recording is amplitude. In referential recording, the larger the deflection associated with a particular focus of activity, the closer the electrode is to the focus. As interelectrode distances are the same for homologous left and right derivations in referential recording, this is the optimal system to use when making amplitude comparisons between the two hemispheres.

When an ear electrode is employed as a reference, ear contamination can present a major problem. This happens when a so-called "active ear" is present. As mentioned briefly in Chapter 11, the contamination results from significant activity occurring in the adjacent temporal region being picked up by the ear reference electrode. An example best describes what happens.

In Fig. 12.10, the focus F is midway between electrode P, the electrode on the earlobe, and electrode Q. Following rule 5, channel 1 shows no deflection because P and Q are located in an equipotential zone. On the other hand, channel 2 shows a deflection, there being a potential difference present between R and P and likewise in the case of channel 3. The deflection is of larger amplitude in channel 3 than in channel 2 because the distance between S and P is greater than the distance between R and P (an application of rule 1). Note that these are downward rather than upward deflections as in the previous examples. This is the result of the fact that grid 2 is negative with respect to grid 1, whereas grid 1 was negative with respect to grid 2 in the earlier examples.

The phenomenon just discussed is referred to as active ear *contamination* of the scalp electrodes because the outputs of channels 1, 2, and 3 show activity that suggests that a surface positive focus is present at electrode S. The fact that in EEG work foci are usually surface negative and only rarely surface positive argues against the latter possibility. That we, indeed, are dealing with a surface negative focus between electrodes P and Q is readily confirmed by connecting P, Q, R, and S in a bipolar chain as in Fig. 12.7(b) and observing the outputs as shown. If the focus were surface positive and located at S instead of surface negative and located at F, as in Fig. 12.7(b), the channel 1, 2, and 3 outputs would be different. We leave it to the reader to determine what these outputs would be like.

Commonly Seen Localizing Patterns

Some of the most commonly occurring localizing patterns are shown in Fig. 12.11. The five examples given illustrate

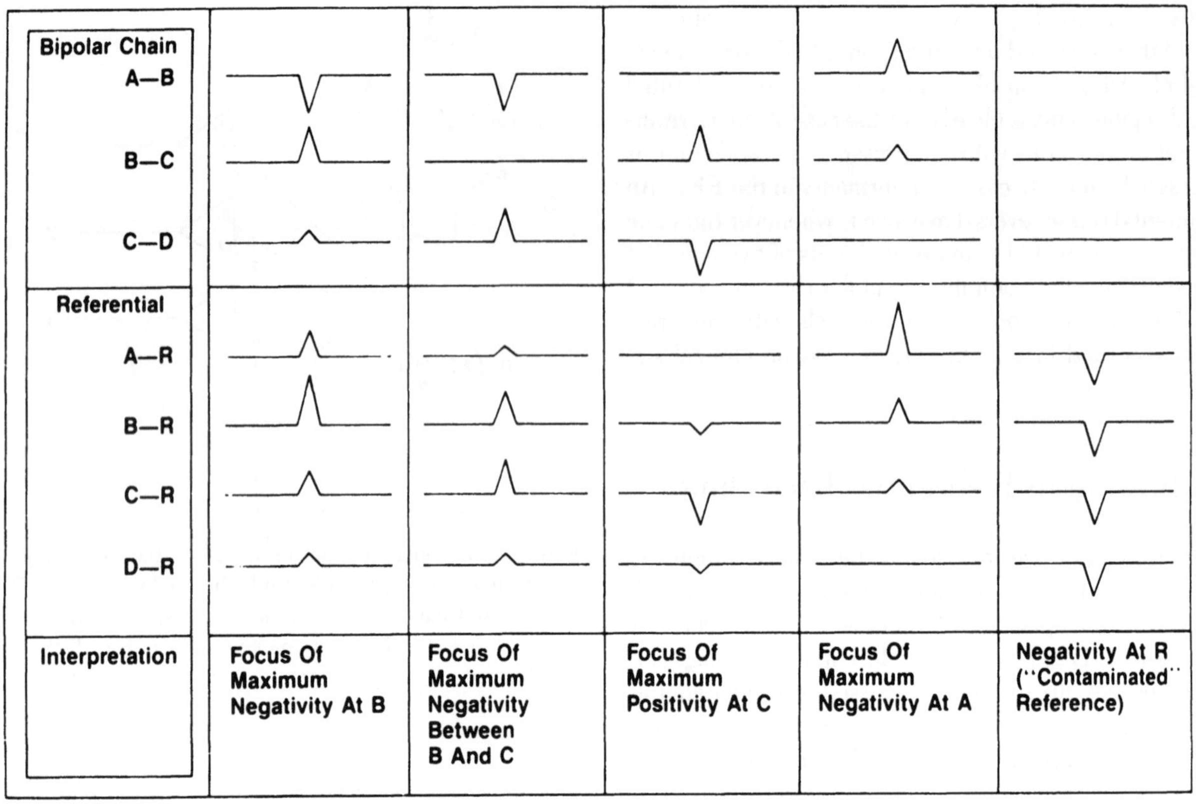

Figure 12.11. Some commonly seen localizing patterns. Simultaneous bipolar and referential recordings. A, B, C, and D represent electrodes placed over F7, T3, T5, and O1. R in referential recording is on the contralateral ear.

the configurations seen in referential as well as in bipolar recording. In each case the interpretation or location of the focus is given at the right. With machines having 18 or more channels, such simultaneous referential and bipolar recording has become practicable and localization of foci of spike activity can be made more quickly and accurately. Note that in confirming by referential recording the location of a focus observed in bipolar recording, or vice versa, we are able to verify its existence.

The last example in Fig. 12.11 shows the localizing patterns when the contralateral ear, which serves as the reference, is contaminated. A different localizing pattern will be seen when the ipsilateral ear is contaminated instead and serves as reference. The localizing patterns for these two different conditions are shown side by side in Fig. 12.12 for comparison. Following conventional EEG practice, the reference electrode is connected to G2. This means that contamination (negativity) at R results in a downward deflection of the tracings, which is a valuable localizing sign.

It is interesting to note that the bipolar localizing patterns for negativity at the ipsilateral ear and for negativity at B (Fig. 12.11) are identical. The difference between the conditions, however, is picked up by the referential recording, a fact that highlights the importance of using both bipolar and referential recording.

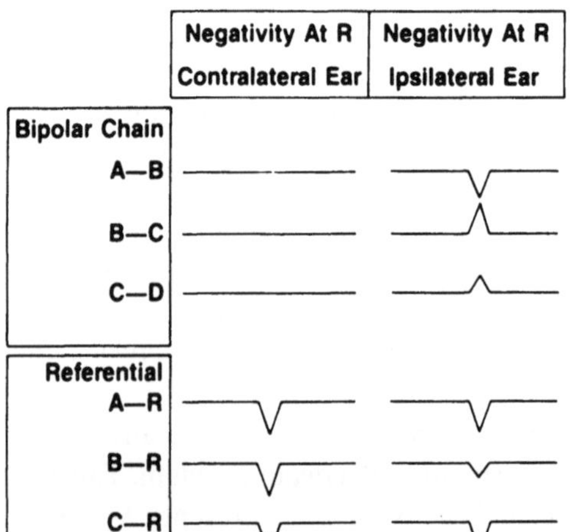

Figure 12.12. Contamination (negativity) at contralateral and ipsilateral ears and the localizing patterns seen in simultaneous bipolar and referential recording. A, B, C, and D represent electrodes placed over F7, T3, T5, and O1.

Contaminated Average Potential Reference

As explained in Chapter 11, the average potential reference (also referred to as averaged common reference) is an attempt to obtain a truly indifferent electrode, that is, a point of zero potential against which the voltages at other electrodes may be compared. In this technique, G2 of each channel is connected to a common point whose voltage is an average of the voltages from all the electrodes. We noted earlier in Chapter 11 that the Fp_1, Fp_2, and Cz electrodes are excluded from the average, and that F7, F8, Fz, and Pz are often also excluded.

The main difficulty with this montage is that a high amplitude potential, as for example a large spike in one electrode, can cause a surge in voltage at G2, which leads to a confusing localizing pattern in many channels. Figure 12.13 shows the effect of having a large spike −200 μV in amplitude present at T5 with spikes of −20 μV at T3 and O1. As there are ten electrodes in the averaged common reference, $^1/_{10}[(-200) + (-20) + (-20)]$ or −24 μV may be expected at G2 of each channel. This leads to a deflection of −176 μV at T5, and virtually no deflection (+4 μV) at T3 and O1. The other channels show a spurious downward deflection due to G2 being 24 μV negative compared with G1. To remedy this situation, the technician has to identify the electrode picking up the high amplitude potential (easily identified because it shows the largest deflection) and remove it from the average. In the present example, once T5 is excluded from the average, the negativity in G2 is reduced to virtually zero so that the downward deflections disappear and T3, T5, and O1 show their actual voltage levels of −20 μV, −200 μV, −20 μV, respectively.

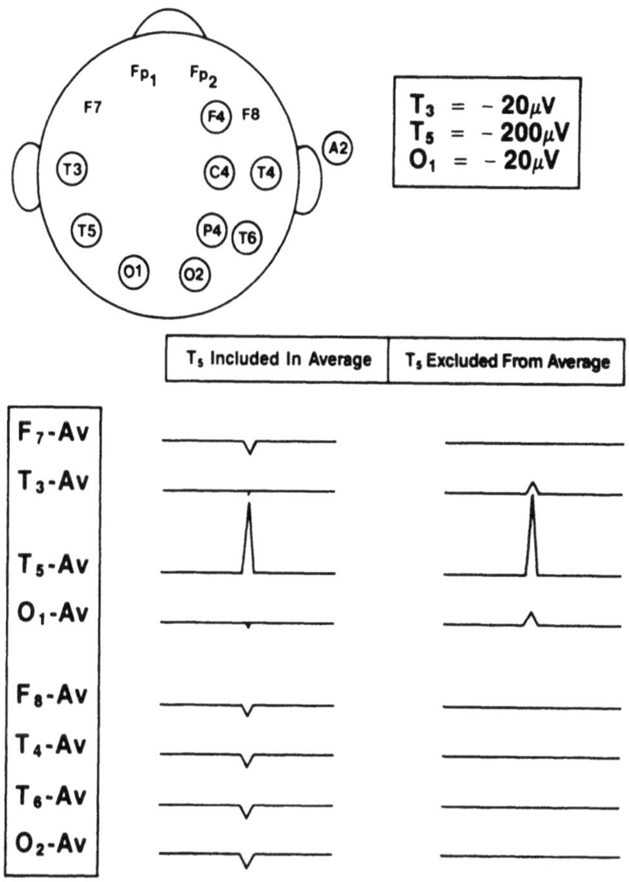

Figure 12.13. Contamination of average potential reference (Av) by spike focus at T5. Circled derivations designate electrodes in the average. Tracings show recordings from electrodes in the temporal chains; all referred to the average potential. The spurious downward deflections disappear when the T5 electrode is removed from the average.

Reference

Helmholtz H: Über einige Gesetze der Vertheilung elektrischer Ströme in körperlichen Leitern mit Anwendung auf die thierischelektrischen Versuche. *Annalen der Physik* 1853; 89:211–233.

Chapter 13
Introduction to EEG Reading

The first encounter with an EEG tracing is somewhat perplexing to the physician and the technician alike as they are likely to be awestruck by the complexity of the record. Often, the initial impression is that the task is too complex to learn. The physician familiar with reading ECGs soon realizes that there is little in common between the analysis of the repetitive complexes of the ECG and the ever changing waveforms of the EEG. At this stage, the prospective electroencephalographer is tempted to exclaim: "This is all Greek and Latin to me." Indeed, learning to read EEGs is not unlike learning to read a foreign language.

Reading EEGs—An Analogy

To read a new language, needless to say, one needs first to learn the alphabet. The alphabet of the EEG consists of the various frequencies and waveforms that comprise the tracing. Just as the letters of the alphabet are combined in different permutations and combinations to form words and then sentences, so the EEG tracings are made up of combinations of waveforms of different frequencies and morphology. To carry the analogy further, it is not enough to be able just to read the words and sentences; one needs to understand quickly the meaning of what is written. In the same way, EEG reading involves analyzing the waveforms and deducing their significance. With experience, one uses a speed reading technique in which a whole page is rapidly scanned for evidences of normal and abnormal phenomena. How successfully this is done depends to a large extent on developing pattern-recognition skills.

Learning to Read

How does one learn to read EEGs? Like any other branch of medicine this involves a continuous process of learning for many months or sometimes even years. Often the initial learning is accomplished through observing an experienced electroencephalographer read EEGs. The next step involves reading under supervision; having seen how an experienced electroencephalographer interprets a record, and having gathered essential information regarding normal and abnormal patterns, the trainee interprets records in the presence of his or her instructor. Ideally, the instructor should regularly quiz the trainee on the various waveforms and artifacts in the tracings, and the trainee should complement this by seeking answers to the questions.

Without at least an elementary knowledge of the basic principles of electricity, neurophysiology, and the technique of recording, it is impossible to learn to read EEGs properly. One needs to know what calibration means, how the various frequency filters work, how various artifacts are identified, and how neurologic disorders produce alterations in electrical activity of the brain. The prospective electroencephalographer also needs to have a thorough working knowledge of the various montages used. All these topics are taken up in considerable detail in this text.

Numerous EEGs need to be seen before one becomes familiar with the wide range of normal variations in different age groups and physiological states. The task becomes even more difficult when tracings of neonates and premature infants have to be interpreted. It may take two or three years of experience in reading before one has acquired reasonable expertise.

Terminology

It is essential to use standard terminology in describing the EEG. The International Federation of Societies for Electroencephalography and Clinical Neurophysiology has proposed definitions for the various terms used in EEG to facilitate communication between different electroencephalographers. In this section, we list the definitions of some of the terms that are most commonly used in EEG reading.[1]

Background Activity

This term denotes the general setting in which changes in frequency, amplitude, or morphology appear. Although the alpha rhythm may be the background activity in the tracings from the posterior regions, it is important to note that the term background activity is not synonymous with alpha rhythm; thus, over the frontal area, the activity may be mostly in the beta frequency band. The background activity may not always be a normal pattern; the term can also refer to abnormal patterns.

Both the background activity and the changes that appear in the features of the tracing are described in terms of frequency, amplitude, wave shape, symmetry, synchrony, location, continuity, and reactivity. It is important to understand the meaning of each of these terms to give a proper description of the EEG.

Frequency

This term refers to the rate at which a particular waveform repeats; it is usually used in the context of rhythmic activity (repeating with regularity). Depending on the frequency, the activity is classified as delta (less than 4 Hz), theta (4 to 8 Hz), alpha (8 to 13 Hz), or beta (more than 13 Hz) activity. Although these terms are ideally restricted to rhythmic activity, they are also often used to describe nonrhythmic or random activity; in this case the frequency of a particular wave will be ascertained by taking the inverse of its duration.

The frequency bands are used to describe the activity irrespective of where it occurs. But the term alpha rhythm is more specifically used to denote the 8 to 13 Hz rhythm occurring during wakefulness over the posterior region of the head; it occurs generally with maximum voltage over the occipital area, is best seen with the patient's eyes closed and under conditions of physical relaxation and relative mental inactivity, and is blocked or attenuated by attention, especially visual attention and mental effort. Sometimes the terms fast and slow activity are used to denote a dominant frequency above or below the alpha band. The term *monorhythmic* is often used when the particular activity shows rhythmic components of a single frequency. When there are multiple frequencies the term *polyrhythmic* is used. The term *periodic* applies to EEG waves or complexes recurring at approximately regular intervals; usually the intervals vary from one to several seconds.

Amplitude

This is expressed in terms of voltage in microvolts based on a peak-to-peak measurement. One needs to know the sensitivity at which a recording was made to determine this. For this purpose, the reader consults the routine calibration at the beginning or end of the record. The amplitude will vary depending on the technique of recording, the bipolar montages with short interelectrode distances giving a smaller amplitude than the referential montages with larger interelectrode distances. Ideally, amplitude should be described in terms of the actual voltage; however, the terms low, medium, and high amplitude are often used. The term low is used when the amplitude is under 20 μV, medium when it falls in the range of 20–50 μV, and high for more than 50 μV. The use of these terms is discouraged owing to lack of uniform criteria.

Attenuation and *blocking* are terms used when there is a reduction in the amplitude of EEG activity, usually in response to some stimulus. The classic example is attenuation of the alpha rhythm in response to eye opening. The term *suppression* is used when little or no electrocerebral activity can be discerned in a tracing. *Paroxysmal activity* is a term denoting activity of much higher amplitude than the background that occurs with sudden onset and offset. It need not necessarily denote an abnormal activity.

Wave Shape or Morphology

Electroencephalographic activity is essentially a mixture of waves of multiple frequencies. The appearance of the waveforms depends on the component frequencies, their relative voltages and phase relationships, and, of course, upon the frequency filters used. The waveforms are also continuously fluctuating in response to stimuli and depend on the state of the patient. Several descriptive terms may be used in this context.

A *transient* is an isolated wave that stands out from the background activity; if it has a sharply pointed peak and the duration is less than 70 ms (less than 2 mm at the paper speed of 30 mm/s) it is called a *spike*; when the duration is between 70 to 200 ms, it is called a *sharp wave*. The term *complex* is used when two or more waves occur together and repeat at consistent intervals; examples are spike and wave complexes and sharp and slow-wave complexes. An activity is described as *monomorphic* when the morphology of subsequent waveforms is similar whereas the term *polymorphic* is used when they are of dissimilar morphology. The description should also include the number of phases. Thus, a wave may be monophasic (positive or negative) or diphasic (positive and negative), triphasic or polyphasic.

[1] For a listing of commonly used terms and their definitions, see the glossary in Appendix 1. The reader may also refer to Chatrian GE, Bergamini L, Dondey M, et al: A glossary of terms most commonly used by clinical electroencephalographers. *Electroencephalogr Clin Neurophysiol* 1974; 37:538–553.

Symmetry

In general, symmetry refers to the occurrence of approximately equal amplitude, frequency, and form of EEG activities over homologous areas on opposite sides of the head.

Synchrony

This term refers to the simultaneous appearance of morphologically identical waveforms in areas on the same side or opposite sides of the head.

Location

Several different terms are used. *Focal* or *localized* are terms used when a particular activity is confined to one particular region of the head. For example, an activity may be localized to frontal, temporal, parietal, or occipital areas. The term *generalized* is used when activity is not limited to one region but occurs over a wide area. An activity is said to be *lateralized* when it is present on one side only.

Continuity

An activity may be described as *continuous* or *intermittent*, depending on the percentage of time it is present. Thus, an activity is called continuous when it occurs without interruption for prolonged periods of time and discontinuous or intermittent when it appears only from time to time.

Reactivity

The term refers to alterations in the amplitude and waveform of activity in response to a stimulus. An example is the attenuation of alpha activity on eye opening.

There are other terms that are used in various special circumstances. These will be taken up in relevant chapters.

Describing the EEG

An adequate and accurate description of the EEG record is important for several reasons. When a clinician would like to compare the EEGs from different laboratories, or when a recent EEG needs to be compared with the findings of an old record that is no longer available, the written description of the records is essential. It may be said that if the description is good, the electroencephalographer can picture the actual EEG record in his mind's eye. To make the description as objective and as accurate as possible, it is important to break down the complex tracing in terms of frequency, voltage, reactivity, synchrony, and distribution.

The various activities occurring in different states of consciousness, namely, wakefulness, drowsiness, and sleep, should be described clearly. One may start with a description of the background activity, which often tends to be the alpha rhythm in the awake subject. Mention should be made about its frequency, amplitude, location, symmetry, and reactivity to eye opening. Next, the features of other rhythmic activities present should be described in similar terms; for example, beta activity in the frontal or central areas.

Intermittent activity should be described in similar terms, including the location and synchrony. The presence of various phenomena such as V (vertex) waves and sleep spindles should be clearly described. If sharp waves, spikes, or other intermittent activity is present, it should be described in terms of location, polarity, and amplitude; how the activity is affected by changes in state should also be noted.

The effect of activation procedures such as hyperventilation and photic stimulation should be described. In the case of hyperventilation, the way in which this procedure influences the background activity and whether it induces other changes should be mentioned. For photic stimulation, the reader should mention whether there is a driving response and, if so, whether it is symmetrical and at what frequency or frequencies it occurs. If there are any specific responses like photoparoxysmal or photomyogenic responses, they should all be called attention to in the description.

Interpreting the EEG

The basic question to ask after completing a visual analysis of the EEG is whether the findings are consistent with the accepted norms for the age and state of the patient. This means that the reader should have a thorough knowledge of the normal variations of EEG patterns in relation to age and state of the patient. Such a judgment is possible only after the reader has seen numerous EEGs and has formed an impression in his or her own mind about normal patterns.

If the EEG is normal, the interpretation ends with a statement to that effect. If an abnormality is present, the next phase of interpretation involves categorization of the abnormality in more specific terms. Categorization should be precise. Some examples are: focal epileptiform abnormality, generalized epileptiform abnormality, focal slowing, generalized slowing, intermittent rhythmic delta activity, polymorphic delta activity, asymmetric alpha rhythm, asymmetric photic driving, asymmetry of sleep spindles or vertex waves. If an abnormality is found to be localized, one needs to specify what area of the brain underlies the abnormality.

The next step in the interpretation is to suggest what kind of changes may be happening in the brain that could account for, or be compatible with, the abnormal EEG pattern. This entails a clear knowledge of the relationship between the various EEG abnormalities and the various disorders that affect the brain. One of the major problems in this aspect of interpretation is that many different types of disorders affecting the brain can give rise to the same type of EEG abnormality so that very often only general comments can be made. These comments may be like the following: "This pattern is suggestive of a diffuse encephalopathy," or "the finding is compatible with a focal structural lesion," or "this finding is suggestive of a focal seizure disorder," or "the abnormality is compatible with a seizure disorder of the generalized type," etc. Following the technical interpretation, it is always useful to provide a clinical correlation on the basis of the patient's clinical history. Often it may be of value to say whether the EEG abnormality seen is consistent with the clinical diagnosis. Table 13.1 gives a brief synopsis of EEG reading.

Table 13.1. EEG Reading—A Synopsis

Steps in EEG Reading	Skill/Knowledge Required
Visually scan the EEG and describe the waveforms in terms of frequency, amplitude, morphology, polarity, symmetry, synchrony, and reactivity.	Familiarity with various EEG waveforms, artifacts, alterations in waveforms at different filter settings, montages, sensitivities and paper speeds.
Determine whether the EEG patterns are compatible with normal patterns for the age and state of the patient.	Thorough knowledge of the normal features of the EEG in various age groups and physiologic states; familiarity with the variations that are accepted to be within the normal range.
If the EEG is abnormal, determine the most likely cause using the type of abnormality, its distribution, and other characteristics as a basis. Look at the available clinical data and decide whether the EEG pattern is or is not consistent with the suspected condition.	Knowledge of the various EEG patterns that accompany different neurological disorders. Familiarity with the diagnostic and prognostic significances of various abnormal patterns.

More on Artifacts—Physiological Artifacts

In general, four different varieties of artifacts are encountered in EEG work: environmental, instrumental, electrode, and physiological. The EEG technician and the physican reading EEGs both need to be familiar with all of them. Needless to say, artifacts in the EEG should be eliminated whenever possible or kept to a minimum. To achieve this goal requires close collaboration between the technician and the physician reading the EEGs.

Environmental, instrumental, and electrode artifacts have been discussed in various earlier chapters, principally the chapters on recording electrodes (Chapter 7) and troubleshooting (Chapter 9). In this section we consider the topic of physiological artifacts.

There are four major sources of physiological artifacts, namely, the heart, the muscles of the head and neck, the eyes, and the skin. The person reading the EEG must be able to recognize these artifacts if they occur in the recording. He or she needs to learn to "read through" the artifacts whenever possible. In this respect, the reader acts like a filter in much the same way as the frequency filters function on the EEG machine. We will take up each of these artifacts in turn. Appendix 6 shows some EEG tracings containing these artifacts.

ECG Artifacts

Because of their regularity and distinctive morphology, ECG artifacts are the easiest to recognize in the EEG tracings. Any uncertainty about whether or not an artifact is an ECG is easily resolved by actually recording an ECG on one channel along with the EEGs. This, of course, needs to be done at the time that the routine EEG is taken. Obviously, a close, harmonious working relationship between the EEG technician and the physician reading the EEGs is essential in resolving such problems.

EMG Artifacts

These are the most common artifacts seen in EEG recordings. Muscle spikes originate mostly from the muscles of the head: the frontalis muscle, the masseters, sternomastoids, and temporal muscles are common sources. Electromyographic artifacts are recognized mainly by their distinctive morphology; the spikes themselves are very sharp and of short duration. In this regard, however, particular attention needs to be given to the settings of the high-frequency filters on the EEG machine. Settings lower than 70 Hz will result in the spikes being rounded off so that they may easily be confused with brain electrical activity in the beta band.

As is the case with ECG artifacts, the EEG technician plays an essential role in the detection of EMG artifacts. While taking the EEG, he/she closely observes the patient, notes various movements, and records their occurrence directly on the EEG record. These notations are especially helpful in recognizing EMG artifacts when the tracings are read and interpreted. Moreover, the technician's efforts in establishing rapport with the patient, and in helping him or her to relax, go a long way toward reducing many EMG artifacts.

Eye-Movement Artifacts

Electrically, the eyes behave very much like batteries rotating in their sockets. This means that electrodes located in the anterior regions of the head can very readily pick up changes in voltage that are correlated with eye movements. The artifacts are quite distinctive and are easily recognized after some experience. To assist recognition, it is helpful in the course of training to record samples of eye movements —up, down, right, and left—using some of the common montages. In this way, technician and reader alike can become more familiar with the spatial distribution of these artifacts.

If there is any question concerning whether or not activity recorded in the EEG is due to eye movement, an electrode should be attached to the patient's cheek directly below the eye. This electrode is connected to grid 1 of one channel; grid 2 of the same channel is connected to the ipsilateral earlobe. The recording from this derivation is compared with the recording from an anteriorly placed derivation in a longitudinal bipolar chain, e.g., Fp_1-F7 or Fp_2-F8. Since the eye falls between these two derivations, eye movements will appear in the two channels as out-of-phase deflections or mirror-image signals. If the deflections in the two channels are in phase, they are not eye movements but may be of cerebral origin.

Galvanic-Skin Artifacts

Like all living tissue, the skin is electrically active. Changes in electrical activity of the skin are usually associated with sweating, although some changes in voltage may be observed between two points on the skin in response to stimulation (the so-called Tarchanoff effect). Generally speaking, changes in skin potential associated with sweating or stimulation are very slow changes. They occur mostly at the very low end of the delta frequency band, making them easy to recognize in the EEG tracings. Sometimes these artifacts can become quite large. When this happens, they usually can be reduced considerably by adjusting the low-frequency filters from the standard 1.0-Hz setting to the 5.0-Hz cutoff point.

Writing the EEG Report

A major complaint sometimes made about the EEG report by some referring physicians (other than neurologists) is that it makes little sense and often does not help in the diagnosis or management of their patients. To some, the report may even seem misleading. For this reason it is important that the report be constructed in two parts: one part deals with the actual description of the EEG findings and their interpretation; the other part contains a clinical correlation that renders the report meaningful to the referring physician.

It is important to begin the report with a brief history and the clinical findings to date (usually available from the physician requesting the EEG and/or from the technologist's worksheet). It is also helpful to mention what the referring physician hopes to find out from the EEG, if it is explicit. The next paragraph should provide descriptive details regarding the testing situation. These should include whether the test was done at the bedside or in the intensive care unit and whether any modifications were made in the electrode connections, as, for example, using a reduced array in a neonate. The use of special electrodes like nasopharyngeal or sphenoidal leads should be noted here. Also mention whether the patient was sleep deprived, and whether any medication was given before or during the test and, if so, what kind and how much.

Next comes the section that describes the EEG and the state of the patient. This portion of the report should be purely descriptive and should not contain any interpretive statements such as normal or abnormal. Following this is the paragraph setting forth the impression. Is the EEG normal or abnormal, and if abnormal, what kind of abnormality was seen?

The last paragraph should attempt to correlate the EEG findings with the clinical picture. Thus, for example, in the case of a seizure disorder in which the EEG is normal, it should be mentioned that the EEG does not support the diagnosis of seizure disorder. But it also may be pointed out that a normal EEG does not necessarily rule out a seizure disorder. In this context, one may suggest further studies such as a sleep-deprived EEG or a repeat EEG using special electrodes. If the EEG in a patient suspected of having a metabolic encephalopathy shows diffuse slowing or frontal intermittent rhythmic delta activity (FIRDA), one may state that this finding is consistent with the clinical diagnosis.

Chapter 14
The Normal EEG

Since the late 1920s when Hans Berger first recorded the EEG in humans, an enormous amount of knowledge concerning the normal EEG has been accumulated. During these years, the EEGs of normal persons have been recorded in numerous situations and under a wide variety of conditions. Tracings have been taken during performance of a virtually endless list of different tasks and activities, as well as during different states of consciousness, from persons over the entire life span. Much of the data concerning the normal EEG need not concern us here. Thus, the clinical application of EEG is concerned mainly with the features of the tracing as they are seen in wakefulness under resting conditions[1] and in sleep. The EEG during specific, clinically significant activation procedures is discussed in Chapter 16.

We begin this chapter with a consideration of the major features of the EEG seen during resting wakefulness. After this, the normal EEG in sleep is taken up. As there are age-related differences present in some features of the normal EEG, separate sections dealing with the normal EEG during maturation and in old age are also included. The reader should understand that a thorough and comprehensive coverage of these topics would require an entire text. For this reason this chapter should be recognized as strictly an elementary introduction to the normal EEG.

Features of the Awake EEG in Adults

The most prominent feature of the normal waking EEG is the posterior dominant rhythm. It was the feature of brain electrical activity that was first described in 1929 by Berger, who named it alpha.

Alpha Rhythm

This is 8- to 13-Hz rhythmic activity that occurs most prominently in the posterior regions and is a conspicuous feature of the EEG in the awake, relaxed adult. In some persons it is so rhythmic—and the waves are so regular—that it looks like the output of a sine wave generator. The alpha rhythm of most adults ranges between 9.5 to 10.5 Hz. In the main, amplitudes are 50 μV or less, and amplitude tends to wax and wane over periods of one to several seconds.

As will be apparent from Fig. 14.1, which is an EEG of an awake, resting adult, the alpha rhythm is best seen when the person's eyes are closed. Opening the eyes results in an attenuation of the alpha rhythm. Reactivity to eye opening is typically used as evidence that the activity is indeed the alpha or posterior dominant rhythm. Mental effort or focusing one's attention can also attenuate the alpha rhythm. An EEG tracing may contain 8- to 13-Hz activity that does not attenuate with eye opening. Although such activity is classified as "alpha" by virtue of its frequency, it is not alpha rhythm and should be distinguished from it.

Occasionally, an increase in amplitude and abundance of alpha activity occurs with attention or eye opening instead of a decrease. This reversal is referred to as a "paradoxical effect" or "paradoxical alpha." It is seen mostly with eye opening in response to stimulation following a brief period of drowsiness (see Fig. 14.2).

The alpha rhythm is commonly of somewhat higher amplitude on the right than the left side. By contrast, frequency normally differs by less than 1 Hz between the two sides. In a small minority of normal individuals (less than 10%), no alpha rhythm is perceptible using conventional recording methods. The significance of this is unknown.

[1] The resting condition also stipulates that the person has not been fasting and that he/she has not consumed stimulants like coffee, tea, or soft drinks before the time the EEG is taken. It is the responsibility of the technician to inform the patient of this at the time the appointment is made for the EEG.

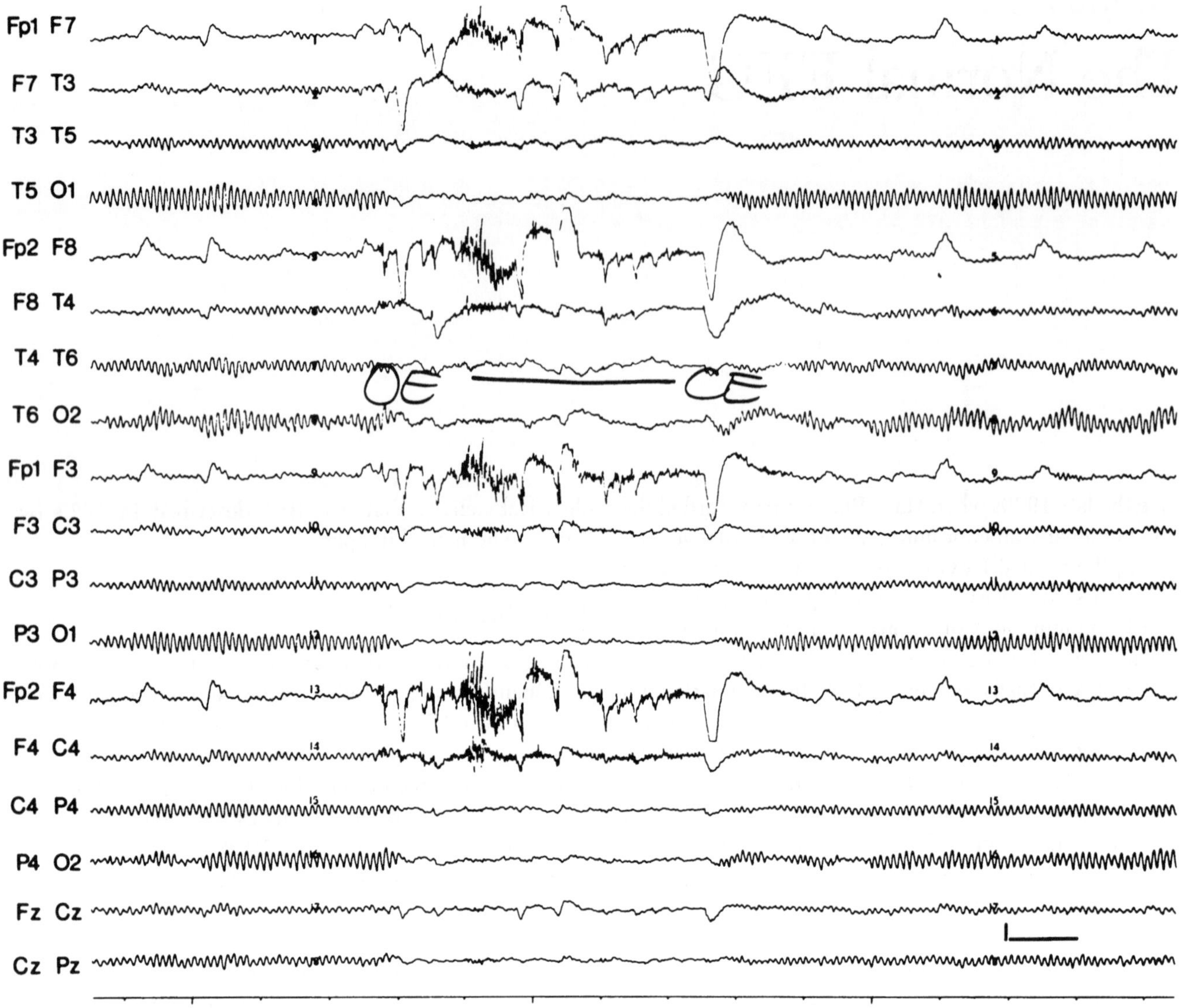

Figure 14.1. Alpha rhythm in an awake, relaxed, and resting young adult. The eyes are closed at the start of the tracing, are opened on command at OE, and closed again at CE. Note that the alpha rhythm is posteriorly dominant; it is markedly attenuated during eye opening but quickly returns to its original level once the eyes are closed. Note also the muscle and movement artifacts associated with eye opening and closing. Frequency of the alpha rhythm is approximately 11 Hz; little variation is seen from second to second. Filters: low frequency = 1 Hz, high frequency = 70 Hz. Calibrations: horizontal = 1 second, vertical = 50 μV.

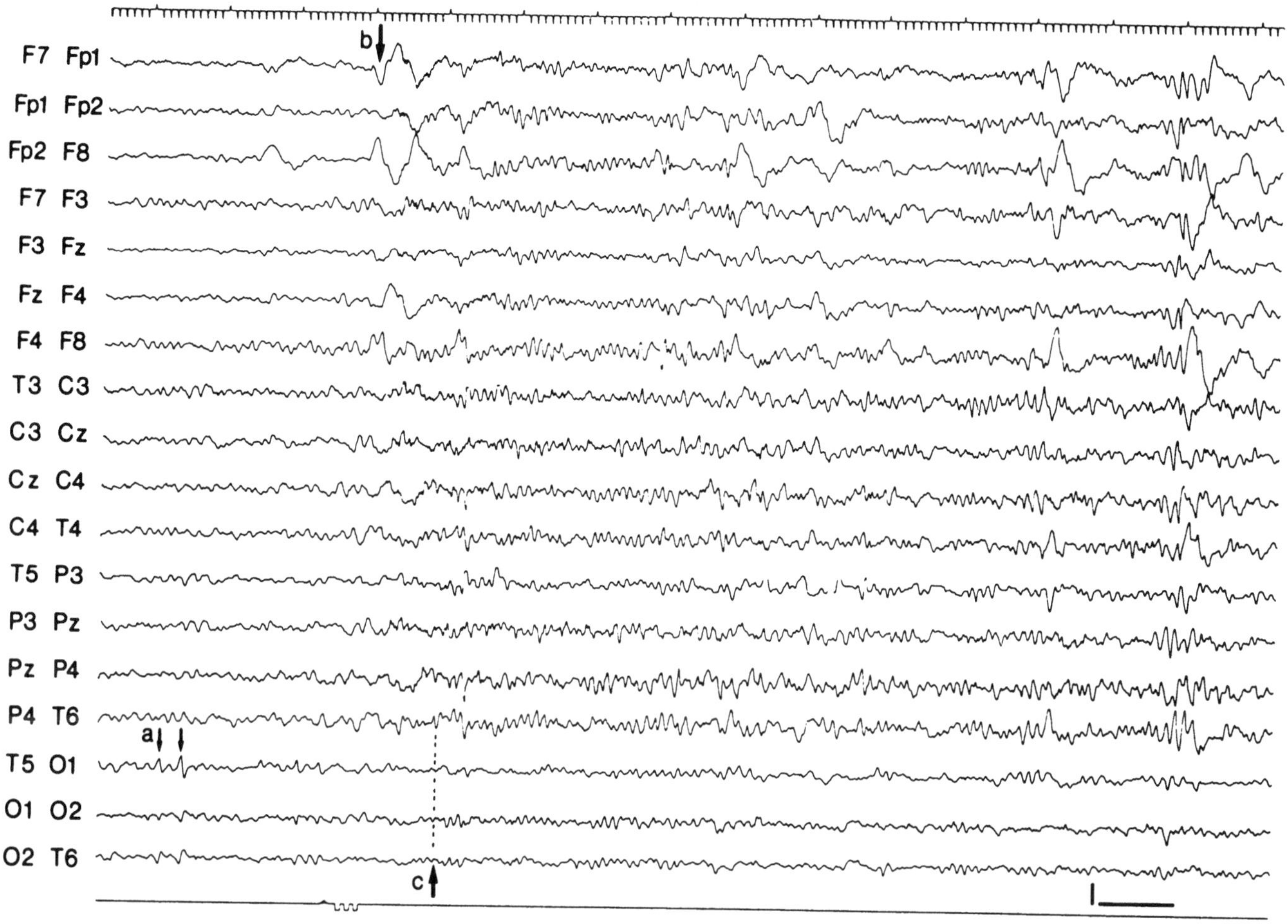

Figure 14.2. Paradoxical alpha, or alpha rhythm occurring in association with eye opening following drowsiness. The two arrows near the start of the tracing at "a" point to POSTS, which are a feature of light sleep. The arrow at "b" indicates artifacts associated with eye opening in response to a noise in the laboratory. Shortly thereafter (arrow at "c"), an alpha rhythm of about 10 Hz is seen in the posterior regions. Filters: low frequency = 1 Hz, high frequency = 70 Hz. Calibrations: horizontal = 1 second, vertical = 50 μV.

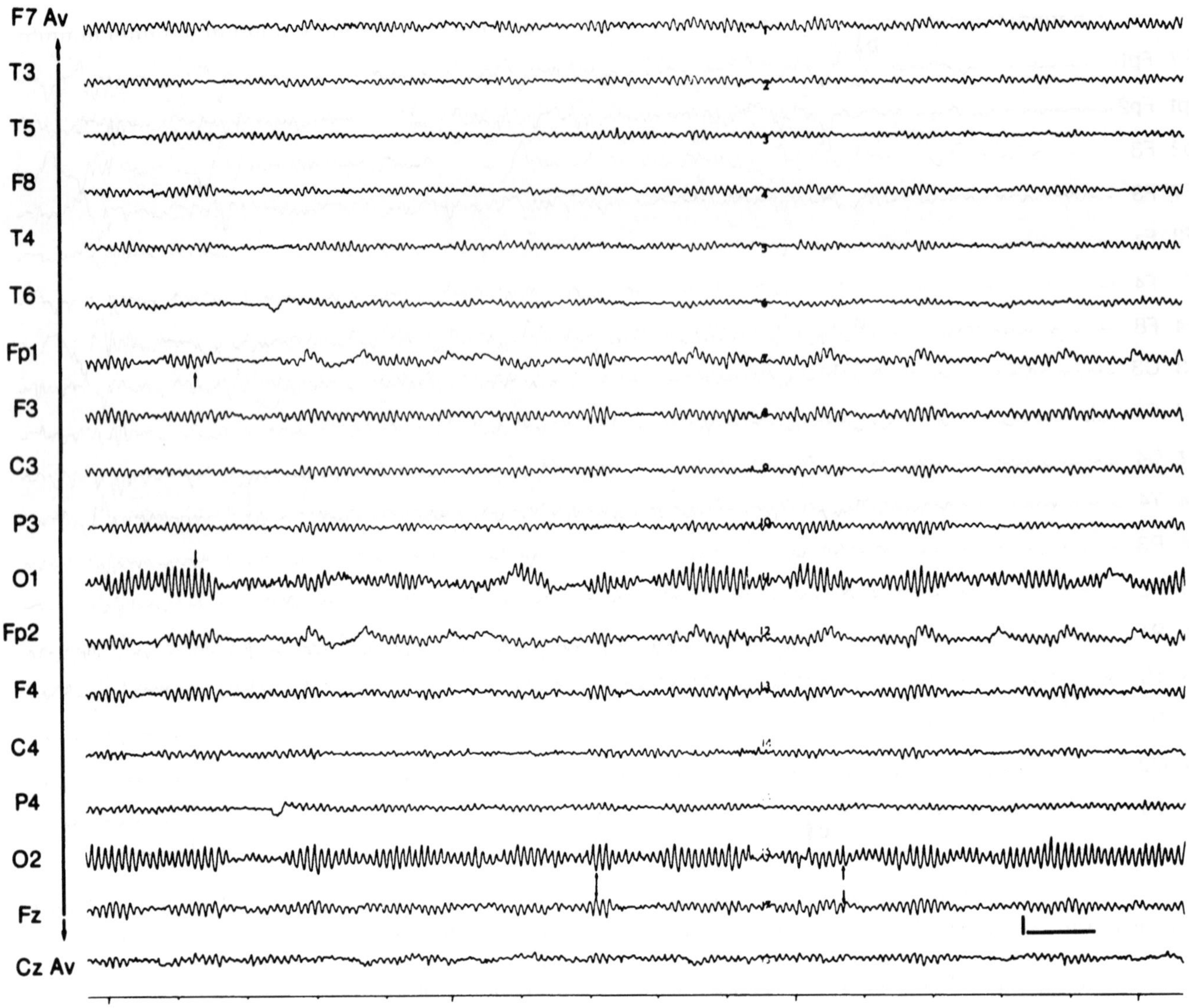

Figure 14.3. Alpha rhythm in widespread distribution as seen in a referential montage using an average potential reference. The person's eyes were closed. The tracing shows persistent, rhythmic activity at 10 to 10.5 Hz that *appears* to be present even in the anterior regions. This peculiar distribution of the alpha rhythm is spurious. It is due to contamination of the average reference, which is apparent from the fact that the waveforms in the anterior leads are 180° out of phase with respect to the waveforms in the posterior regions. Arrows point to some of these waves. This should not be confused with alpha coma (see Chapter 15). Filters: low frequency = 1 Hz, high frequency = 70 Hz. Calibrations: horizontal = 1 second, vertical = 50 μV.

Occasionally, the alpha rhythm occurs in widespread distribution, extending to the central and temporal regions. When this happens and an average potential reference is being used, it is important to recognize that the reference, which is common to all channels, may become significantly contaminated with alpha activity. The result is a record that gives a false impression concerning the distribution of the activity. Indeed, the record may show an alpha rhythm in all derivations, even in the anterior regions of the head. Such is illustrated in Fig. 14.3. Note, particularly, that the alpha activity appearing in the anterior regions is of opposite phase to that in the posterior areas. This readily identifies it as spurious and resulting from contamination of the average reference. The lesson gained is a general rule that the EEG technician should know and follow: avoid the average reference when a feature of the EEG is widespread in distribution and of high amplitude.

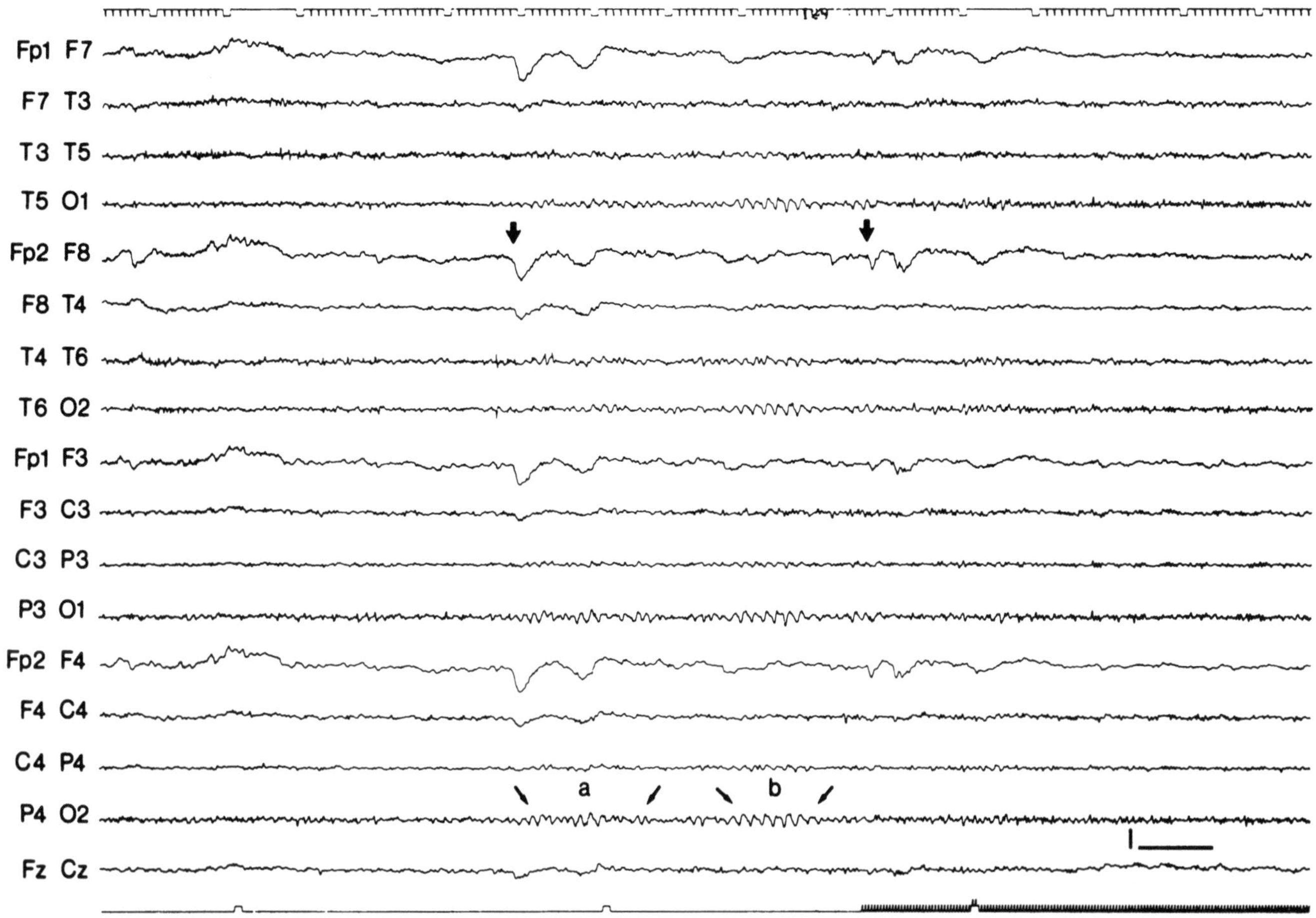

Figure 14.4. Slow alpha variant rhythm at 6 Hz. The person's eyes closed at the first vertical arrow and opened again at the second. A mixture of 6-Hz and 12-Hz rhythmic activity is seen in the posterior region on either side at "a." Rhythmic activity at "b" is the 6-Hz alpha variant alone; it disappears with eye opening and the commencement of photic stimulation at 24 flashes per second. Filters: low frequency = 1 Hz, high frequency = 70 Hz. Calibrations: horizontal = 1 second, vertical = 50 µV.

Alpha Variant

In some persons the posterior dominant rhythm shows some interesting variations. These features of the waking EEG, which are termed alpha variants, consist of rhythmic activity like the alpha rhythm but of frequencies that are faster or slower. They occur mostly when the person's eyes are closed. Slow alpha variant rhythms range between 3.5 and 6 Hz, and generally alternate or are intermixed with the ordinary alpha rhythm. Often, the frequency of the slow variant is harmonically related to the frequency of the alpha rhythm present. Fast alpha variant rhythms have a frequency range of 14 to 20 Hz; as in the case of the slow variants, the fast activity alternates or is intermixed with the ordinary alpha rhythm.

Alpha variant rhythms react to stimulation in the same way as the ordinary alpha rhythm. Thus, they are attenuated or blocked by eye opening and by mental effort. Figure 14.4 shows an alpha variant at 6 Hz that is intermixed, at times, with an alpha rhythm of 12 Hz. The significance of alpha variant rhythms, if any, is unknown.

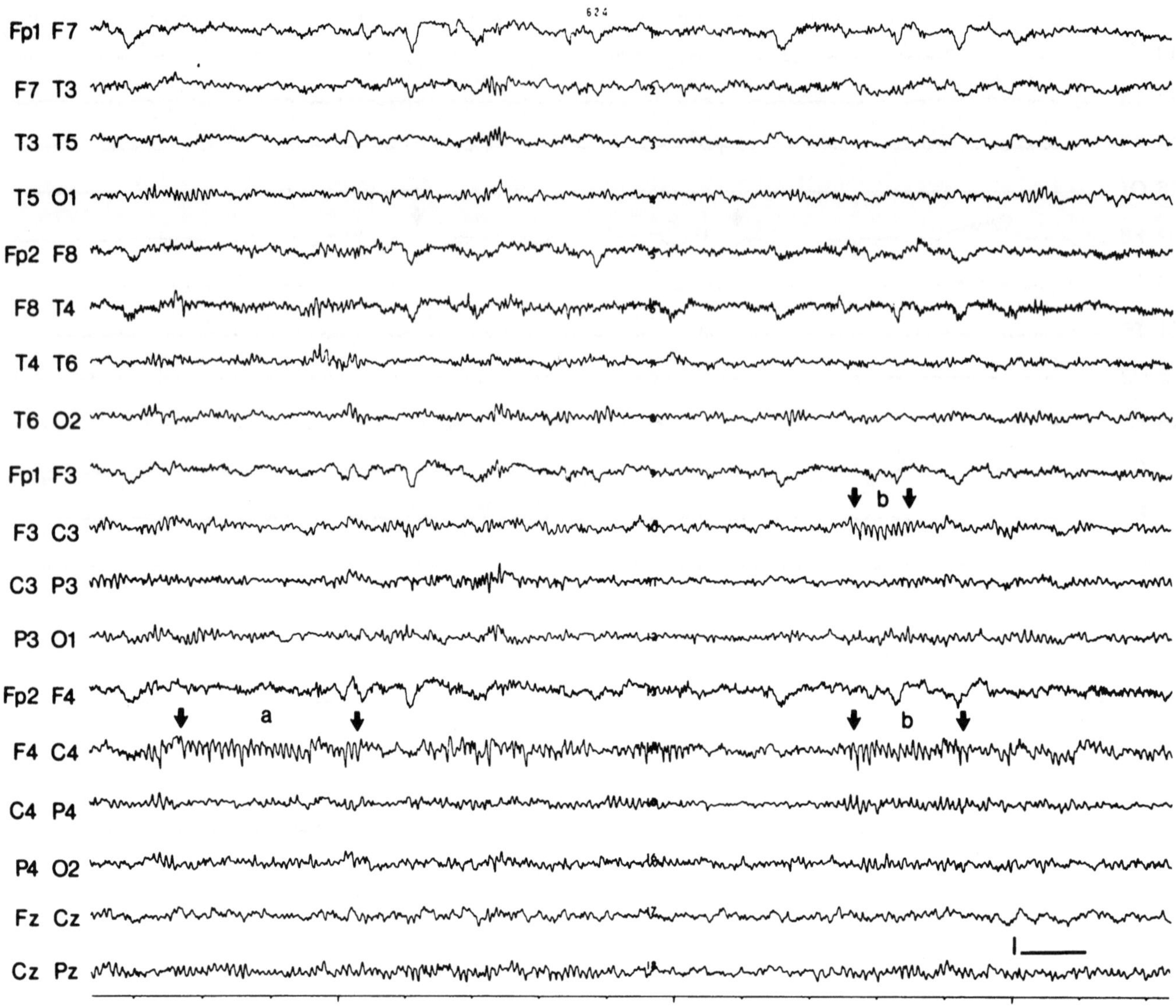

Figure 14.5. A mu rhythm at about 11 Hz. Between the arrows at "a," the mu activity is present almost solely in the centroparietal region on the right side. At "b" it occurs on both sides but is asymmetrical, being of somewhat higher amplitude and more persistent on the right. The eyes were closed throughout the recording. Filters: low frequency = 1 Hz, high frequency = 70 Hz. Calibrations: horizontal = 1 second, vertical = 50 μV.

Mu Rhythm

Although the mu rhythm resembles the alpha rhythm in both frequency and amplitude, the similarity ends there. Ranging in frequency from 7 to 11 Hz, the mu rhythm is composed of arch-like or comb-shaped waves that occur over the central or centroparietal regions of the head on either side. The activity may be bilaterally symmetrical and synchronous, or asymmetrical and asynchronous, as seen in Fig. 14.5. Because the mu and alpha rhythms frequently occur at the same time in a tracing, the two may be confused. As shown in Fig. 14.6, however, they are easily separated. Thus, whereas the alpha rhythm is attenuated by eye opening, the mu rhythm is unchanged by this maneuver. The fact that the two rhythms are independent of each other is further documented in Fig. 14.7, where an alpha rhythm is observed both in the absence and presence of mu activity.

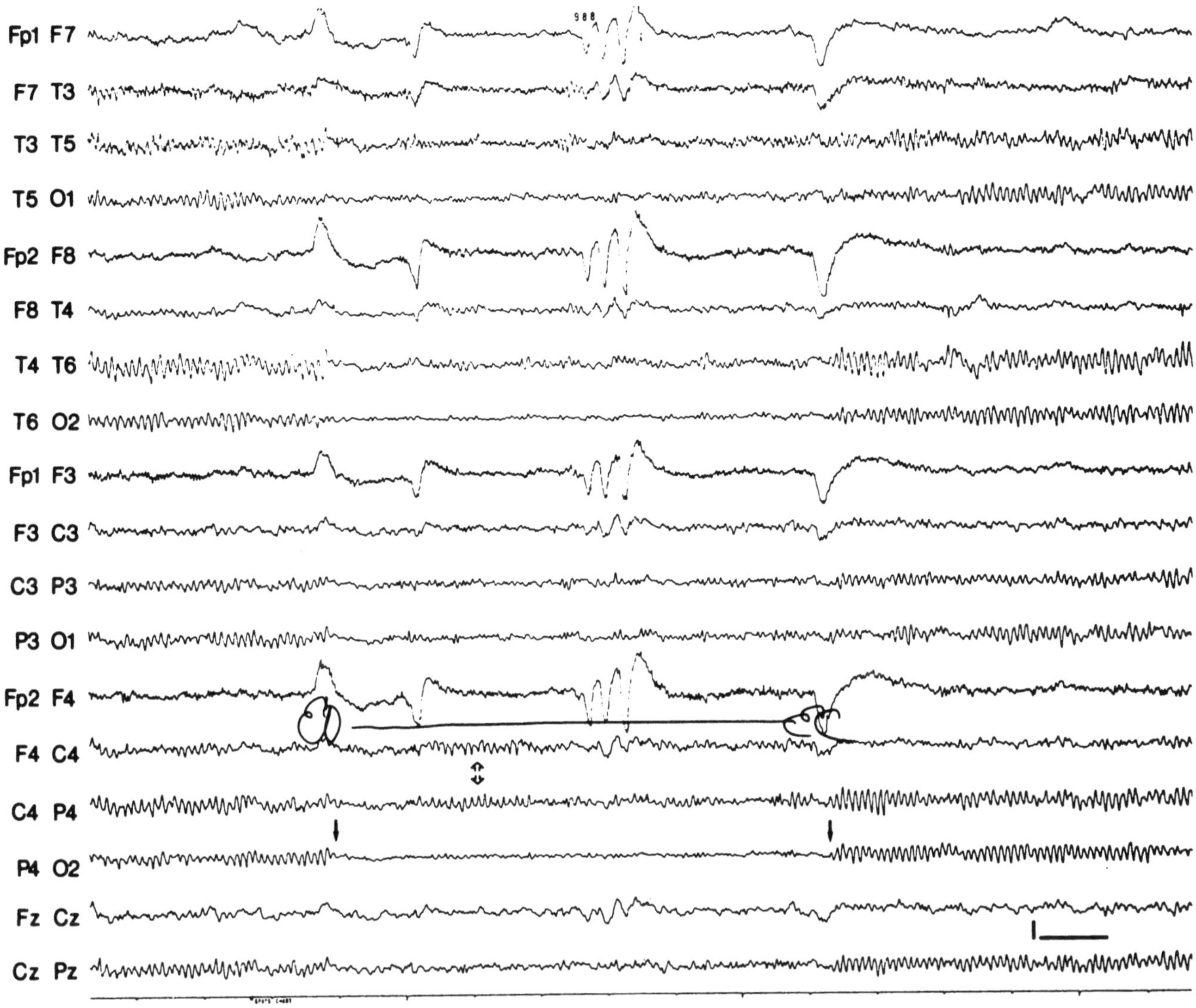

Figure 14.6. Mu rhythm during attenuation of the alpha rhythm in response to eye opening. The eyes were opened at the first solid arrow from the left, and closed again at the second solid arrow. During eye closure the alpha rhythm was greatly attenuated, but the mu rhythm is conspicuously present on the right side (open arrows). Note the sharp, comb-like character of these waves. Filters: low frequency = 1 Hz, high frequency = 70 Hz. Calibrations: horizontal = 1 second, vertical = 50 μV.

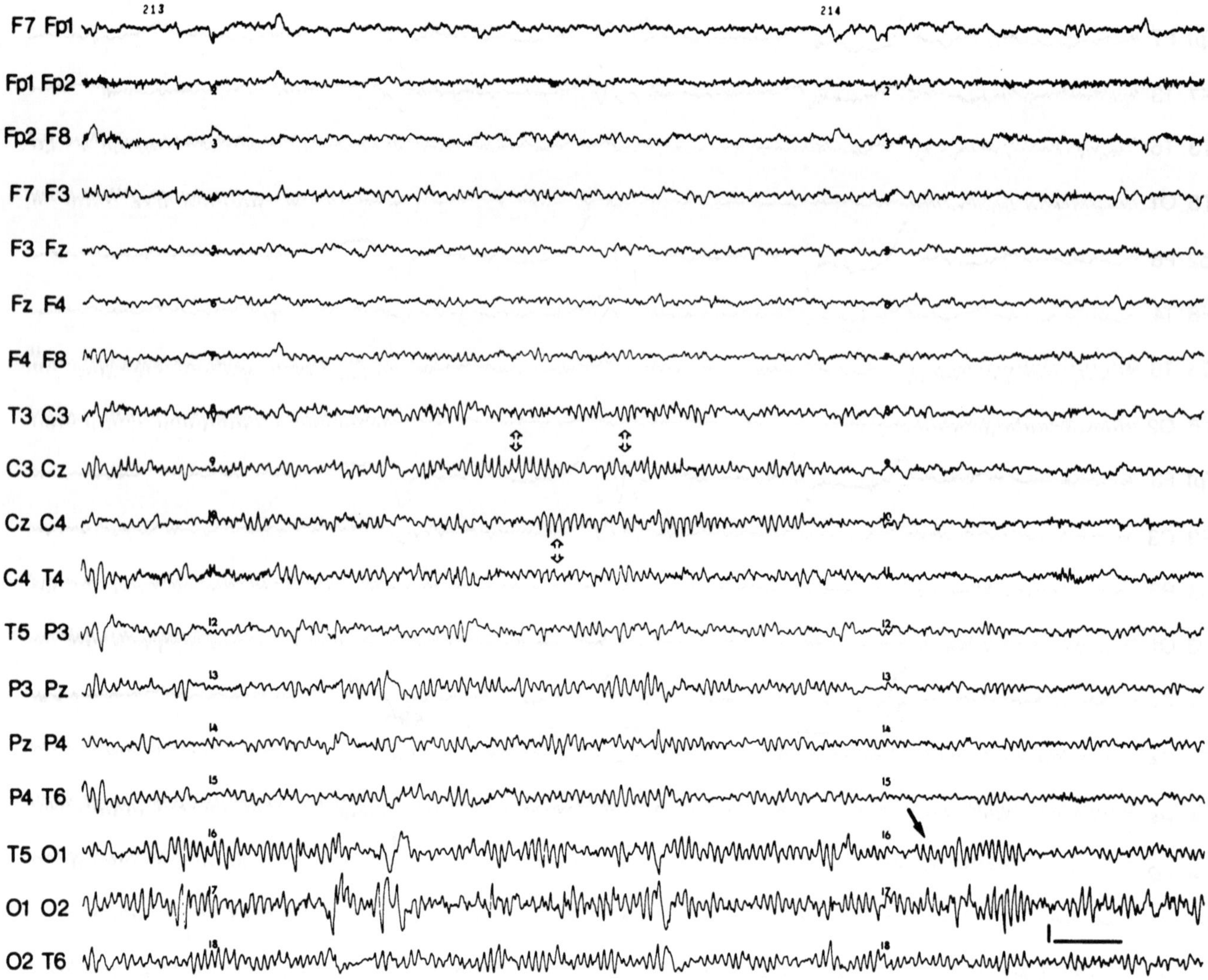

Figure 14.7. A mu rhythm in the central region on either side (open arrows). The mu activity seen here is neither bilaterally synchronous nor symmetrical. The subject's eyes were closed throughout the recording; note that an alpha rhythm is seen concomitantly with the mu rhythm and also when it is absent (solid arrow). Filters: low frequency = 1 Hz, high frequency = 70 Hz. Calibrations: horizontal = 1 second, vertical = 50 μV.

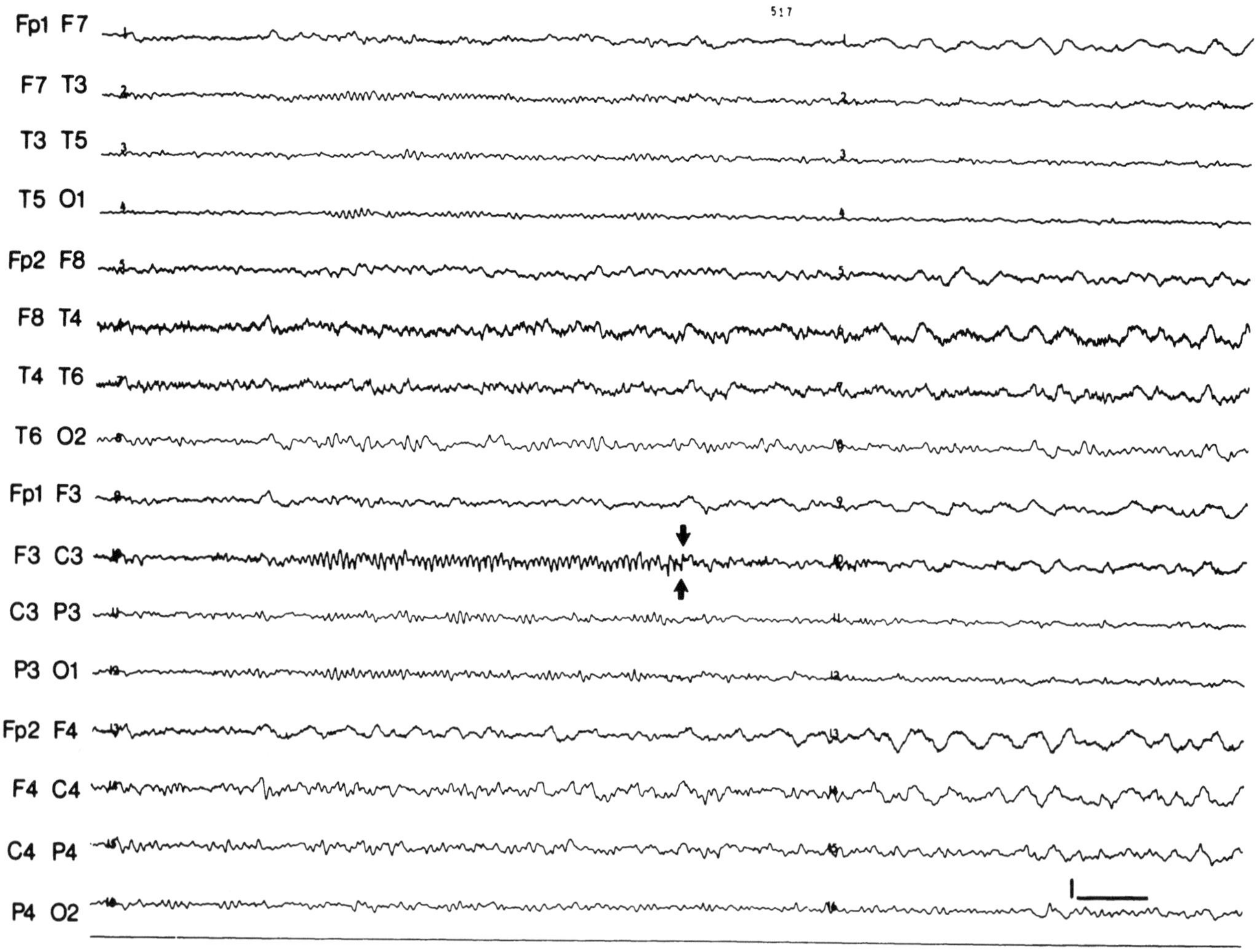

Figure 14.8. Blocking of the mu rhythm resulting from muscular contraction on the contralateral side. The tracing shows a very marked reduction in amplitude of the mu rhythm in the *left* central region in response to the subject clenching the *right* fist beginning at the arrows. This is a normal response. The slow activity, which is seen on both sides but occurs more frequently and is of considerably greater amplitude on the right, is not normal. Filters: low frequency = 1 Hz, high frequency = 70 Hz. Calibrations: horizontal = 1 second, vertical = 50 μV.

Although mu rhythms are not reactive to eye opening, they are responsive to activity of the motor system. Movement of the extremities or muscle contraction, such as clenching the fist, on one side of the body results in the attenuation or blocking of a mu rhythm present on the contralateral side. Figure 14.8 shows how a prominent left-sided mu rhythm is blocked by having the person clench the right fist.

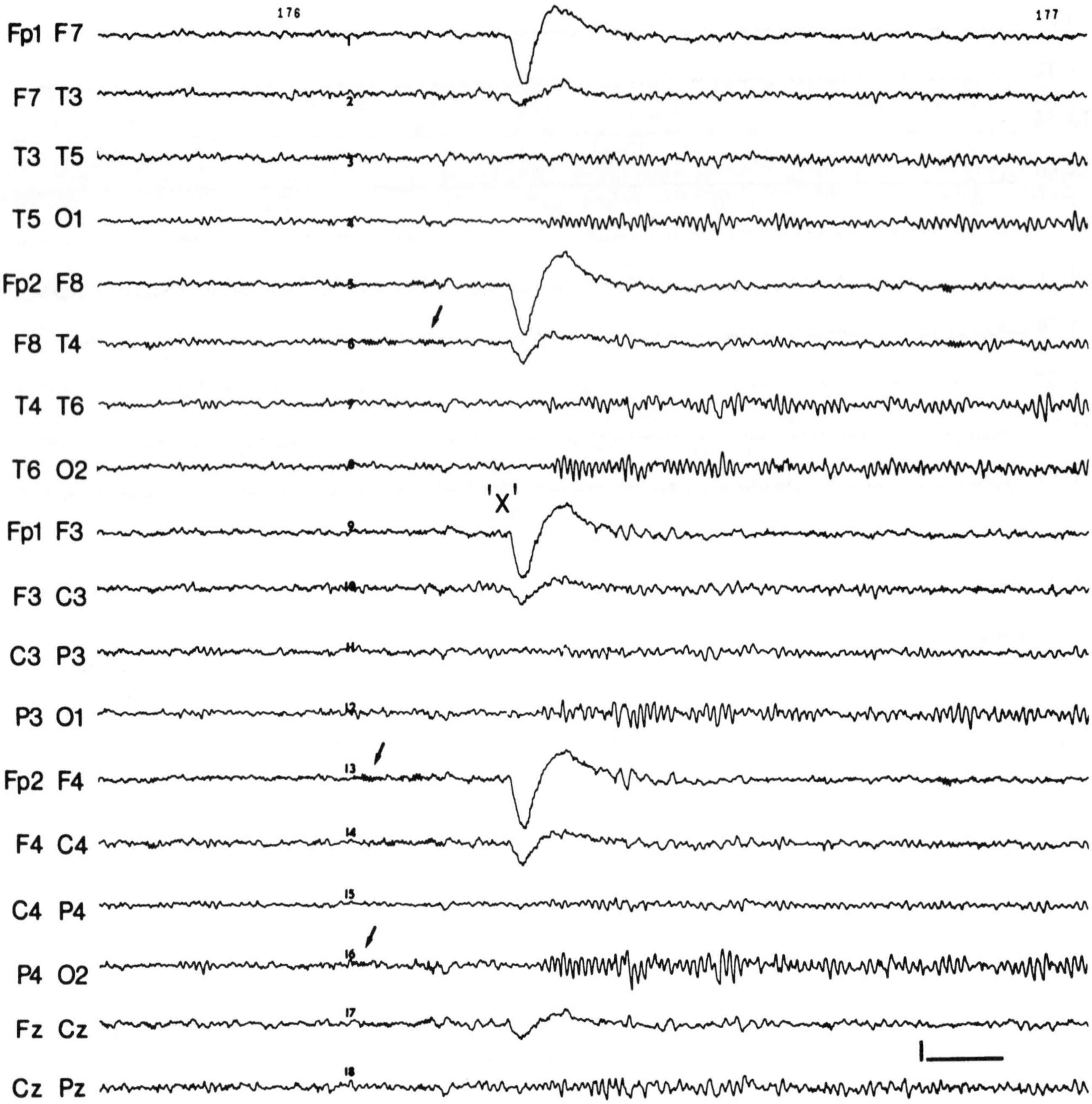

Figure 14.9. Beta activity with eyes open and then closed starting at "x." Arrows point to low amplitude (less than 20 μV) beta activity at about 22 to 24 Hz. Note that with the eyes opened, the beta activity is present in most derivations, being visible in the posterior regions as well an anteriorly. However, when the eyes are closed, the alpha rhythm masks the beta activity so that beta is best seen in tracings from the anterior regions. Filters: low frequency = 1 Hz, high frequency = 70 Hz. Calibrations: horizontal = 1 second, vertical = 50 μV.

Beta Activity

Although activity with frequencies higher than 13 Hz is common in the waking EEG of adults (and children as well), amplitudes are normally only 20 μV or less in more than 90% of the cases. For this reason, beta activity, which usually consists of frequencies greater than 13 Hz and up to 35 Hz, is not a prominent feature of the waking EEG. Although beta activity is best seen in tracings from the anterior regions, it is commonly present in the posterior regions as well, albeit masked by the alpha rhythm. Because of this, it is easier to appreciate the distribution of beta activity when the person's eyes are opened. Figure 14.9 shows beta activity in a tracing with the eyes open and closed.

The frequency of beta activity should be the same on both sides. Amplitude, however, may display an asymmetry,

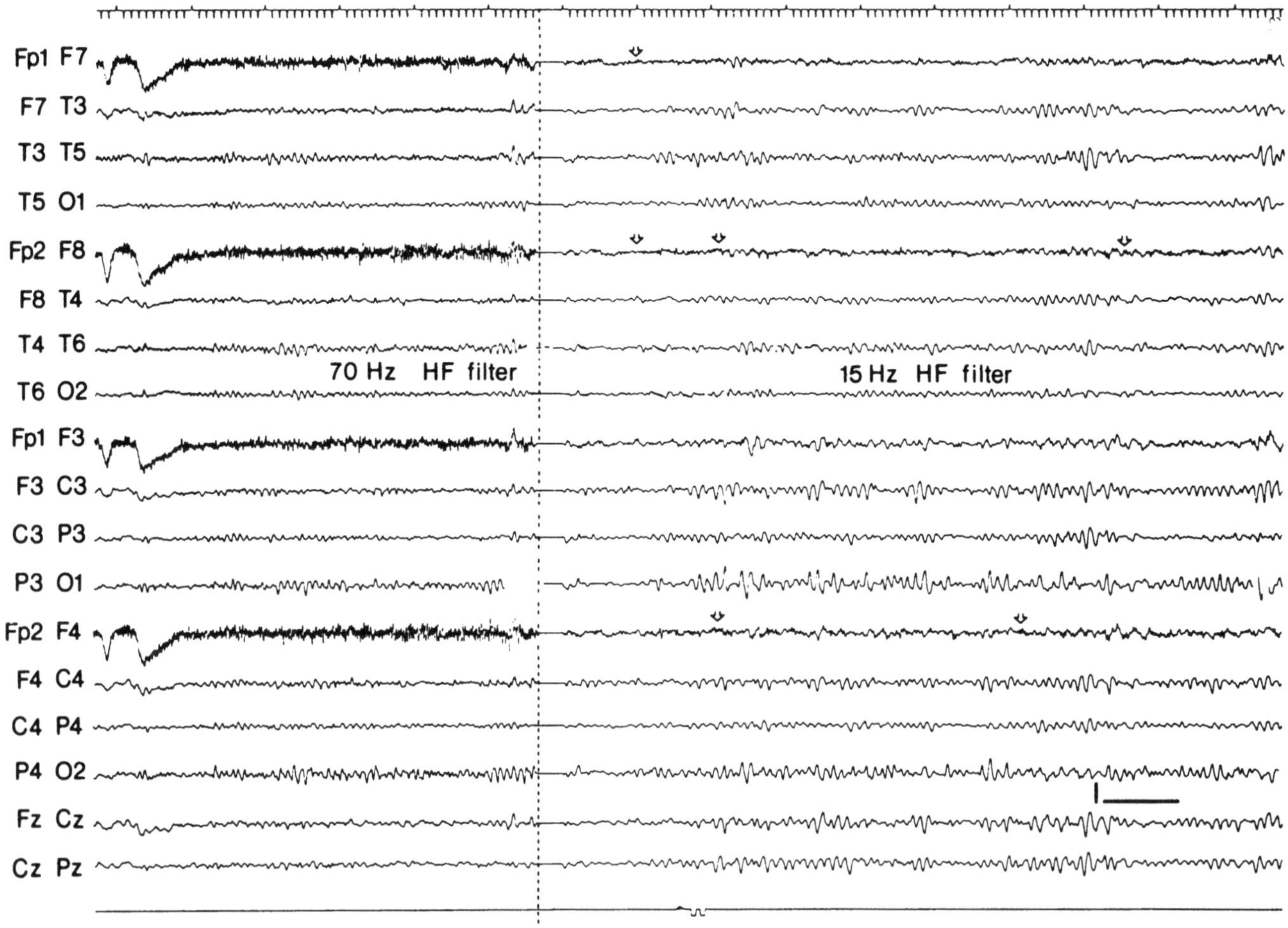

Figure 14.10. The effect of high-frequency filtering on muscle artifacts. Prominent muscle spikes were recorded in the anterior leads with the high-frequency filter at the standard 70-Hz setting. Switching to the 15-Hz setting removed the spikes, but traces of low-amplitude fast activity were left. This activity looks like beta (open arrows), but is really muscle activity with the spikes rounded off by filtering. Calibrations: horizontal = 1 second, vertical = 50 μV.

with minor asymmetries being attributable to differences in skull thickness on the two sides. Consistent, major differences in amplitude between the two sides—differences in excess of 35%—are considered to be abnormal.

When examining a tracing for evidences of beta activity, the electroencephalographer needs to be especially aware of the high-frequency filtering that was used in taking the EEG. The reason for this will be apparent from a perusal of Fig. 14.10. As noted elsewhere in the text, the recommended high-frequency-filter setting for routine EEG work is 70 Hz. Occasionally, however, muscle artifacts may be so numerous and persistent that the technician's only recourse is to drop the setting to 35 Hz or, rarely, even to 15 Hz. When this is done, any muscle spikes that remain in the recording will show a rather marked change in character. Figure 14.10, in which the filter setting was abruptly changed from 70 to 15 Hz, illustrates this. Note that with the filter set at 15 Hz, the sharply pointed muscle spikes are no longer seen. Instead, there is rhythmic, 20- to 30-Hz activity present in the anterior leads that looks a great deal like beta activity. In reality, however, this activity consists of low-amplitude muscle spikes that have been rounded off by the action of the filtering. To avoid mistaking filtered muscle spikes for beta activity, it is essential to keep a close watch on the high-frequency-filter settings used in taking the EEG.

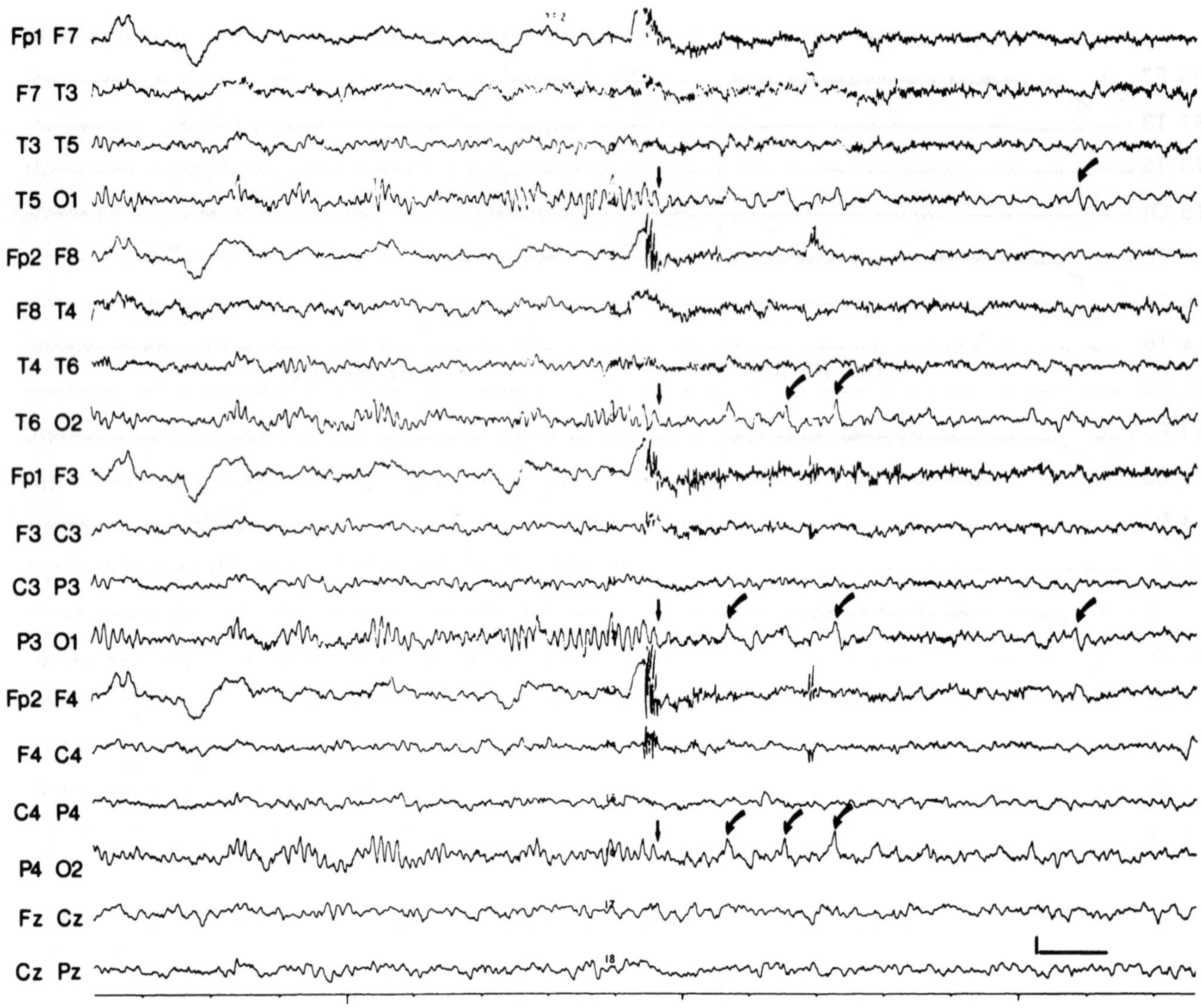

Figure 14.11. Lambda waves. The subject's eyes were closed and the tracing shows a prominent alpha rhythm at 8 to 9 Hz, which is blocked when the eyes were opened at the vertical arrows. The eyes remained open throughout the rest of the record. Curved arrows point to lambda waves. Filters: low frequency = 1 Hz, high frequency = 70 Hz. Calibrations: horizontal = 1 second, vertical = 50 μV.

Theta Activity

This includes activity having frequencies of 4 Hz to less than 8 Hz. Although a small amount of 6- to 7-Hz random activity is present in the background of the waking EEG of most young adults, theta activity in any but trace amounts is not considered to be normal. As theta activity is a normal accompaniment of drowsiness, it is important to avoid mistaking theta activity of drowsiness for waking theta activity. In this regard, the technician's observations and notations concerning the subject's behavior while the EEG was taken are important in helping to make the distinction; they are essential in cases where the posterior dominant rhythm is absent or not readily discernible in the tracing. Theta activity of drowsiness is discussed and illustrated in the next major section of the present chapter.

Lambda Waves

These are sharp transients that occur in the occipital regions when the eyes are open and the person is engaged in visual exploration. Durations of the transients vary considerably and are reported to range from 100 to 250 ms; amplitudes are usually under 100 μV. Lambda waves are predominantly positive at the occipital electrodes with respect to other areas. They may be fairly symmetrical on the two sides, as in Fig. 14.11, or quite asymmetrical; or they may be present only on one side.

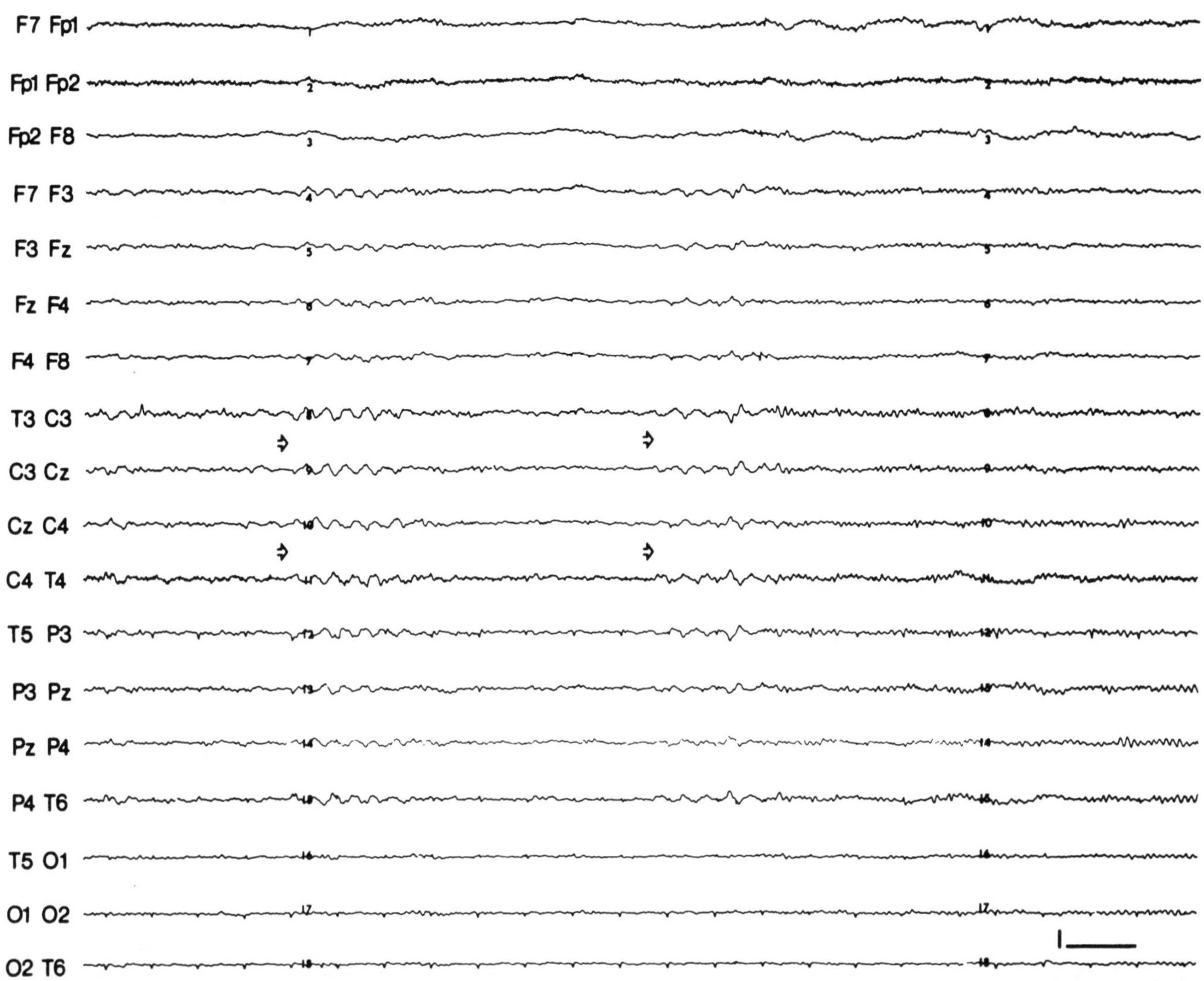

Figure 14.12. Rhythmic theta and delta activity during drowsiness. The slow activity (open arrows) is mainly in the central regions. The episode of drowsiness is followed after about 10 seconds by partial awakening during which time the alpha rhythm appears. Note the low-amplitude ECG artifacts. Filters: low frequency = 1 Hz, high frequency = 70 Hz. Calibrations: horizontal = 1 second, vertical = 50 µV.

Lambda waves should not be confused with positive occipital sharp transients of sleep (POSTS), which occur during sleep, nor mistakenly interpreted as a focus of abnormal activity. Lambda is easily distinguished from these other EEG features by the fact that the waves promptly disappear when the eyes are closed. This, again, highlights how important it is for the EEG technician to be a careful observer; for if he/she should fail to note that the activity occurred only with the eyes open, the interpretation could become uncertain.

Features of the EEG During Drowsiness and Sleep in Adults

The transition from the awake to the drowsy state, or stage I sleep, is marked by some profound changes in the background activity of the EEG. The transition may be gradual or it may be very abrupt. The most prominent change is the disappearance of the posterior dominant (alpha) rhythm. In some persons this is preceded first by a perceptible slowing in the frequency of alpha. With the alpha rhythm gone, the background becomes dominated by theta activity, which occurs in generalized distribution but is commonly most prominent in central or fronto-central regions. The theta activity varies in amplitude from 10 to about 50 µV; it may be rhythmic as in Fig. 14.12, or semirhythmic and/or irregular as in Fig. 14.13. At times, some slower activity may also be intermixed. It is not unusual for periods of drowsiness to alternate back and forth with periods of wakefulness, at which time the alpha rhythm returns. Two such cycles are seen in Fig. 14.13.

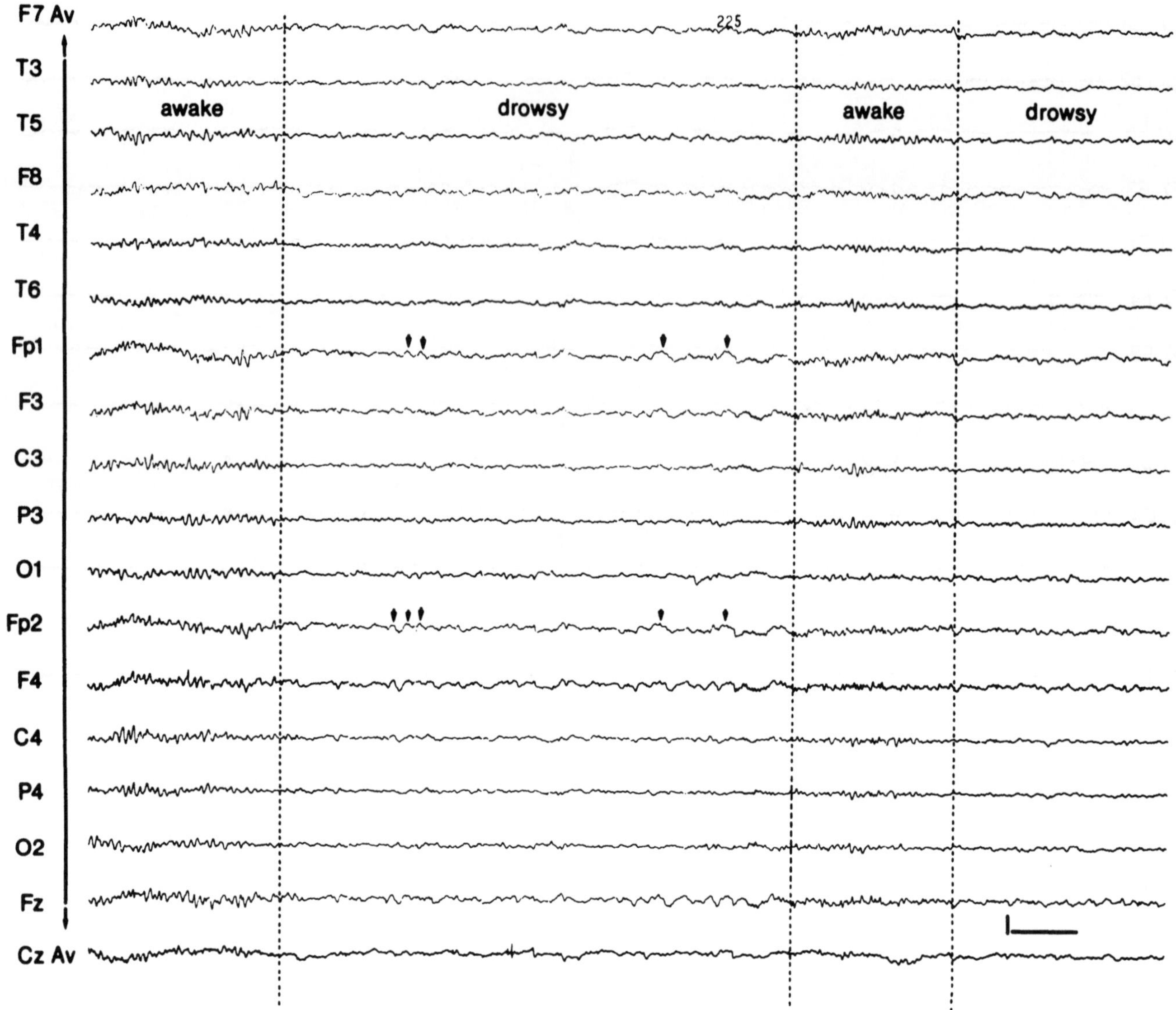

Figure 14.13. Tracing showing rapid shifts between an awake and a drowsy EEG. The record begins with an alpha rhythm at approximately 11 Hz, which abruptly disappears after about 2.5 seconds. A 7-second interval of drowsiness follows during which time 15- to 35-μV theta (arrows on left) and delta (arrows on right) activity is present. There is a return to wakefulness that lasts about 2 seconds before the tracings again revert back to drowsiness. Filters: low frequency = 1 Hz, high frequency = 70 Hz. Calibrations: horizontal = 1 second, vertical = 50 μV.

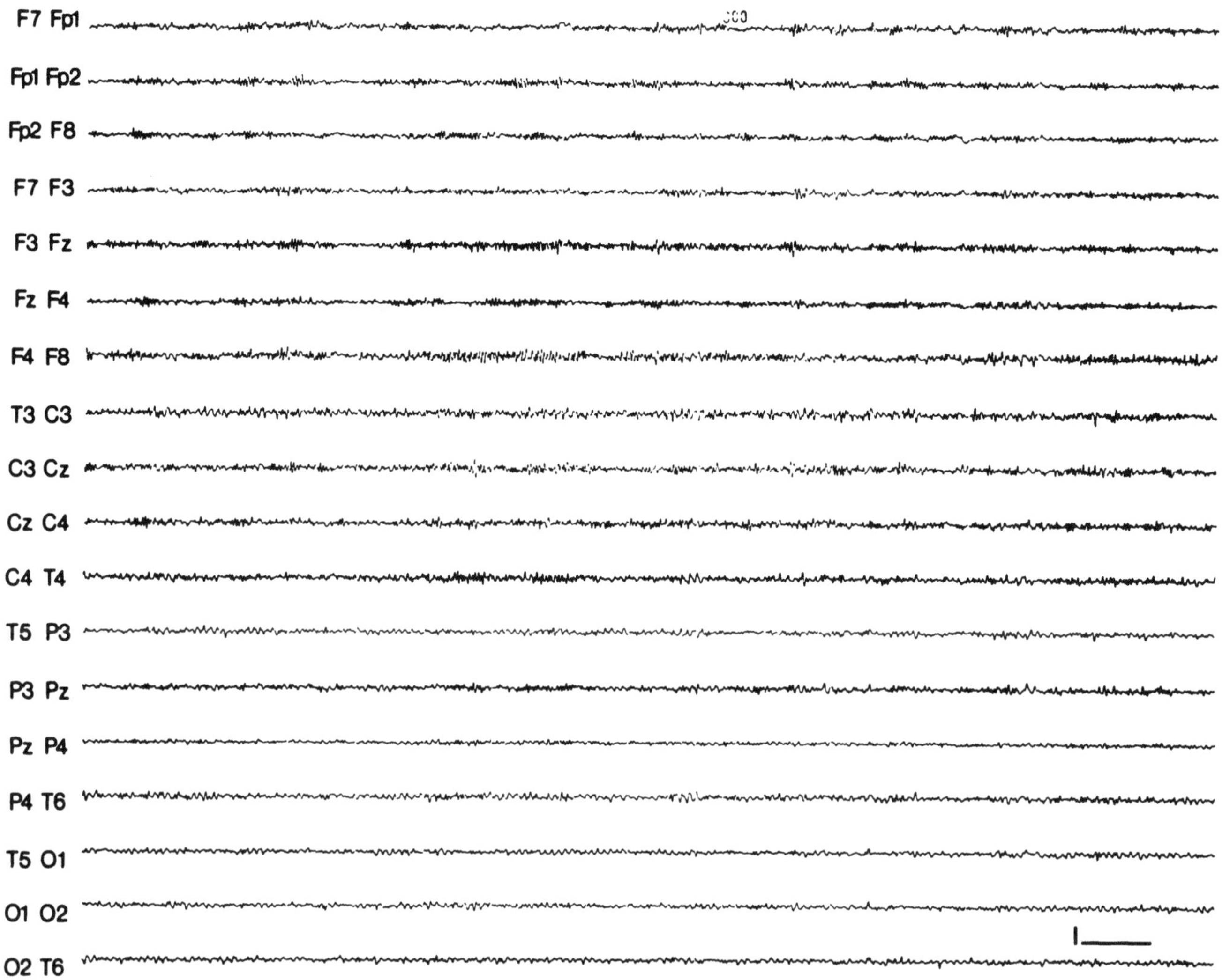

Figure 14.14. Diffuse beta activity during drowsiness following administration of chloral hydrate to promote sleep. The 20- to 24-Hz rhythmic activity is most prominent in the frontal regions where it sometimes has an amplitude of 40 μV. Filters: low frequency = 1 Hz, high frequency = 70 Hz. Calibrations: horizontal = 1 second, vertical = 50 μV.

Another change in the background activity that occurs in the transition from the awake to the drowsy state concerns the beta activity. Beta activity in the range of 18 to 25 Hz usually, but not always, increases in amplitude with drowsiness; this has been termed ***subvigil beta***. At times, the beta activity appears in bursts of short duration; these are referred to as beta spindles (see Fig. 14.15). When chloral hydrate is administered to promote sleep during the EEG test, beta activity may become widespread and quite prominent, sometimes attaining amplitudes in excess of 50 μV. Figure 14.14 shows diffuse, persistent beta activity during drowsiness following administration of chloral hydrate.

As a person goes from the drowsy state, or stage I sleep, into stage II sleep, we find that the EEG displays a number of distinctive, easily recognized features. These are taken up in turn.

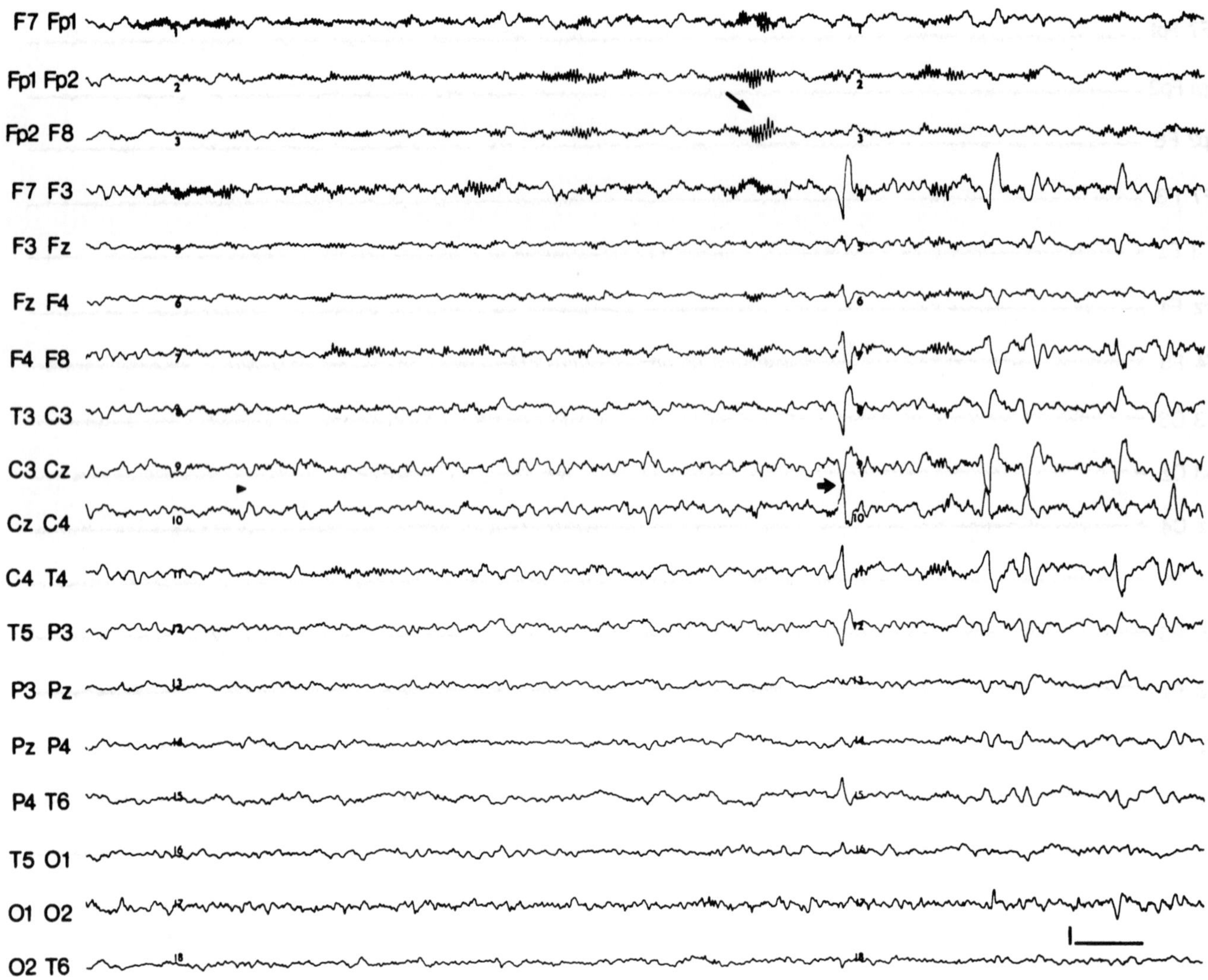

Figure 14.15. Early stage II sleep showing the initial appearance of V waves. In the first 2 seconds, the tracing shows central and frontocentral theta activity that is characteristic of drowsiness. Directly thereafter an incompletely formed V wave (marked by the triangle) is seen. About 8 seconds later, a typical V wave occurs (horizontal arrow), closely followed by several other V waves. Note the phase reversal in channels 9 and 10, which indicates that the focus of the wave is at Cz. The diagonal arrow points to one of several beta spindles present in the tracing. Filters: low frequency = 1 Hz, high frequency = 70 Hz. Calibrations: horizontal = 1 second, vertical = 50 μV.

Vertex Waves

Also referred to as V waves or vertex sharp transients, this feature of the EEG is most prominent in stage II sleep. The waves are aptly named, as their focus lies at Cz, the vertex. When the waves are of large amplitude—100 μV and larger is not uncommon—they also are picked up in the C3 and C4 electrodes. Their fields frequently spread to the frontocentral regions and sometimes even extend to the parietal areas. V waves usually are diphasic, but occasionally may be triphasic as well; the initial deflection is negative, and this is followed by a lower-amplitude, positive phase. Figure 14.15 shows the emergence of V waves in the early phase of stage II sleep.

V waves are bilaterally synchronous and essentially symmetrical on the two sides, although some shifting asymmetries are not uncommon. Thus, the amplitude may be somewhat higher on one side at one time, and then higher on the other side at another time. The waves can assume a variety of different forms, as will be seen in Fig. 14.16. Sometimes they appear as sharp waves, and at other times they fit the definition of spikes. The particular morphology has no clinical significance. As mentioned later in this section, a V wave may be followed by a sleep spindle.

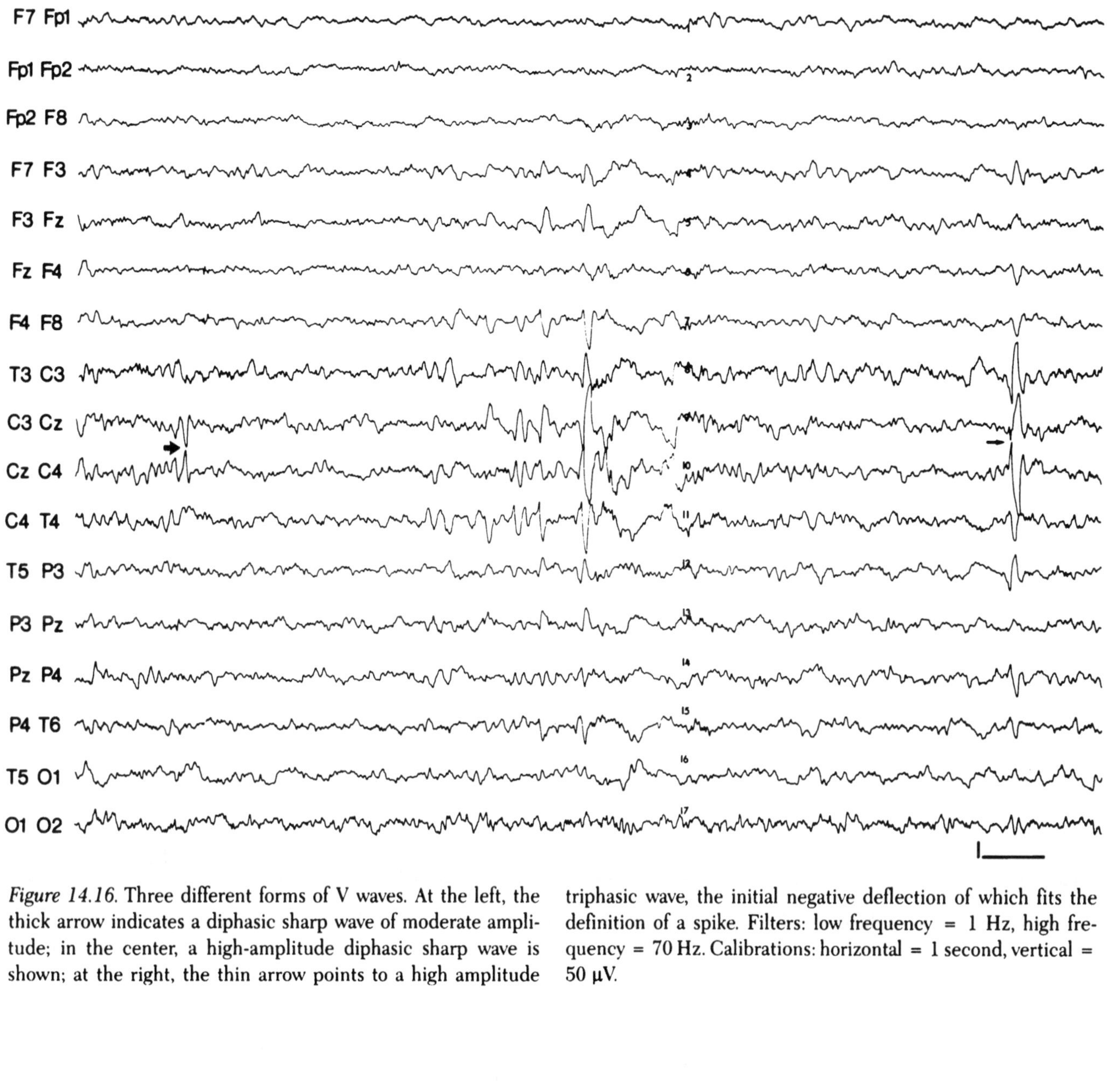

Figure 14.16. Three different forms of V waves. At the left, the thick arrow indicates a diphasic sharp wave of moderate amplitude; in the center, a high-amplitude diphasic sharp wave is shown; at the right, the thin arrow points to a high amplitude triphasic wave, the initial negative deflection of which fits the definition of a spike. Filters: low frequency = 1 Hz, high frequency = 70 Hz. Calibrations: horizontal = 1 second, vertical = 50 μV.

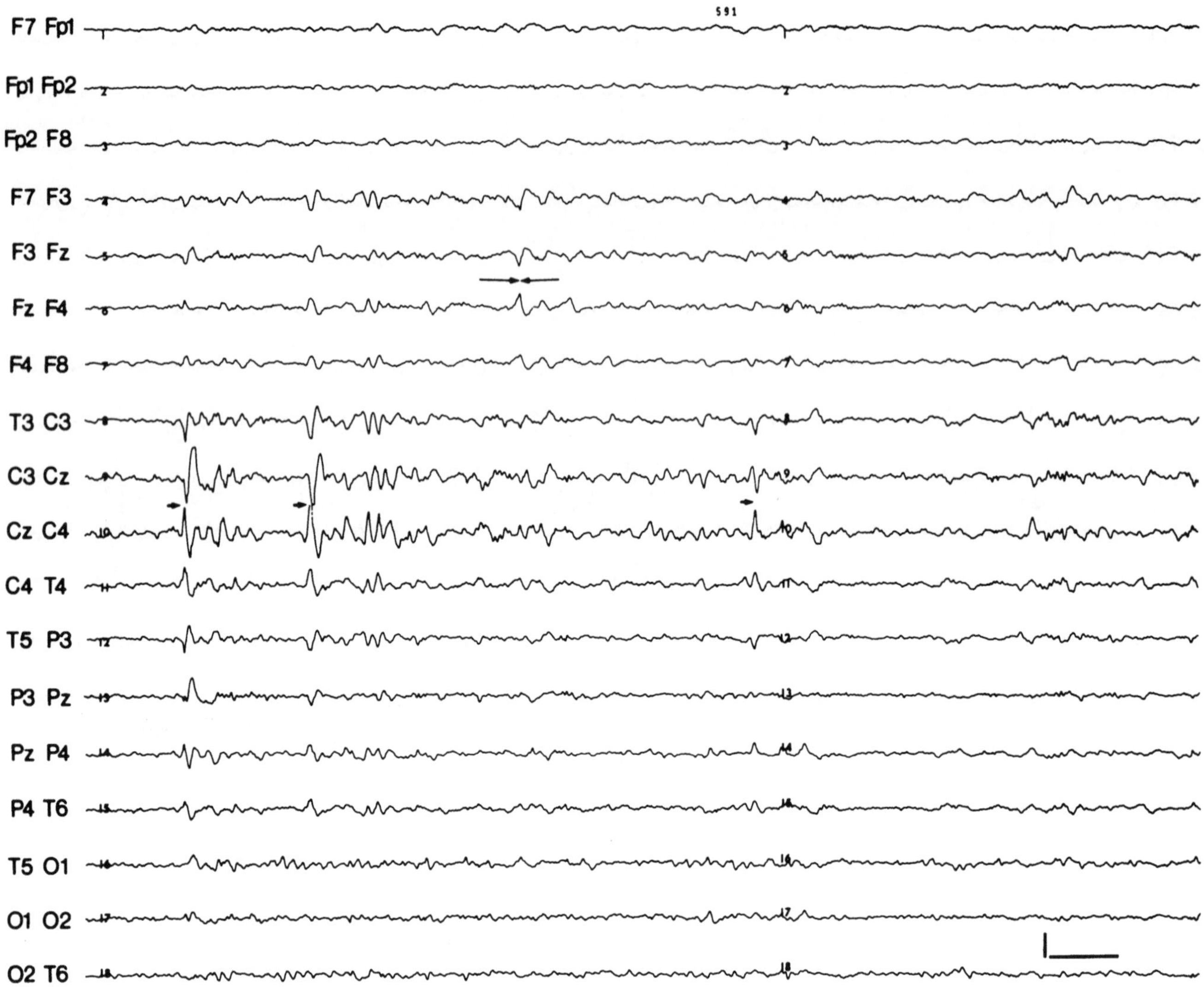

Figure 14.17. Stage II sleep showing an F wave (opposing arrows) and several V waves (single arrows). Note that the focus of the F wave is in the frontal region at the midline, while the V waves have their focus at Cz. Filters: low frequency = 1 Hz, high frequency = 70 Hz. Calibrations: horizontal = 1 second, vertical = 100 μV.

F Waves or Frontal Waves

Occasionally, a sharp transient not unlike a V wave appears in the frontal regions at the midline without a corresponding wave present at the vertex (Fig. 14.17). Such transients are sometimes referred to as F waves. Their amplitude is usually less than 100 μV. V waves and F waves have the same significance; they are normal features of the EEG during stage II sleep.

K Complex

This is yet another feature of stage II sleep that is similar to the V wave. The K complex is a slow-wave transient, it is commonly diphasic, and amplitude is generally a maximum at the vertex. This is a large-amplitude wave, with amplitudes running as high as several hundred microvolts. A sleep spindle (see below) may immediately follow or be associated with the K complex. As seen in Fig. 14.18, a K complex may last for nearly a second; but at times, the duration can be somewhat longer. K complexes can occur apparently spontaneously. They also can occur in response to sudden sensory stimulation such as an unexpected, loud noise in the EEG laboratory.

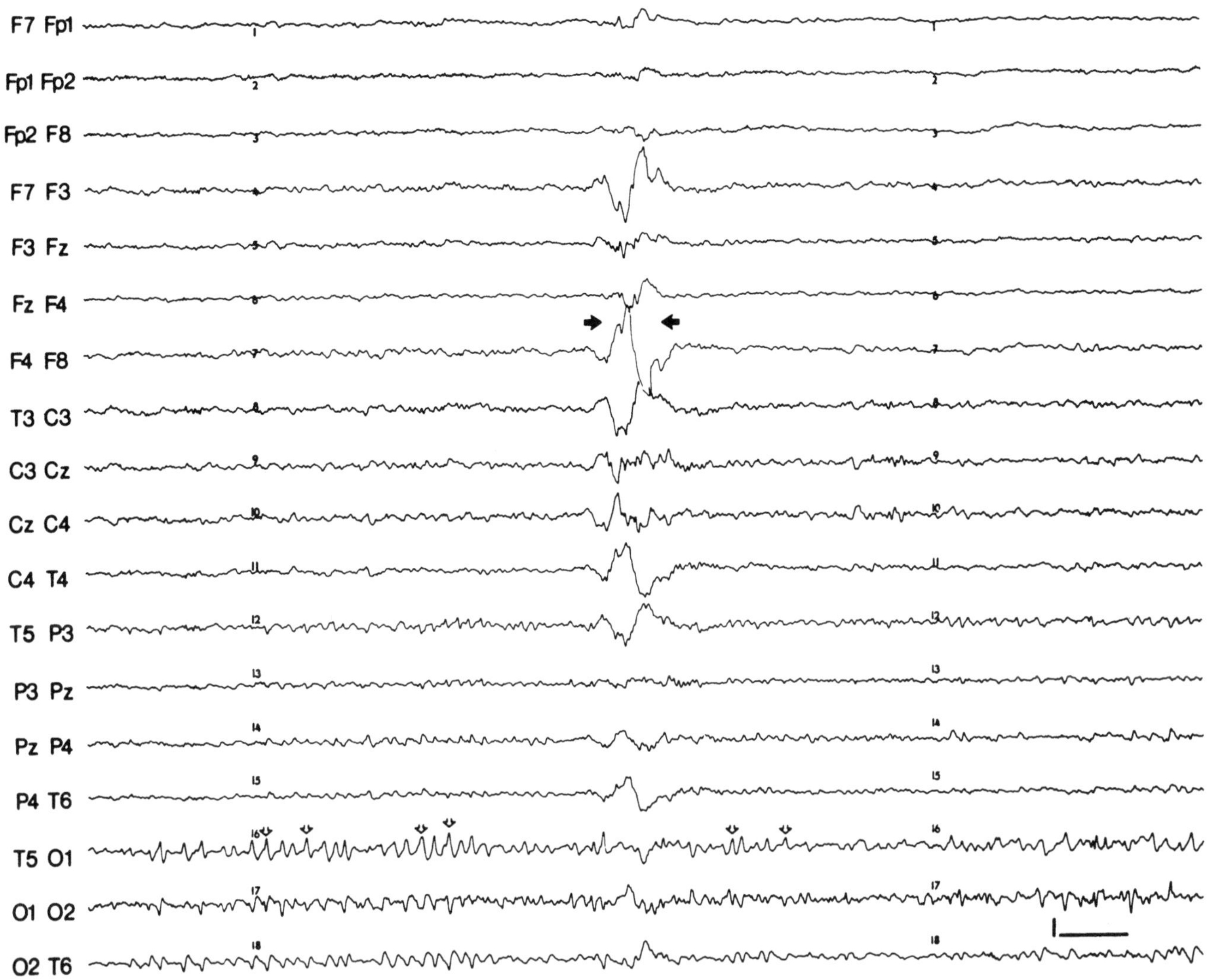

Figure 14.18. Stage II sleep showing a K complex between the solid arrows. Amplitude of the wave appears to be larger in the frontal than in the central region; duration is markedly longer than a typical V wave. Open arrows point to some of the numerous POSTS also seen in the tracing. Filters: low frequency = 1 Hz, high frequency = 70 Hz. Calibrations: horizontal = 1 second, vertical = 50 μV.

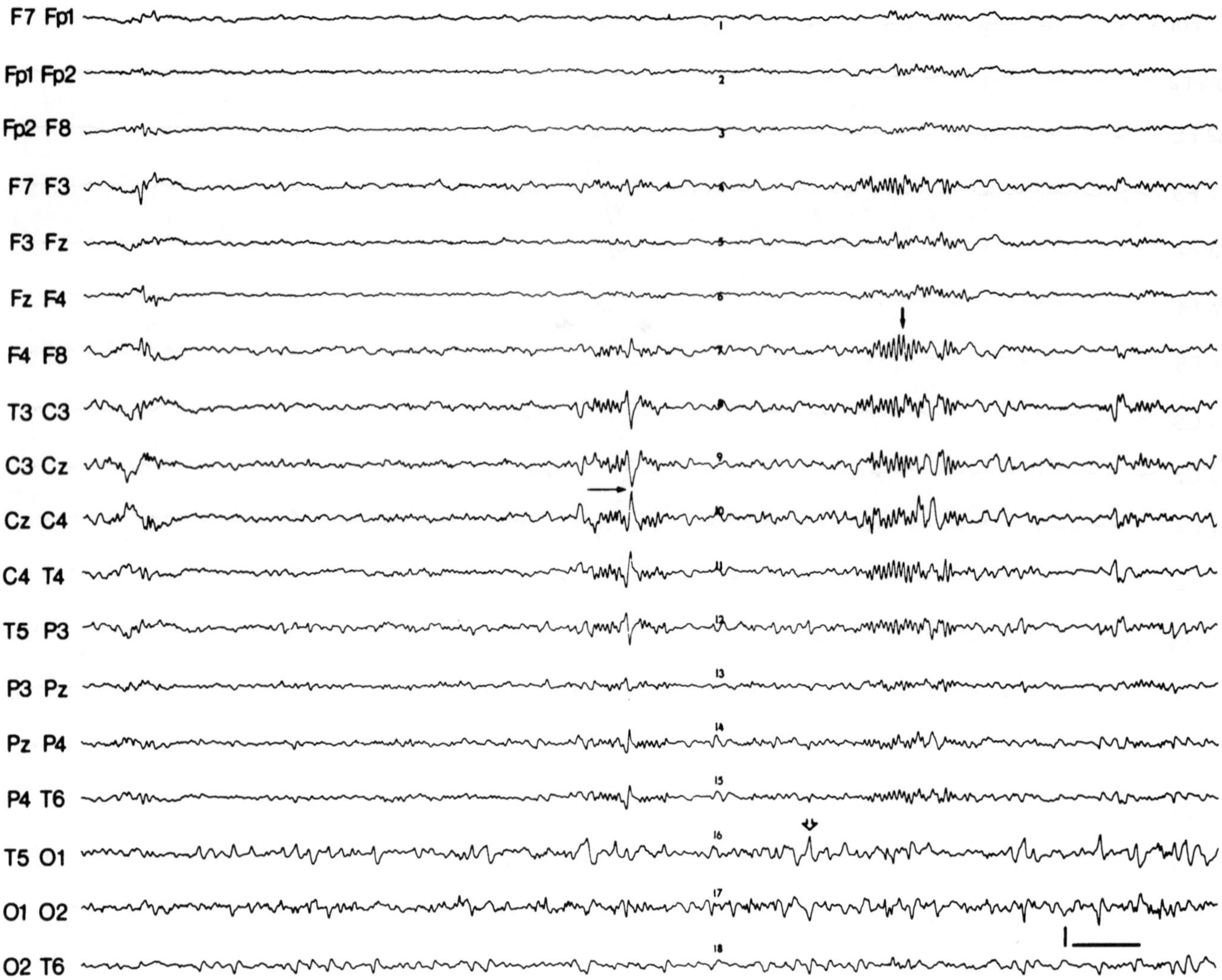

Figure 14.19. Stage II sleep in which a sleep spindle (vertical closed arrow), a V wave (horizontal arrow), and POSTS (vertical open arrow points to one of the POSTS) are all present in the same recording. The sleep spindle has a frequency of about 14 Hz and an amplitude of approximately 50 μV. Note that a sleep spindle also accompanies the V wave. The sleep spindles are widely distributed, but appear to be of higher amplitude in the frontocentral regions. Filters: low frequency = 1 Hz, high frequency = 70 Hz. Calibrations: horizontal = 1 second, vertical = 50 μV.

Sleep Spindles

These are bursts of very rhythmic activity at 11 to 15 Hz that are seen in stage II and in the early phase of stage III sleep. Duration and amplitude are both variable; amplitude is generally less than 50 μV but occasionally may exceed 100 μV. Sleep spindles generally occur in widespread distribution. Commonly, they are of higher amplitude in the central regions, as is seen in Fig. 14.19, but sometimes an anterior dominance is noted instead (Fig. 14.20). Sleep spindles in adults should be bilaterally synchronous and essentially symmetrical. However, it is not unusual to observe some shifting asymmetries in which amplitude is alternately somewhat higher on one side, and then higher on the other during the course of the recording. Sleep spindles are sometimes preceded by a V wave, as may be observed in Fig. 14.20.

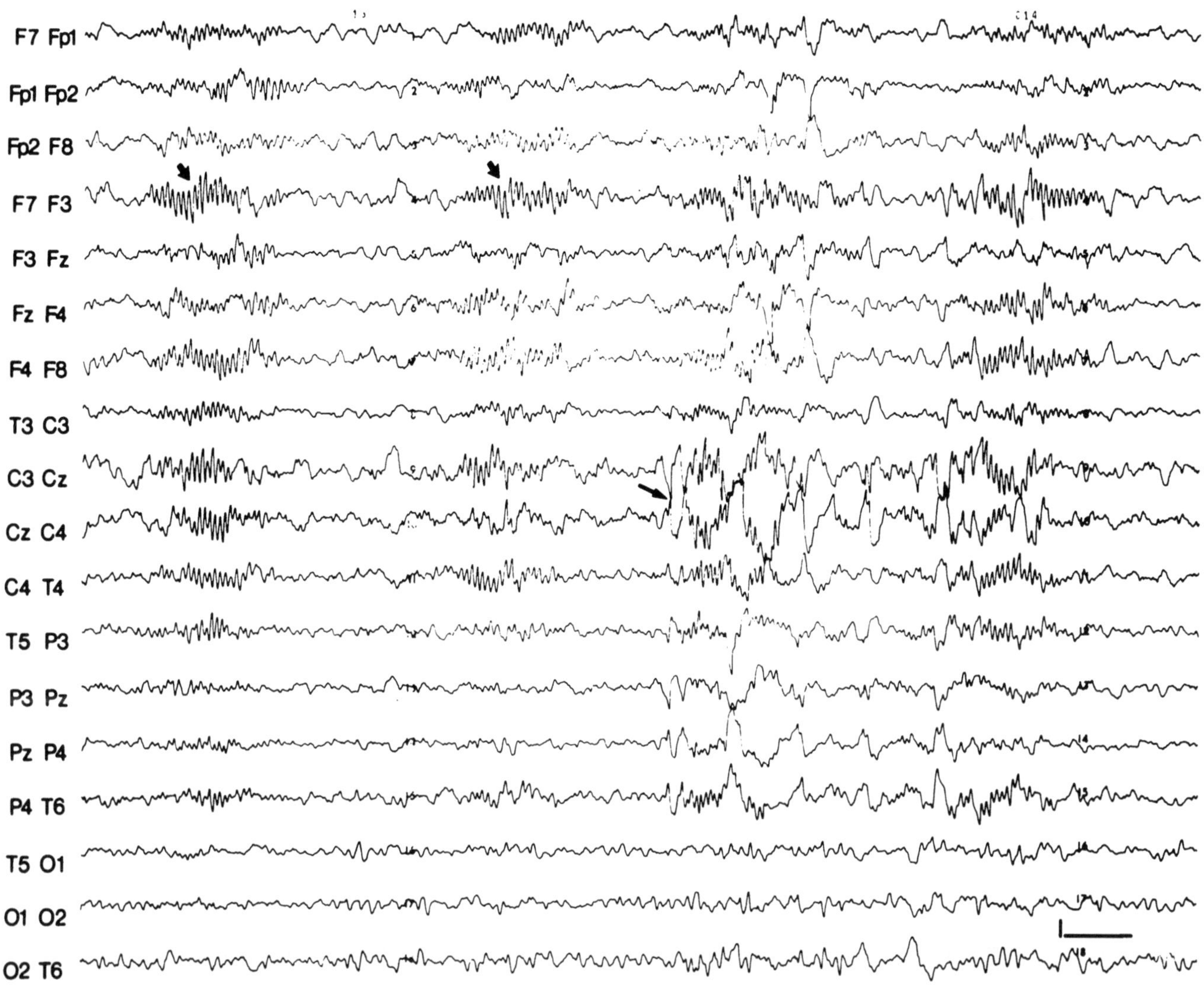

Figure 14.20. Sleep spindles and V waves in stage II sleep. The sleep spindles in the tracing appear to have their highest amplitudes in the frontal region (thick arrows), sometimes attaining amplitudes greater than 100 μV. The thin arrow points to a V wave that is directly followed by a sleep spindle. Filters: low frequency = 1 Hz, high frequency = 70 Hz. Calibrations: horizontal = 1 second, vertical = 50 μV.

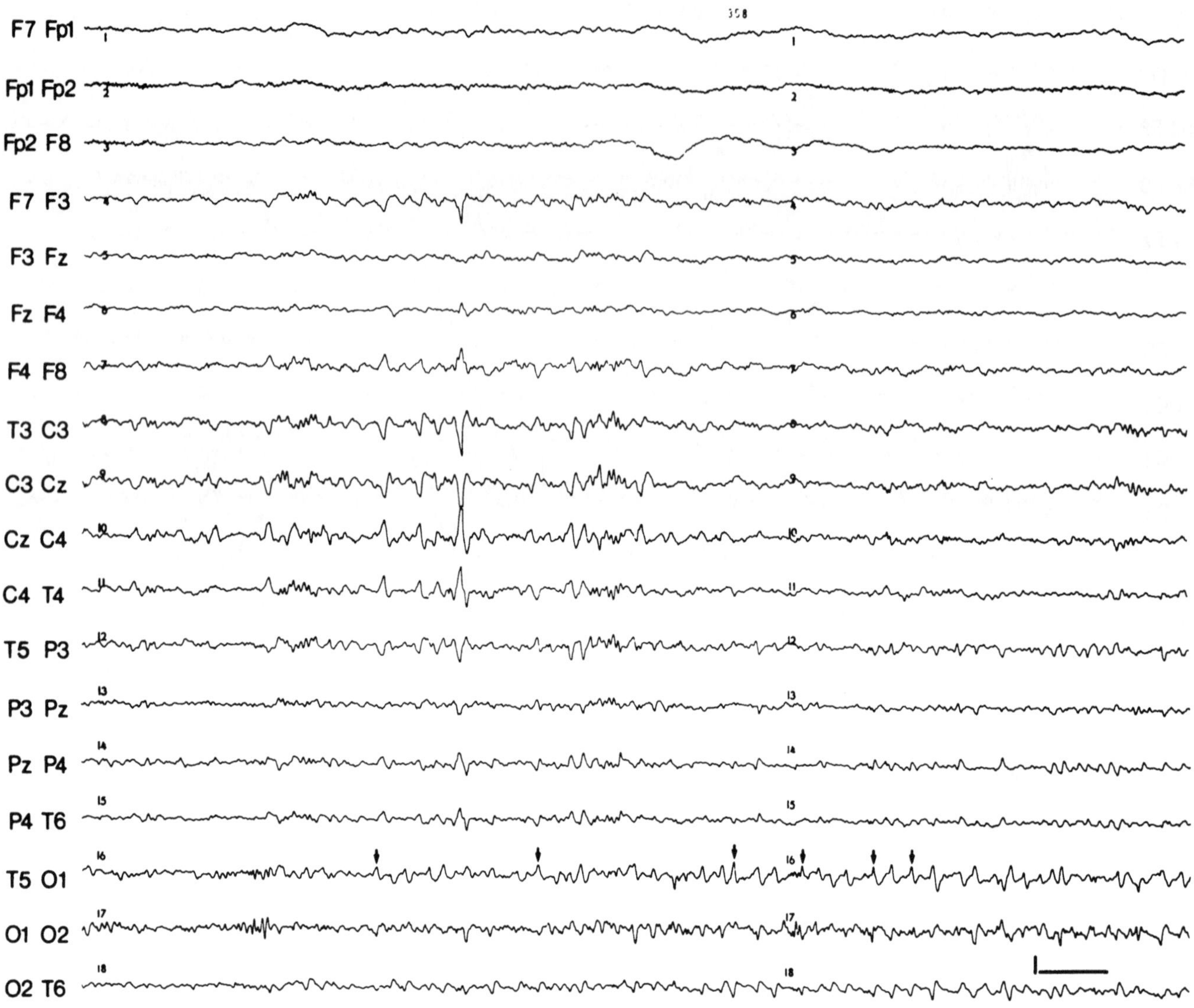

Figure 14.21. Stage II sleep showing POSTS, a number of which are indicated by the arrows. These sharp transients commonly occur in runs as seen in the figure. The phase reversals apparent in the bottom three channels indicate that the POSTS are positive relative to the voltages at other areas. Filters: low frequency = 1 Hz, high frequency = 70 Hz. Calibrations: horizontal = 1 second, vertical = 50 μV.

POSTS

This very descriptive acronym stands for positive occipital sharp transients of sleep. These transients, which do look somewhat like the posts of a fence, are seen in stage II sleep. They occur over the occipital regions on either side, being positive relative to other areas. POSTS occur singly or, more commonly, in runs; sometimes as many as four or five may be seen in a single second. Figure 14.21 is typical of their appearance. Whereas amplitudes are usually 50 μV or less, POSTS may attain amplitudes in excess of 100 μV, as seen in Fig. 14.22. Note in this figure that the POSTS are very sharp indeed and in some instances would fit the definition of a spike. For this reason it is important that they not be mistaken for a focus of abnormal activity.

POSTS are often bilaterally synchronous. At the same time they are commonly asymmetrical on the two sides. Amplitude differences as large as 60% may be seen and are considered to be normal. This, also, can sometimes make their distinction from a focus of abnormal, sharp-wave activity difficult.

The EEG in Deeper Stages of Sleep

In a routine clinical EEG, the sleep recording is normally limited to tracings of stages I and II sleep only. When deeper stages of sleep are observed, the person is usually aroused. In most cases, recordings from the deeper stages of sleep are of little diagnostic value. Nevertheless, it is useful in a general text to mention briefly the other stages of sleep so that the reader will be able to recognize them and distinguish them from stages I and II. The term slow-wave sleep is used to denote stages III and IV.

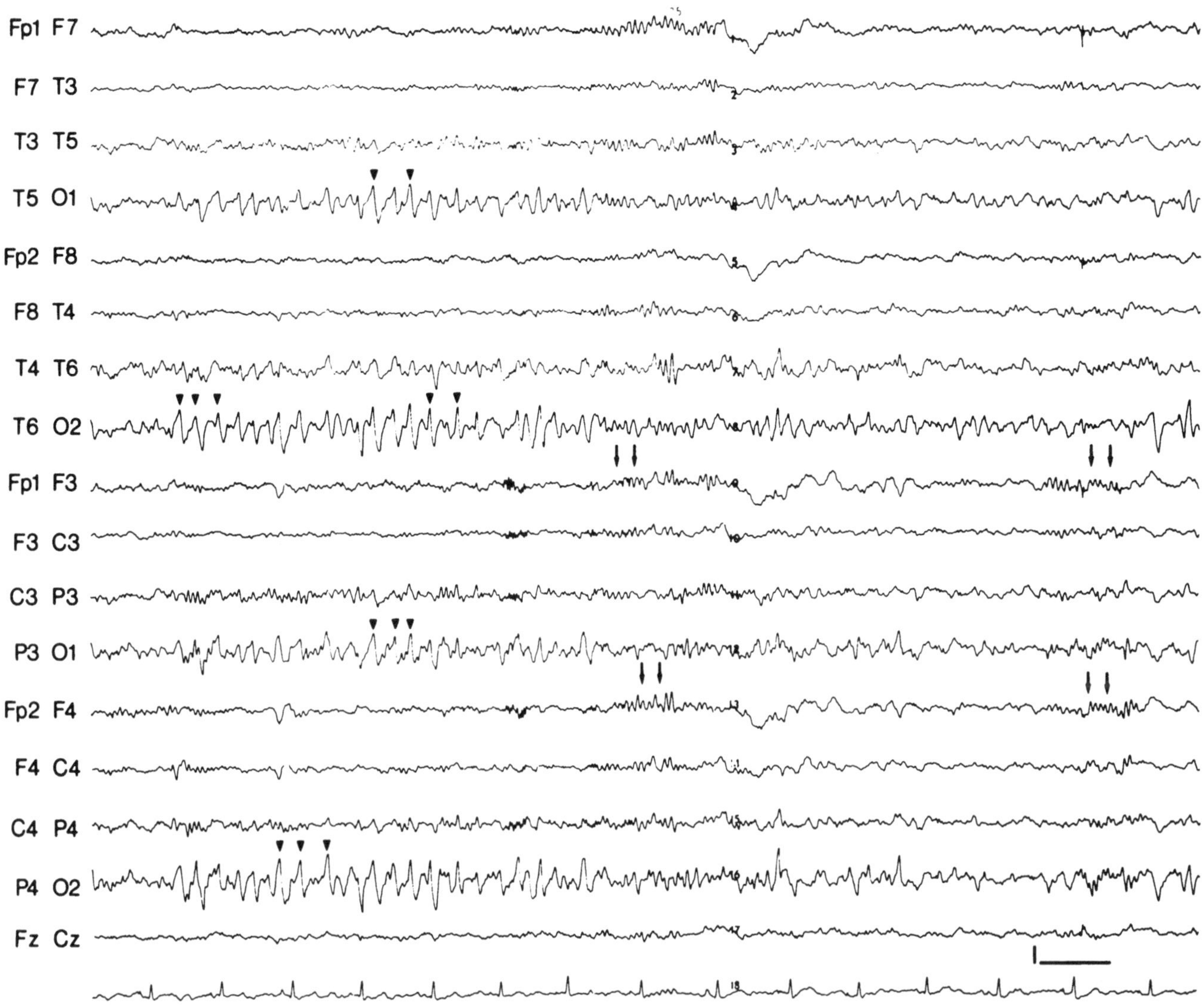

Figure 14.22. Stage II sleep in which numerous, high-amplitude POSTS are present. A number of these sharp transients are indicated by arrowheads. Note that some of the POSTS are very sharp, indeed, and look like spikes. The double arrows point to sleep spindles, another feature of stage II sleep. The channel at the bottom shows the ECG. Filters: low frequency = 1 Hz, high frequency = 70 Hz. Calibrations: horizontal = 1 second, vertical = 50 μV.

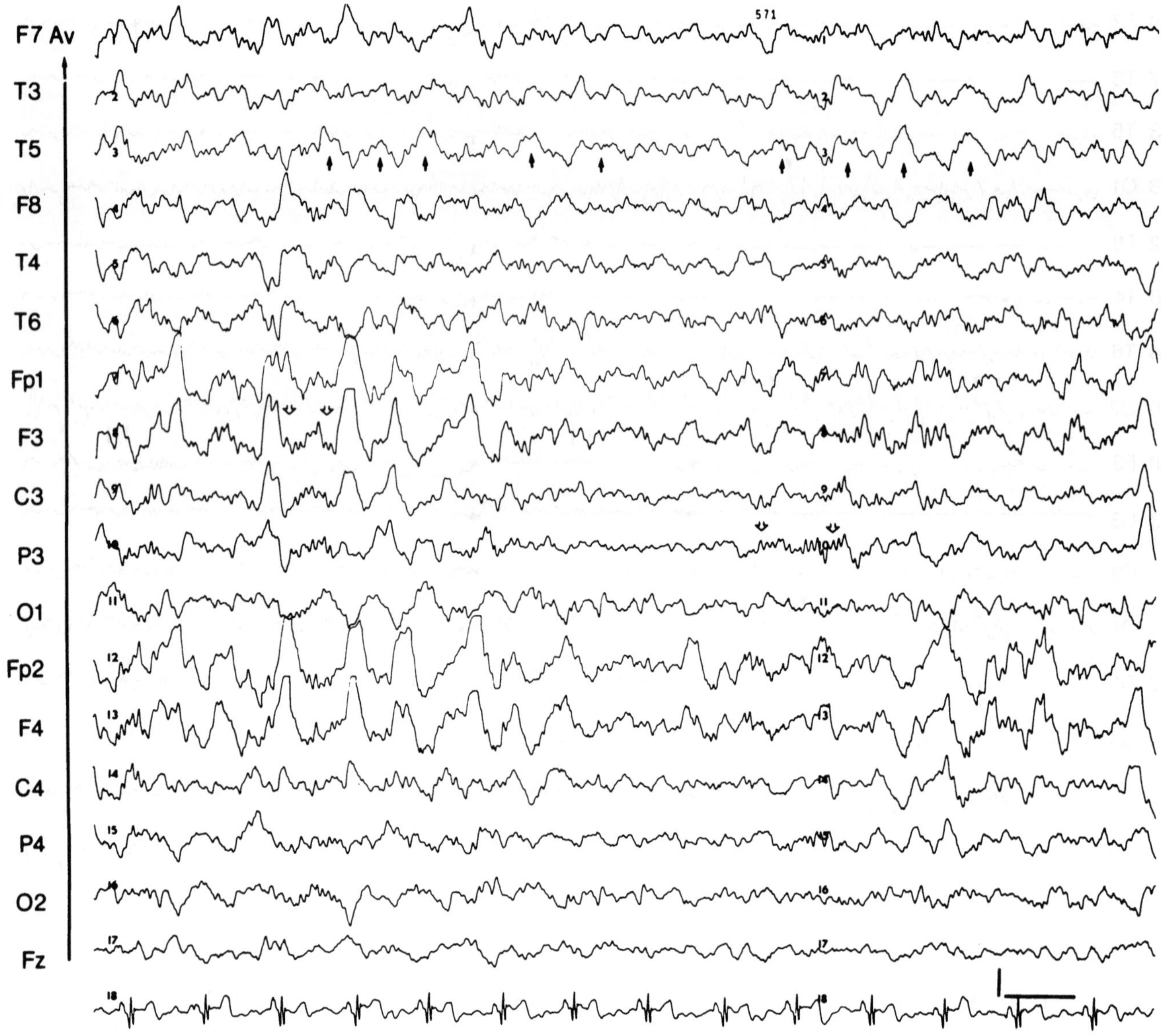

Figure 14.23. Stage III sleep. The solid arrows point to some of the rhythmic delta waves, which in stage III sleep constitute more than 20% of the record. Open arrows indicate sleep spindles. The channel at the bottom shows an ECG. Filters: low frequency = 1 Hz, high frequency = 70 Hz. Calibrations: horizontal = 1 second, vertical = 100 μV.

Stage III

This stage is characterized by a background that consists of irregular and semirhythmic theta activity mixed with mostly rhythmic delta activity, i.e., activity of frequencies less than 4 Hz. The delta activity, which is the hallmark of the deeper stages of sleep, has amplitudes in excess of 100 μV. In stage III sleep, delta activity comprises more than 20% of the record. Occasionally, a few sleep spindles are mixed in with this activity. These tend to disappear as the amount of time spent in stage III sleep increases. Figure 14.23 shows a segment of stage III sleep.

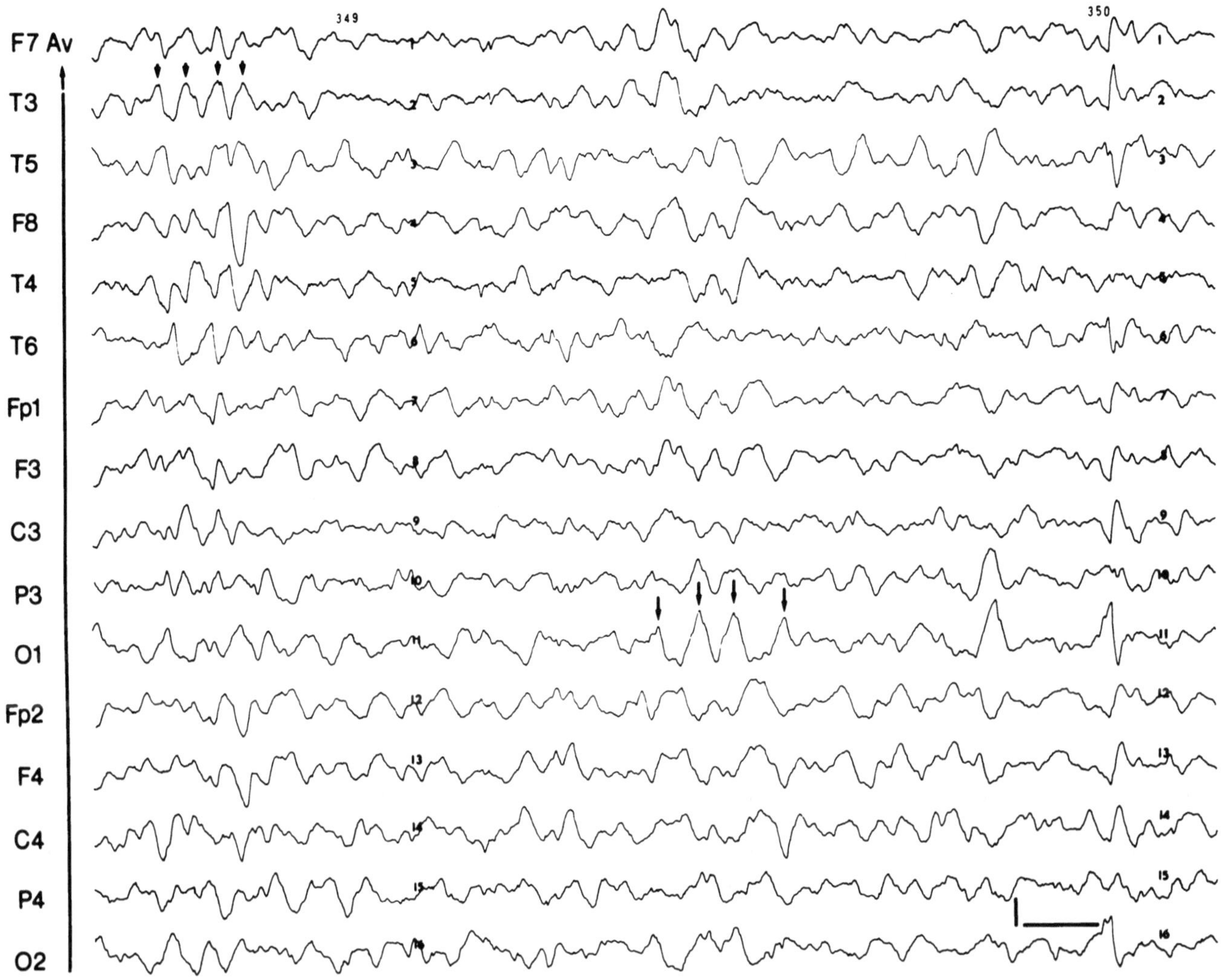

Figure 14.24. Stage IV sleep. The arrows point to rhythmic, high amplitude delta waves, which in stage IV sleep constitute more than 50% of the record. Some of the waves have amplitudes in excess of 300 μV. Filters: low frequency = 1 Hz, high frequency = 70 Hz. Calibrations: horizontal = 1 second, vertical = 150 μV.

Stage IV

This is a deeper stage of sleep. The EEG consists of high-amplitude, rhythmic theta and delta activity. The delta activity, which may show amplitudes of 300 μV or greater, comprises more than 50% of the recording during stage IV sleep. Sleep spindles disappear during this stage. Figure 14.24 is a short recording of stage IV sleep.

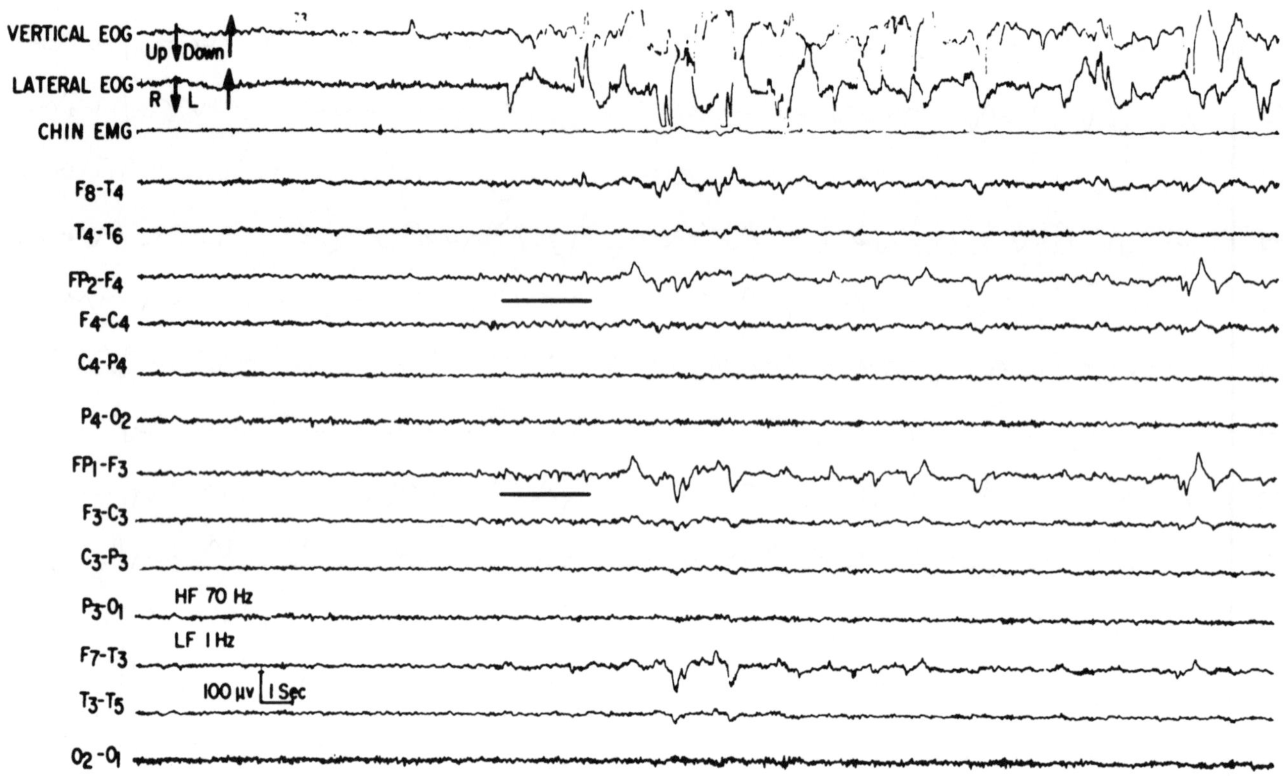

Figure 14.25. The EEG in REM sleep. The tracing includes recordings of vertical (channel 1) and horizontal (channel 2) eye movements and muscle activity from the chin (channel 3). The frontal pole to frontal derivations on both sides (Fp_2-F4 and Fp_1-F3) show a cluster of saw tooth waves (underlined) that initiate an episode of REMs seen in channels 1 and 2. Note the absence of chin muscle activity in channel 3. The EEG consists mostly of low-amplitude beta activity in diffuse distribution. Note the compressed time scale. (Reprinted from Matsuo F, Gaskin JA: Unexpected REM sleep episodes in standard EEG laboratory recording. *Am J EEG Technol* 1986; 26:33–40, by permission of authors and publisher.)

REM Sleep

The EEG during rapid eye movement (REM) sleep is strikingly different from the tracings seen in the other stages of sleep. The background is paradoxically similar to that observed during wakefulness with the eyes open. Figure 14.25 shows a segment of recording obtained while the subject was in REM sleep. It documents the presence of REMs that are the distinguishing feature of this interesting stage of sleep. Such recordings are of special importance in polysomnography, particularly for the diagnosis of conditions like narcolepsy. The interested reader should consult the specialized texts that are available as well as the periodical literature for more information about REM sleep and its clinical significance.

Age-Related Differences: The EEG in Relation to Maturation

It is impossible in an elementary, general text of this kind to cover the wide range of differences that the EEG displays during the course of growth and development. Particularly extensive are the differences evident in the first days, weeks, and months of life. From the standpoint of technology, special montages are needed in taking a neonate's EEG. In dealing with the EEG in neonates and young infants, therefore, the reader should consult the specialized texts that are available for guidance in these matters. Nevertheless, in the routine practice of clinical electroencephalography, children often need to be tested. This section, therefore, is meant to serve as an introduction to the EEG in children. The material, however, is limited to a sampling of the major differences that are seen between the EEGs of children and adults.

Waking Activity

The most striking difference that meets the eye in EEGs from awake children is the character of the posterior dominant rhythm. Generally speaking, the posterior dominant rhythm is slower and of higher amplitude in children than in adults. Insofar as frequency is concerned, the posterior dominant rhythm in many young children does not even qualify as an alpha rhythm. Thus, frequency may be less than the 8-Hz lower limit that defines the alpha rhythm. This will be apparent from a glance of Fig. 14.26, where the posterior dominant rhythm is only 6 to 7 Hz in the

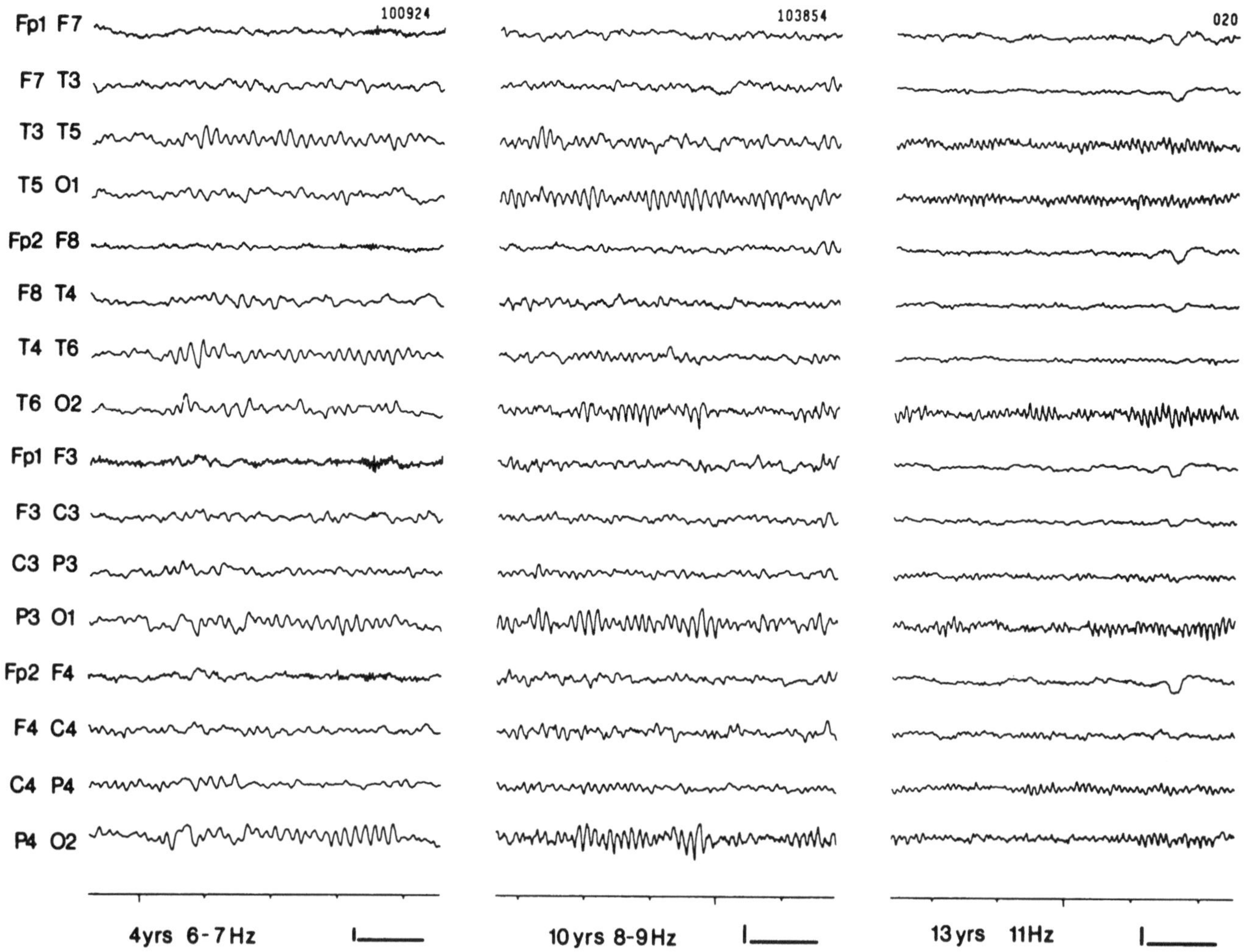

Figure 14.26. Changes in frequency of the posterior dominant rhythm with age during growth and development. Subjects' eyes were closed. Note that between the ages of 4 and 13 years, frequency increases by about 5 Hz. Note also the higher amplitude of the waves in the 4-year-old child. Filters: low frequency = 1 Hz, high frequency = 70 Hz. Calibrations: horizontal = 1 second, vertical = 50 μV.

4-year-old child. In younger children, it is even slower (Fig. 14.30). Nevertheless, such activity is, indeed, the posterior dominant rhythm as it occurs most prominently in the posterior regions and attenuates with eye opening.

The tracing from the 10-year-old child in Fig. 14.26 shows a posterior dominant rhythm of 8 to 9 Hz. The frequency of the waves in this child's tracing satisfies the definition of alpha. Note in the case of the 13-year-old child, the posterior dominant rhythm has reached 11 Hz. In other words, age and the frequency of this feature of the EEG are directly related. But at what age should a child's posterior dominant rhythm fit the definition of alpha rhythm? At what age should it have a frequency of at least 8 Hz? Unfortunately, the question cannot be answered with absolute certainty because the posterior dominant rhythm shows a great deal of variability in normal children. However, a convenient rule of thumb is that it should be at least 8 Hz by 8 years of age. This conservative rule is based on the finding that more than 95% of normal 8-year-old children have a posterior dominant rhythm of 8 Hz or more.

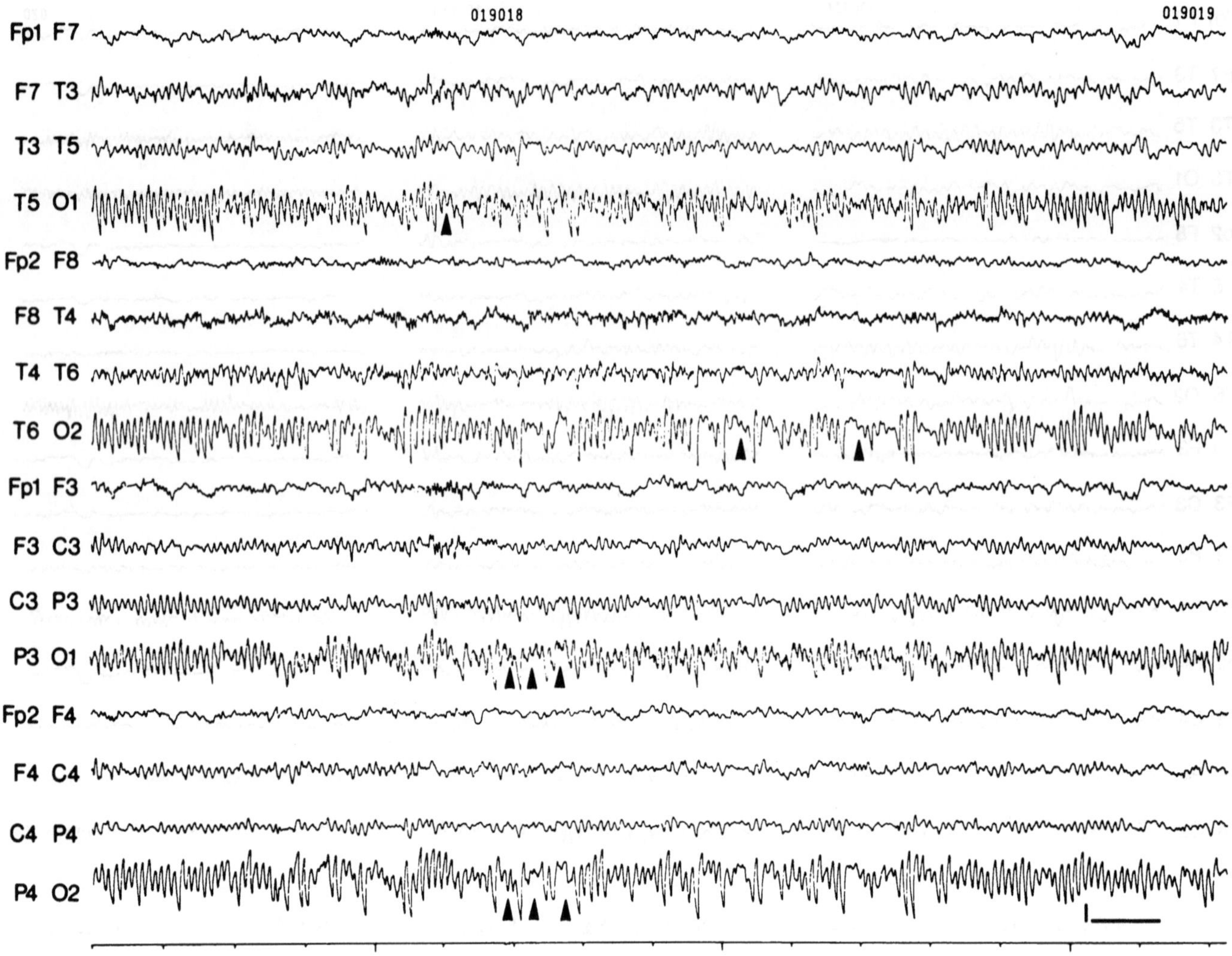

Figure 14.27. Posterior slow waves of youth in a 10-year-old child. The subject's eyes were closed. Arrowheads mark some of the more prominent waves that are seen to fuse with the posterior dominant rhythm of about 10.5 Hz. Filters: low frequency = 1 Hz, high frequency = 70 Hz. Calibrations: horizontal = 1 second, vertical = 50 μV.

Tracings taken from the posterior derivations show other differences in normal children. The most notable are the slow transients that have been termed *posterior slow waves of youth*. They are recorded with the person's eyes closed. Figure 14.27 illustrates these interesting waves, which are shaped like the sails of a schooner. Note that these waves are fused with the waves of the posterior dominant rhythm with which they are intermixed; their amplitude is similar to the amplitude of the alpha rhythm. Although Fig. 14.27 shows them to be present only in the occipital region, at times they also are seen in the posttemporal and parietal regions.

Posterior slow waves of youth occur most commonly in the EEGs of children 8 to 14 years of age. They may be present in younger children and in older adolescents, but they are rare in normal adults over 21 years of age. Although these waves are frequently bilaterally synchronous and symmetrical, they may be asynchronous and/or asymmetrical as well. Posterior slow waves of youth attenuate or block with eye opening and disappear during drowsiness, along with the alpha rhythm. They frequently are accentuated and/or become more numerous during hyperventilation.

Although only traces of theta activity are seen in the waking EEGs of normal young adults, frontocentral and central theta activity is quite common in normal children. Figure 14.28 shows evidence of significant amounts of frontocentral theta activity in a normal 5 year old. At times, paroxysmal bursts of rhythmic theta activity—mostly in frontal and frontocentral regions—are also seen in children. Such activity, which may have amplitudes in excess of 100 μV, should not be confused with similar activity—termed hypnagogic hypersynchrony—that occurs during drowsiness. The significance of the high-amplitude, rhythmic theta activity sometimes present in wakefulness is unknown; however, it is not considered to be abnormal.

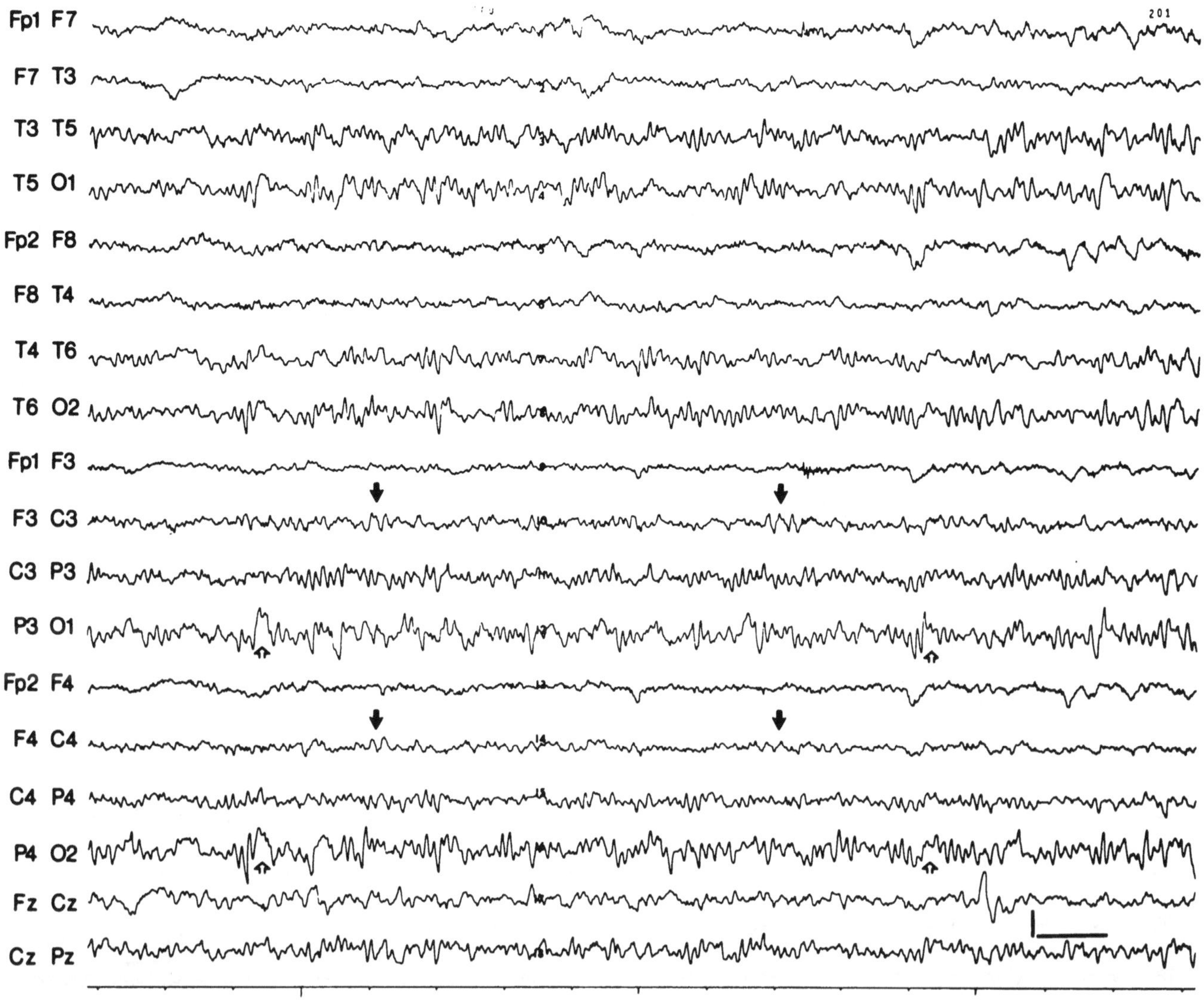

Figure 14.28. Frontocentral theta activity (solid arrows) in a 5-year-old child. The subject's eyes were closed. The tracing also contains a number of posterior slow waves of youth, some of which are marked by open arrows. The posterior dominant rhythm is variable and has a range of about 7.5 to 10 Hz. Such variability is not uncommon in children. Filters: low frequency = 1 Hz, high frequency = 70 Hz. Calibrations: horizontal = 1 second, vertical = 100 μV.

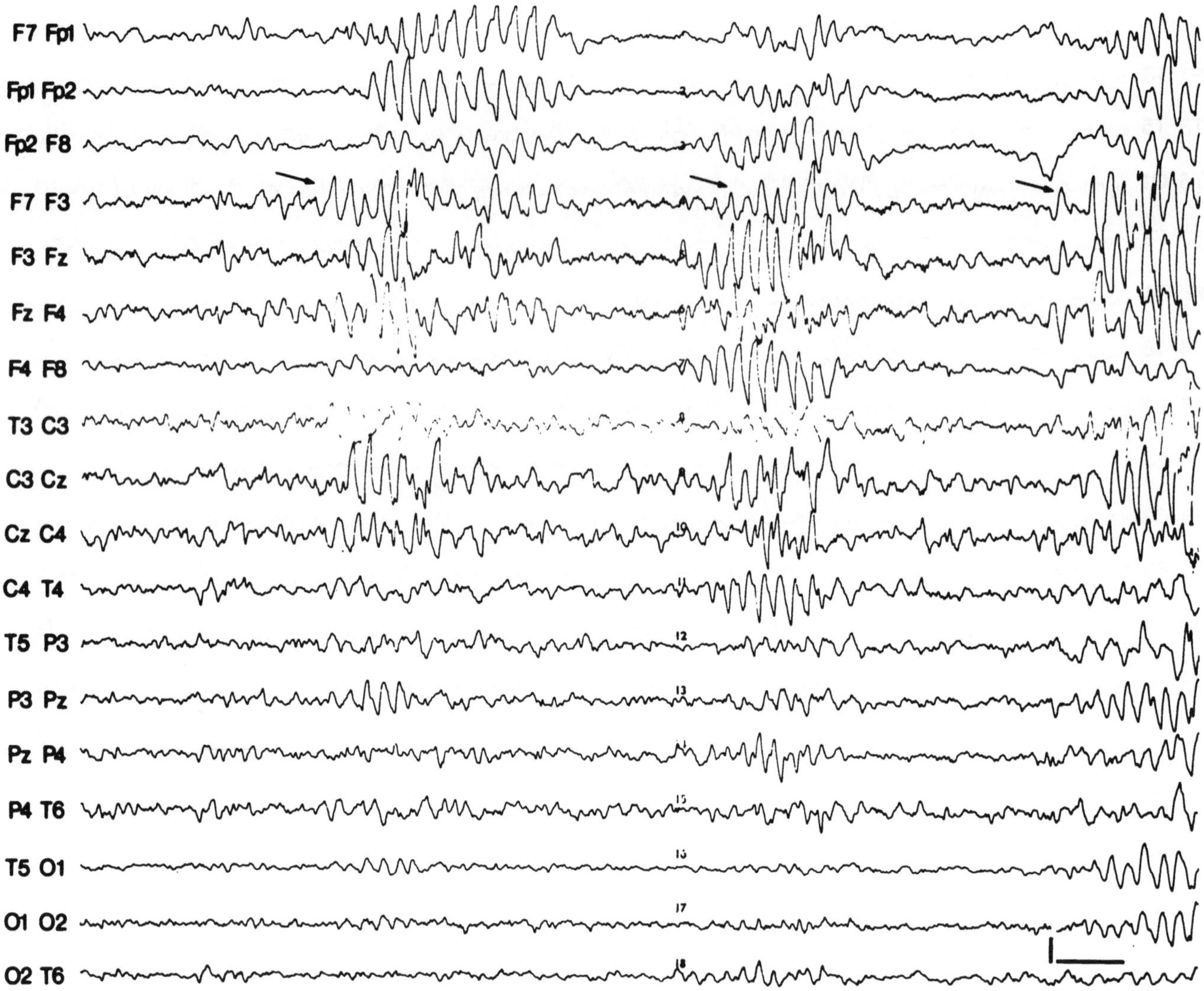

Figure 14.29. Hypnagogic hypersynchrony in a 4-year-old child. The arrows indicate three bursts of paroxysmal, high-amplitude rhythmic activity at about 4 Hz. The subject's eyes were closed throughout. This is a normal finding. Filters: low frequency = 1 Hz, high frequency = 70 Hz. Calibrations: horizontal = 1 second, vertical = 100 μV.

Activity During Drowsiness and Sleep

Although drowsiness in normal children, like drowsiness in adults, is often signaled by an overall reduction in amplitude of the background, the EEG in the majority of infants 3 to 11 months old and a significant percentage of children 1 to 6 years of age displays a unique, high-amplitude phenomenon during drowsiness. This phenomenon is known as ***hypnagogic hypersynchrony***. It consists of paroxysmal, high-amplitude bursts of slow (3 to 6 Hz), very rhythmic activity. Having amplitudes sometimes in excess of 300 μV, this activity may occur in widespread distribution but is more commonly centrally or frontocentrally dominant (see Fig. 14.29). When intermixed with fast activity that may be present at the same time, these paroxysmal bursts may falsely give the impression of spike and wave discharges. Despite their ominous appearance, these waves are a normal feature of the EEG in childhood.

As already mentioned, hypnagogic hypersynchrony can occur in the very young. Figure 14.30 shows hypnagogic hypersynchrony alternating with a wakeful pattern in a 14-month-old baby. The phenomenon is a normal feature up to age 15 years, but is rarely seen after 11 years.

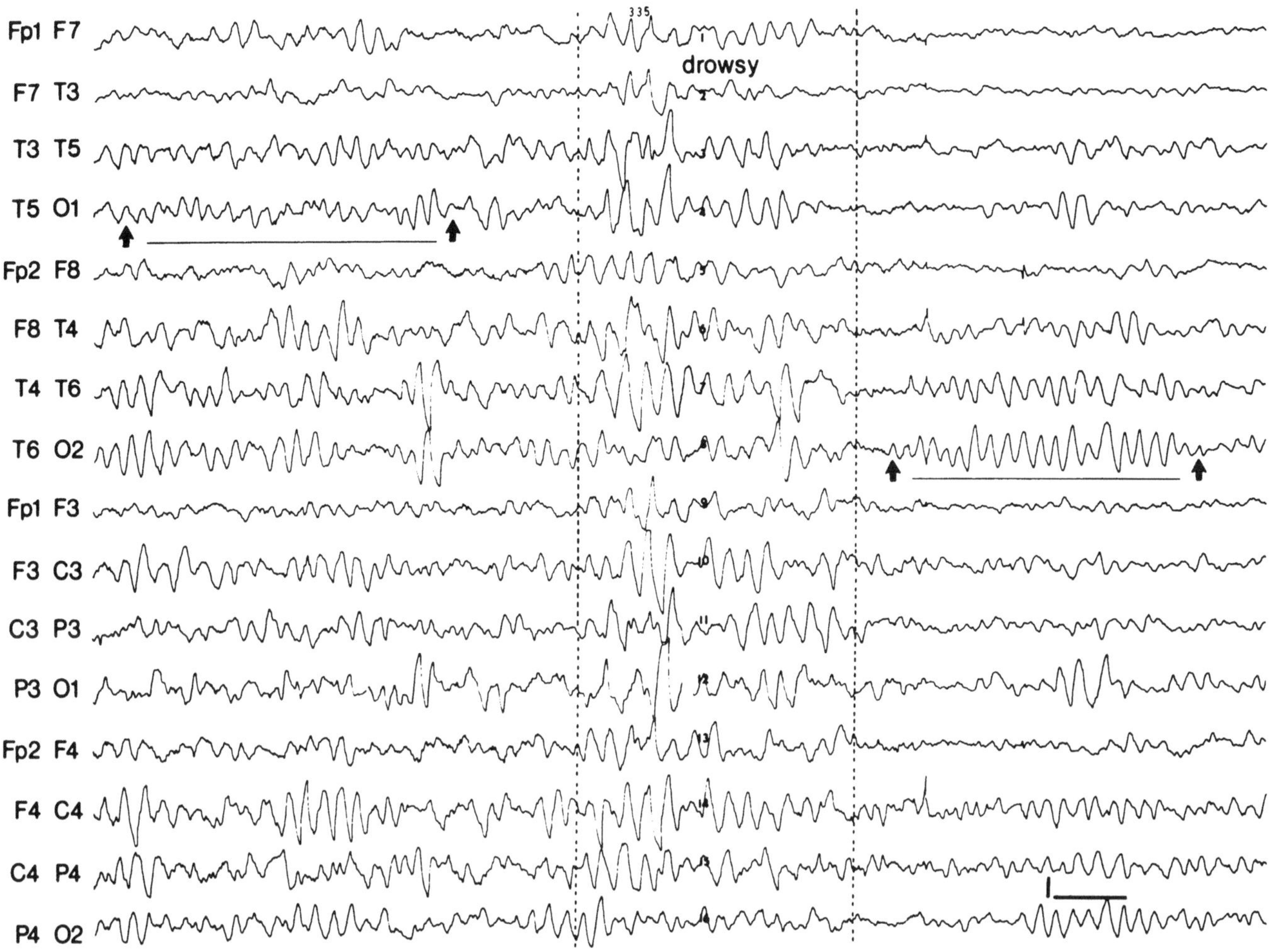

Figure 14.30. Alternating wakeful and drowsy states in a 14-month-old child. The child's eyes were closed throughout the recording. At the left, a posterior dominant rhythm of about 4 to 4.5 Hz (underlined and marked by arrows) is present. This is replaced by a burst of generalized, high-amplitude, somewhat slower activity (hypnagogic hypersynchrony). Following the brief drowsy episode, the posterior dominant rhythm returns (underlined and marked by arrows). Filters: low frequency = 1 Hz, high frequency = 70 Hz. Calibrations: horizontal = 1 second, vertical = 50 μV.

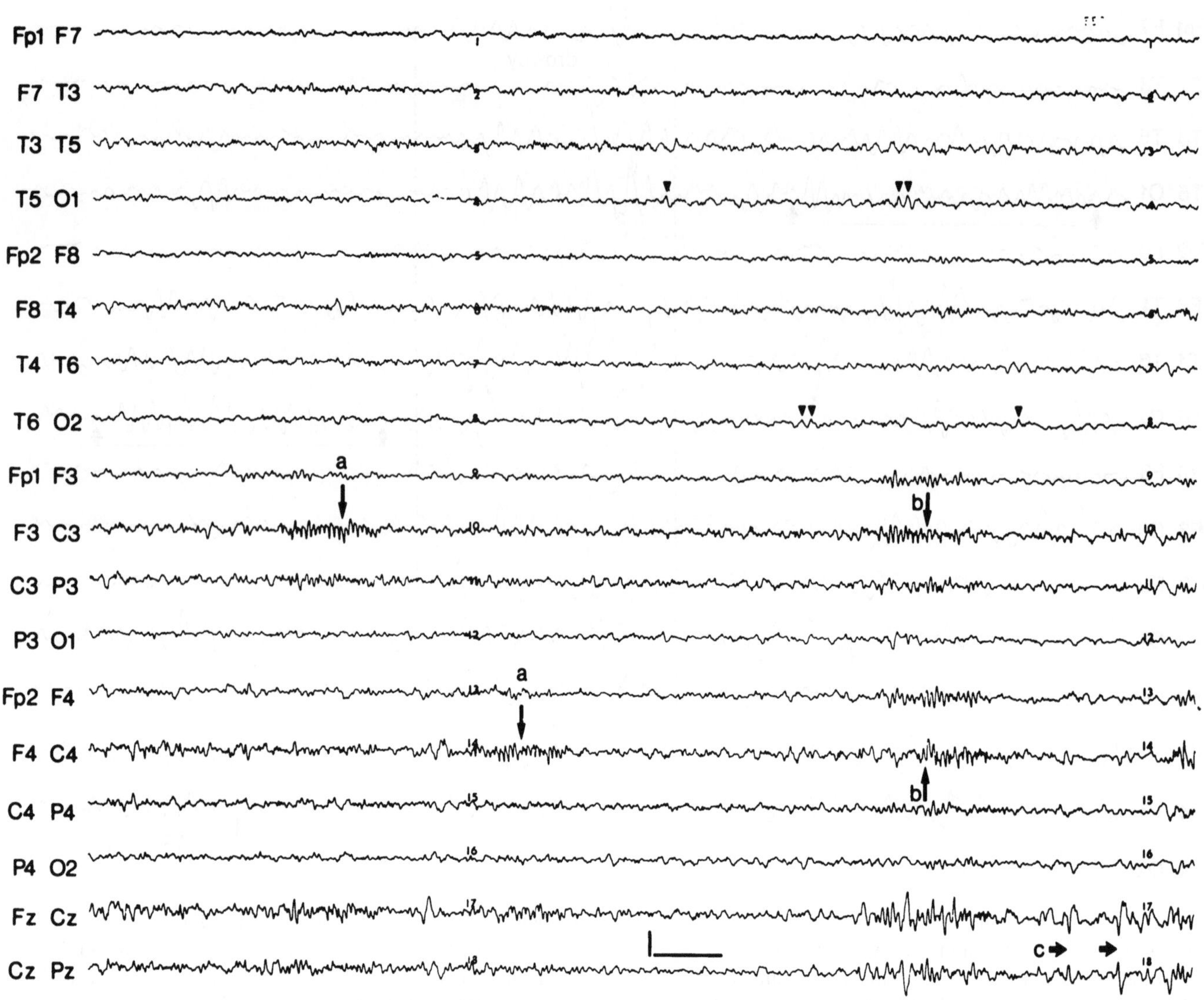

Figure 14.31. Asynchronous sleep spindles in a 1-year-old child. The arrows at "a" mark sleep spindles on left and right sides that are asynchronous. At "b" the spindles are partly synchronous. The arrows at "c" point to vertex waves, and the arrowheads note several POSTS that are also features of stage II sleep. Filters: low frequency = 1 Hz, high frequency = 70 Hz. Calibrations: horizontal = 1 second, vertical = 100 μV.

The principal features of stage II sleep taken up earlier in our discussion of the adult EEG are generally applicable to children as well. Some of these features emerge and acquire adult characteristics very early in life; others acquire adult characteristics somewhat later. The sleep spindle is an example of the latter. Thus, sleep spindles, which first appear at 6 to 8 weeks postterm, are bilaterally asynchronous and continue to show some asynchrony during the first year of life (Fig. 14.31). Some degree of asynchrony may continue and is not considered to be abnormal until age 2 to 2.5 years. Details concerning the other EEG features of sleep in young children may be found by consulting the more advanced, specialized texts that are available.

Arousal from sleep in children may show a phenomenon that is akin to hypnagogic hypersynchrony that was discussed earlier. This phenomenon or feature of the EEG in childhood is termed *hypnopompic hypersynchrony*, and a sample is shown in the tracing of Fig. 14.32. This figure nicely illustrates the rapid shifts that may occur from sleep to drowsiness, and back again to sleep.

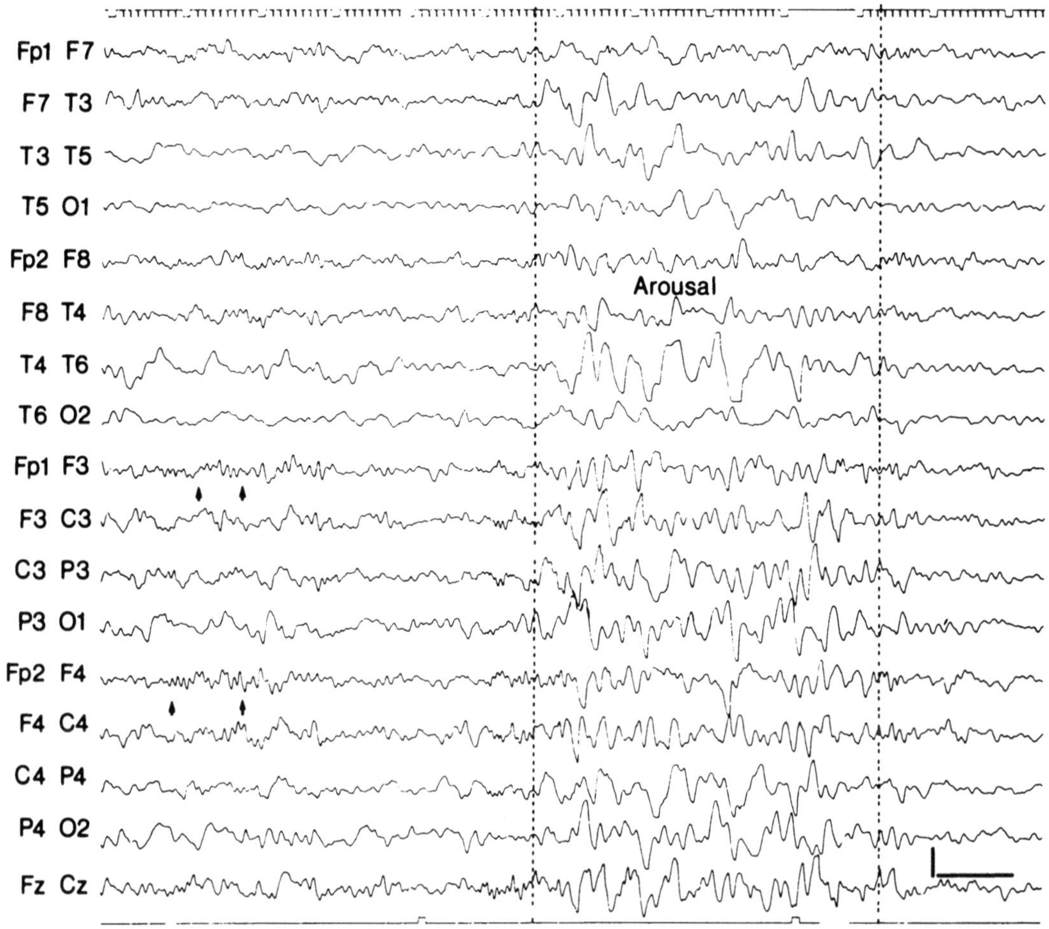

Figure 14.32. Arousal from sleep in an 8-year-old child. The stage II sleep—note the sleep spindles at arrows—is interrupted by a paroxysmal burst of high-amplitude, rhythmic slow activity lasting about 4 seconds. The paroxysmal activity is known as hypnopompic hypersynchrony. Filters: low frequency = 1 Hz, high frequency = 70 Hz. Calibrations: horizontal = 1 second, vertical = 150 μV.

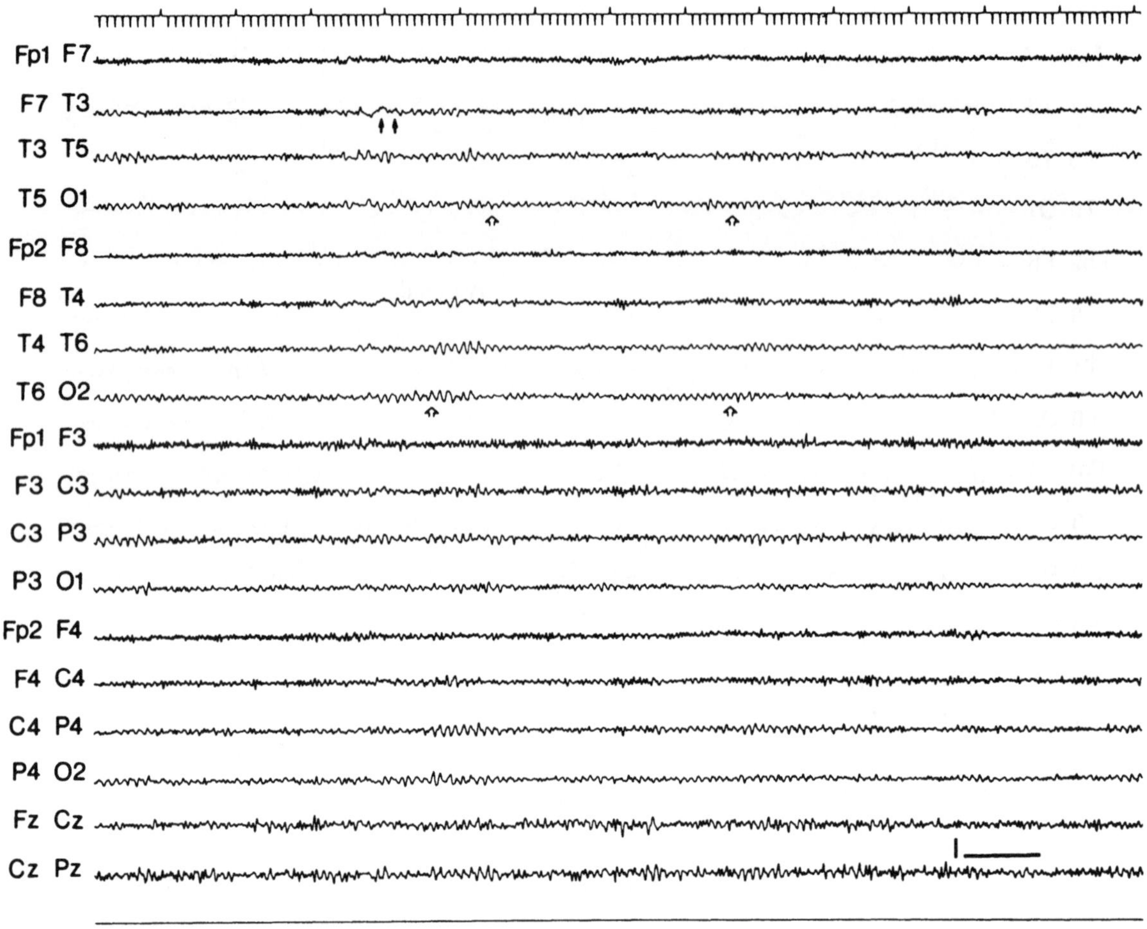

Figure 14.33. Awake, eyes-closed EEG in a 60-year-old person. The open arrows indicate the posterior dominant rhythm at about 10 Hz. The two solid arrows point to a single episode of temporal slowing on the left side. Filters: low frequency = 1 Hz, high frequency = 70 Hz. Calibrations: horizontal = 1 second, vertical = 50 μV.

Age-Related Differences: The EEG in Old Age

There is some evidence that suggests that the frequency of the posterior dominant rhythm declines with advancing age. The clinical value of this finding is somewhat dubious, however, since thoroughly healthy individuals in advanced years may not show such a change. Moreover, since EEG slowing is commonly associated with dementia, it is not clear whether the reported slowing is an accompaniment of normal aging or an early sign of pathology.

The same problem obtains in the case of temporal slow activity, which is undoubtedly the feature most commonly observed in the EEGs of older persons. This activity is most often seen in the temporal region on the left side, but it also may occur intermittently or simultaneously on both sides. Temporal slow activity is episodic, irregular, and mostly of low amplitude (20 to 30 μV). The episodes may consist of one to three or four waves. Figure 14.33 shows a single episode of very mild temporal slowing in a 60 year old. Such rare instances of temporal slowing at this age are considered to be normal.

In older individuals the episodes become more frequent and may be of higher amplitude. Figures 14.34 and 14.35, which are tracings from apparently asymptomatic 78 and 80 year olds, respectively, illustrate this. Considering the patient's advanced age, the EEG in Fig. 14.34 may be considered normal, assuming of course that such episodes of temporal slowing occur only occasionally throughout the record. The EEG in Fig. 14.35, however, illustrates a borderline case. Interpretation in such cases is difficult, and, at best, the final judgment concerning normality or abnormality is tenuous. Obviously, more research into the EEG of old age is needed before such clinical judgments can be sharpened.

The normal EEG in hyperventilation and in photic stimulation is discussed in the chapter on activation procedures.

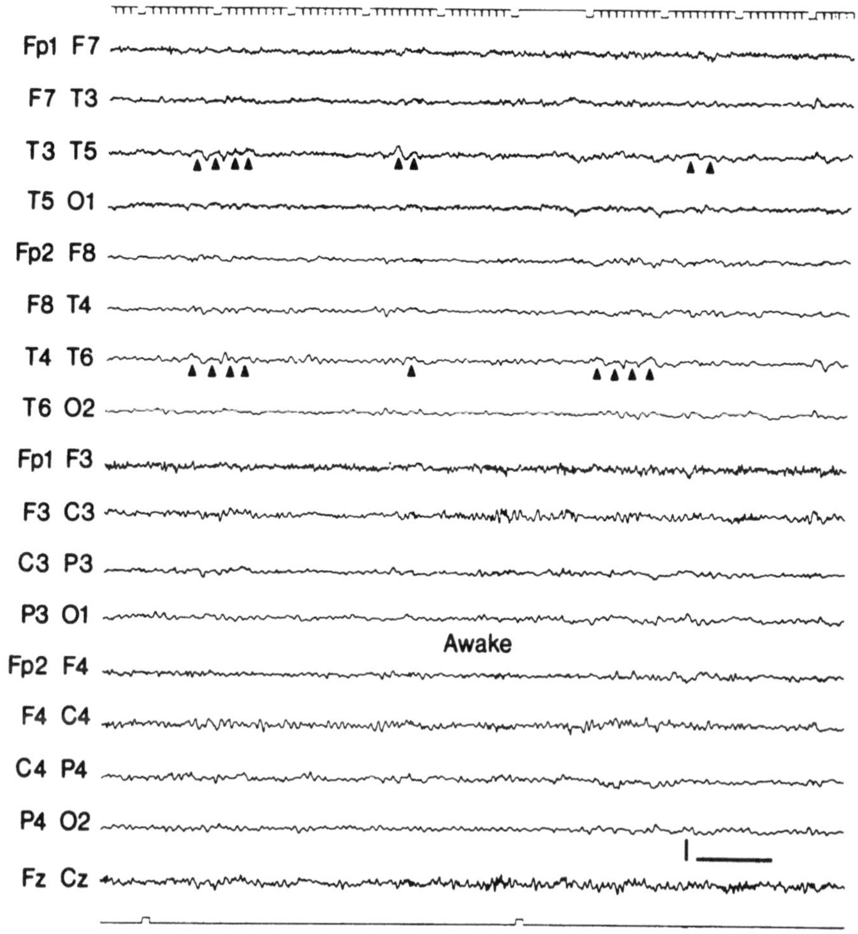

Figure 14.34. Temporal slowing on both sides in a 78-year-old person. The arrowheads indicate three episodes of slowing, which consist of irregular, low-amplitude theta and delta activity. Filters: low frequency = 1 Hz, high frequency = 70 Hz. Calibrations: horizontal = 1 second, vertical = 50 μV.

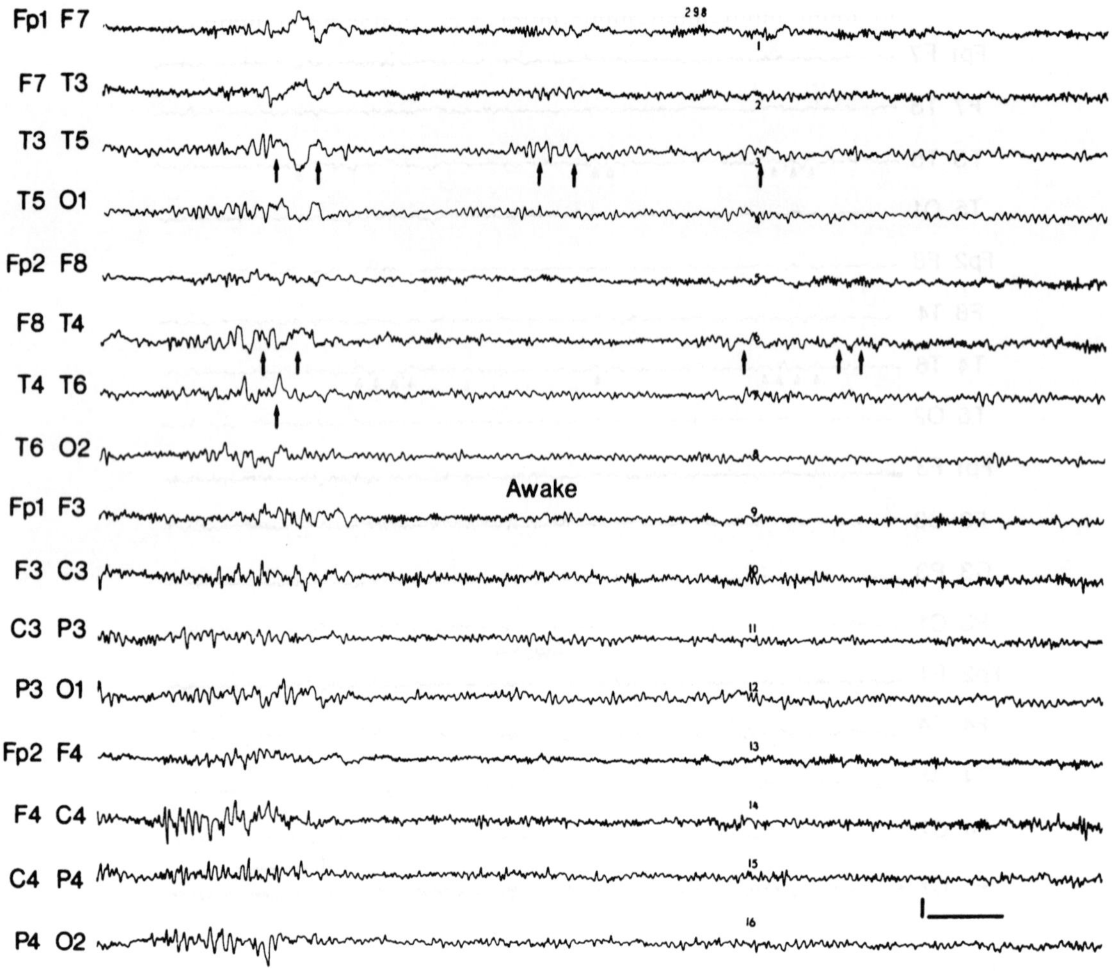

Figure 14.35. Temporal slowing in an 80-year-old person. The slowing (arrows) is more severe and more extensive on the left side, where it extends to F7. Filters: low frequency = 1 Hz, high frequency = 70 Hz. Calibrations: horizontal = 1 second, vertical = 50 μV.

Chapter 15
Abnormal EEG Patterns

A perusal of Chapter 13, Introduction to EEG Reading, should convince the reader that a thorough knowledge of the different normal and abnormal EEG patterns is essential to gain proficiency in EEG interpretation. This knowledge can be gained only by going over a large number of EEG tracings and becoming familiar with the various physiological and pathological alterations commonly and uncommonly seen in the EEG. This chapter, therefore, takes up the descriptive features of individual abnormal EEG patterns and briefly mentions their significance. Later, in a chapter on clinical correlations (Chapter 21), the specific disorders of the central nervous system (CNS) that lead to the abnormal EEG patterns are highlighted, and the EEG is discussed from the standpoint of its diagnostic and prognostic implications. The reader will note some amount of repetition and overlap between the two chapters; but this, hopefully, will serve to reinforce the important aspects of the topic.

General Considerations

The term abnormal EEG patterns refers to patterns of activity that are judged to be outside the normal range. These judgments are based on current knowledge accumulated through the evaluation of EEGs of large numbers of neurologically normal individuals of different ages by many electroencephalographers. In the chapter on the normal EEG (Chapter 14), we saw that one needs to take into account the age and state of the patient to correctly interpret a particular EEG pattern. Thus, for example, a pattern that is normal for a drowsy patient may be considered abnormal if the patient is fully awake. Similarly, a pattern that is normal for a neonate may be quite abnormal for an older child. These examples underscore the importance of the technologist's notations regarding age and state of the patient on the EEG tracing. Without such information, EEG interpretation and judgments of normality or abnormality may be of doubtful clinical value.

The abnormal EEG patterns may be broadly divided into five categories: abnormalities of the background rhythms, abnormal sleep patterns, focal or generalized abnormal slow activity, paroxysmal epileptogenic abnormalities, and abnormal periodic paroxysmal patterns. It should be obvious, of course, that there is some overlap between some of the categories; thus, for example, there are times when slow activity may appear in paroxysms.

Abnormalities of the Background Rhythms

Alternations in rate, rhythm, distribution, symmetry, amplitude, or reactivity of the background activity may occur during various CNS disorders. The alterations may involve one or more of the physiological rhythms, namely, the alpha, beta, or mu rhythms.

Alpha Rhythm

In the awake adult an alpha rhythm of less than 8 Hz is abnormal. Since a number of clinical conditions can produce slowing of the alpha rhythm, the slowing is considered a nonspecific abnormality. Thus, bilateral slowing of the alpha rhythm may be seen in metabolic, toxic, and infectious encephalopathies of diverse etiology. It is also a consistent finding in patients with dementia irrespective of the underlying cause. The degree of slowing often parallels alteration in the mental status of the patient. It should also be noted that the alpha rhythm slows down in patients with hypothyroidism and can become normal when a euthyroid state results from adequate treatment. Figure 15.1 shows an "alpha rhythm" of 7 Hz in an adult patient who was found to have toxic serum levels of phenytoin.

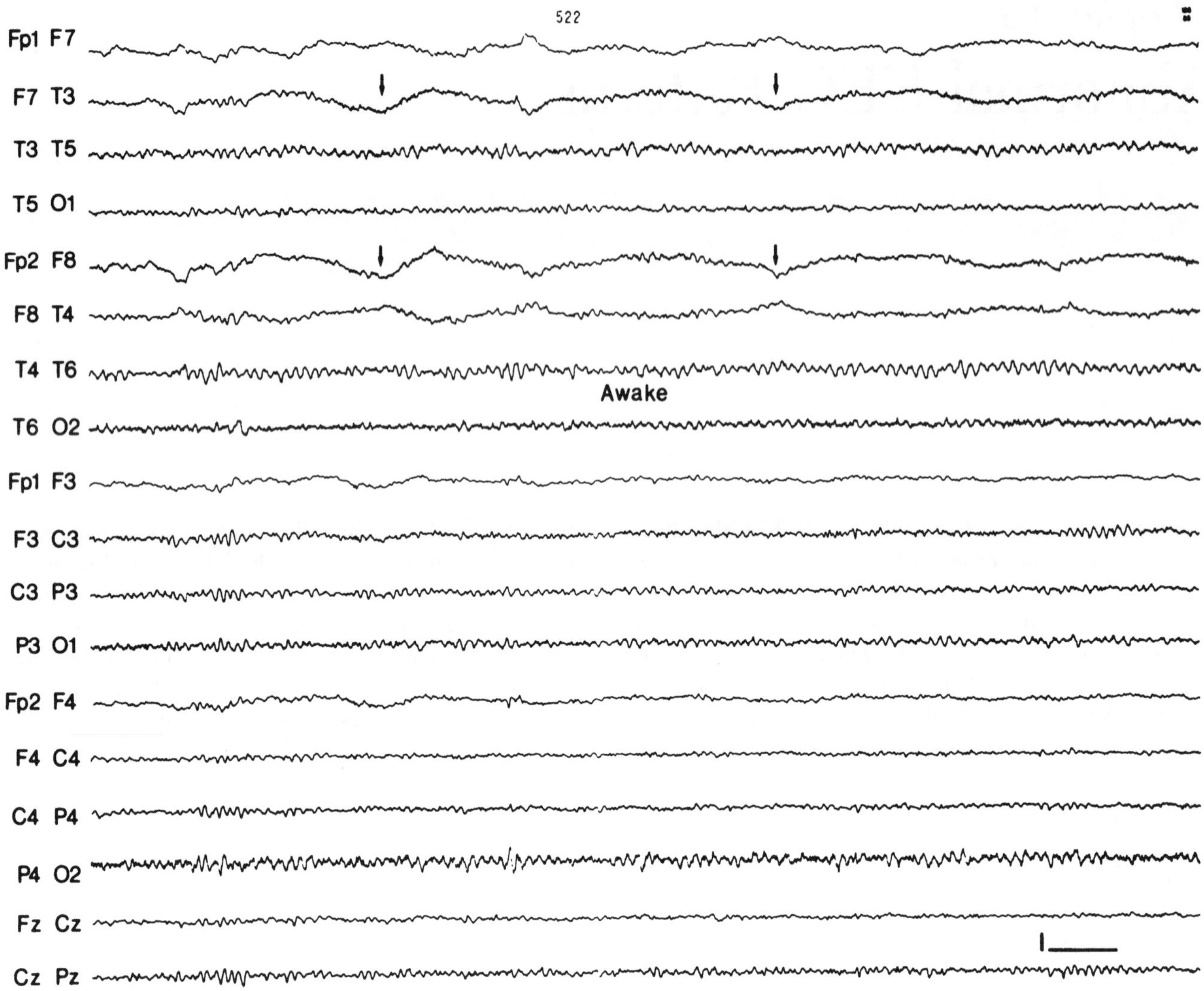

Figure 15.1. Posterior dominant rhythm of 7 Hz in a 30-year-old epileptic with a serum phenytoin level of 24 μg/mL. The arrows point to lateral eye movements. Filters: low frequency = 1 Hz, high frequency = 70 Hz. Calibrations: horizontal = 1 second, vertical = 50 μV.

Asymmetrical slowing of the alpha rhythm with a consistent difference of greater than 1.5 Hz between the two sides is abnormal and should suggest the possibility of a lesion on the slower side. However, such a finding does not necessarily indicate the presence of a lesion in the occipital lobe itself; asymmetrical slowing of the alpha rhythm is known to occur even with lesions that are more anteriorly located.

A difference in amplitude of the alpha rhythm between the two sides is considered significant if it exceeds 50%. Since the alpha rhythm in most normal persons is of higher amplitude on the right side, even a 35% decrease on the right side may be significant. Figure 15.2 shows a marked amplitude asymmetry of the alpha rhythm, which is of much lower amplitude on the right side compared with the left. The computed tomography (CT) scan showed a right-sided subdural hematoma. Lesions that involve the cerebral cortex—especially in the posterior regions—or that cause accumulation of fluid between the brain and the recording electrode, as in the case of subdural or epidural hematoma and scalp edema, may lead to attenuation of alpha rhythm ipsilaterally.

Abnormalities may also occur in the distribution of the alpha rhythm. Normally, it is distributed in the occipital, parietal, and, to some extent, posterior temporal areas; however, activity may occur over widespread areas, including the frontal regions. Such an alpha pattern is abnormal and is seen in alpha coma, which may result from a number of conditions such as brain stem infarct or cerebral anoxia, or it may be a drug effect. In this context, it is worth reiterating that an alpha rhythm may appear spuriously in the frontal areas when using an average potential reference (see Fig. 14.3, Chapter 14), and this should not be mistaken for an alpha coma pattern.

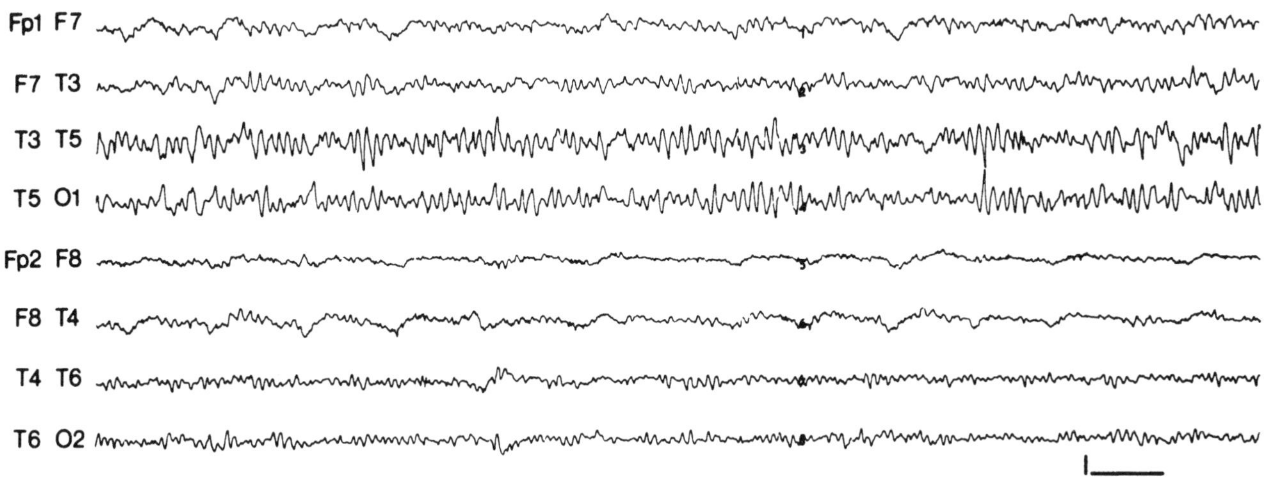

Figure 15.2. Asymmetrical and slow (7–7.5 Hz) alpha rhythm in a 14-year-old adolescent having a right-sided subdural hematoma. Filters: low frequency = 1 Hz, high frequency = 70 Hz. Calibrations: horizontal = 1 second, vertical = 50 μV.

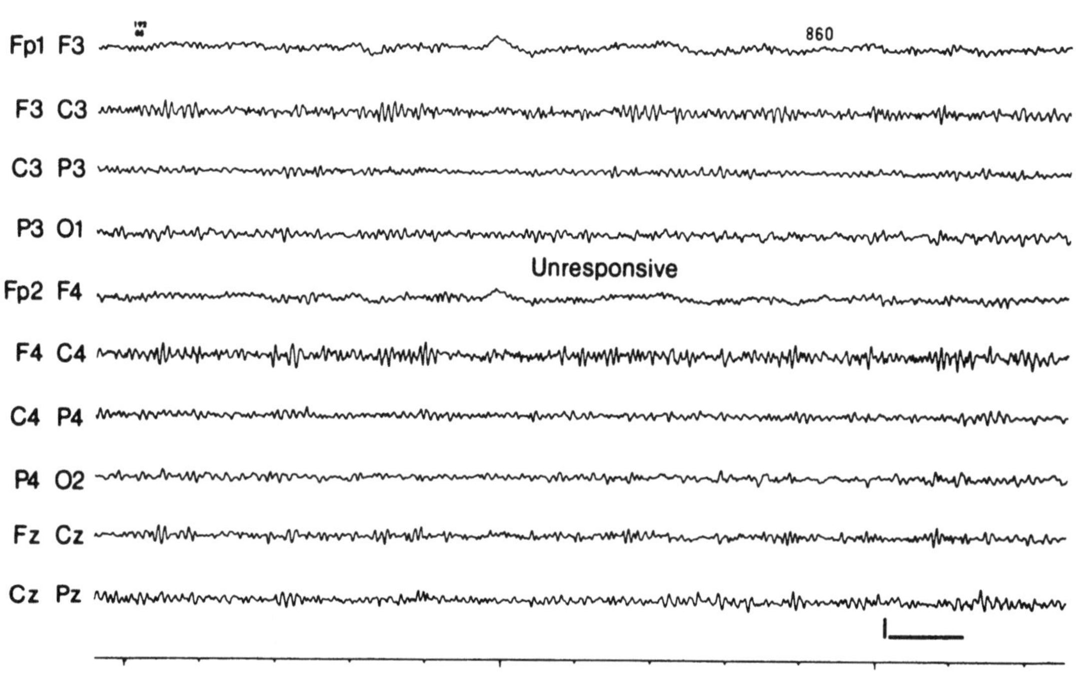

Figure 15.3. Alpha coma pattern in a 60-year-old patient suffering from a lower brain stem infarct. The patient was unresponsive to verbal commands and external stimuli. Note the diffusely distributed alpha rhythm. Passive eye opening and external stimulation did not evoke any change in the alpha rhythm. Filters: low frequency = 1 Hz, high frequency = 70 Hz. Calibrations: horizontal = 1 second, vertical = 50 μV.

Lack of reactivity to eye opening is a significant finding, particularly if consistently demonstrated on one side. This may be an early finding in occipital lobe lesions. A total lack of reactivity of the alpha rhythm is a feature that may be seen in alpha coma, particularly in cases of diffuse cerebral anoxia due to cardiac arrest. In the case of alpha coma resulting from lower brain stem lesions, there may be some degree of reactivity. This contrasts with findings in psychogenic unresponsiveness where reactivity to eye opening is normal. Figure 15.3 shows an alpha coma pattern in a 60-year-old patient with a low brain stem infarct. The reader will find further discussion on the alpha coma pattern in Chapter 21.

A focal increase in amplitude and/or frequency of the alpha rhythm is known to occur in patients with structural lesions, particularly tumors; but this is quite an uncommon finding. Remember that a localized increase in amplitude may also be seen over a skull defect.

Asymmetrical photic driving in the alpha frequency band, if quite consistent, may be abnormal. Rarely, the driving may be more pronounced on the side of the lesion. See Chapter 16 for further details on photic stimulation.

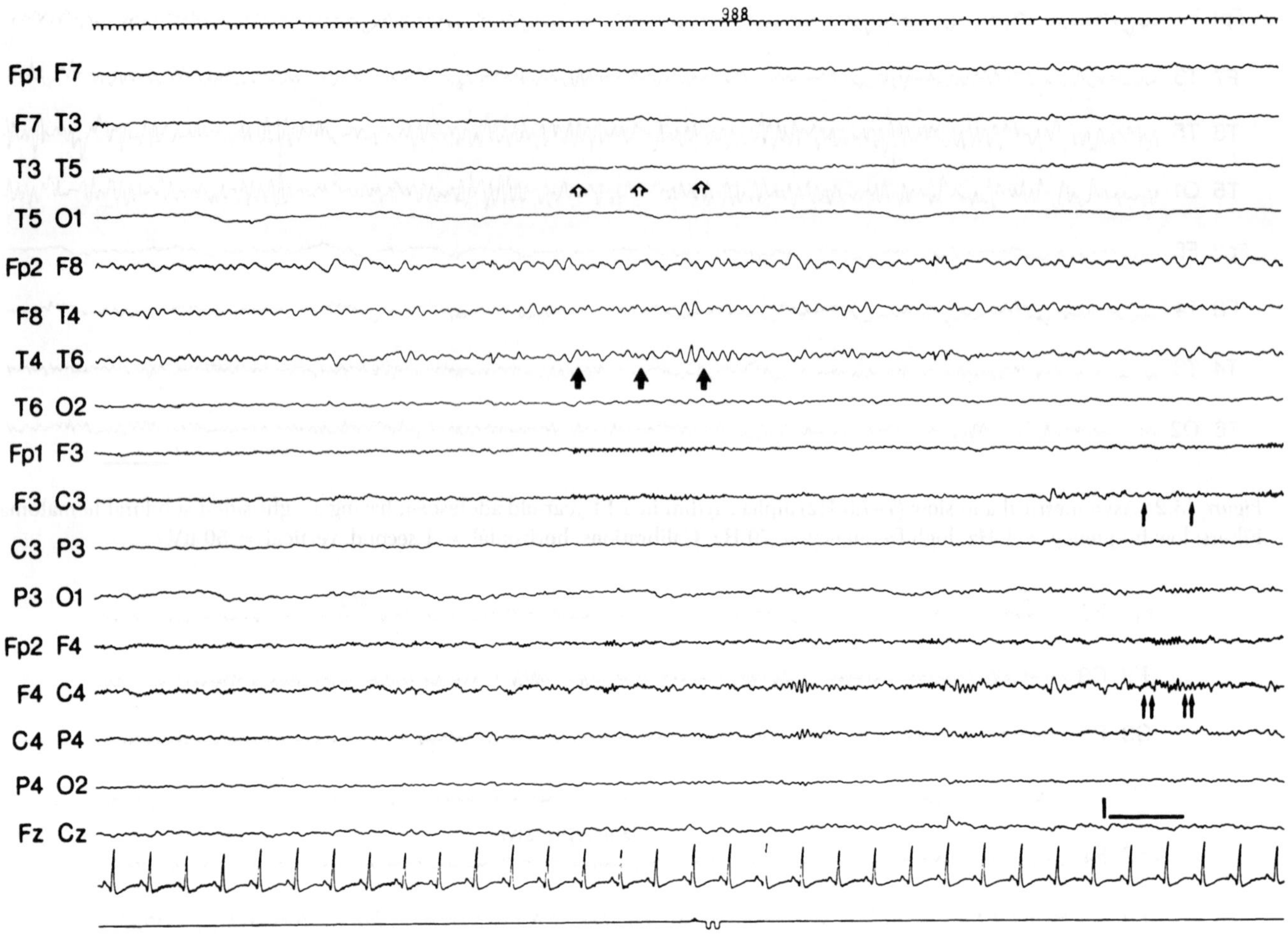

Figure 15.4. Asymmetry of beta activity in a 30-year-old patient with a subdural hematoma on the left side. The tracing also shows a marked attenuation of slow background activity on the left (open arrows) as compared with the right side (solid arrows). The thin arrows point to rhythmic beta activity, which has a higher amplitude on the right side (double arrows). Filters: low frequency = 1 Hz, high frequency = 70 Hz. Calibrations: horizontal = 1 second, vertical = 50 μV.

Beta Activity

Both attenuation and accentuation of beta activity may be abnormal. Beta attenuation is often seen in patients with destructive cortical lesions or when there is an abnormal collection of fluid between the cortex and the recording electrodes. Figure 15.4, which is from a patient with a left subdural hematoma, shows attenuation of alpha as well as beta activity on that side.

An excessive amount of beta activity, especially in diffuse distribution, is usually a drug effect. Many drugs can produce an increase in beta activity; barbiturates are the most common. Figure 15.5 shows a marked increase in beta activity in a patient who is on barbiturates for epilepsy. A focal increase in beta activity is most often seen over a skull defect. The skull bones act as high-frequency filters; in the presence of a bone defect, therefore, beta activity may become quite prominent.

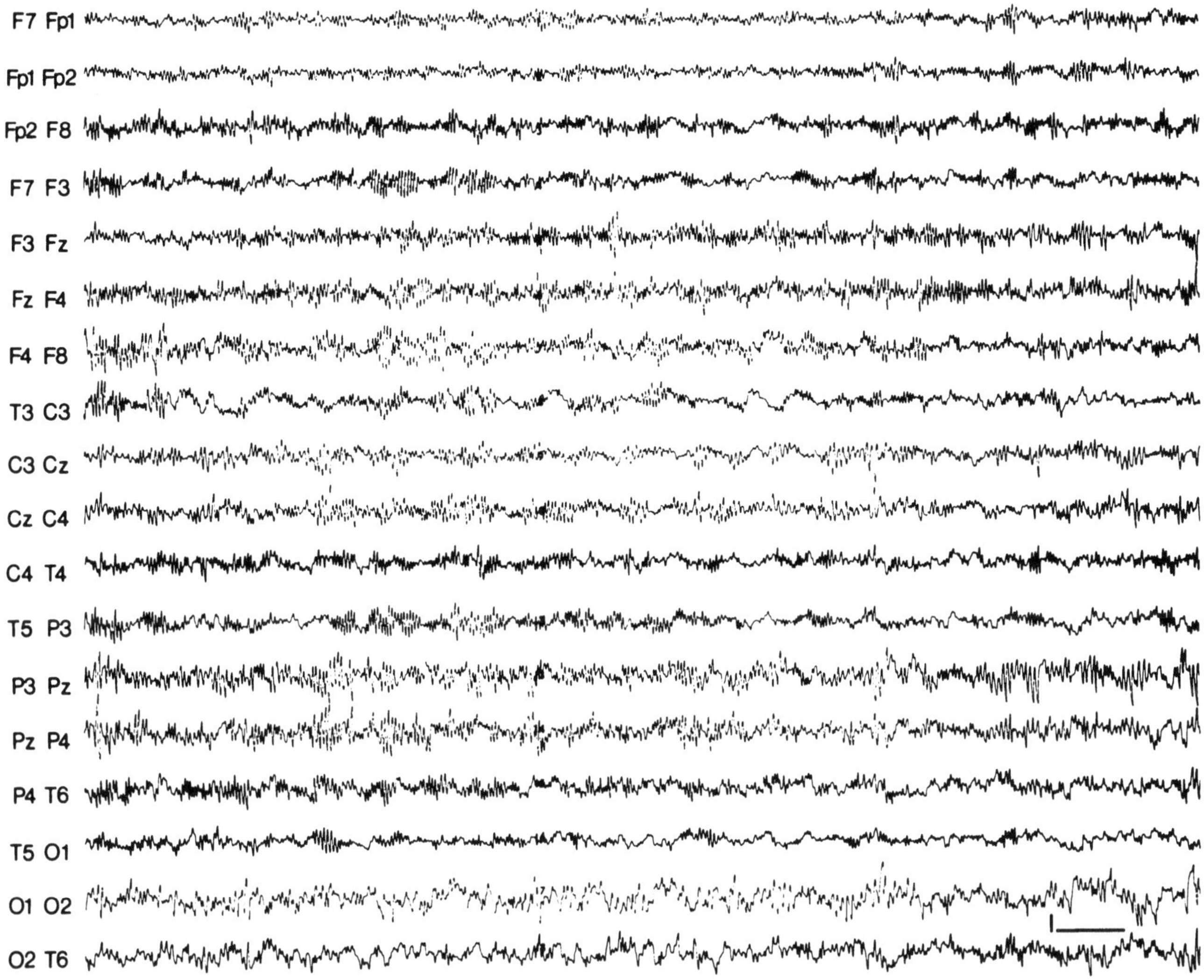

Figure 15.5. Excessive beta activity in a 5-year-old child who is on phenobarbital for control of generalized seizures. Note the diffusely distributed, high-amplitude beta activity that sometimes forms obvious spindles. Filters: low frequency = 1 Hz, high frequency = 70 Hz. Calibrations: horizontal = 1 second, vertical = 50 μV.

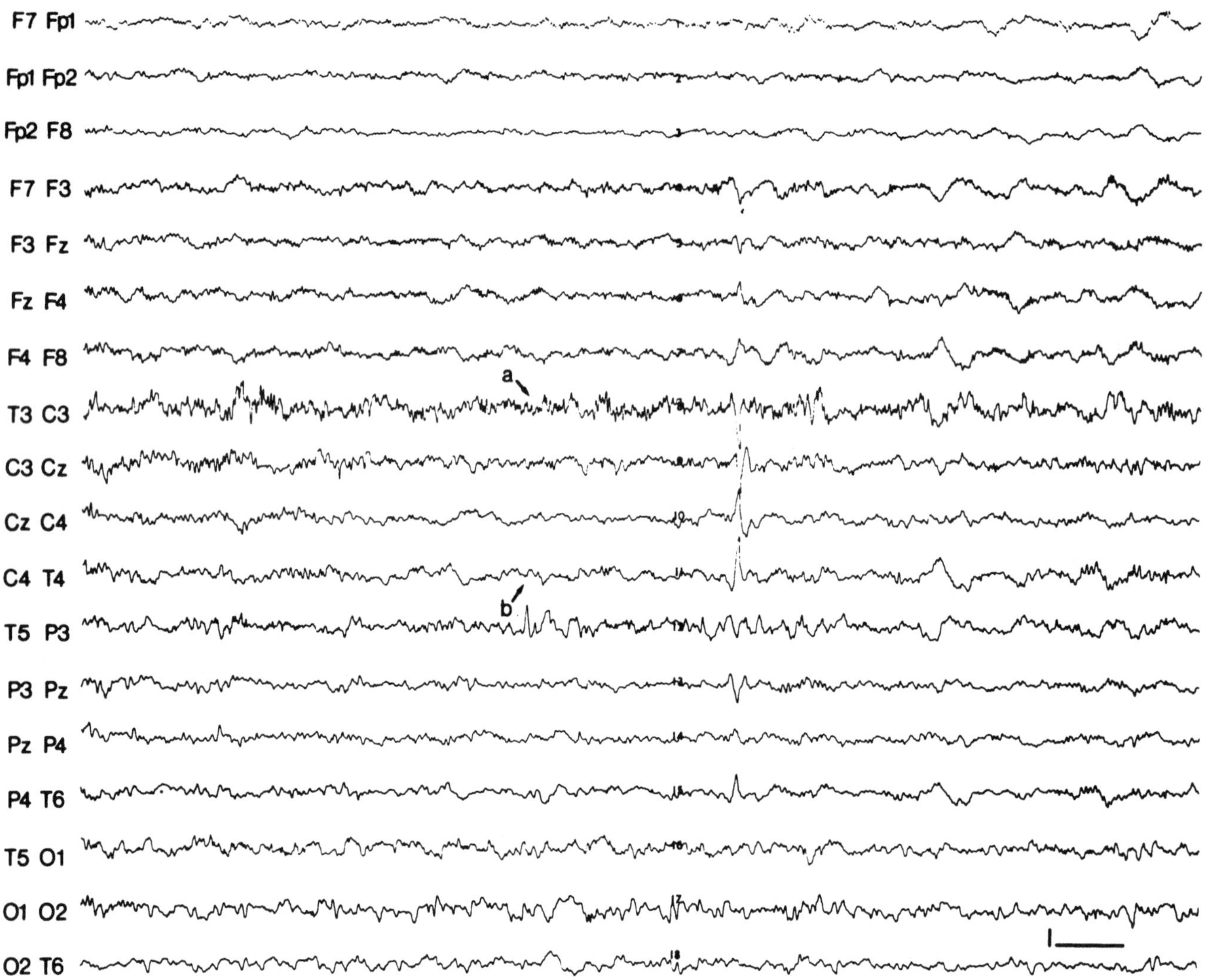

Figure 15.6. Focal increase in beta activity over the left midtemporal and central areas in a patient with an underlying skull defect. The arrow at "a" points to prominent beta activity on the left side, and the arrow at "b" indicates the lower amplitude beta in the homologous area on the right. Note the vertex wave. Filters: low frequency = 1 Hz, high frequency = 70 Hz. Calibrations: horizontal = 1 second, vertical = 50 μV.

Figure 15.6 shows a focal increase in beta activity in the left midtemporal and central areas in a patient with a skull defect from a craniotomy. Figure 15.7 shows a similar increase in beta activity over the right frontal, central, and temporal areas in a patient who had previous cranial surgery.

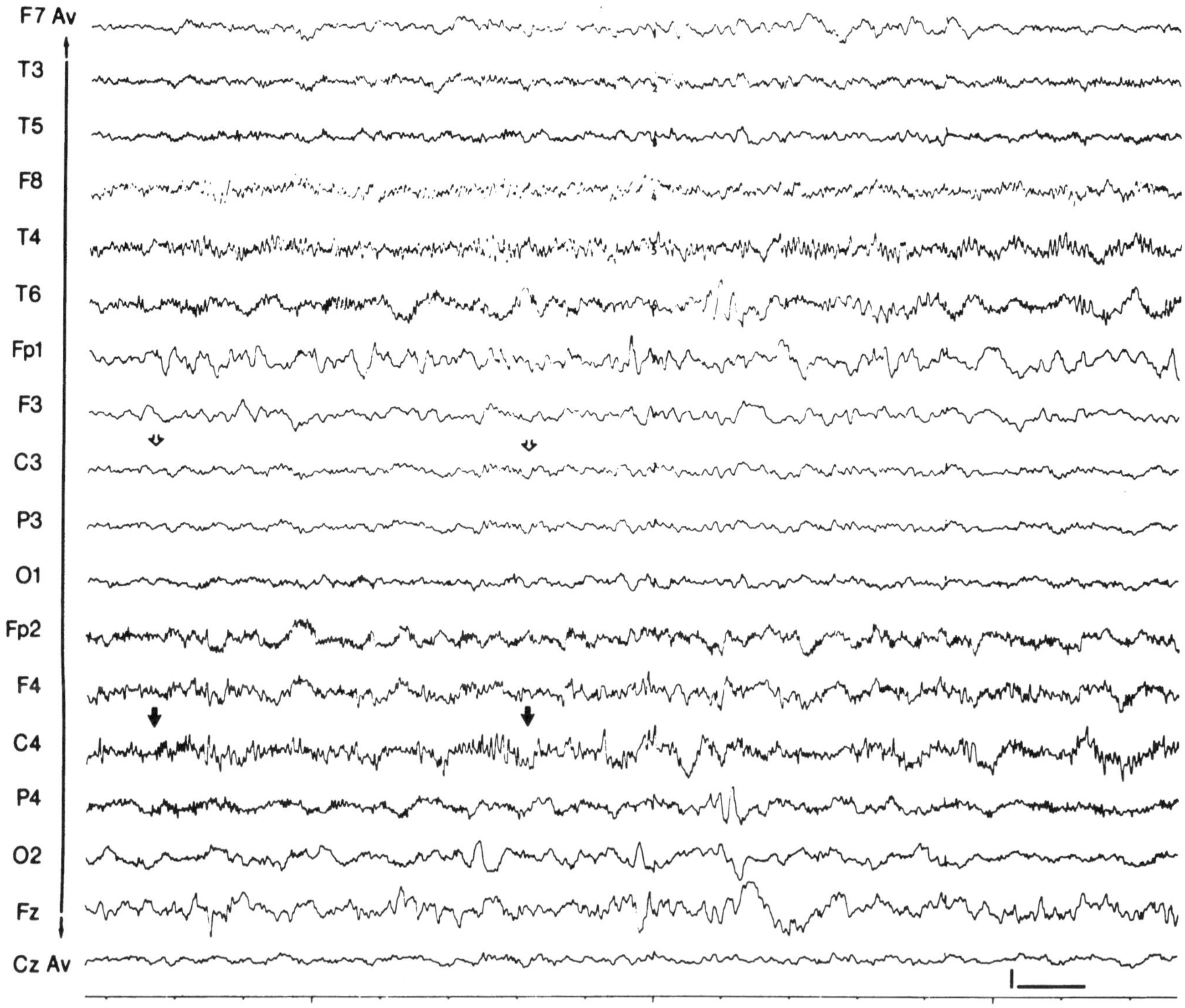

Figure 15.7. Increased amplitude of beta activity in the right frontal, central (solid arrows) and temporal areas compared with the homologous areas on the left side (open arrows). The 40-year-old patient had a history of craniotomy. Filters: low frequency = 1 Hz, high frequency = 70 Hz. Calibrations: horizontal = 1 second, vertical = 50 μV.

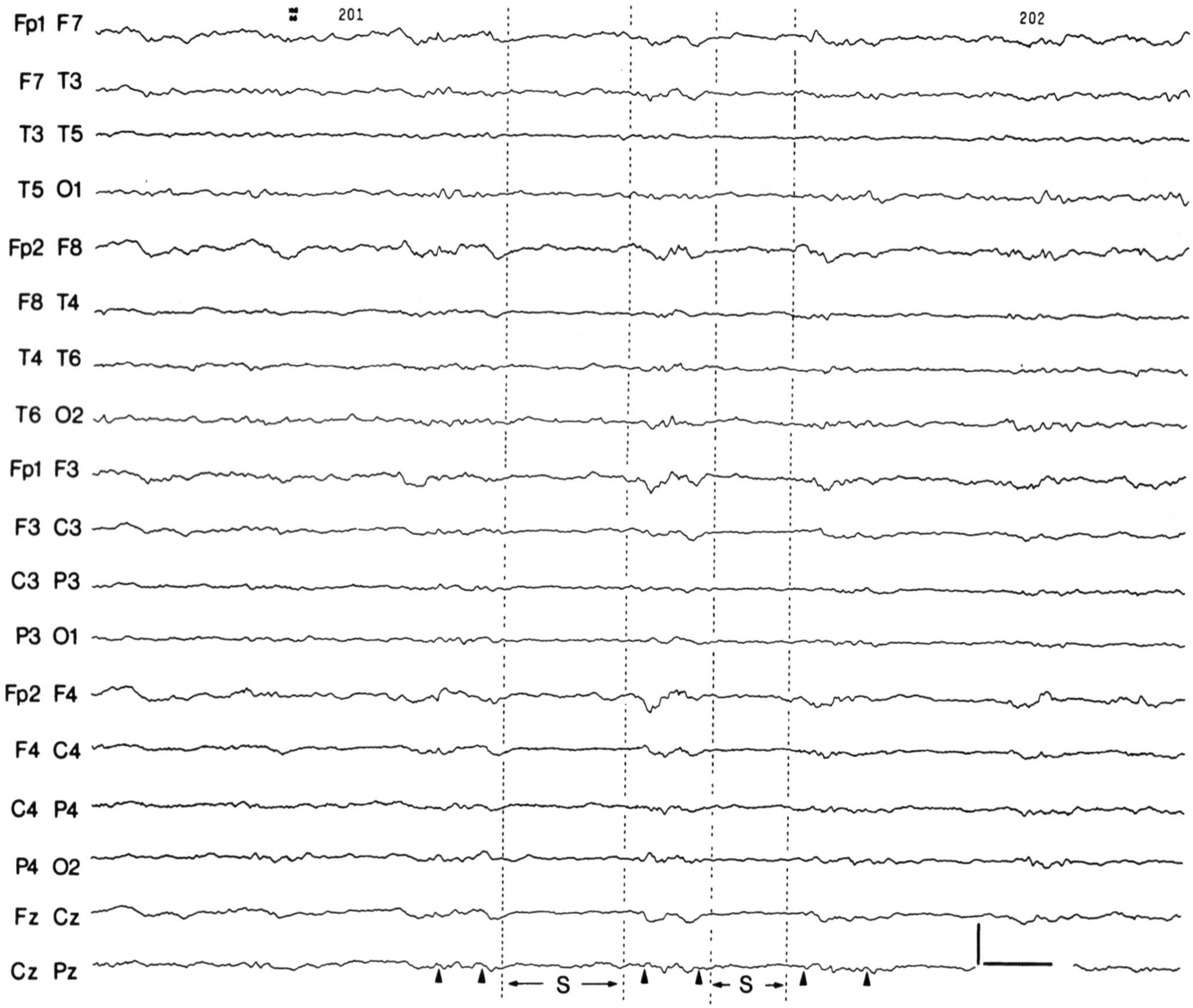

Figure 15.8. Burst-suppression pattern in a patient who had suffered from cerebral hypoxia following cardiopulmonary arrest. Areas marked "S" indicate epochs of suppression. Arrowheads indicate low-voltage theta activity that occurs during the bursts. Filters: low frequency = 1 Hz, high frequency = 70 Hz. Calibrations: horizontal = 1 second, vertical = 50 μV.

Mu Rhythm

Asymmetry of the mu rhythm may not always be abnormal as some degree of shifting asynchrony and asymmetry is well-known to occur in normal persons. However, if it is consistently absent on one side but present on the other, the finding might suggest an abnormality on the side that mu activity is absent. A consistent difference in the frequency of mu activity between the two sides is also considered abnormal. Higher amplitude mu rhythm on one side has been known to occur over skull defects.

Other Abnormalities of Background Activity

Marked attenuation of background rhythms is abnormal, whether it is generalized or localized. As mentioned in Chapter 14, however, in a small percentage of otherwise normal individuals the alpha rhythm is absent and the waking EEG may show mostly low amplitude, fast activity.

When the background activity in a tracing appears greatly attenuated, the technician has to double-check a number of points. These include checking the impedance of the electrodes, making sure that the amplifier sensitivity settings are correct, and verifying that the cable from the jack box is indeed connected to the EEG machine. It should be recognized that when this cable is disconnected, the tracing may show some artifacts resulting from the amplifier inputs being open; these artifacts could be mistaken for brain activity. When background activity is of very low amplitude, it is necessary to record at a higher sensitivity like 5, 3, or even 2 μV/mm. If no activity greater than 2 μV/mm is present in the tracing, the possibility of brain death (cerebral cortical death) may be considered if the clinical picture warrants it; under such circumstances,

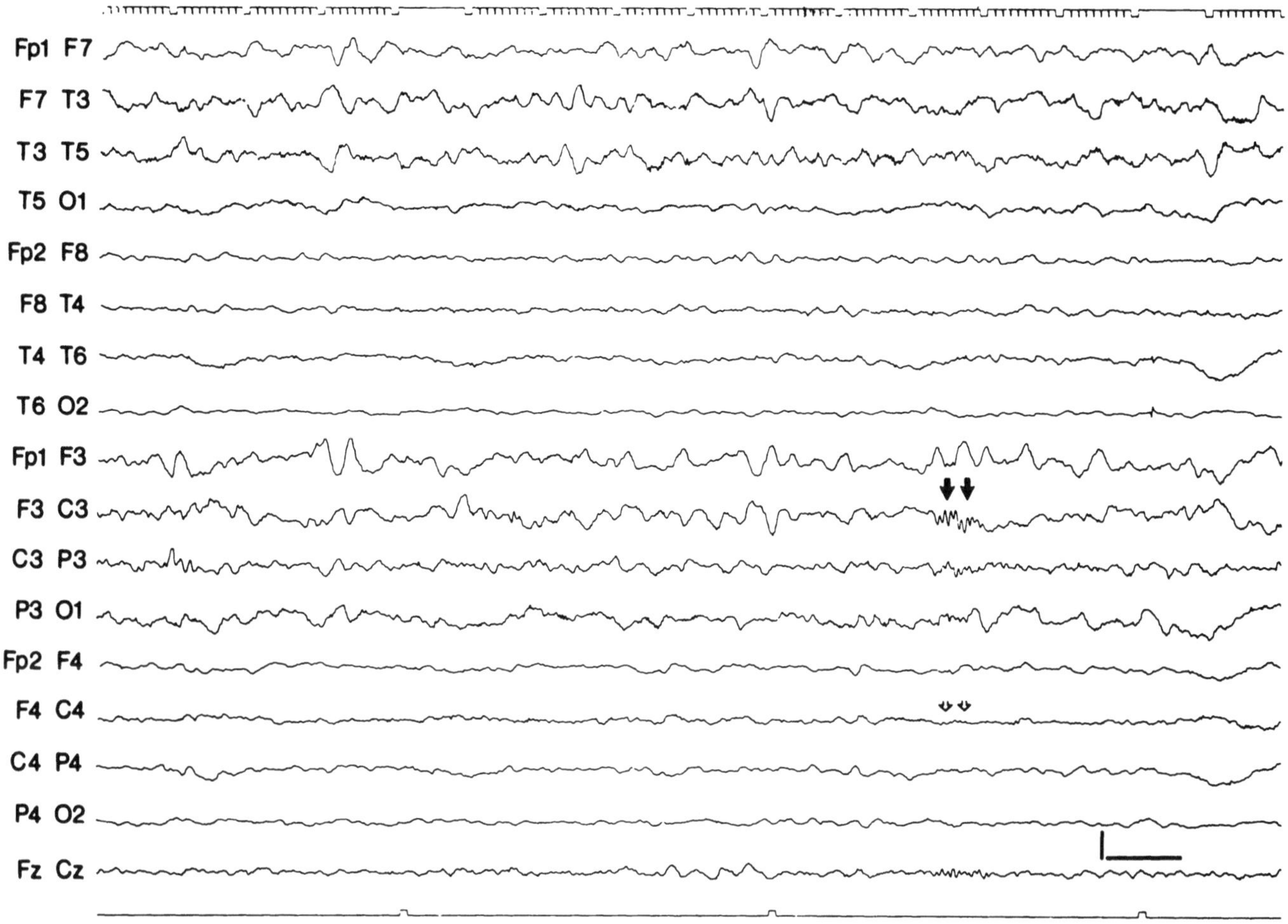

Figure 15.9. Asymmetry of sleep spindles in an 8-year-old child with left hemiplegia accompanying Sturge-Weber syndrome. A CT scan showed atrophy and gyral calcification involving the right side. The closed arrows point to a sleep spindle in the left frontocentral area, and the open arrows show attenuated spindles in the homologous area. There is also a decrease in amplitude of the background activity on the right side. Filters: low frequency = 1 Hz, high frequency = 70 Hz. Calibrations: horizontal = 1 second, vertical = 100 μV.

the EEG has to be taken according to specific guidelines stipulated by the American EEG Society (1986) for establishing electrocerebral silence. These include, among other things, using double-distance electrode placements, documenting the effects of external stimulation, and recording at a sensitivity of 2 μV/mm. Figure 15.8 is an example of extreme suppression of electrical activity following cerebral hypoxia. The only activity seen is very low amplitude theta, which occurs between the areas of suppression.

Abnormal Sleep Patterns

Amplitude asymmetry of sleep spindles is suggestive of a lesion on the side with the lower amplitude. This may happen both in structural lesions and also when there is abnormal collection of fluid between the brain and the recording electrode, as in subdural hematoma. Figure 15.9 shows asymmetry of sleep spindles in a patient with left hemiplegia accompanying Sturge-Weber syndrome. By contrast, sleep spindles may appear with higher amplitude over a skull defect.

The V waves may also be asymmetrical in amplitude. The presence of consistently asymmetrical V waves indicates a structural lesion, a subdural hematoma or effusion on the side with the lower amplitude. Sometimes the V waves may appear larger on the side of a skull defect. Figure 15.10 shows asymmetrically large V waves on the right side due to a skull defect.

Another group of sleep-pattern abnormalities includes disorders of sleep architecture. A person normally goes through stages I and II sleep before the first phase of REM sleep occurs, usually 90 minutes after the onset of sleep. But the REM phase can occur at the onset of sleep, and this abnormality is a feature of narcolepsy. Polysomnographic studies are needed to evaluate such sleep disorders.

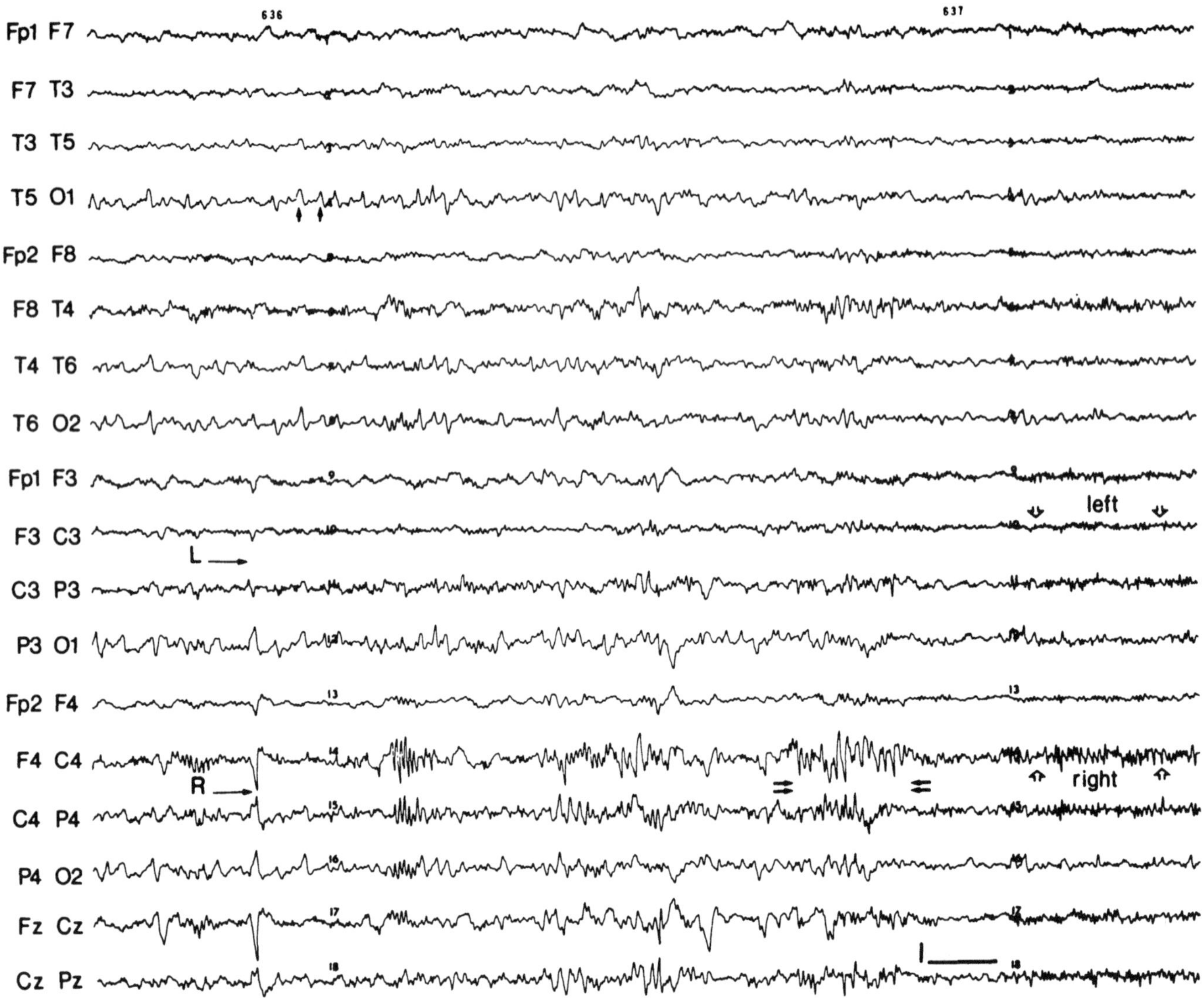

Figure 15.10. Asymmetrical V wave and beta activity in a 40-year-old patient with a skull defect on the right side. Note the higher amplitude of the V wave on the right (R) as compared with the left (L). The open arrows indicate the asymmetrical beta activity, which is of higher amplitude on the right. The horizontal double arrows mark an epoch of breach rhythm. Small vertical arrows point to POSTS. Filters: low frequency = 1 Hz, high frequency = 70 Hz. Calibrations: horizontal = 1 second, vertical = 50 μV.

Abnormal Slow Activity

One of the commonest abnormalities in EEG is the occurrence of abnormal slow activity. It must be understood that certain forms of slow activity are entirely normal. These include the delta activity that occurs in stages III and IV sleep, the theta activity that is present in the background activity of children during wakefulness, and also the delta activity that may be seen during hyperventilation. The presence of a consistent asymmetry in amplitude and/or frequency of such activity between the two sides is considered abnormal.

The distinction between normality and abnormality is less precise in the case of theta than delta activity. With theta activity, asymmetries in amplitude and frequency may be more significant than the mere presence of the activity. This is due to the variable occurrence of theta activity in drowsiness and in the waking state of normal young and very old persons.

Abnormal slow activity may occur either intermittently or more or less continuously, in which case it is termed persistent. It can be generalized, focal, or lateralized. We take these up in turn.

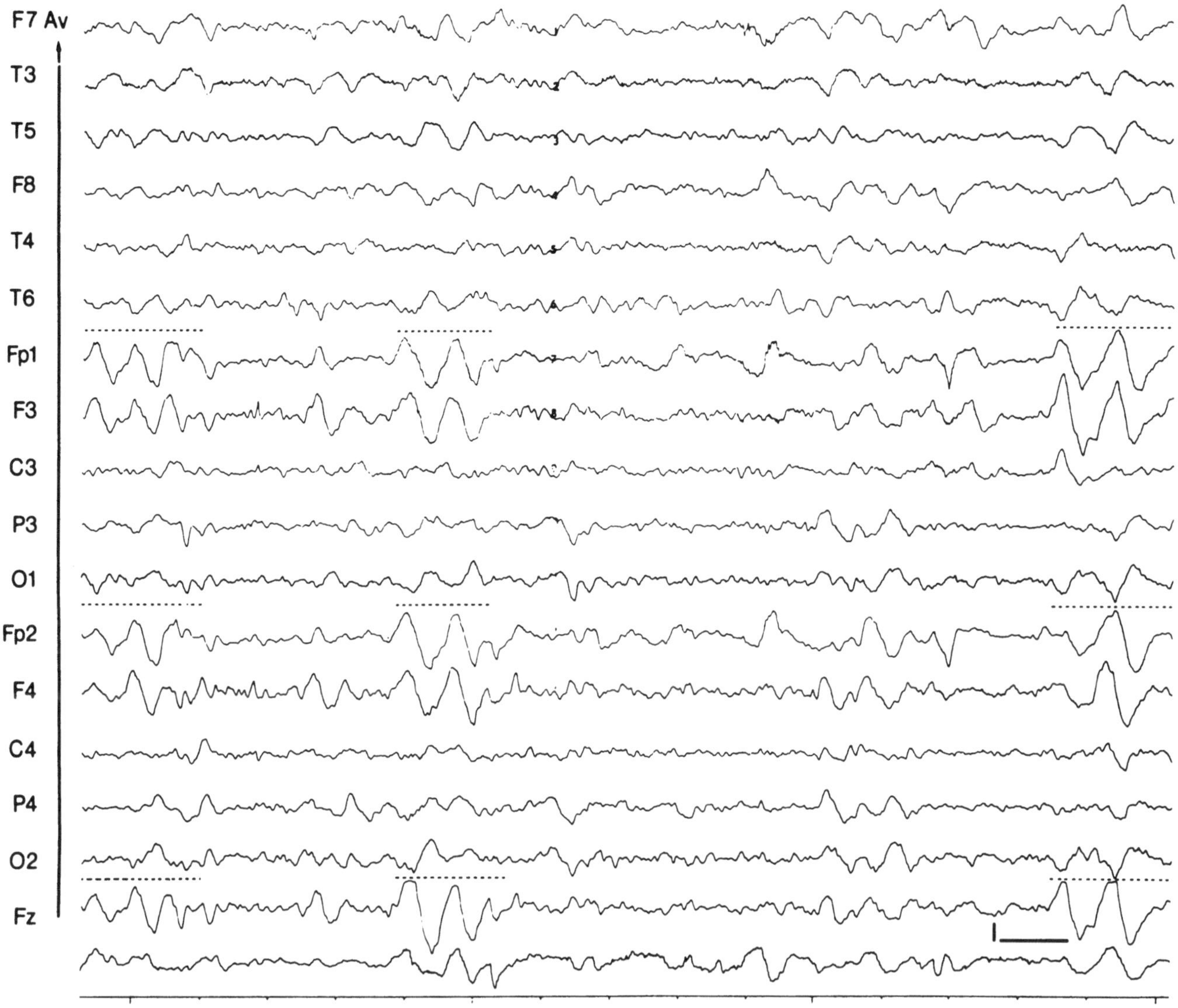

Figure 15.11. Frontal intermittent rhythmic delta activity (FIRDA) in a 40-year-old patient with chronic renal failure. Note the 1.5- to 2-Hz frontally dominant rhythmic, high-amplitude delta activity occurring in approximately 2-second epochs (dotted lines). The bottom channel, which is an eye electrode referred to the ipsilateral ear, shows activity that is *in phase* with the delta waves, thus confirming that the waves represent cortical activity and not eye activity. The patient appeared to be confused during the recording. Filters: low frequency = 1 Hz, high frequency = 70 Hz. Calibrations: horizontal = 1 second, vertical = 50 μV.

Generalized Intermittent Slow Activity

This is a common and easily identified abnormal EEG pattern. It consists of intermittent rhythmic, usually monomorphic, slow activity most commonly occurring in the delta frequency band. The acronym IRDA (intermittent rhythmic delta activity) is often used for this pattern. The activity is characteristically bilateral and synchronous, showing frontal (FIRDA, see Fig. 15.11) or occipital dominance (OIRDA, see Fig. 15.12); rarely, it may be most prominent over the temporal areas. The dominance seems to be age related, OIRDA being more common in children.

Intermittent rhythmic delta activity is usually of high amplitude; it stands out from the background and often has a frequency of 2 to 3 Hz. The component waves are often sinusoidal or saw-toothed, and have a rapid upstroke and slower downstroke. Each epoch of delta activity may last from 1 to 3 seconds and may repeat every few seconds. Both the frequency of occurrence of the epochs and their amplitude may increase during hyperventilation or drowsiness. Intermittent rhythmic delta activity is often attenuated by eye opening and other alerting stimuli. It may disappear during deeper stages of sleep, but reappear during the REM stage. Since eye-movement

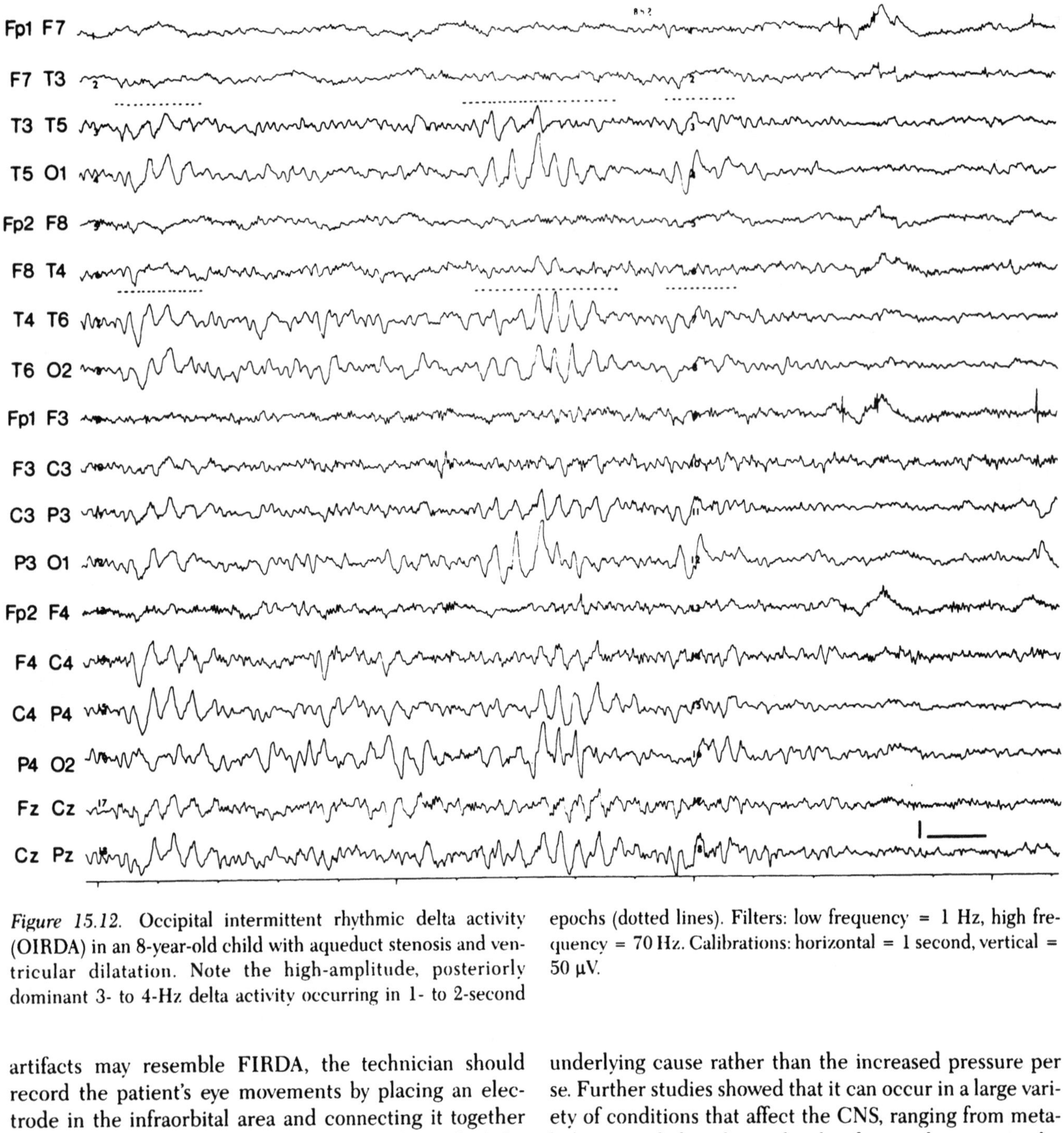

Figure 15.12. Occipital intermittent rhythmic delta activity (OIRDA) in an 8-year-old child with aqueduct stenosis and ventricular dilatation. Note the high-amplitude, posteriorly dominant 3- to 4-Hz delta activity occurring in 1- to 2-second epochs (dotted lines). Filters: low frequency = 1 Hz, high frequency = 70 Hz. Calibrations: horizontal = 1 second, vertical = 50 μV.

artifacts may resemble FIRDA, the technician should record the patient's eye movements by placing an electrode in the infraorbital area and connecting it together with the ipsilateral ear electrode to one channel of the EEG machine. If the waveform recorded on this channel is in phase with the waveforms on the other channels, eye-movement artifacts are excluded.

Although IRDA is easily identified, some difficulty arises with regard to its precise clinical correlation. It was once believed that IRDA resulted from increased intracranial pressure. However, the observation that IRDA is not seen as a feature of benign increased intracranial tension changed this concept; the IRDA may be related to the underlying cause rather than the increased pressure per se. Further studies showed that it can occur in a large variety of conditions that affect the CNS, ranging from metabolic encephalopathy to localized mass lesions. At the present time, IRDA is considered to be a totally nonspecific abnormality reflecting disturbance in cerebral function as a result of diffuse encephalopathies of metabolic, toxic, infectious, or traumatic origin, or focal encephalopathies such as cerebrovascular accidents, and intracranial tumors. In focal encephalopathies, FIRDA may occur with higher amplitude on the side of the lesion, but this is not always a reliable finding. For this reason, IRDA is considered to be of little localizing value.

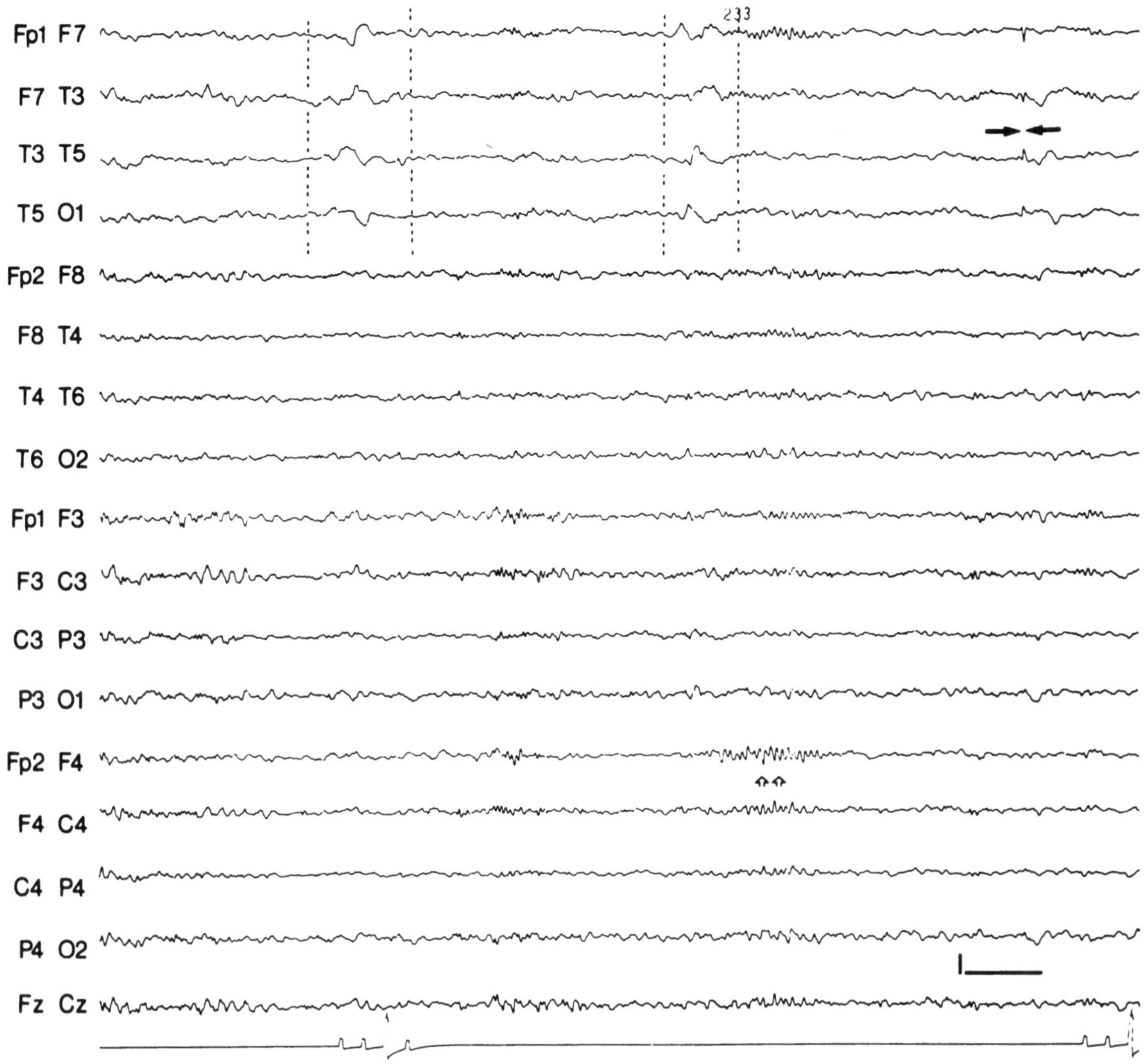

Figure 15.13. Intermittent, left-sided delta activity (between the dotted lines) and focal spike discharge at T3 (solid arrows) in a 50-year-old patient with a left temporal lobe glioma. The open arrows point to sleep spindles that occur at higher amplitude on the right than the left side. Filters: low frequency = 1 Hz, high frequency = 70 Hz. Calibrations: horizontal = 1 second, vertical = 50 μV.

Since IRDA was initially associated with the presence of mass lesions is deep midline regions like the third ventricle, the subfrontal regions, or the posterior cranial fossa, the terms "distant rhythm" or "projected rhythm" have been used. But this is no longer favored because there is little difference between the intermittent delta activity of such mass lesions and that resulting from diffuse encephalopathies.

Intermittent rhythmic delta activity often accompanies decrements in mental status of the patient such as lack of attention, confusion, and obtundation. In focal lesions the presence of IRDA also is often accompanied by such changes in mental status, suggesting that a diffuse cerebral dysfunction may be accompanying the focal lesion.

Focal and Lateralized Intermittent Slow Activity

These abnormalities have the same features as generalized intermittent delta activity except that they are limited to one area or to one side of the brain. When consistently present, such activity should arouse the suspicion of an underlying structural lesion (Fig. 15.13), although the correlation is not as consistent as in the case of persistent polymorphic delta activity. The underlying cause may be a mass lesion, trauma, or ischemia; or the activity may even be a postictal phenomenon. Figure 15.14 is an example of lateralized intermittent rhythmic slow activity in a patient with a subcortical infarct.

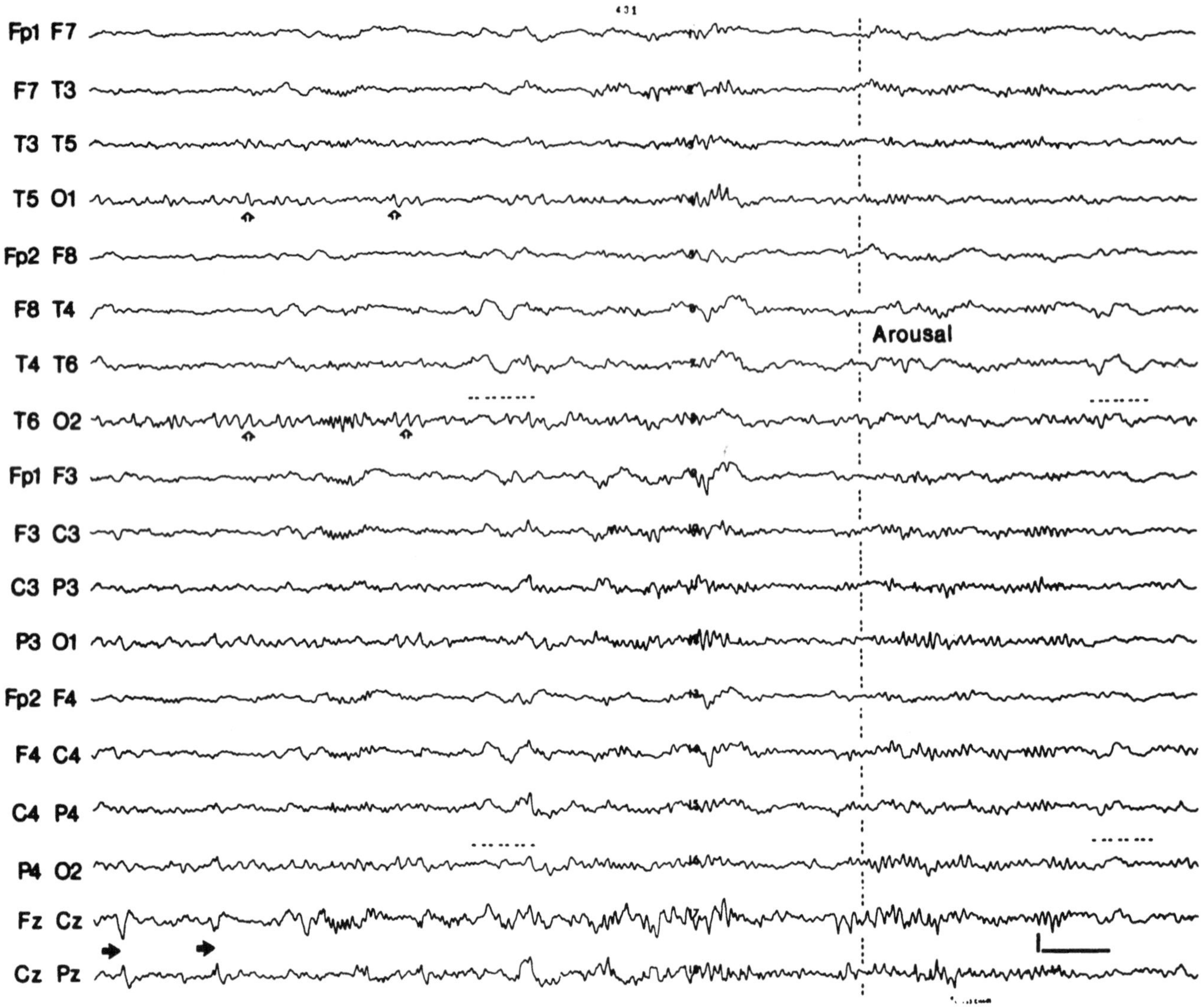

Figure 15.14. Lateralized intermittent rhythmic delta activity occurring during both sleep and wakefulness in a 50-year-old patient with a right subcortical infarct. Note the 2- to 3-Hz rhythmic delta activity occurring intermittently (dotted lines) on the right side. Vertical arrows point to POSTS. Horizontal arrows point to V waves. Filters: low frequency = 1 Hz, high frequency = 70 Hz. Calibrations: horizontal = 1 second, vertical = 50 μV.

Persistent Slow Activity

Persistent slow activity is usually in the delta frequency band. It may occur as monomorphic, rhythmic waves or as polymorphic, arrhythmic waves. It may be generalized, lateralized, or focal. In the case of the rhythmic delta activity, the waveforms resemble each other and maintain a somewhat constant frequency. On the other hand, the polymorphic delta activity (PDA) usually has a frequency of 0.5 to 3 Hz, and the waveforms change in frequency, amplitude, and morphology (shape) in a continuous fashion. In other words, no two succeeding waves appear to be quite alike. Polymorphic delta activity tends to show no reactivity to stimulation and persists both during wakefulness and sleep. Even hyperventilation may not have much effect on PDA.

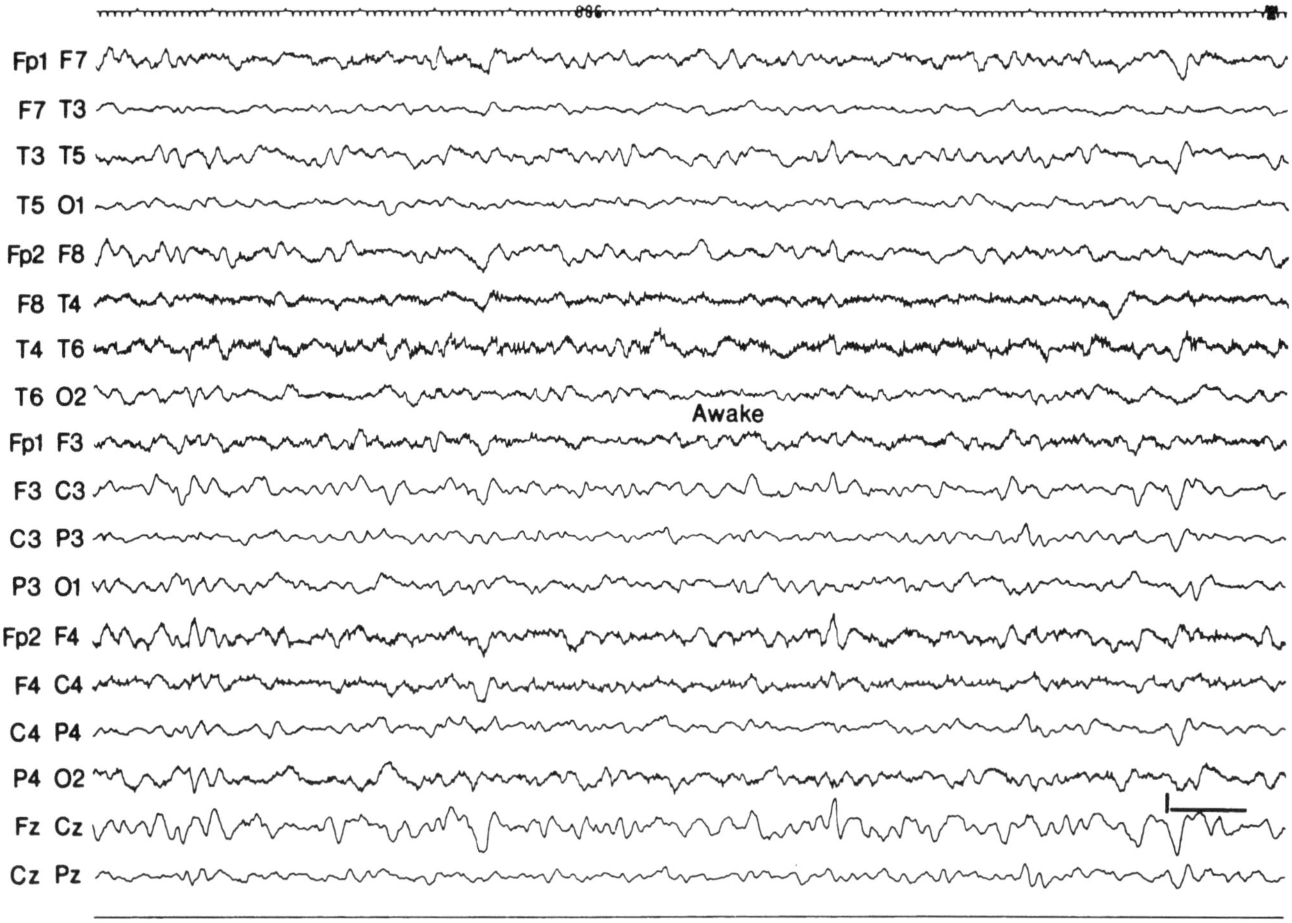

Figure 15.15. Diffuse slow activity in a 20-year-old patient presenting with fever, headache and vomiting. Although the patient appeared awake at the time of recording, he was confused and disoriented. Cerebrospinal fluid showed normal glucose, with 48 lymphocytes and 6 polymorphs per cubic millimeter. A diagnosis of viral encephalitis was made. Note the muscle artifacts in most of the channels. Filters: low frequency = 1 Hz, high frequency = 70 Hz. Calibrations: horizontal = 1 second, vertical = 50 μV.

Figure 15.15 is an example of generalized, persistent slow activity consisting of a mixture of theta and delta activity that is both rhythmic and arrhythmic. The patient was diagnosed to have viral encephalitis. Figure 15.16 shows the occurrence of polymorphic delta activity on the right side in a patient with an underlying glioma. When the delta activity is of very low amplitude and of very low frequency, it may easily be overlooked. The technician should be vigilant about such patterns as he/she can carry out certain maneuvers—such as changing the low-frequency-filter setting to 0.3 to 0.1 Hz, and/or using a slow paper speed—to make the slow activity more prominent.

Continuous PDA, especially when it is focal, is indicative of an underlying structural lesion unless proved otherwise. The finding often correlates well with other tests like the CT scan and Magnetic resonance imaging (MRI), but there are certain situations where the CT may be negative as, for example, in recent infarct or contusion. For the exact localization of the lesion, the frequency of the waveform is a better indicator than the amplitude; thus, the area showing the slowest activity is the most likely site of the lesion. The amplitude is often higher in the immediately surrounding areas. Sometimes the tracing from the area overlying a destructive lesion may be of such low amplitude that it appears to be flat, whereas the surrounding areas show large amplitude PDA.

It is now believed that deafferentation of the cortex by a lesion that interrupts the thalamocortical afferents is the underlying mechanism in the genesis of PDA. Polymorphic delta activity is most likely to be associated with acute destructive lesions, but it gives no clues as to the specific etiology of the lesion.

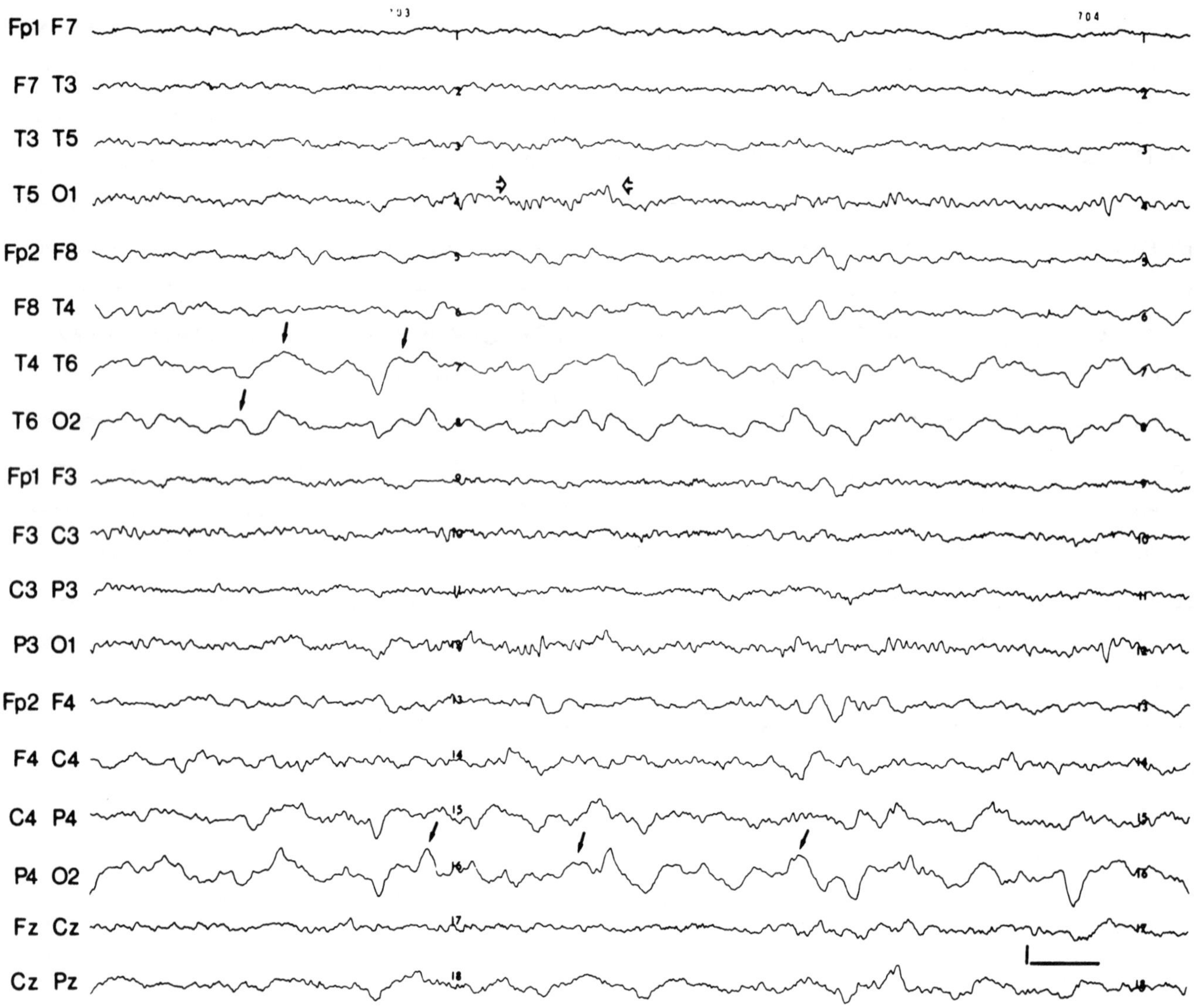

Figure 15.16. Polymorphic delta activity on the right side in a patient with a large glioma. The open arrows indicate an intact alpha rhythm on the left side, which contrasts with the right side where none is present. The slanted arrows point to some of the delta transients. Filters: low frequency = 1 Hz, high frequency = 70 Hz. Calibrations: horizontal = 1 second, vertical = 50 μV.

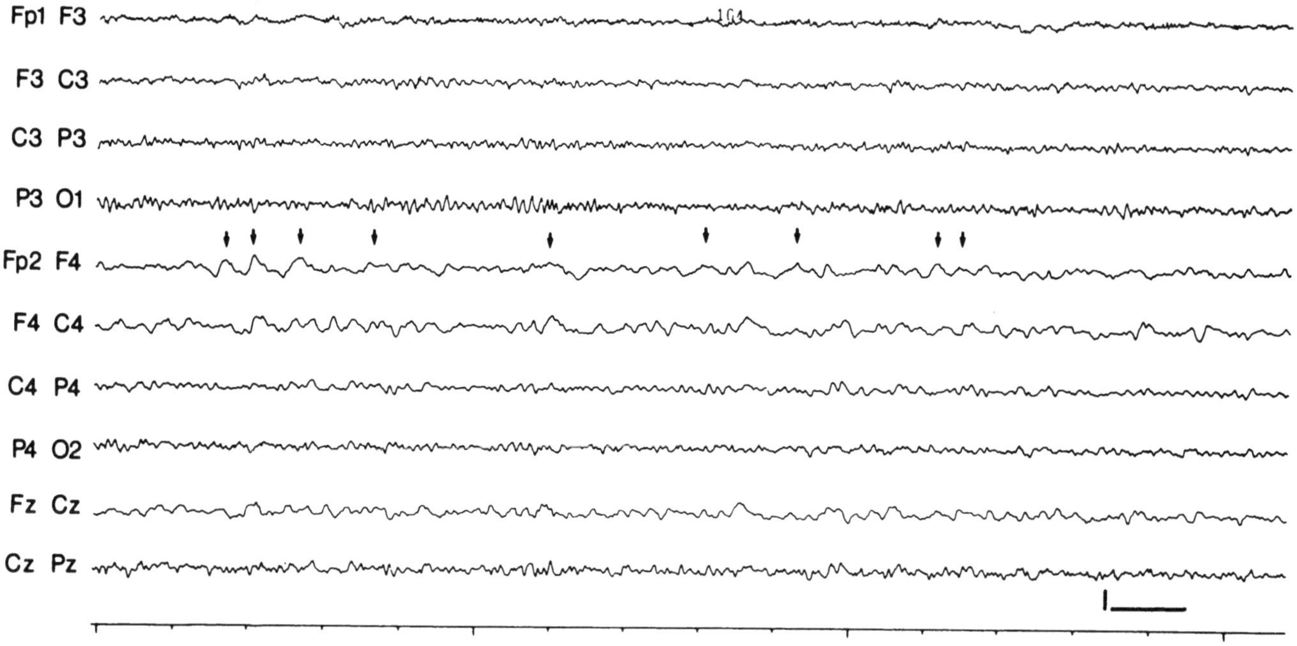

Figure 15.17. A lateralized mixture of rhythmic and polymorphic delta activity in a 60-year-old patient with a right frontal infarct involving both the cortex and the subcortical areas. The vertical arrows point to some of the delta waves. Filters: low frequency = 1 Hz, high frequency = 70 Hz. Calibrations: horizontal = 1 second, vertical = 50 μV.

Sometimes PDA and rhythmic delta activity may coexist in the same tracing as in Fig. 15.17. This may conceivably depend on different degrees of involvement of the cortical and subcortical areas by the lesion.

Paroxysmal Epileptogenic Abnormalities

The term paroxysmal refers to activity that shows changes in amplitude or frequency, which occur with sudden onset and offset, and that stands out distinctly from the ongoing background activity. While certain patterns of activity such as V waves, lambda activity, and hypnagogic hypersynchrony are also paroxysmal by this definition, the term is usually reserved for abnormal activity. Broadly speaking, paroxysmal abnormalities may be classified into epileptogenic abnormalities and periodic patterns. Abnormal periodic paroxysmal patterns are taken up later in the chapter.

There has been considerable debate about the correct terminology to use in describing patterns of activity that are associated with seizure disorders. The term epileptogenic abnormalities is usually used to indicate those transients or patterns of activity that are known to have a definite correlation with seizure disorders. These abnormalities may be further classified into interictal (in between seizures) phenomena and ictal (during a seizure) phenomena. The interictal phenomena, which are termed epileptiform patterns or discharges, include spike discharges and sharp waves with or without accompanying slow waves. The ictal discharges or electrographic seizure patterns, on the other hand, include a number of patterns that are known to accompany clinical seizures.

Epileptiform Discharges

Spike Discharges

A spike discharge is defined as a transient that is clearly distinguished from the background activity, has a pointed peak at a paper speed of 30 mm/s, and has a duration of 20 to 70 ms; it usually is surface negative and is of variable amplitude (Chatrian GE, Bergamini L, Dondey M et al, 1974). Ordinarily, a spike stands out from the background activity because of its distinct appearance and/or size; but when the amplitude is small, it may be difficult to identify, especially when there is a considerable amount of beta activity in the background. This is one reason why it is better to avoid using medications for promoting sleep when taking an EEG, as most such drugs cause diffuse beta activity. As mentioned earlier in Chapter 14, use of the 15-Hz high-frequency filter should also be avoided because the sharp-pointed character of a spike is lost and its amplitude becomes markedly attenuated. When this happens, a spike may be indistinguishable from beta activity or muscle artifacts.

Sharp-wave Discharges

A sharp wave is defined as a transient that clearly stands out from the background activity, has a pointed peak at a paper speed of 30 mm/s, and has a duration of 70 to 200 ms;

amplitude is variable, and like spikes, sharp waves usually are surface negative (Chatrian GE, Bergamini L, Dondey M, et al, 1974). Since there is little distinction between spikes and sharp waves from the standpoint of their potential for epileptogenicity, the terms are used interchangeably in the ensuing discussion.

Polyspikes or Multispikes

Spike discharges are usually monophasic or biphasic. The term multispike is used when several spikes comprise a single waveform. As with spike discharges, multispike discharges may also be accompanied by slow waves.

Spike and Wave Complexes

Another epileptiform pattern is the spike-wave complex or sharp and slow-wave complex. Each spike or sharp wave is accompanied by a slow wave, usually of the same polarity. Spike-wave complexes may take different forms, the classical example being the three per second spike-wave complex of absence seizures.

Detection of Epileptiform Discharges

When a spike or sharp-wave discharge is noted in an EEG, the following questions should go through the minds of both technician and electoencepalographer. Is it indeed a spike, or is it an artifact? Where is it coming from? Is it a physiological phenomenon that looks like a spike such as a V wave, a POSTS, or lambda wave?

To ensure that a particular transient is a true spike, and to establish its exact location, the technician may have to carry out certain maneuvers. Figure 15.18 shows what looks like a spike discharge that is confined to a single electrode, namely, F4. It does not appear to have an electrical field, as it is not seen in C4 or F8. Such being the case, one needs to prove that it is not simply an electrode artifact. To do this, an additional electrode is placed between the incriminated one and its neighbor so that any electrical field associated with the transient can be documented. In this instance, the electrode at F4′ showed a similar transient, thereby confirming that the transient observed in F4 is not an electrode artifact.

When a spike is noted in an electrode at the end of a bipolar chain, the localization can be made easier if the montage is revised and the electrode is placed in the middle of the chain. As explained in the chapter on localization, it is sometimes useful to combine bipolar and referential recording on the same page to show the exact localization of a spike more accurately (see Fig. 12.11). If there is a question as to whether a particular discharge is an abnormal spike or a normal transient, other measures are necessary. For example, midline spikes may be confused with V waves, so a waking record is essential whenever midline spikes are suspected. If the spikes do occur in the waking EEG, one is dealing with midline spikes. Sometimes ECG artifacts may look like spikes; when this happens, it is essential to record the ECG simultaneously on the same chart with the EEG. It is important to keep in mind that genuine spike discharges are often followed by slow waves, whereas ECG and other artifacts are not. The role of activation procedures in enhancing epileptiform activity is discussed in Chapter 16.

Parameters of Epileptiform Discharges

The size of the spike is only rarely a reliable criterion for the assessment of its epileptogenicity. There are but few such examples. Thus, it may be noted that small sharp spikes (see later in this chapter) which are of low amplitude and of very short duration do not usually have any correlation with seizure disorders. Similarly, the very small spikes present in 6-Hz spike and wave discharges (phantom spike and wave) are also known to have little relevance from the point of view of epileptogenicity. It is important in this context to remember that amplitude is not a reliable measure in bipolar chains (Chapter 11).

Polarity of a spike may provide clues as to its potential for epileptogenicity. Thus, negative spikes are more common by far than positive spikes and are more significant from the point of view of epileptogenicity. Positive spikes like the 14- and 6-Hz bursts that are taken up later are considered to be of little clinical significance. On the other hand, positive Rolandic spikes, which are seen in premature infants and are known to occur in association with intraventricular or periventricular hemorrhage, are an exception.

Location of a spike is also relevant from the standpoint of correlation with seizure disorders. Thus, Rolandic spikes and spikes arising from the occipital areas are said to be less well correlated with seizure disorders than anterior temporal or frontal spikes.

Pathophysiology of Spike Generation

Epileptiform activity is believed to be the result of paroxysmal discharges occurring synchronously in a large aggregate of neurons. Such discharges could conceivably give rise to spikes or sharp waves in the scalp EEG, particularly if the neuronal aggregate happened to be in an area covered by the field of the electrode and the dipole happened to be in the appropriate orientation. At the cellular level the corresponding abnormality consists of bursts of spontaneous depolarization of the neuronal membrane, which are termed paroxysmal depolarization shifts. Although a paroxysmal depolarization shift is considered to be the cellular hallmark of a hyperexcitable neuron, the hypersynchrony (several neurons discharging simultaneously) probably results from abnormalities occurring at the synaptic level. In other words, abnormalities of the

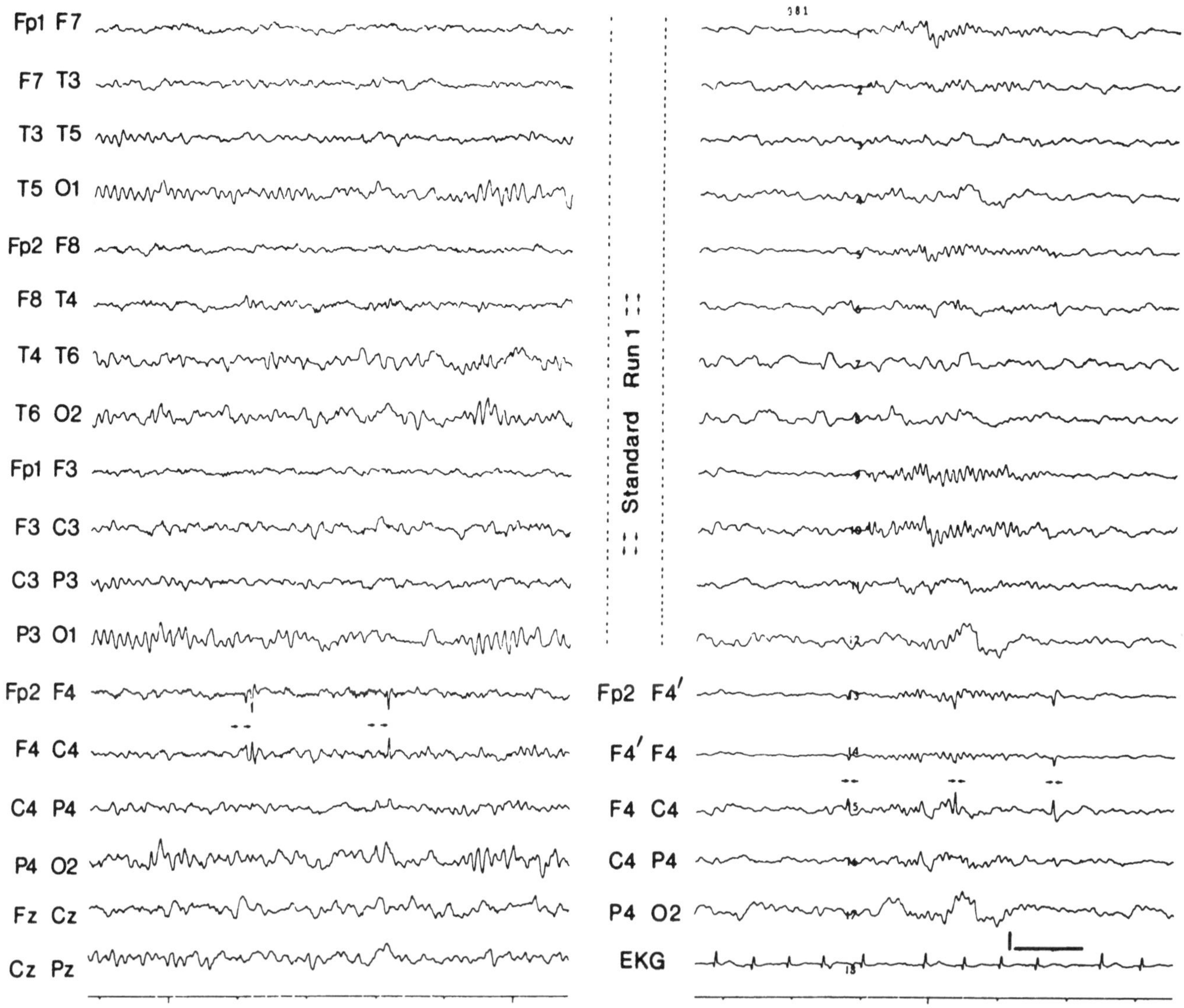

Figure 15.18. Focal spike discharges from the midfrontal area on the right side. In the left half of the tracing, a standard bipolar montage was used. Horizontal arrows denote spike discharges with a phase reversal at F4. As there was no definite electrical field for the spike, the technician added another electrode (F4′) between Fp_2 and F4. The revised montage in the right half of the tracing shows that the spike has a definite electrical field and, hence, is not an electrode artifact. The horizontal arrows show a phase reversal that places the focus between F4′ and F4, but closer to F4. Use of the additional electrode should be a standard procedure when spikes or sharp waves appear restricted to a single electrode. Filters: low frequency = 1 Hz, high frequency = 70 Hz. Calibrations: horizontal = 1 second, vertical = 50 μV.

neuronal membrane, as well as abnormalities at the synapses that lead to hypersynchronization, are necessary for the production of interictal as well as ictal discharges.

Let us now consider some of the commonly seen interictal epileptiform abnormalities. Broadly speaking, two types of interictal patterns are noted, namely, focal abnormalities and generalized abnormalities.

Focal Epileptiform Abnormalities

Interictal focal epileptiform abnormalities manifest themselves as focal spikes or sharp waves, or focal spike and wave discharges. Their presence helps to confirm the clinical suspicion of a focal or partial seizure. In the following, we consider some of the common focal spike patterns.

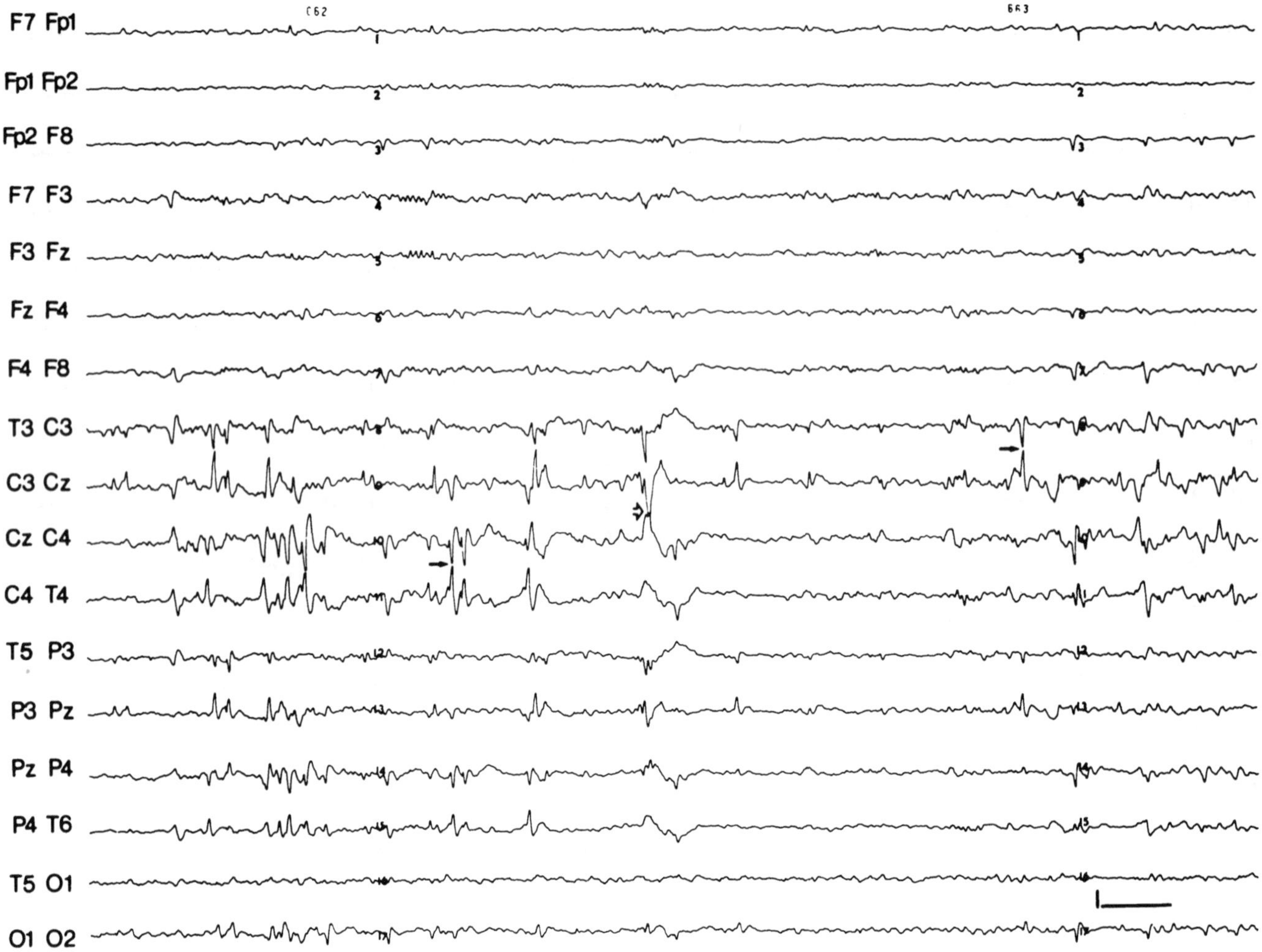

Figure 15.19. Focal spikes arising independently from C3 and C4 during sleep in a 10-year-old girl with benign Rolandic epilepsy. The closed arrows point to phase reversals at C3 and C4. The open arrow points to a V wave. Filters: low frequency = 1 Hz, high frequency = 70 Hz. Calibrations: horizontal = 1 second, vertical = 50 μV.

Rolandic spikes, which are distinguished by their occurrence over C3 and/or C4, may be unilateral, bilateral, or may shift from side to side. They may be seen simultaneously over the parietal area (centroparietal spikes) or in the midtemporal area (centromidtemporal or Sylvian spikes) as well. Rolandic spikes are easily distinguished by their location and by their tendency to be greatly accentuated during drowsiness and sleep. They are usually abundant in number. A slow wave usually comes after the spike. Rolandic spikes should be distinguished from asymmetrical V waves and from isolated components of the mu rhythm.

Since centroparietal spikes occur in a number of patients with cerebral palsy and other central motor dysfunction, their presence does not always indicate a seizure disorder. Central and midtemporal spikes are correlated with benign Rolandic epilepsy of childhood in about two out of three cases. The reader will find more detail about this in Chapter 21 on clinical correlations. Again, it must be pointed out that such spikes may be seen occasionally in asymptomatic children. Figures 15.19 and 15.20 are examples of Rolandic spikes occurring in patients with benign Rolandic epilepsy of childhood, each showing slightly different spike morphology. Figure 15.21 shows spikes from the left central area in a patient with focal seizures occurring several months following a left cerebral infarct.

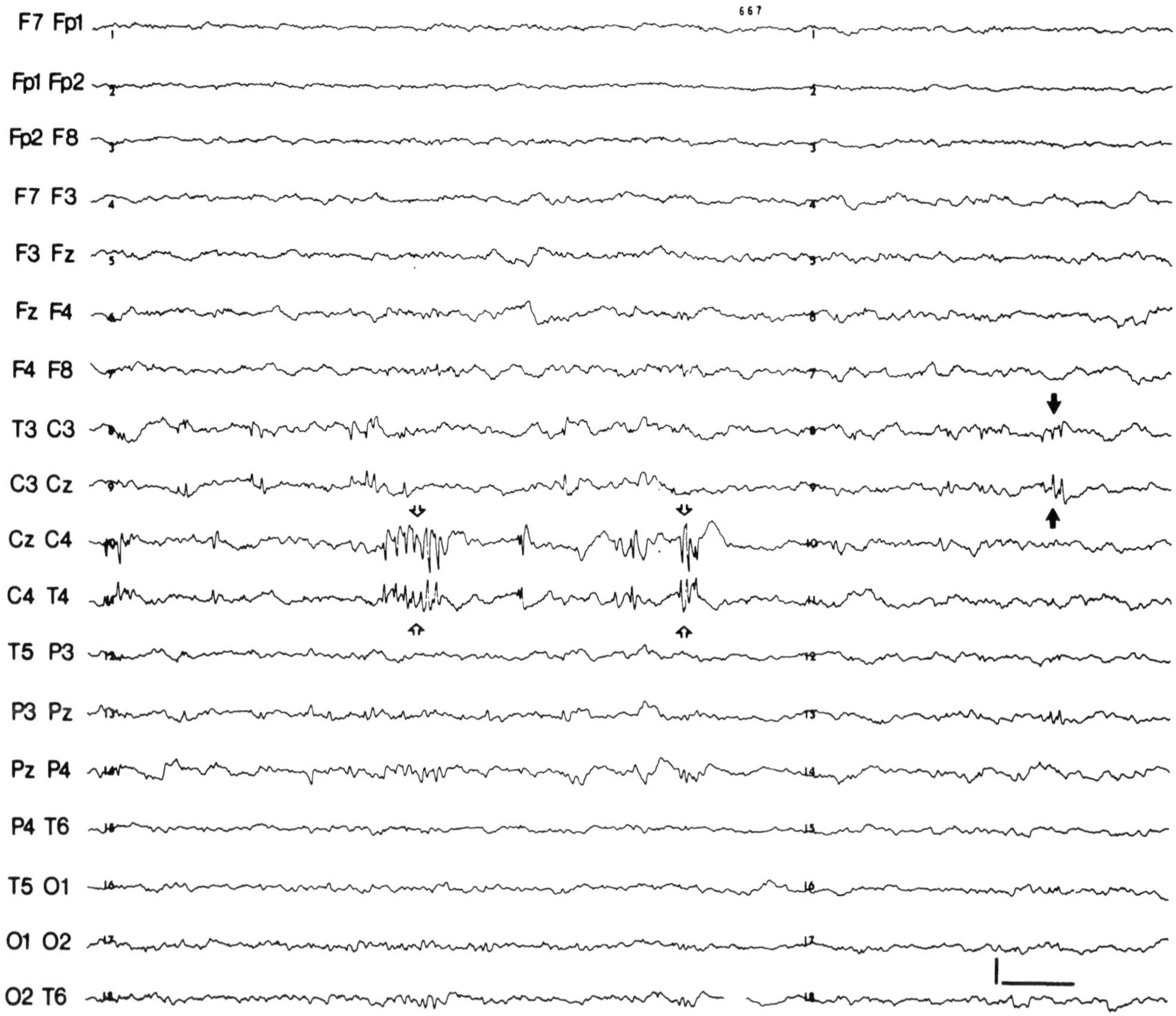

Figure 15.20. Rolandic spikes occurring in clusters during sleep. Open arrows point to spikes arising from C4, and closed arrows point to independent spikes from C3. The patient was a 12-year-old child with nocturnal seizures characterized by facial twitching and salivation. Filters: low frequency = 1 Hz, high frequency = 70 Hz. Calibrations: horizontal = 1 second, vertical = 150 μV.

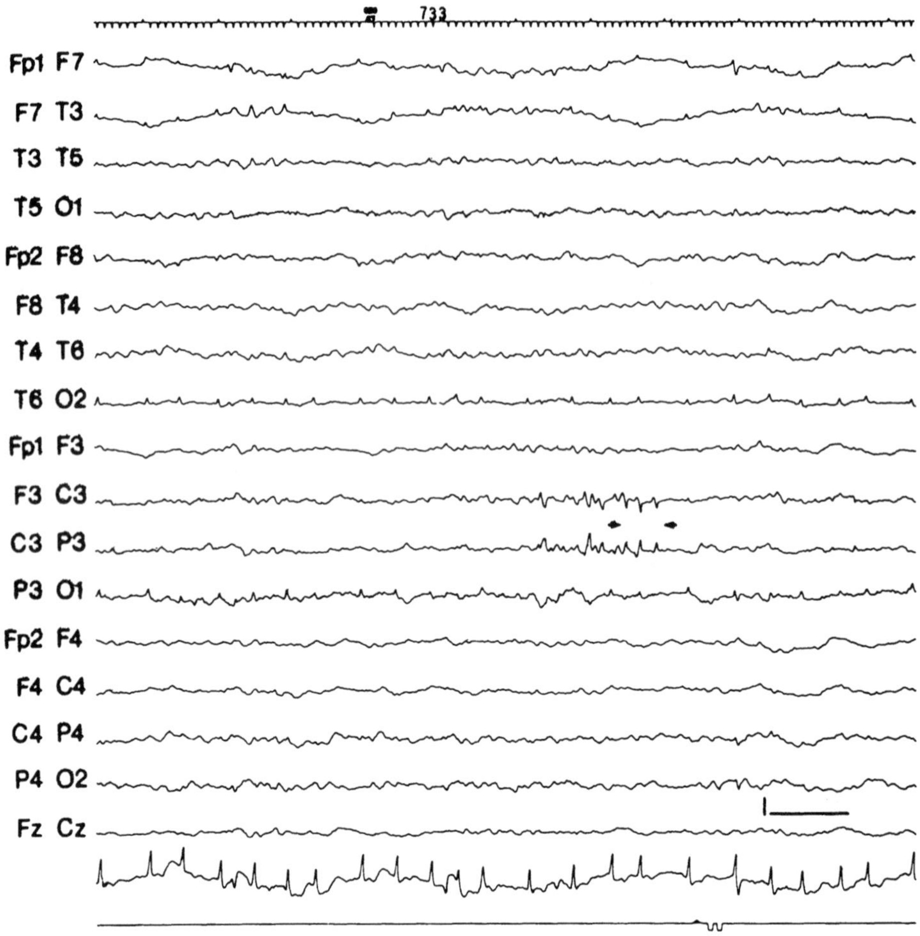

Figure 15.21. Left-sided Rolandic spikes in a 60-year-old woman with recent onset of focal motor seizures involving the right upper extremity and face. The horizontal arrows point to phase reversals occurring at C3. The CT scan showed an old infarct in the left frontoparietal area. Filters: low frequency = 1 Hz, high frequency = 70 Hz. Calibrations: horizontal = 1 second, vertical = 50 μV.

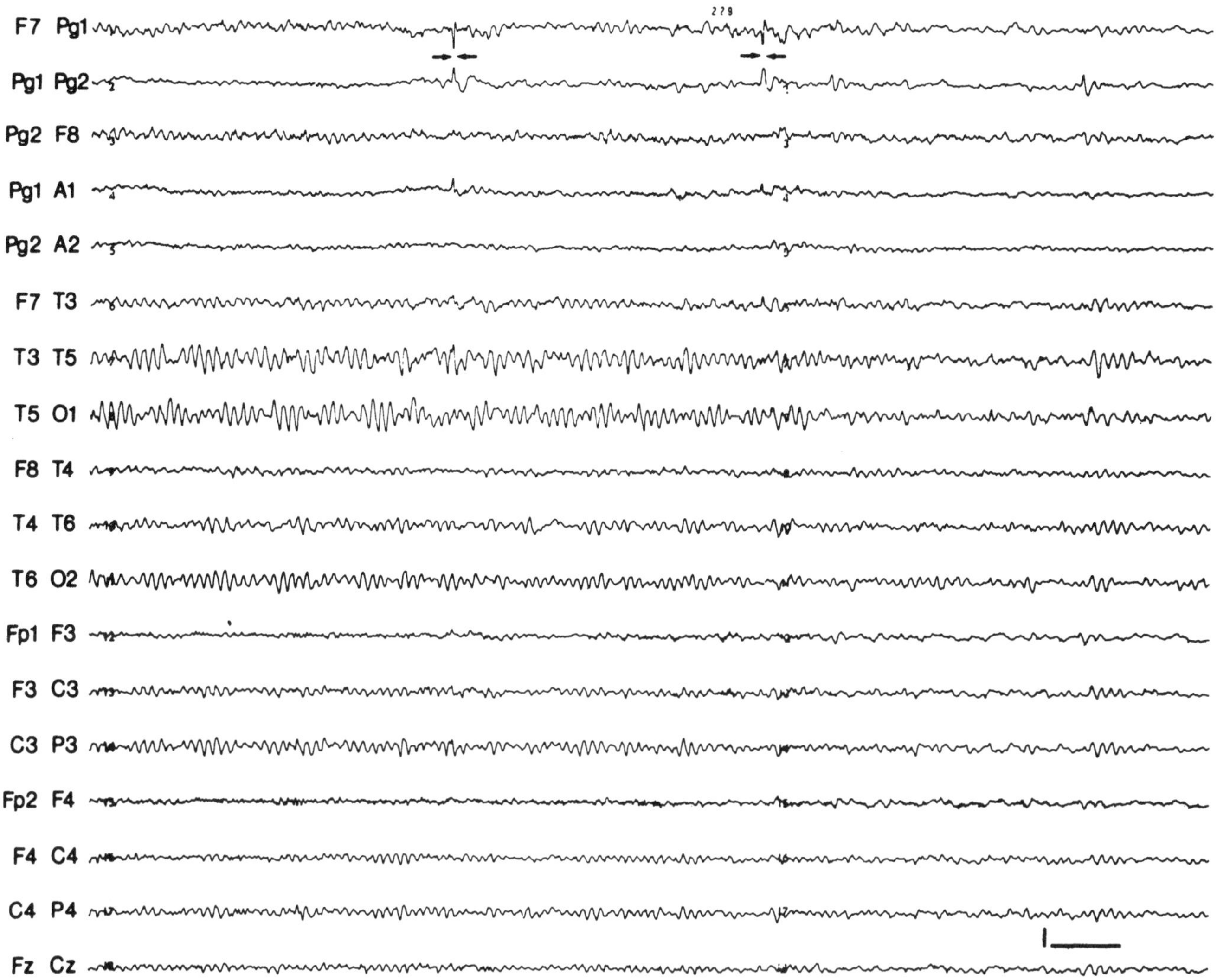

Figure 15.22. Mesiotemporal spikes arising from the left side (horizontal arrows) picked up by nasopharyngeal lead Pg_1 in a 40-year-old man with a history of recurrent episodes of confusion lasting for about five minutes—preceded, each time, by an apparent smell of gasoline. The CT scan was suggestive of a medial sphenoidal wing meningioma. Note that the spikes are hardly visible in the scalp recordings. Filters: low frequency = 1 Hz, high frequency = 70 Hz. Calibrations: horizontal = 1 second, vertical = 50 µV.

Spikes that are localized to the inferior frontal (F7, F8) and/or anterior temporal electrodes (T1, T2) are also potentiated by drowsiness and sleep. These discharges are highly correlated with complex partial epilepsy. They should not be confused with small sharp spikes, wicket spikes, or sharply pointed theta activity of drowsiness. Detection of spikes arising from the inferior and medial aspects of the temporal lobe is facilitated by the use of nasopharyngeal or sphenoidal electrodes. Figure 15.22 shows an example of spike discharges appearing in the nasopharyngeal electrodes almost exclusively.

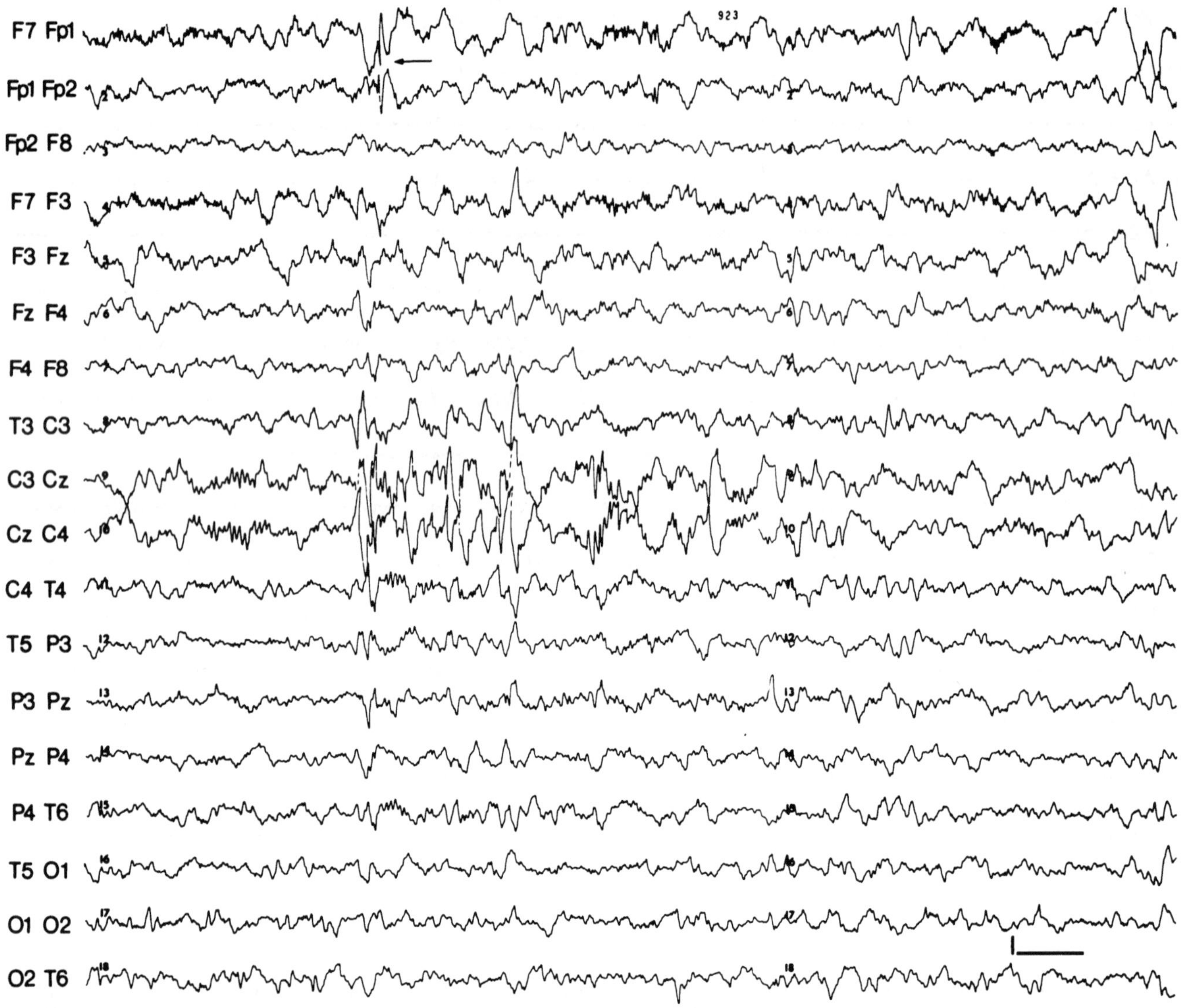

Figure 15.23. Focal spike and wave discharge arising from the left frontopolar area in a 10-year-old boy who developed generalized tonic-clonic seizures 3 months following a head injury. The arrow points to the slow-wave component of the discharge. The patient was asleep, as shown by the vertex waves and sleep spindles. A CT scan showed bilateral contusion in the frontal pole. Filters: low frequency = 1 Hz, high frequency = 70 Hz. Calibrations: horizontal = 1 second, vertical = 50 μV.

Spikes originating in the frontal lobe (Figs. 15.18, 15.23) are also considered to have high epileptogenicity. Such patients may present with adversive, focal motor, complex partial or secondarily generalized seizures. These discharges should be distinguished from asynchronous F waves, small sharp spikes, and frontalis muscle spikes.

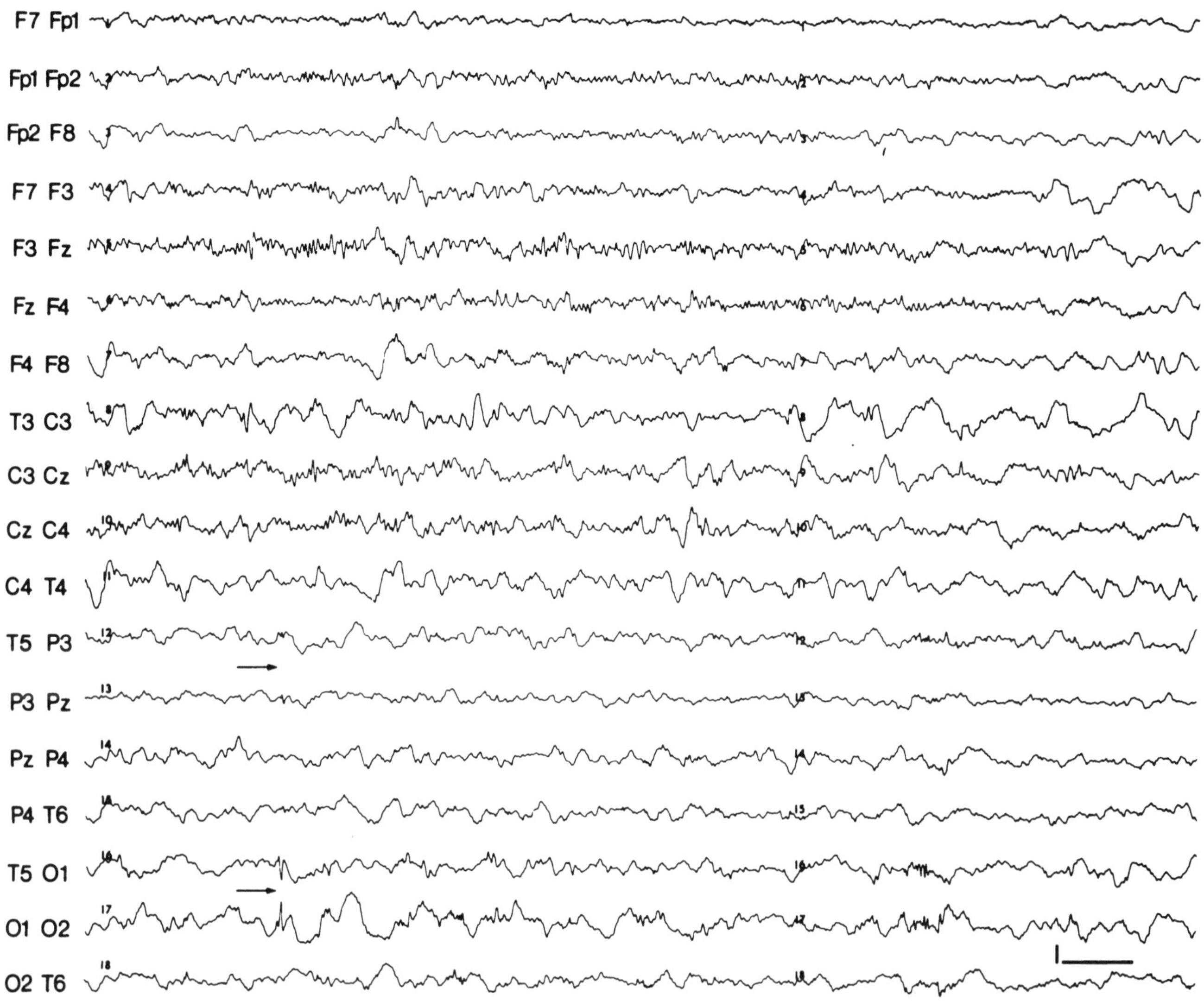

Figure 15.24. Focal spikes from left occipital area in a 5-year-old girl with a history of a single generalized seizure. The horizontal arrows point to phase reversals occurring at O1 and P3. Filters: low frequency = 1 Hz, high frequency = 70 Hz. Calibrations: horizontal = 1 second, vertical = 50 μV.

Occiptal spikes (Fig. 15.24), which are most often seen in young children, may be unilateral or bilateral; they are not generally considered to be highly epileptogenic. Since they are seen at the end of the chain of electrodes in the longitudinal bipolar montage, a coronal chain including O1 and O2 may bring them out better. They have to be distinguished from lambda waves, lambdoid activity, and "spiky" alpha. Occipital spikes that are of very short duration and have a "needle-like" appearance are known to occur in children with early-onset blindness. When occipital spikes occur in later age groups, they are often epileptogenic and may be associated with an underlying structural disorder. They are also known to occur in children with a form of benign focal epilepsy similar to Rolandic epilepsy.

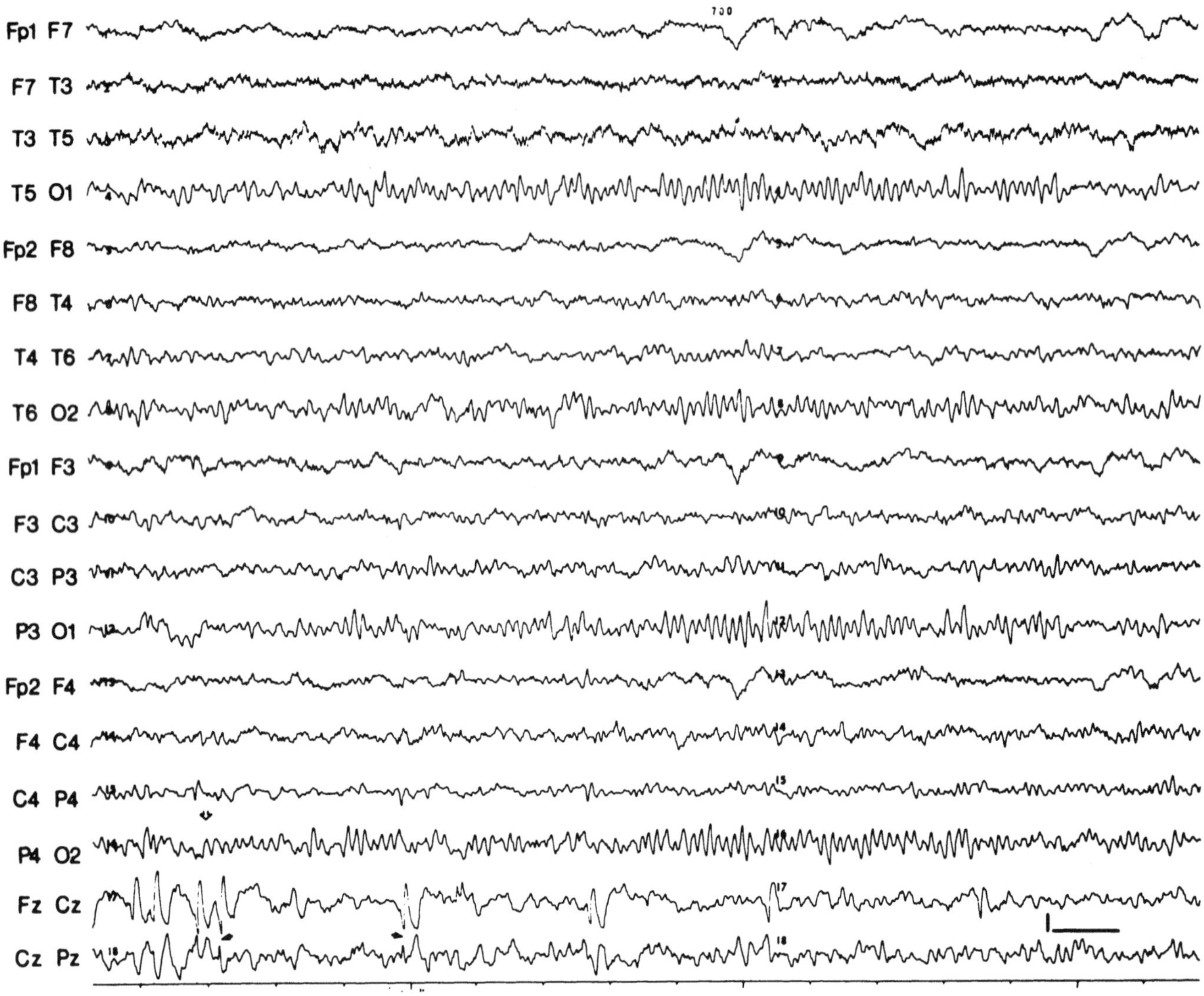

Figure 15.25. Midline spikes in an awake 6-year-old child with recent onset of tonic-clonic seizures. The horizontal arrows point to two of the spikes, which show phase reversals at Cz. The vertical open arrow marks the start of a continuing alpha rhythm over the occipital area, which indicates that the patient was awake. Filters: low frequency = 1 Hz, high frequency = 70 Hz. Calibrations: horizontal = 1 second, vertical = 50 μV.

Midline spikes (Figs. 15.25, 15.26) may easily be missed unless coronal montages, including the midline electrodes (Fz, Cz, and Pz) are used. They should be distinguished from V waves, but the distinction may be difficult unless they are also seen in a waking record. Midline spikes have a more restricted electrical field than V waves, which have fields that extend into the parasagittal areas. These discharges may occur in patients with seizures characterized by posturing of the arms, adversive movements, speech arrest, and sometimes bladder dysfunction.

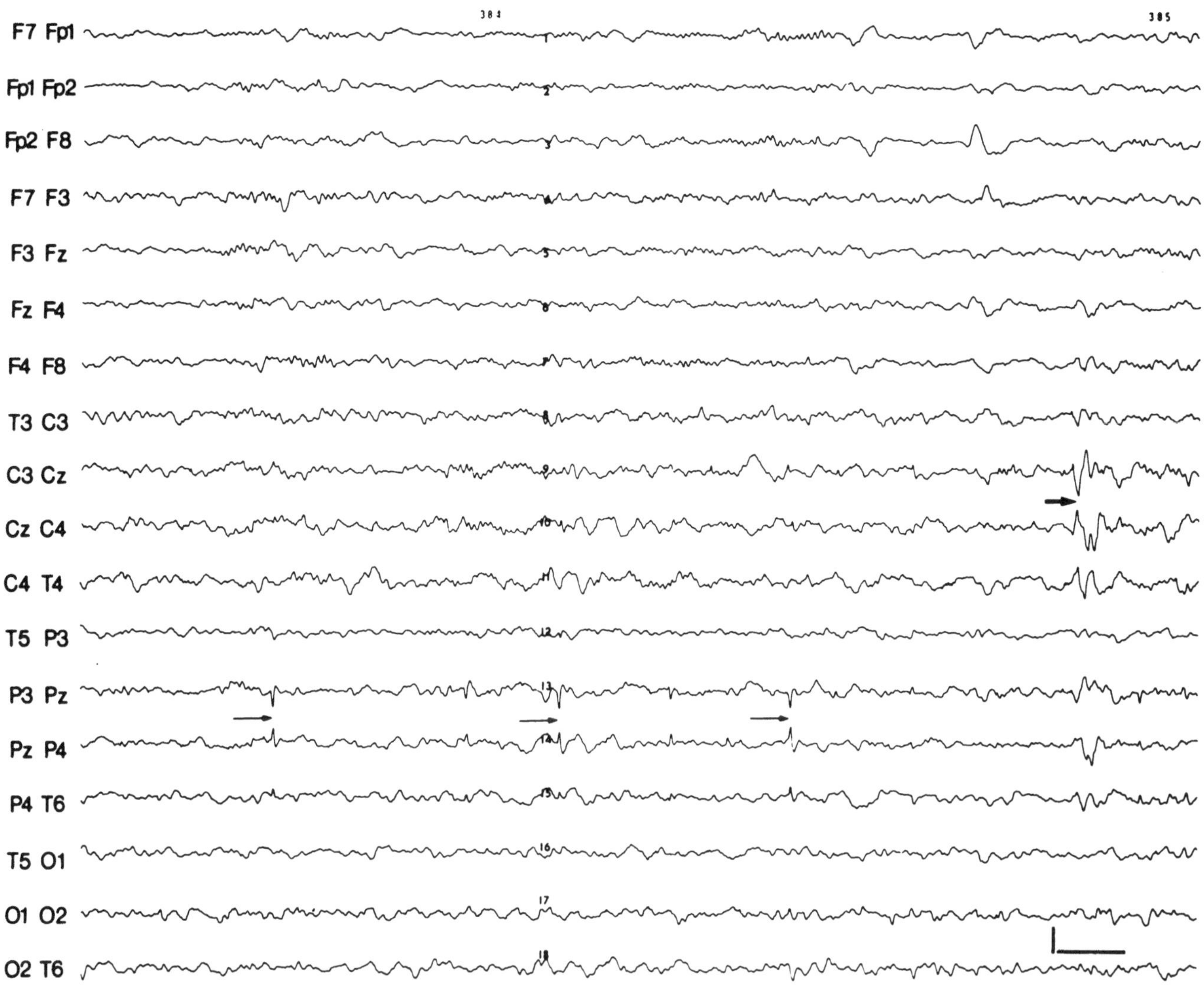

Figure 15.26. Midline spikes arising during sleep from Pz (thin arrows) in a 9-year-old child with generalized tonic-clonic seizures. The thick arrow points to a V wave. Filters: low frequency = 1 Hz, high frequency = 70 Hz. Calibrations: horizontal = 1 second, vertical = 200 μV.

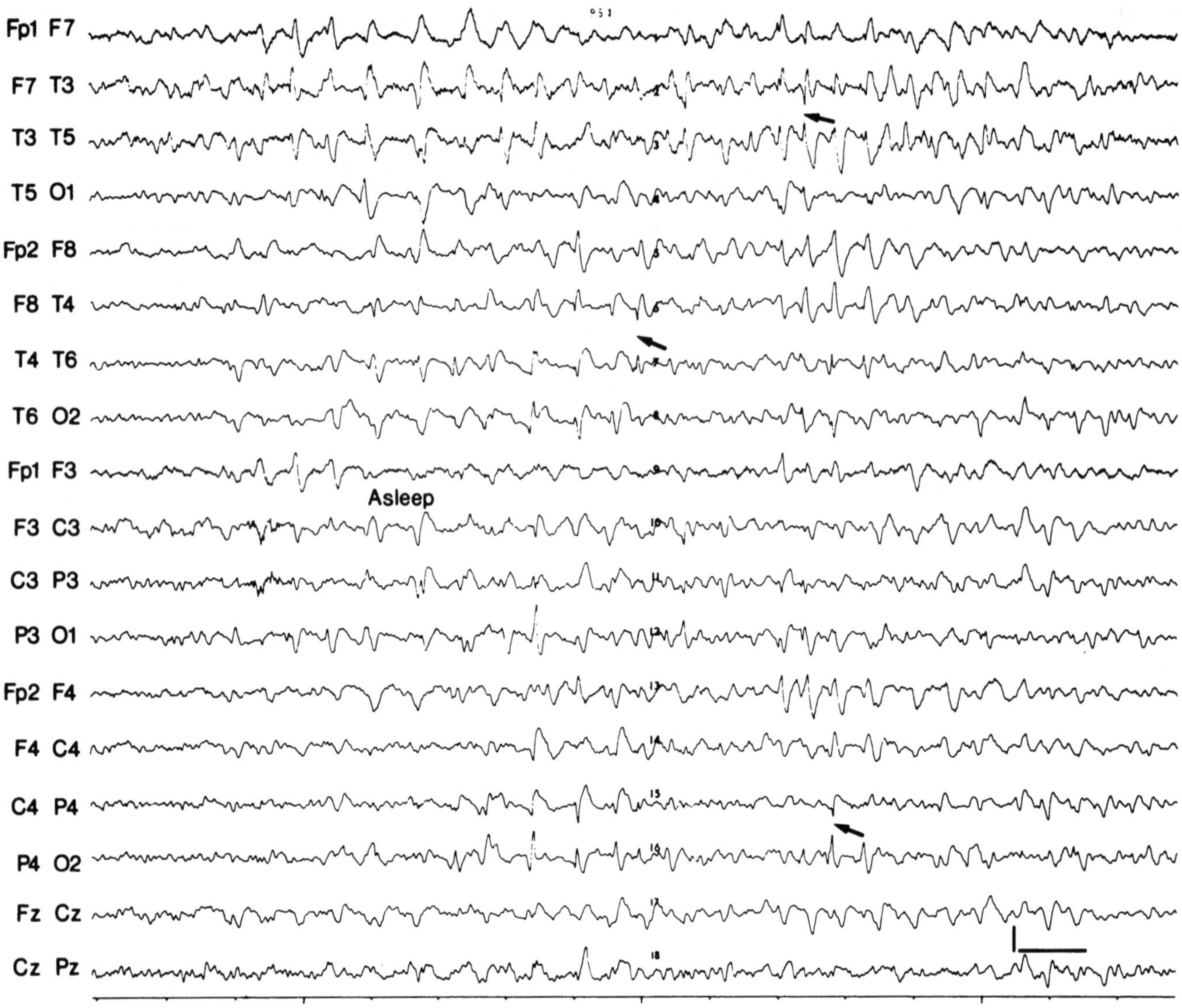

Figure 15.27. Multifocal spike discharges in an 8-year-old, mentally subnormal girl with a history of tonic-clonic as well as myoclonic seizures that were poorly controlled with anticonvulsants. The arrows point to some of the spike discharges that arise multifocally from either side. Filters: low frequency = 1 Hz, high frequency = 70 Hz. Calibrations: horizontal = 1 second, vertical = 100 μV.

Spike or sharp-wave discharges may arise independently from multiple sites in the same hemisphere or from both sides. The terms multifocal spikes or "multiple independent spike foci pattern" are used when there are spikes originating from three or more noncontiguous electrode placements, with a minimum of one focus in each hemisphere. Such a pattern correlates very highly with a seizure disorder, usually of the generalized type and sometimes accompanied by mental subnormality. Figure 15.27 is an example of multifocal spikes.

Generalized Epileptiform Abnormalities

Unlike the focal epileptiform patterns, these are discharges that appear simultaneously on both sides. There are a number of patterns that show a significant degree of clinical correlation.

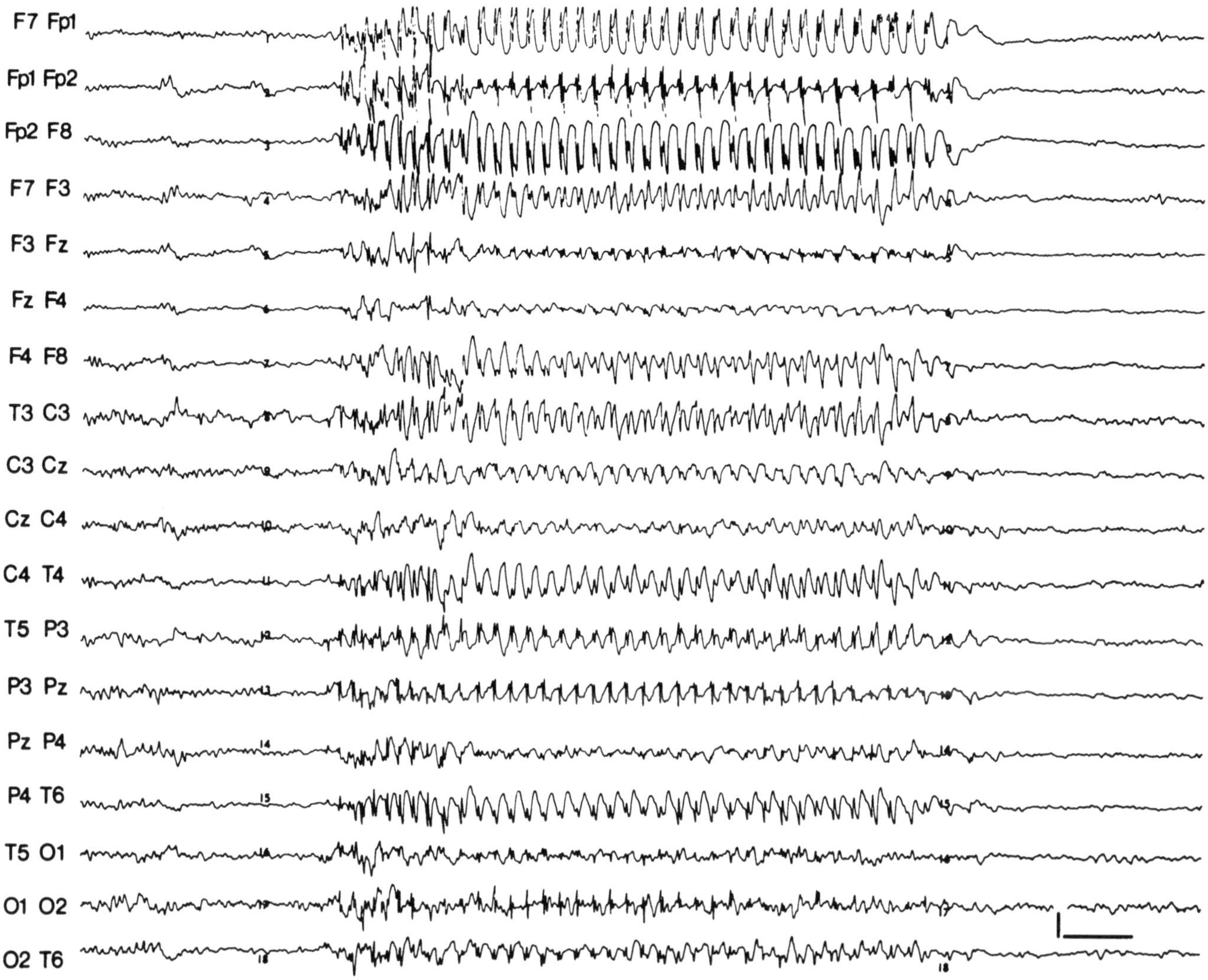

Figure 15.28. Three per second spike and wave discharges typical of absence in a 10-year-old boy with staring spells. The generalized bisynchronous, frontally dominant spike and slow-wave complexes occur at a frequency of about 3.5 to 4 Hz. The patient was unresponsive during this period. Filters: low frequency = 1 Hz, high frequency = 70 Hz. Calibrations: horizontal = 1 second, vertical = 150 μV.

3-Hz Spike and Wave Discharges

This pattern is classically seen in absence seizures. It consists of bilaterally synchronous and symmetrical complexes, each made up of a high-amplitude, surface-negative spike and wave. The complexes repeat at a rate of about 3 Hz and appear in a generalized distribution (Fig. 15.28); maximum amplitude is usually in the frontal and rarely in the occipital areas. At the onset of the discharges, repetition rate may be faster (4 Hz), whereas toward the end it may become slower (2.5 Hz). The discharges usually last for 3 to 4 seconds and are often precipitated by hyperventilation (see Fig. 16.4). When prolonged, they are considered to be ictal phenomena. During sleep the discharges may get fragmented and appear as spikes or multispikes.

The 3-Hz spike and wave pattern is one of the easiest EEG patterns to recognize. Certain physiological patterns that may resemble it are hyperventilation-induced delta activity—especially when the waves appear sharp or are "notched"—and paroxysmal hypnagogic hypersynchrony. Certain artifacts like hiccup, and artifacts associated with rhythmic rocking of a baby by its mother while the EEG is taken, may also resemble this discharge. The 3-Hz spike and wave pattern is readily distinguishable from the 6-Hz spike and wave discharge, the latter having very small spikes and short-duration discharges. When the onset of the 3-Hz pattern is asynchronous, or when the frequency is less than 2.5 Hz, the possibility of secondary bilateral synchrony (see later) or slow spike and wave should be considered.

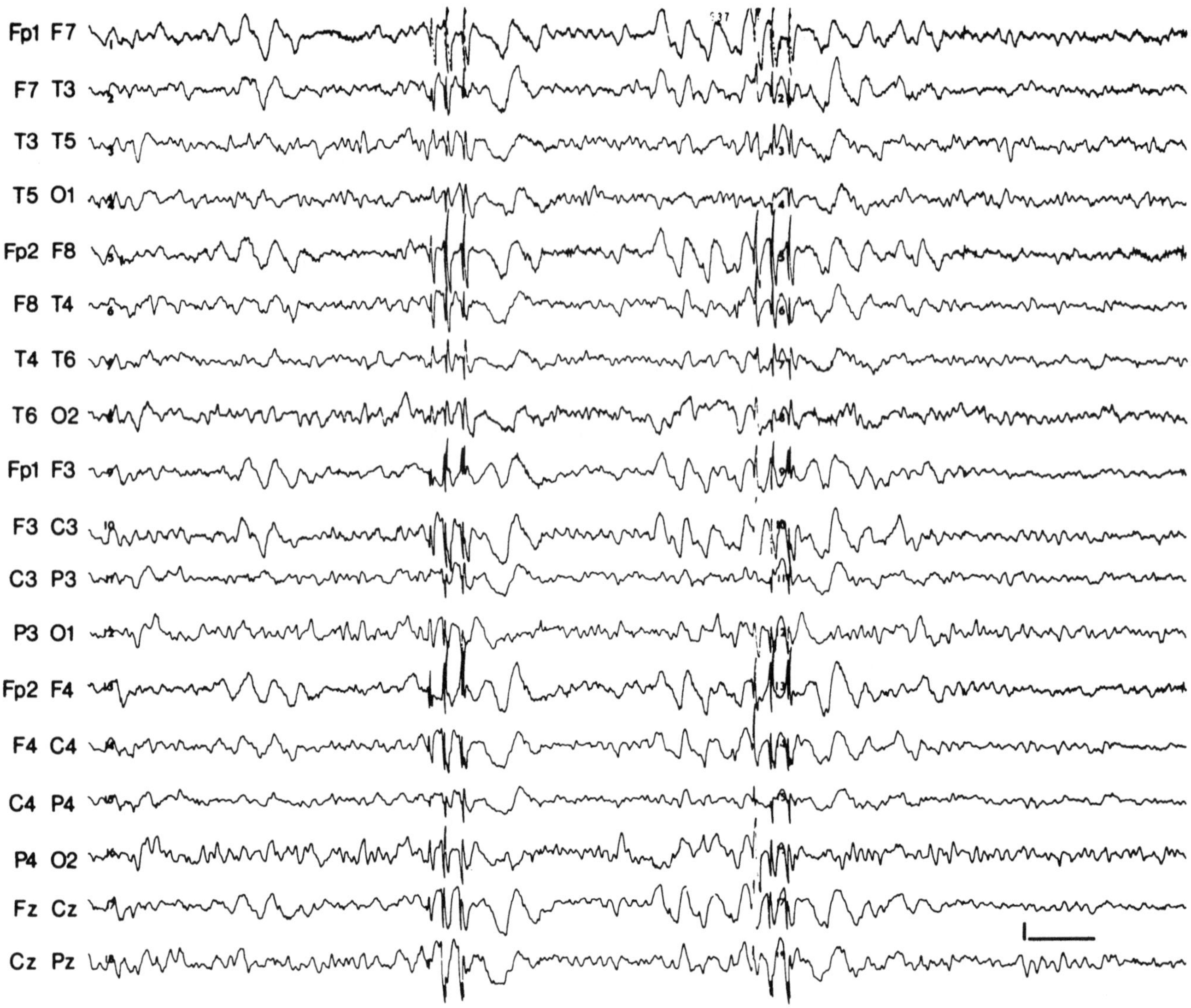

Figure 15.29. Generalized, atypical fast spike and wave pattern. The frontally dominant, 4- to 5-Hz spike and wave paroxysms occurred during wakefulness in a 26-year-old patient with a history of generalized tonic-clonic seizures and occasional myoclonic jerks from the 10th year of life. The patient's mother had experienced similar seizures up to age 25 years. Filters: low frequency = 1 Hz, high frequency = 70 Hz. Calibrations: horizontal = 1 second, vertical = 50 μV.

Generalized Atypical Fast Spike and Wave Discharges

This pattern lacks the typical appearance and repetition rate of 3-Hz spike and wave discharges. It consists of 3- to 5-Hz spike and wave discharges that show variations both in repetition rate and morphology. The waveform may be a mixture of spike and wave complexes and multispike and wave complexes. Figure 15.29 shows an example of this pattern. Note the variation in the appearance of the discharges, with some of them showing biphasic spikes and others multispikes.

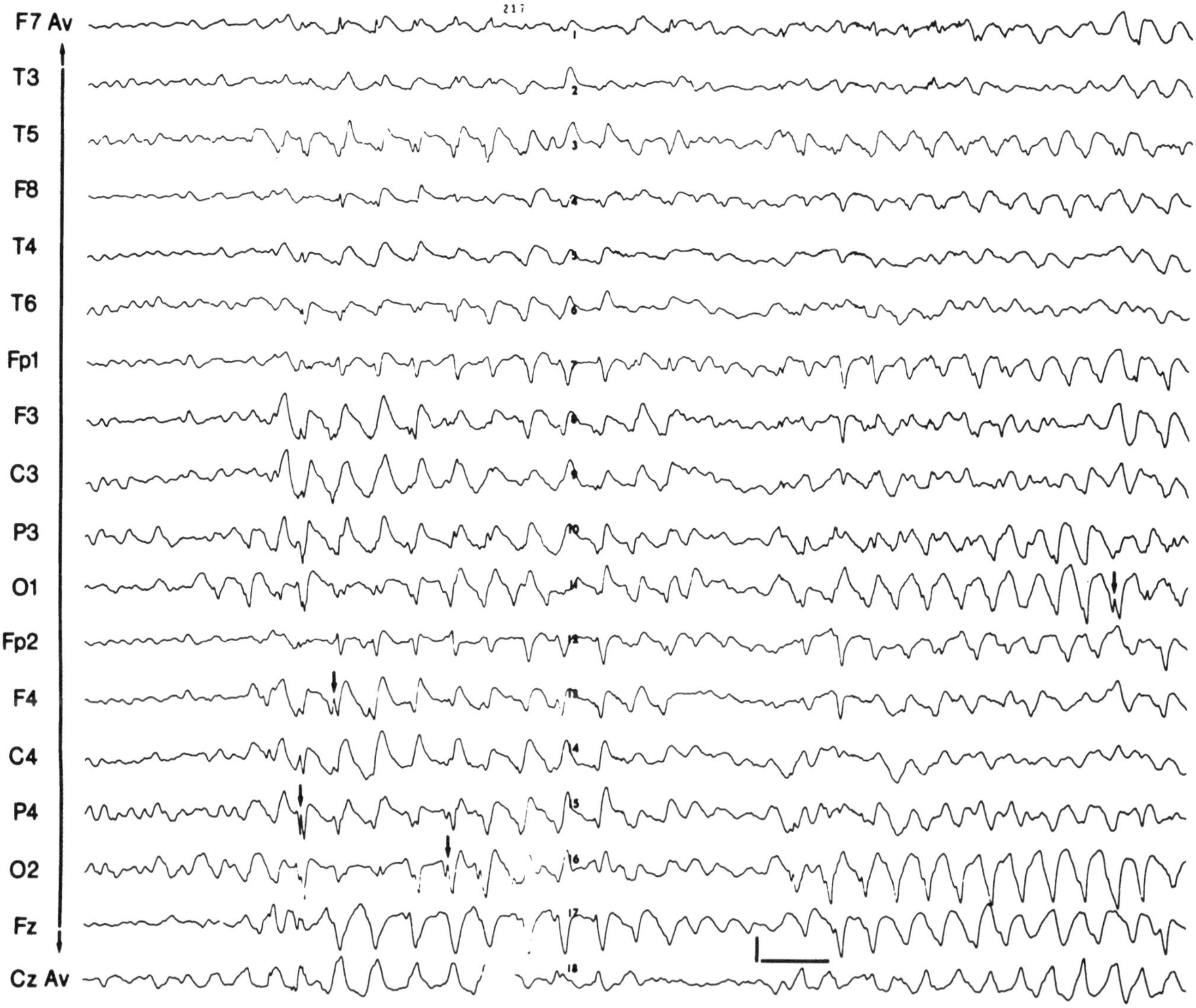

Figure 15.30. Generalized sharp- and slow-wave complexes (sharp waves indicated by arrows) at 1.5 to 2 Hz in a 5-year-old boy with Lennox-Gastaut syndrome. The patient was mentally subnormal and hyperactive and had seizures that were resistant to anticonvulsants. The seizures were tonic or myoclonic and led to frequent falls. Filters: low frequency = 1 Hz, high frequency = 70 Hz. Calibrations: horizontal = 1 second, vertical = 150 μV.

Generalized Slow Spike and Wave Discharges

This pattern consists of sharp and slow-wave complexes occurring at a rate of less than 2.5 Hz (Fig. 15.30). It also has been called "petit mal variant." The discharges are bilaterally synchronous and generalized, although fluctuating asymmetry is not uncommon. Each epoch may last for several seconds, usually without clinically obvious seizures. The background activity is abnormally slow. Unlike the case of classic 3-Hz spike and wave, there is little evidence of activation during hyperventilation. The pattern is often correlated with intractable seizures and mental subnormality (see Lennox-Gastaut syndrome, Chapter 21).

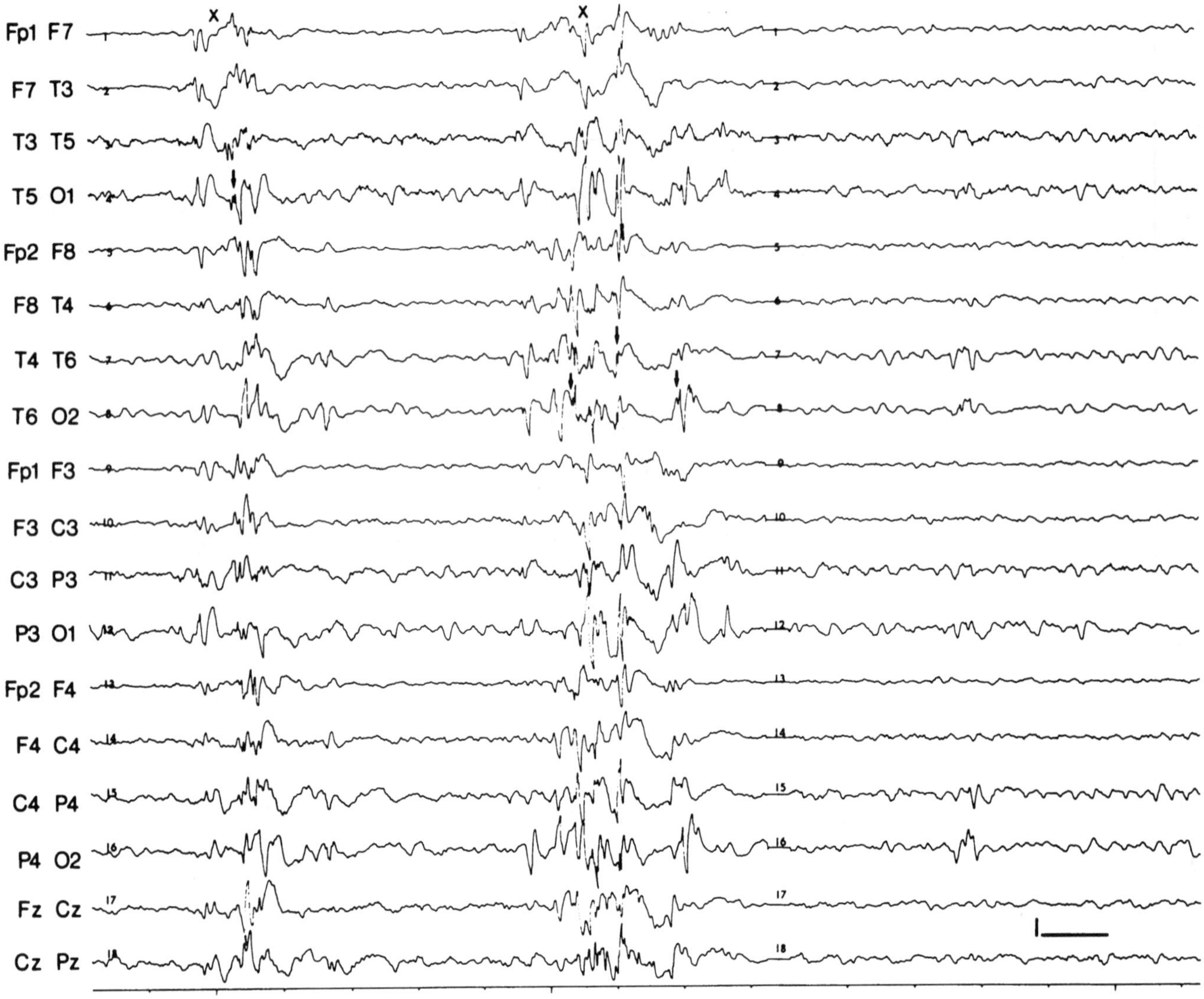

Figure 15.31. Generalized multispike and wave complexes in a 3-year-old child with recurrent tonic-clonic as well as myoclonic seizures involving mostly the upper extremities. The patient had a documented history of cerebral hypoxia during the perinatal period. The arrows point to the multispikes. Myoclonic jerks are indicated by the "X's." Filters: low frequency = 1 Hz, high frequency = 70 Hz. Calibrations: horizontal = 1 second, vertical = 50 μV.

Generalized Multispike and Wave Discharges

Generalized multispikes or multispike and wave complexes may occur as an interictal phenomenon in patients with primary generalized epilepsy (often with myoclonus) and in Lennox-Gastaut syndrome. They may be seen in awake but more often in sleep recordings. Sometimes the discharges are accompanied by myoclonic jerks. Generalized or lateralized multispikes may also occur in diffuse encephalopathies of different etiology. Figures 15.31 and 15.32 show generalized and lateralized multispike discharges.

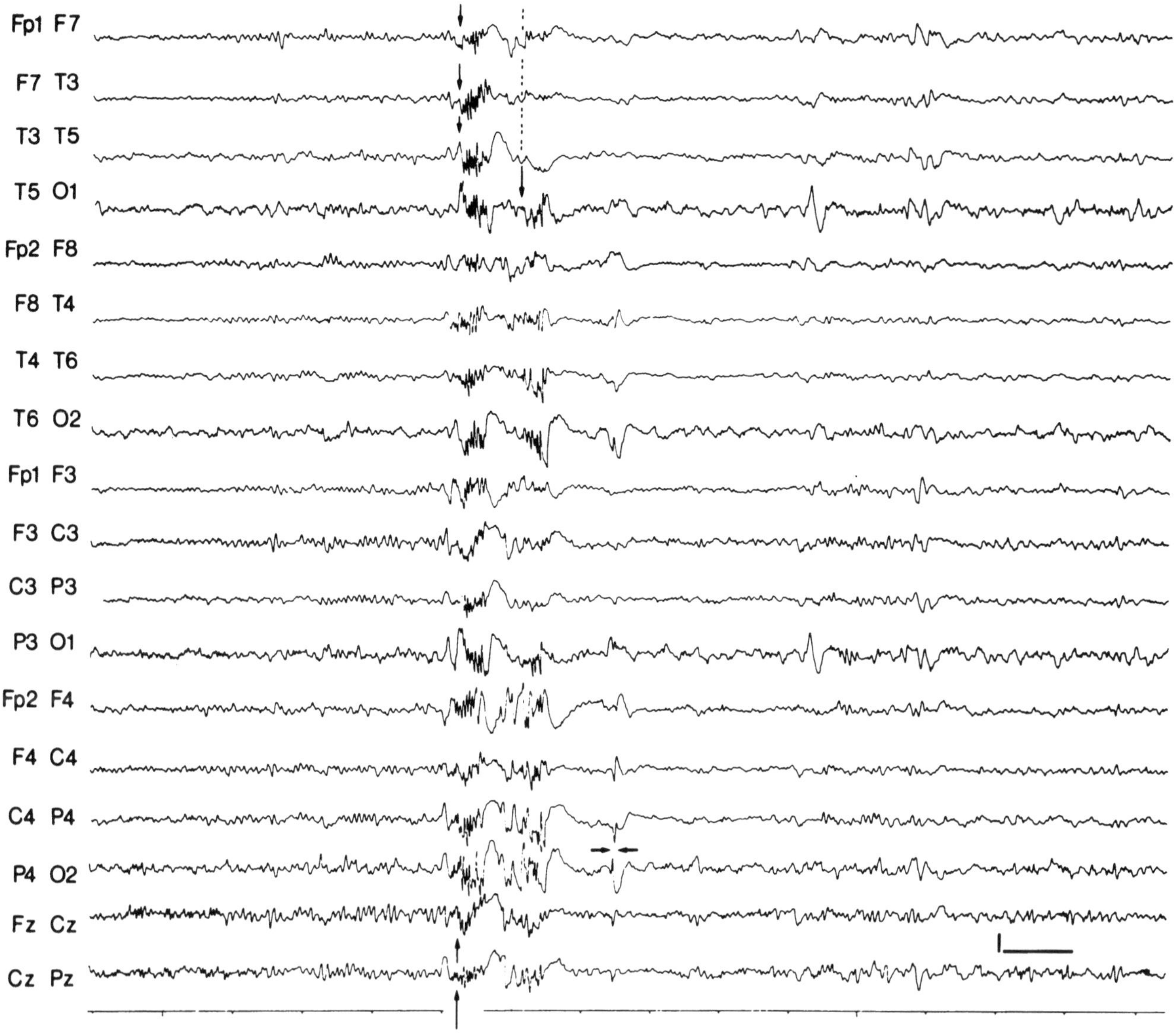

Figure 15.32. Generalized (solid arrow) as well as lateralized (arrow with dotted tail) multispike discharges in a 12-year-old patient with tonic-clonic, myoclonic, and left focal motor (upper extremity and face) seizures from age 3 years following a febrile illness diagnosed as viral encephalitis. The horizontal arrows point to a focal spike showing a phase reversal at P4. Filters: low frequency = 1 Hz, high frequency = 70 Hz. Calibrations: horizontal = 1 second, vertical = 50 μV.

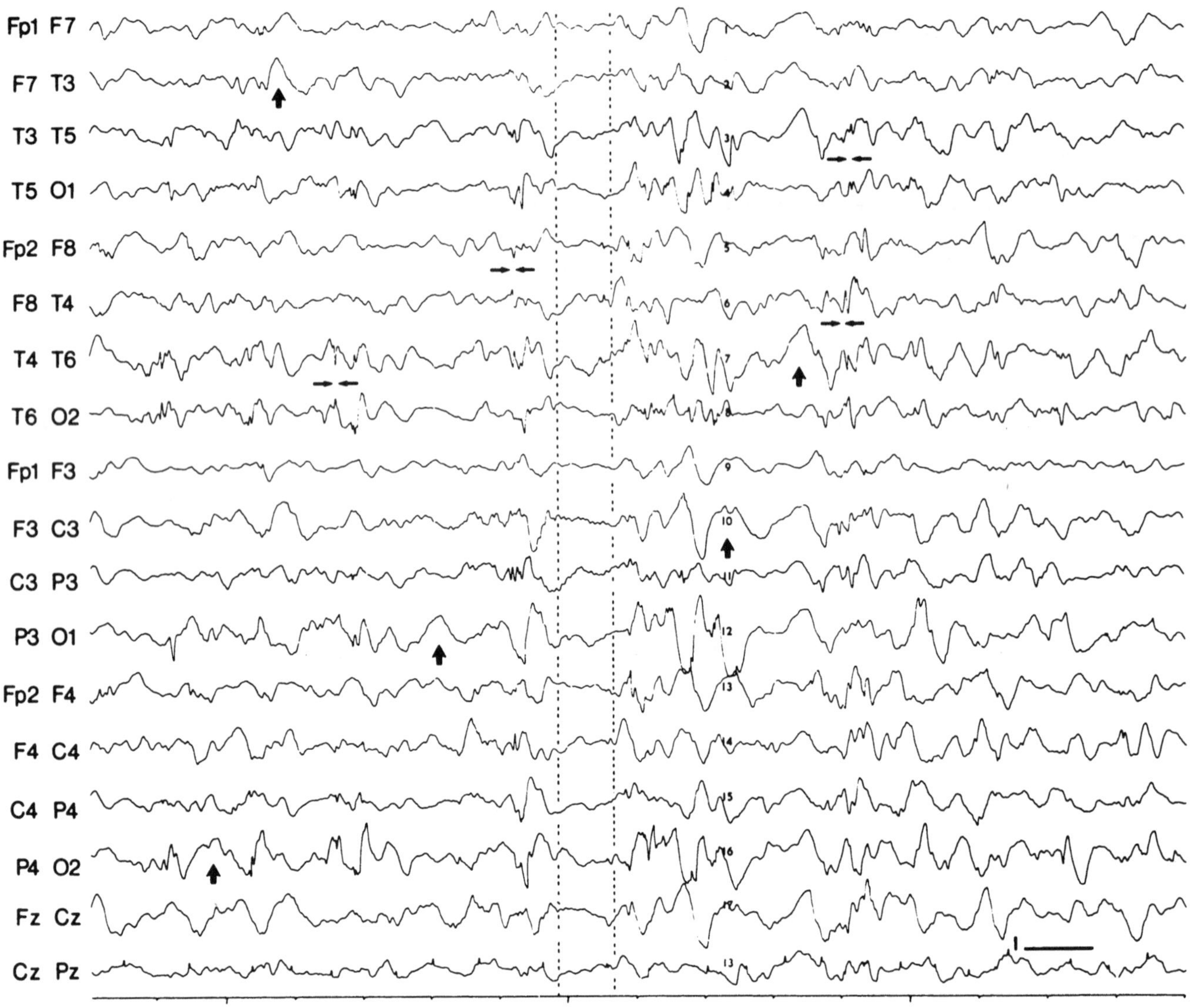

Figure 15.33. Hypsarrhythmic pattern in an 8-month-old baby with infantile spasms. Note the chaotic background activity containing high-amplitude, irregular delta transients (vertical arrows) and multifocal spike discharges indicated by horizontal arrows. A brief interval of electro-decrement can be seen between the dotted lines. The patient made a jerky movement that involved the trunk and upper extremities at the beginning of this interval. Filters: low frequency = 1 Hz, high frequency = 70 Hz. Calibrations: horizontal = 1 second, vertical = 100 μV.

Hypsarrhythmia

This pattern consists of continuous, high-amplitude 1- to 3-Hz irregular, disorganized background activity and shifting spike foci (Fig. 15.33). It occurs in most patients with infantile spasms. During sleep, the activity may become discontinuous. The pattern as described is the interictal pattern; when spasms occur, there is an abrupt attenuation of the background activity (electro-decrement) and rhythmic, low-amplitude fast activity follows.

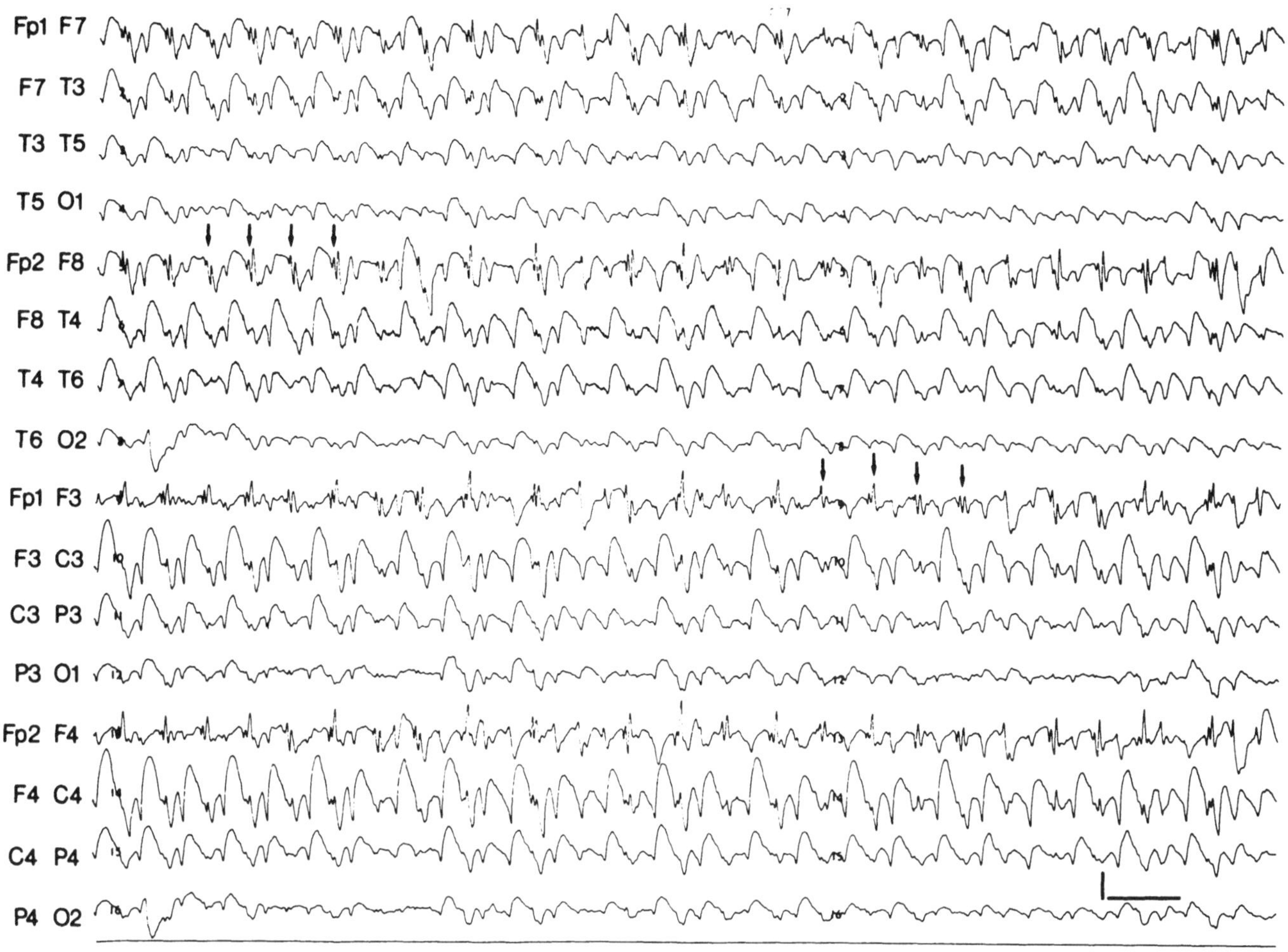

Figure 15.34. Generalized 1.5- to 2.5-Hz continuous spike and wave discharges (absence status/spike-wave stupor) in a 32-year-old patient who appeared confused. The patient had a history of staring spells and tonic-clonic seizures from age 6 years and had discontinued anticonvulsants seven days before the recording was taken. Filters: low frequency = 1 Hz, high frequency = 70 Hz. Calibrations: horizontal = 1 second, vertical = 100 μV.

Ictal Patterns

The ictal patterns often consist of either a change in the frequency and amplitude of the interictal patterns, or a totally different pattern that may be activity in the alpha, beta, or slower frequency bands. The characteristic features of an ictal pattern are abrupt onset and an evolution characterized by changes in frequency, morphology, and amplitude. However, a nonevolving ictal pattern is also known to occur in generalized seizures, the most common example being that seen in absence. Here the ictal and interictal patterns are quite similar, the difference being that the ictal pattern lasts longer and may be of higher amplitude. The onset and offset are abrupt, with a quick return of normal background activity. In the case of absence status, there is continuous spike and wave activity at 2 to 4 Hz (Fig. 15.34). As pointed out in Chapter 21, the distinction between interictal and ictal pattern in a nonconvulsive generalized seizure is sometimes nebulous owing to the difficulty in establishing that the patient has indeed had a clinical seizure within the short time that the discharges last.

A progressive ictal pattern may be seen in both generalized and focal seizures. There is usually an attenuation of the preceding interictal activity and even of the background activity, followed by a rhythmic discharge that may take several forms. In the case of generalized seizures, there initially is diffuse fast activity (10 to 25 Hz) that progressively slows down and increases in amplitude. At the onset, the spike component of this activity may not quite be obvious and may look like rhythmic alpha or beta activity. Different terms such as fast paroxysmal rhythms, generalized paroxysmal fast activity, and rhythmic spikes have been used to denote this activity. Since it is a common feature at the beginning of a grand mal seizure, the term "grand mal pattern" was originally used. Such discharges are particularly common in Lennox-Gastaut syndrome when the child is experiencing a generalized tonic or akinetic seizure.

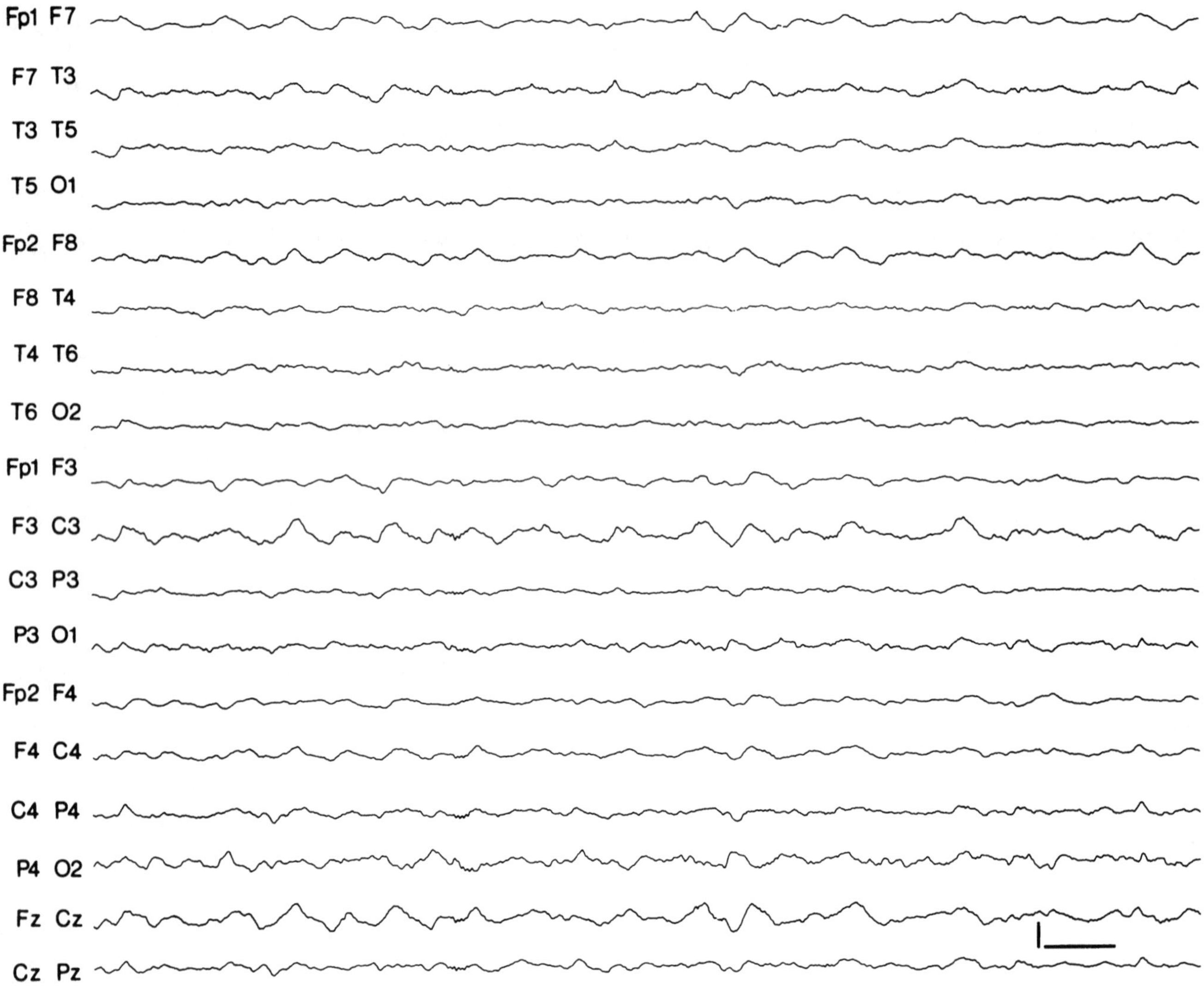

Figure 15.35. Generalized delta activity in the postictal phase of a grand mal seizure in a 20-year-old patient with primary generalized epilepsy. The patient was unresponsive at this time. Filters: low frequency = 1 Hz, high frequency = 70 Hz. Calibrations: horizontal = 1 second, vertical = 100 μV.

In the case of a generalized tonic-clonic or grand mal seizure, the fast-frequency activity correlates with the tonic phase; this is followed by progressive slowing and the appearance of rhythmic generalized spikes that are bilaterally synchronous. These discharges correspond to the clonic jerks. Soon the spike discharges become less and less frequent and cease. This is followed by generalized attenuation or slowing of the background activity (Fig. 15.35). In the case of convulsive status, the same sequence of events may repeat at short intervals, or there may be continuous rhythmic generalized spiking (Fig. 15.36). Generalized multispike and wave discharges may be ictal phenomena and are often accompanied by myoclonic jerks. Figure 15.37 is an example of such a tracing.

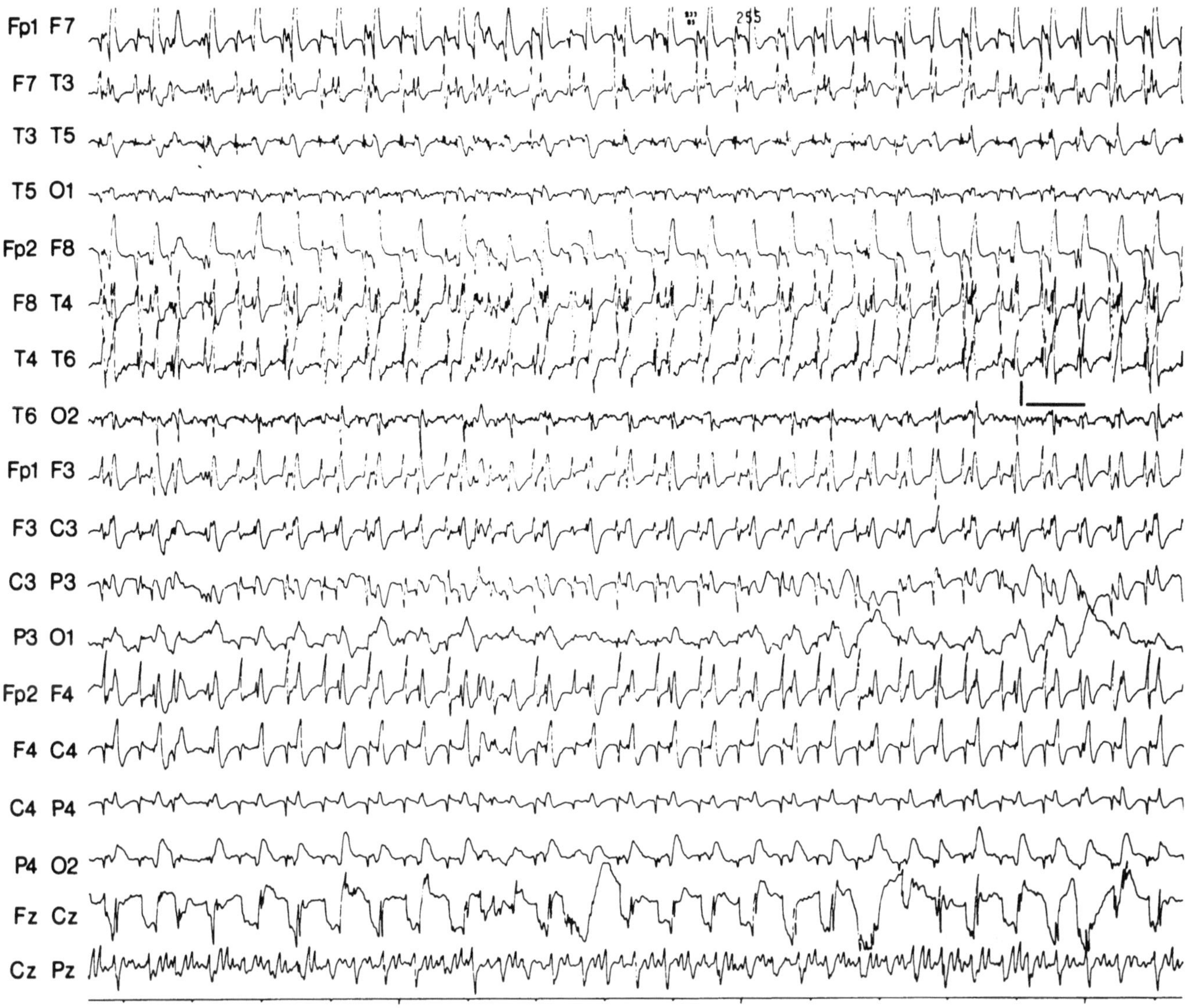

Figure 15.36. Generalized spike discharges occurring continuously at about 2- to 3- Hz in a 30-year-old patient who was unresponsive, had frequent grand mal seizures, and was diagnosed as convulsive status epilepticus. This tracing was taken during an interval between two convulsions. Intravenous diazepam suppressed the spikes. Filters: low frequency = 1 Hz, high frequency = 70 Hz. Calibrations: horizontal = 1 second, vertical = 100 μV.

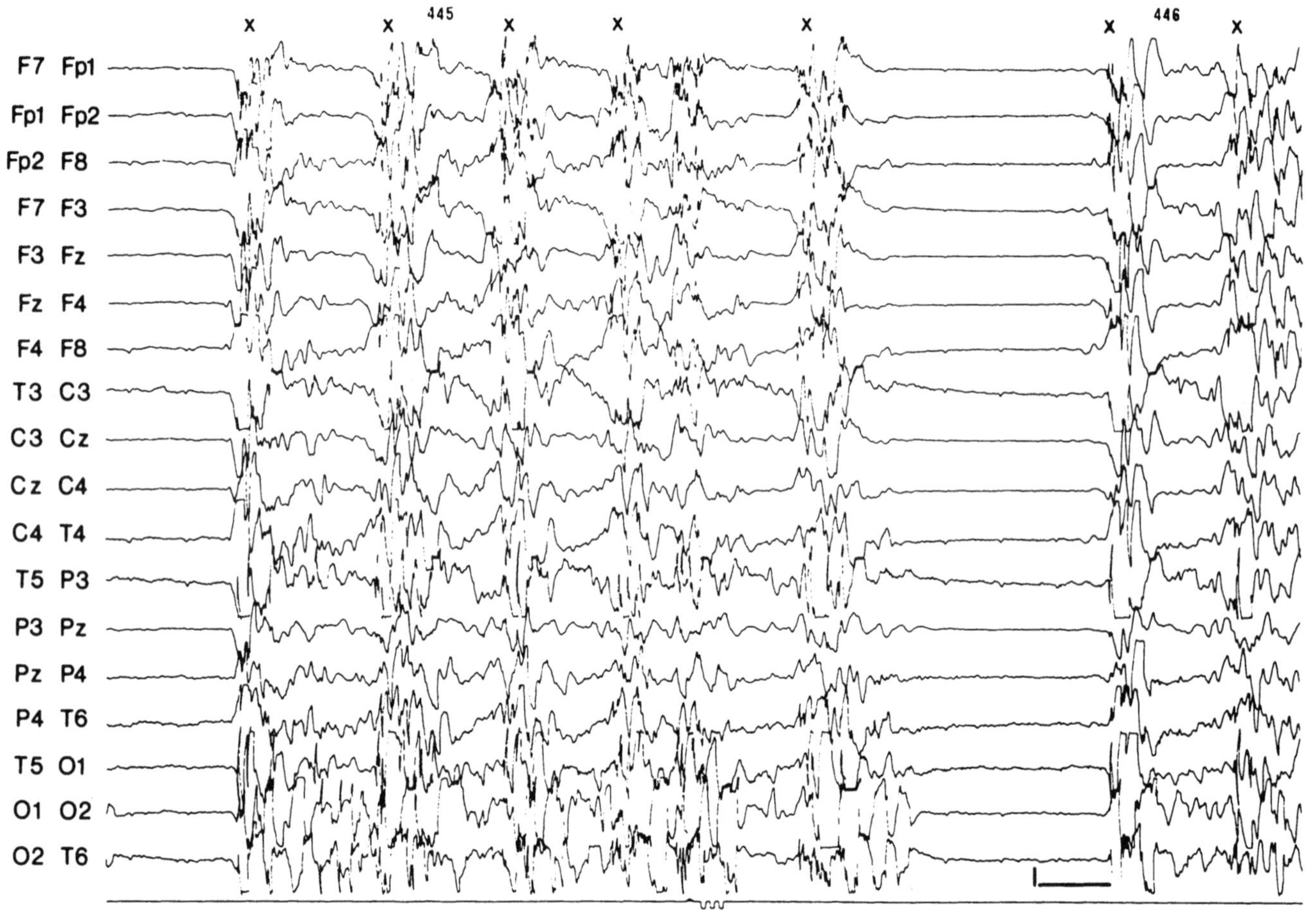

Figure 15.37. Bursts of bilaterally synchronous spikes, multispikes, and spike and wave discharges in a 60-year-old patient following cardiopulmonary arrest and resuscitation. Note the periods of suppression of the background between the bursts. The "X's" mark the occurrence of myoclonic jerks. Filters: low frequency = 1 Hz, high frequency = 70 Hz. Calibrations: horizontal = 1 second, vertical = 50 μV.

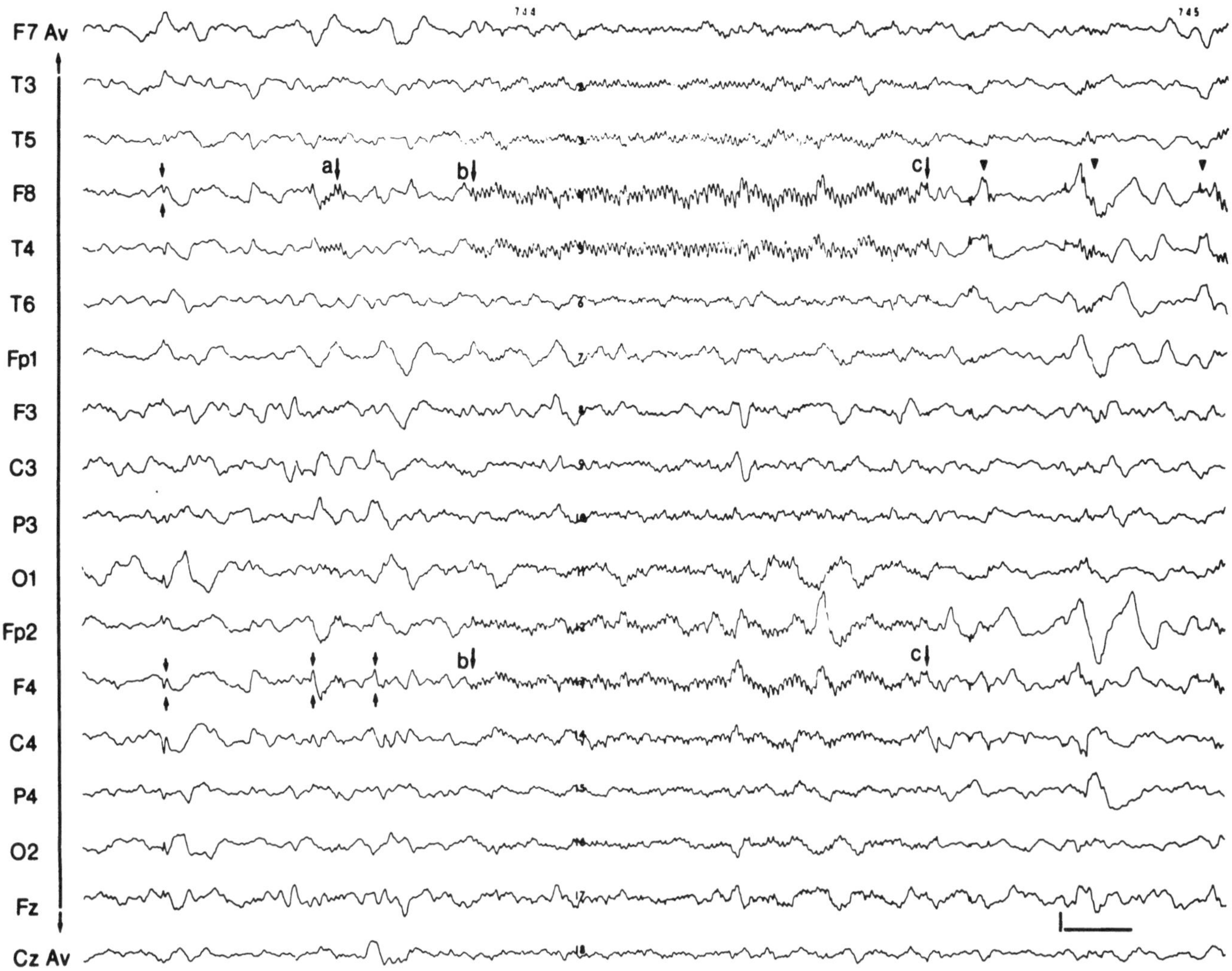

Figure 15.38. Focal subclinical electrographic seizure arising from the right inferior frontal and temporal areas in a 45-year-old patient with brain metastasis from melanoma. The patient presented with recurrent episodes of speech arrest, confusion, and clonic movements of the right side of the face and upper extremities. Short vertical arrows point to focal spikes at F4 and F8. A brief cluster of rapidly occurring spikes can be seen at arrow "a" over F8 and T4. A run of similar rapidly occurring, medium amplitude spikes is seen over F8, T4 and F4, beginning at arrow "b" and ending at arrow "c." The triangles point to sharply contoured delta transients that occur in the same area. No behavioral changes could be detected during the rapid spiking. Filters: low frequency = 1 Hz, high frequency = 70 Hz. Calibrations: horizontal = 1 second, vertical = 50 μV.

In the case of focal seizures, there usually is a fast-frequency discharge over the site of origin of the epileptiform activity. This is followed by a slower rhythmic discharge that may take the form of spikes or sharp waves followed, in turn, by postictal slowing (see Figs. 15.38–15.40). In seizure discharges originating in the temporal lobe, particularly the medial part, surface recordings may show only the cessation of the previously occurring focal discharges followed by the rhythmic, slow postictal activity. It has been suggested that the fast-frequency activity occurring at the onset may be picked up if sphenoidal electrodes are used.

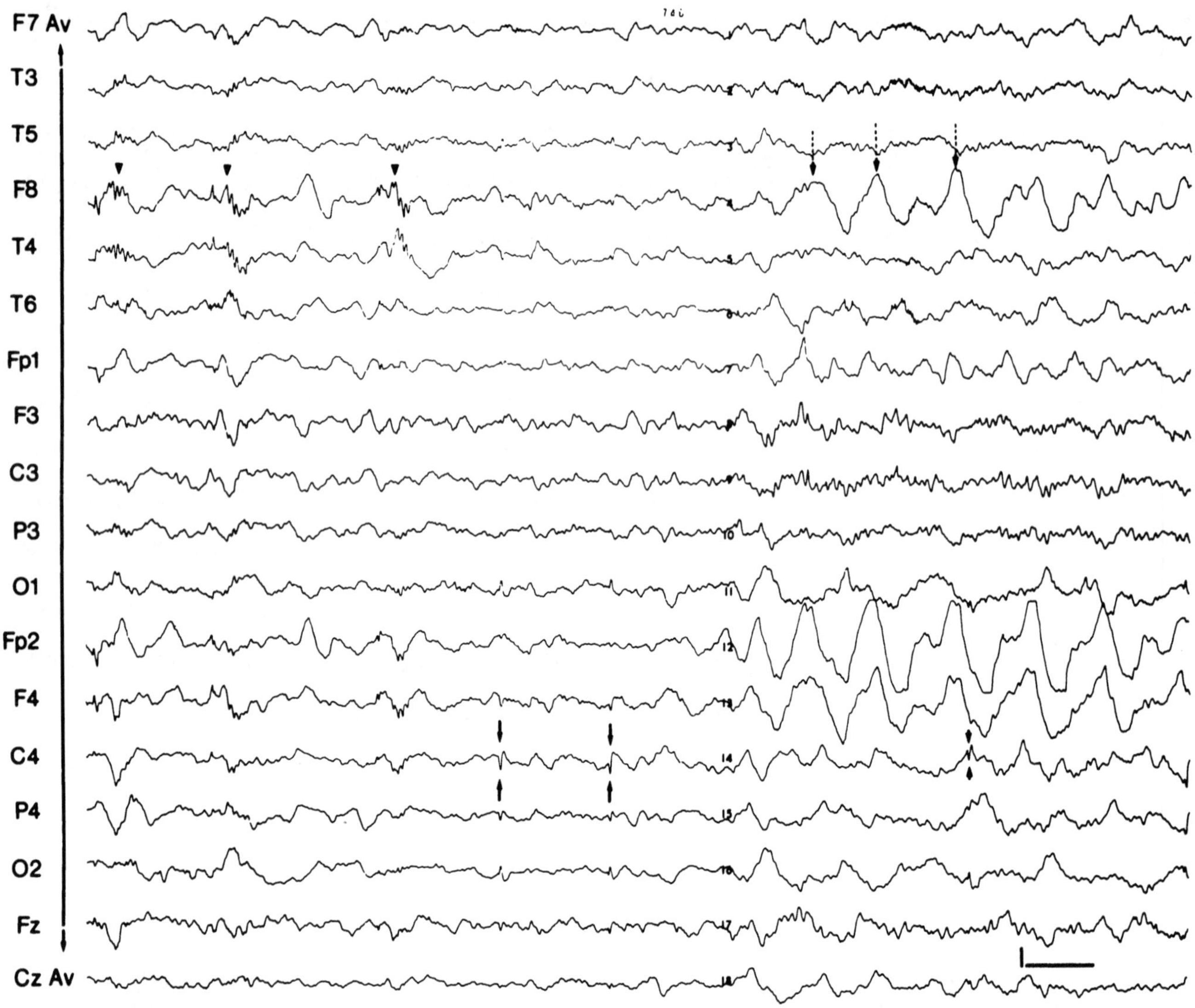

Figure 15.39. Rhythmic focal delta activity occurring as a postictal phenomenon in the right inferior frontal and frontopolar areas. Tracings are from the same patient as in Fig. 15.38. The arrows with dotted tail point to some of the high-amplitude delta waves. Brief clusters of spikes (arrowheads) preceded the delta activity. Long vertical arrows point to focal spikes occurring on the same side. The rhythmic delta activity disappeared after 10 seconds. Filters: low frequency = 1 Hz, high frequency = 70 Hz. Calibrations: horizontal = 1 second, vertical = 50 µV.

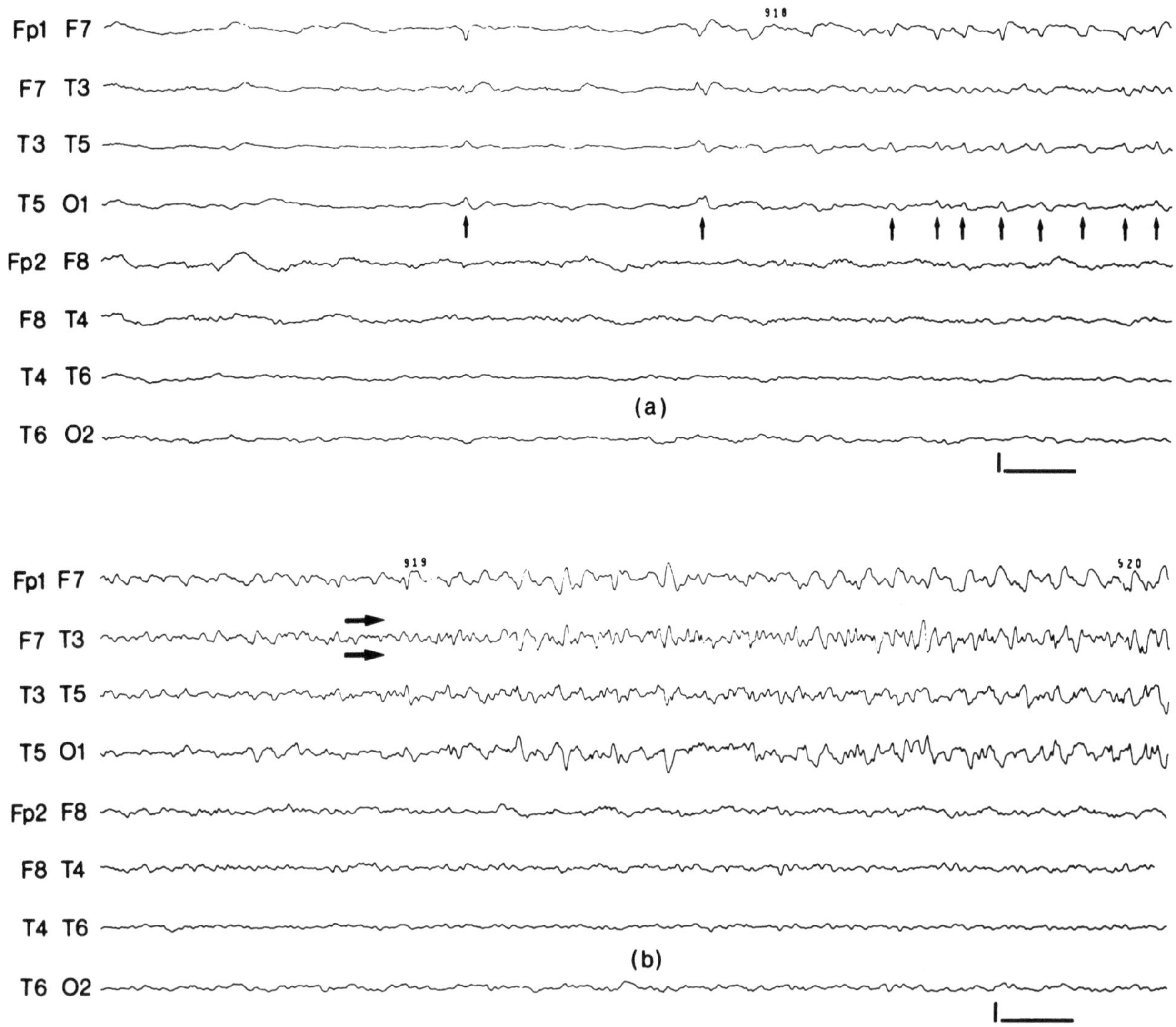

Figure 15.40. Focal electrographic seizure arising from the left temporal area in a 38-year-old patient with an underlying glioma. In (a) the vertical arrows point to sharp-wave discharges that show an equipotential zone between F7 and T3. Immediately following, in (b) there is a rapid buildup of activity involving the whole left temporal area. This activity lasted for approximately 60 seconds and was followed by a period of postictal focal slowing. During the ictus the patient appeared very confused, sat up in bed, and showed some bizarre movements of the right upper extremity. Filters: low frequency = 1 Hz, high frequency = 70 Hz. Calibrations: horizontal = 1 second, vertical = 50 μV.

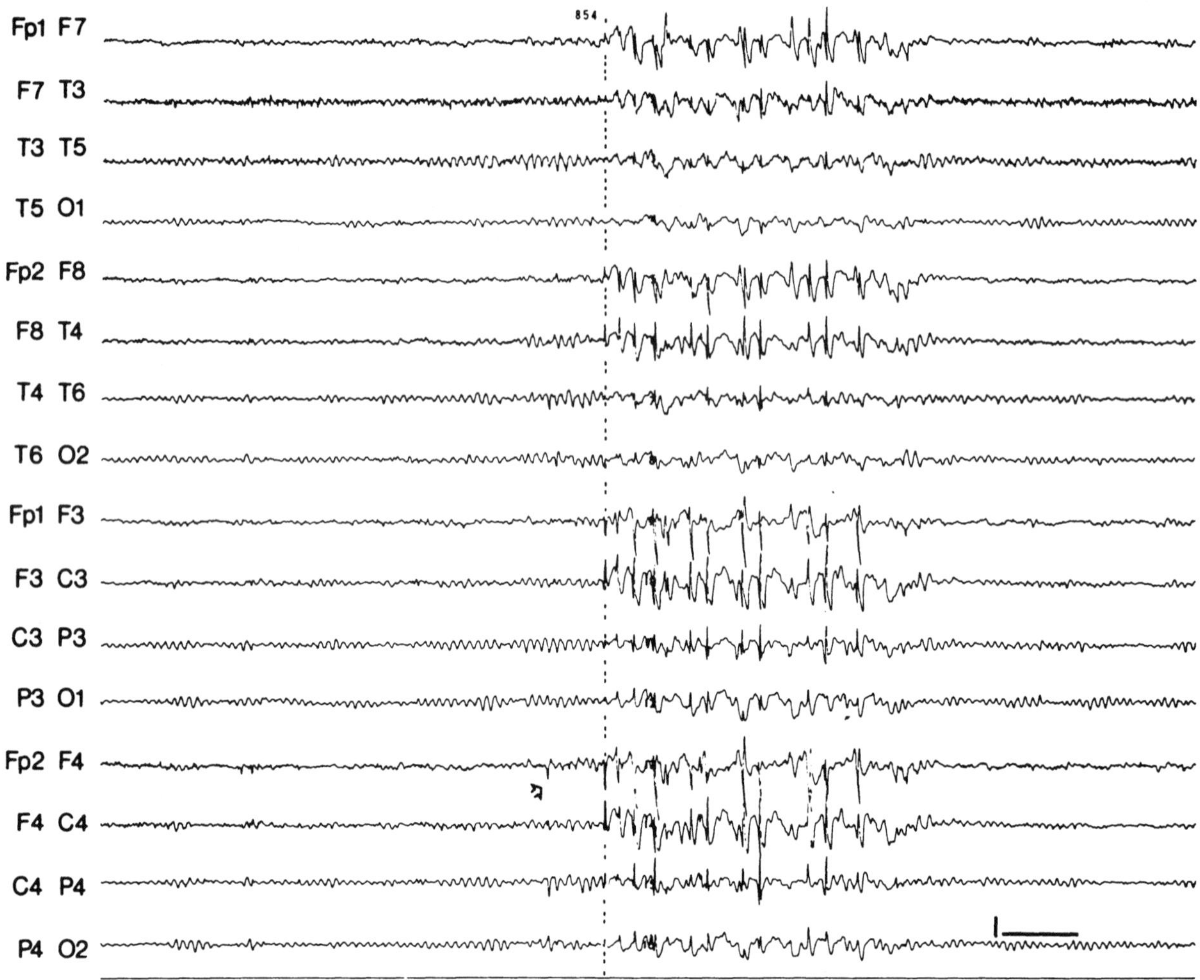

Figure 15.41. Secondary bilateral synchrony in a 40-year-old patient with new onset tonic-clonic seizures. The patient had a history of head trauma 2 years prior to the first seizure. Note the bilaterally synchronous discharge of spike-wave complexes starting at the dotted line. The open arrow shows a spike on the right side, with negativity at F4, preceding the bilaterally synchronous discharge. Filters: low frequency = 1 Hz, high frequency = 70 Hz. Calibrations: horizontal = 1 second, vertical = 50 μV.

Secondary Bilateral Synchrony

Seizure discharges that originate focally may spread rapidly and involve both sides, so that the resulting seizure may be of the generalized type. The EEG pattern, which consists of generalized bilaterally synchronous discharges, often holds clues to the focal onset. This may be in the form of a focal spike (as in Fig. 15.41) or fast activity preceding the bisynchronous discharge. Amplitude or frequency differences in the discharge or in the postictal slowing between the two sides are a less reliable clue.

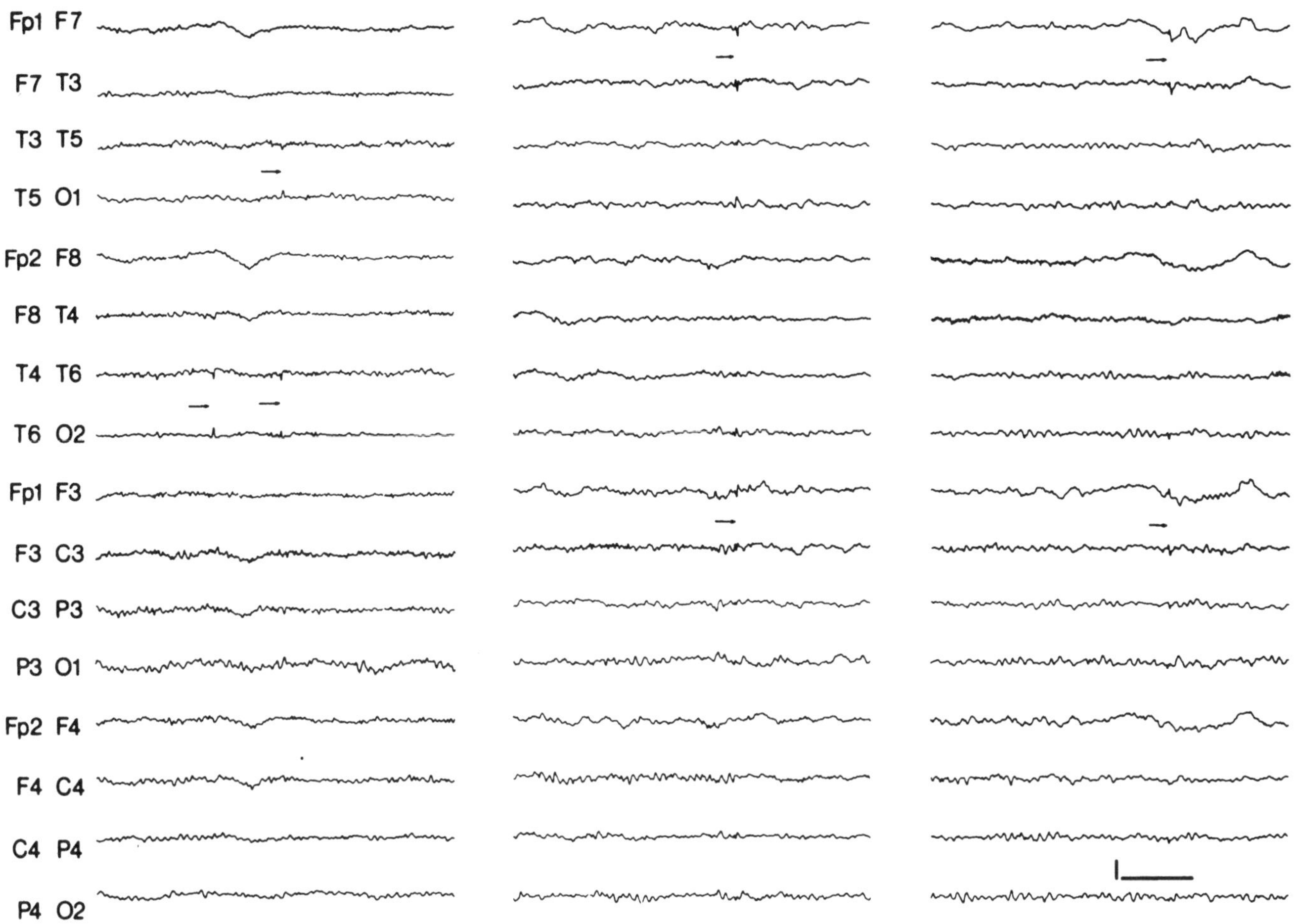

Figure 15.42. Small sharp spikes (SSS), also known as benign epileptiform transients of sleep (BETS), in a 30-year-old patient with headaches and depression. The arrows indicate the low-amplitude, short-duration spikes that occur independently in the temporal areas on left and right sides; some of them show a more widespread distribution, involving the inferior and midfrontal areas as well. Filters: low frequency = 1 Hz, high frequency = 70 Hz. Calibrations: horizontal = 1 second, vertical = 50 µV.

Epileptiform Patterns of Doubtful Significance

These patterns have been denoted "pseudoepileptiform" abnormalities, since they are of doubtful or no clinical significance. Nevertheless, they do resemble epileptiform patterns. It is important, therefore, to differentiate them from patterns that do show significant clinical correlation with epilepsy.

Small Sharp Spikes

Small sharp spikes (SSS) are usually less than 50 µV in amplitude and less than 50 ms in duration (Fig. 15.42). Because of their duration and size, they may look like muscle spikes; but they often are followed by a slow wave. The slow wave is usually of smaller amplitude than the spike. Small sharp spikes may be monophasic, biphasic, or even multiphasic. They are often seen over temporal and frontal areas, but may be more widespread. They are usually bilateral, showing a shifting laterality from time to time; occasionally they are synchronous. Small sharp spikes are seen during drowsiness but disappear as the person goes into deeper sleep. These discharges are also referred to as benign epileptiform transients of sleep (BETS). They occur mainly in adulthood and are not usually seen in children.

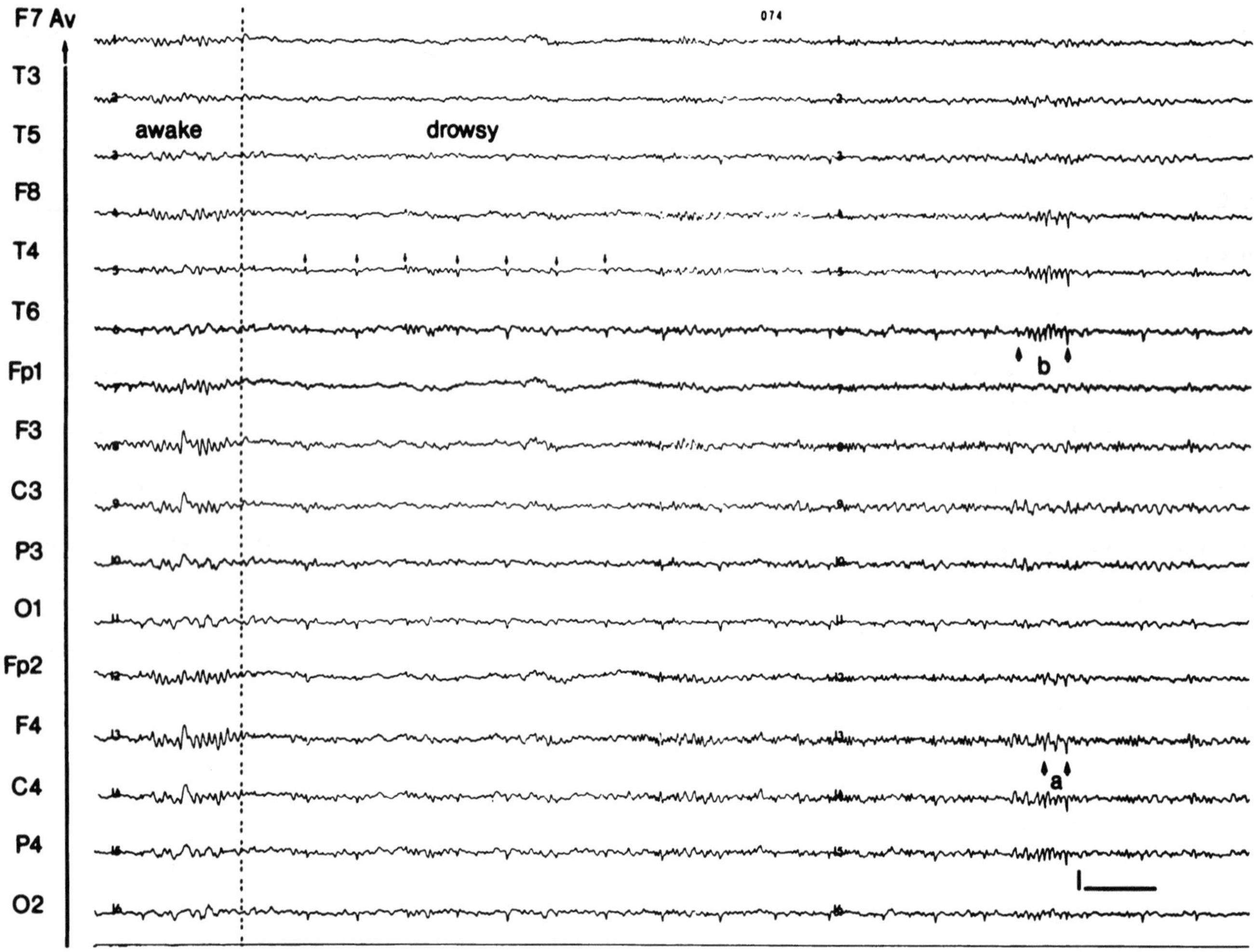

Figure 15.43. Tracing showing 14- and 6-Hz positive spikes in a 16-year-old patient with behavior problems. Arrows at "b" mark off an interval of 14-Hz spikes showing positivity at T6 and T4, and to a lesser extent at F8. Arrows at "a" point to 6-Hz spikes at F4 and to a lesser extent at Fp_2. The small vertical arrows point to ECG artifacts. Filters: low frequency = 1 Hz, high frequency = 70 Hz. Calibrations: horizontal = 1 second, vertical = 50 μV.

14- and 6-Hz Positive Spikes

Positive spikes occurring at a frequency of 14 and 6 Hz have been a source of controversy for many years. The pattern consists of bursts of comb-shaped waves (Fig. 15.43) having a frequency of 13 to 17 Hz and/or 5 to 7 Hz. The pattern is seen generally over the posterior temporal region and adjacent areas on one or both sides of the head during light sleep. The sharp peaks of the waves are surface positive. The pattern is best demonstrated by using contralateral ear-reference recording. Due to the comb-like shape of the waves, the term ctenoids has also been used. Although the pattern is best seen during light sleep, it sometimes may occur during wakefulness.

Fourteen- and 6-Hz positive spikes have been reported in a number of disorders like headache, dizziness, abdominal pain, and behavioral problems, but there seems to be no definite correlation to epilepsy. It is worth mentioning that 14- and 6-Hz positive spikes have also been reported in patients with Reye's syndrome, particularly when the patients are comatose. The significance of this finding is unclear.

Wicket Spikes

The term is derived from the waveform's appearance, which is similar to an arch. Wicket spikes occur either in brief runs or as isolated transients. They are not accompanied by a slow wave. They occur over the temporal areas and may be unilateral or bilateral. Wicket spikes are seen best during stages I and II sleep, although rarely they may occur during wakefulness. They are seen mostly in older persons, with a reported incidence of about 1%. No clinical correlation with epilepsy has been documented.

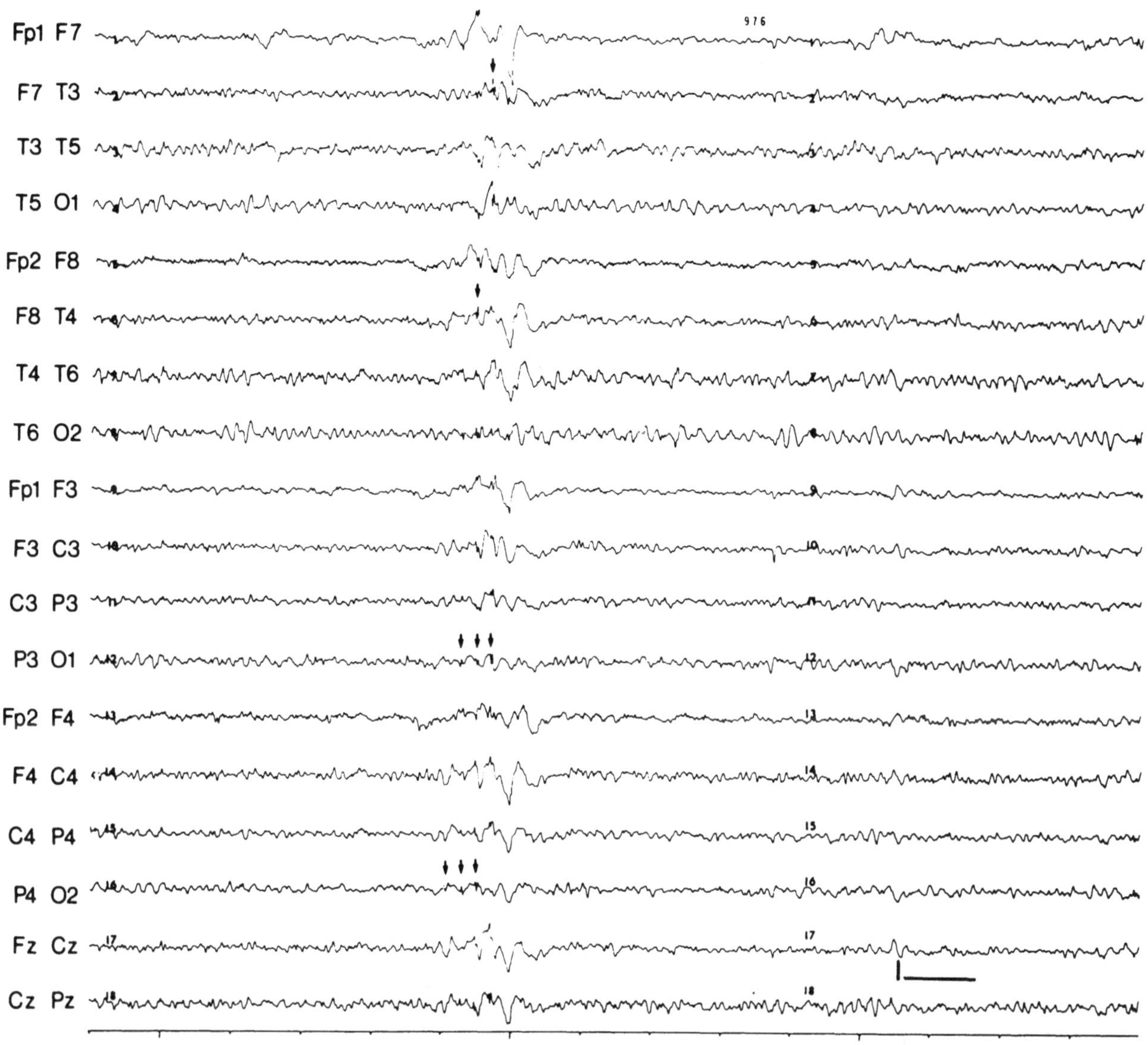

Figure 15.44. Six Hz (phantom) spike and wave discharge in a 25-year-old patient with recurrent headaches. Arrows point to the small-amplitude spikes. Filters: low frequency = 1 Hz, high frequency = 70 Hz. Calibrations: horizontal = 1 second, vertical = 50 μV.

Breach Rhythm

This consists of 6- to 12-Hz sharply pointed waves that appear in clusters, at higher amplitude than background, in the proximity of a skull defect (see Fig. 15.10). The waveforms are often intermixed with beta activity, and their morphology may resemble a mu rhythm. They may show a variable response to motor activity of the contralateral limb. This pattern is not considered to be epileptiform.

6-Hz Spike and Wave

Otherwise known as phantom spike and wave discharges, this pattern consists of brief (½ to 1 second) bursts of 5- to 7-Hz spike and wave complexes. As seen in Fig. 15.44, the spikes are of very low amplitude and are much smaller than the waves. The discharge, which is usually bilateral with frontal or occipital dominance, may occur during wakefulness or sleep, but it is seen mostly during drowsiness. The reported incidence varies from 0.5% to 2.5%. The pattern occurs in children as well as adults.

There has been some controversy as to the significance of this pattern. The occipitally dominant discharges are said to be very poorly correlated with epilepsy, whereas the frontally dominant discharges, according to some authors, may carry some degree of potential for epileptogenicity.

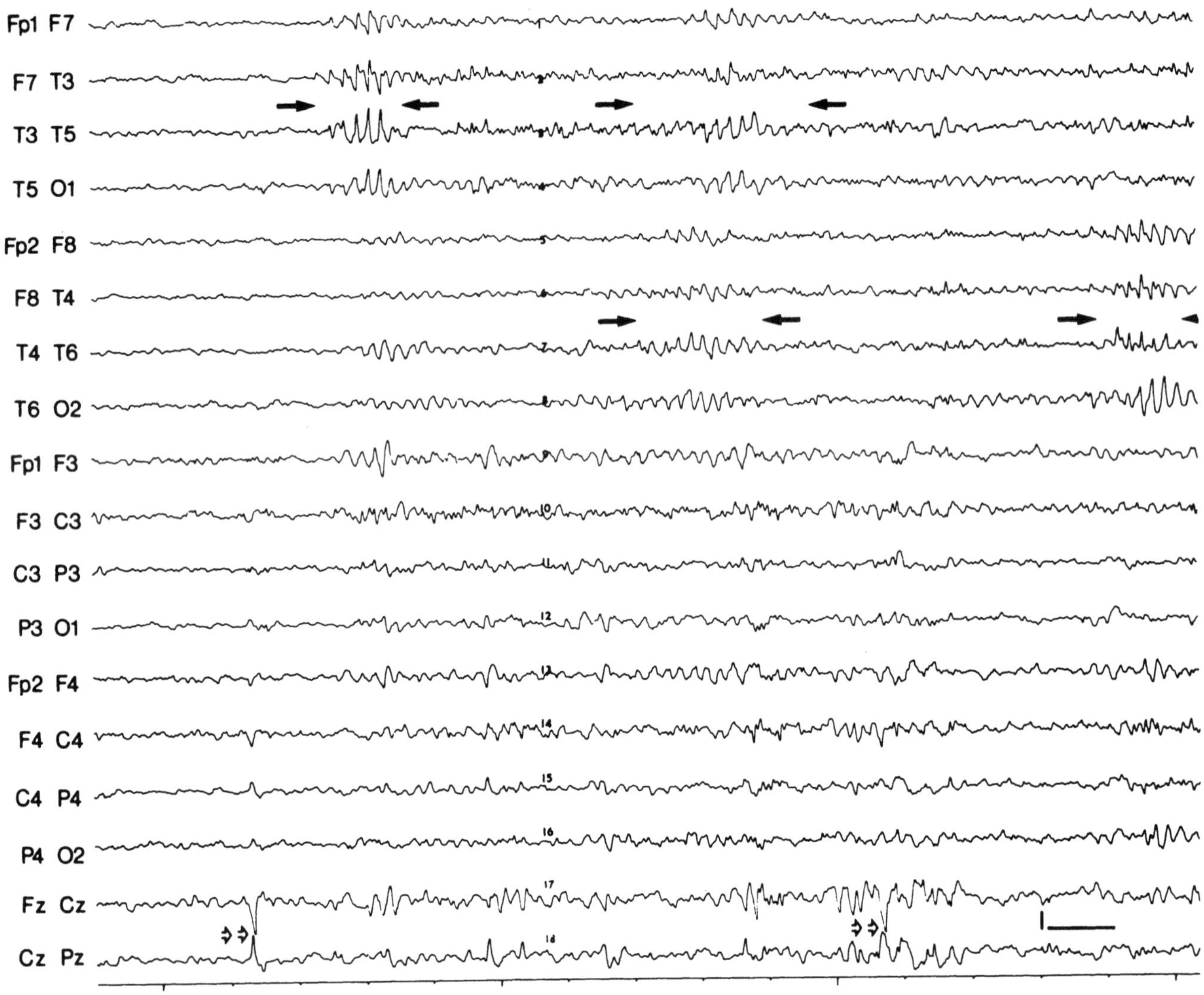

Figure 15.45. Rhythmic midtemporal theta activity in a 35-year-old patient with recurrent headaches and one episode of syncope. Note the sharply contoured 5- to 5.5-Hz waves occurring independently over the left and right midtemporal areas (horizontal arrows) during stage II sleep. The open arrows indicate V waves. Filters: low frequency = 1 Hz, high frequency = 70 Hz. Calibrations: horizontal = 1 second, vertical = 50 μV.

Rhythmic Temporal Bursts of Drowsiness

Otherwise known as rhythmic midtemporal theta of drowsiness (RMTD) or psychomotor variant, this pattern consists of runs of sharply contoured theta activity that is often notched at the top (Fig. 15.45). Rhythmic midtemporal theta of drowsiness may last for several seconds without a significant change in the frequency of the waves. The waves occur most commonly in the midtemporal area, although the activity may spread to the more anterior and posterior temporal areas and to the parietal regions. The RMTD pattern is usually bilateral and may show some shifting laterality. It is seen in both children and adults. No correlation has been found with seizure disorders.

Subclinical Rhythmic Electroencephalographic Discharge of Adults (SREDA)

This pattern is reported to occur mostly during drowsiness in older adults. Often the discharge starts with a sharp or slow wave of high amplitude and is followed by a build up of sharp waves to a sustained frequency of 4 to 7 Hz. The activity may last for several seconds to minutes and may end abruptly. The most common areas involved include the temporal and parietal regions, but the activity may become more widespread. Distinction from an ictal discharge may be difficult, but the lack of a progressive change in discharge frequency is considered to be a differentiating feature. No definite correlation has been reported with seizure phenomena.

Other Related Phenomena

There are a number of other waveforms that may resemble epileptiform activity but that should be distinguished from it. These include, to mention a few, paroxysmal forms of hypnagogic hypersynchrony; asynchronous "spiky" sleep spindles of early childhood; and lambda, lambdoid, and mu rhythms. Transients that resemble spikes or sharp waves occur in the first few weeks after birth. These patterns pose interpretive problems that are beyond the scope of this text.

Abnormal Periodic Paroxysmal Patterns

These are defined as stereotyped recurrences of paroxysmal complexes at relatively fixed intervals (Kuroiwa Y and Celesia G, 1980). They should be present throughout the entire tracing or a major portion of it. The discharges should stand out from the background. They may be composed of slow waves, sharp waves, or sharp and slow-wave complexes. Although they may appear to be epileptiform, they are not necessarily associated with a chronic seizure disorder. They often indicate severe encephalopathy and may or may not be associated with clinical seizures. Some of these patterns may suggest a specific diagnosis when taken in conjunction with the clinical picture, and for this reason it is important to recognize them.

The discharges may be generalized, lateralized, or even focal. Generalized periodic paroxysmal patterns are seen classically in subacute sclerosing pan encephalitis (SSPE), Jakob-Creutzfeldt disease (JCD), and herpes simplex encephalitis (HSE). Electroencephalographic tracings with a burst-suppression pattern may also appear periodic, especially when the bursts occur at regular intervals. Lateralized and focal periodic paroxysmal patterns are seen in acute destructive lesions involving one hemisphere. These particular patterns are taken up in turn.

Generalized Periodic Paroxysmal Patterns

SSPE

The EEG in SSPE may be sufficiently specific to suggest the diagnosis. The periodic discharges consist of high-amplitude (100 to 1,000 μV) complexes—each consisting of one or two slow waves—that recur at intervals of several seconds. The interval between the complexes may have a range of 4 to 14 seconds. Sharp waves may occur along with these large delta transients, and each complex may last from 0.5 to 3 seconds. The activity is often frontocentrally dominant, and for this reason it may be confused with eye-movement artifacts. Myoclonic jerks may accompany the discharges, and when present, they are time locked to the periodic discharges. Initially, the background activity between the complexes may be normal; but as the disease progresses, the background shows slowing, disorganization, and epileptiform abnormalities.

Jakob-Creutzfeldt Disease

The periodic complexes consist of sharp waves of 100- to 300-ms duration that occur at intervals of 0.5 to 2 seconds (Fig. 15.46). Note that these complexes occur at a much faster rate than those in SSPE. The interval between them remains relatively constant in a given patient, although it may vary from patient to patient. Sometimes the complexes have a triphasic configuration. The background activity is severely disorganized in advanced cases of the disease. The periodic discharges may show a temporal relationship to the myoclonic jerks that occur. This interesting pattern is said to be present in as high as 90% of patients with moderately advanced JCD.

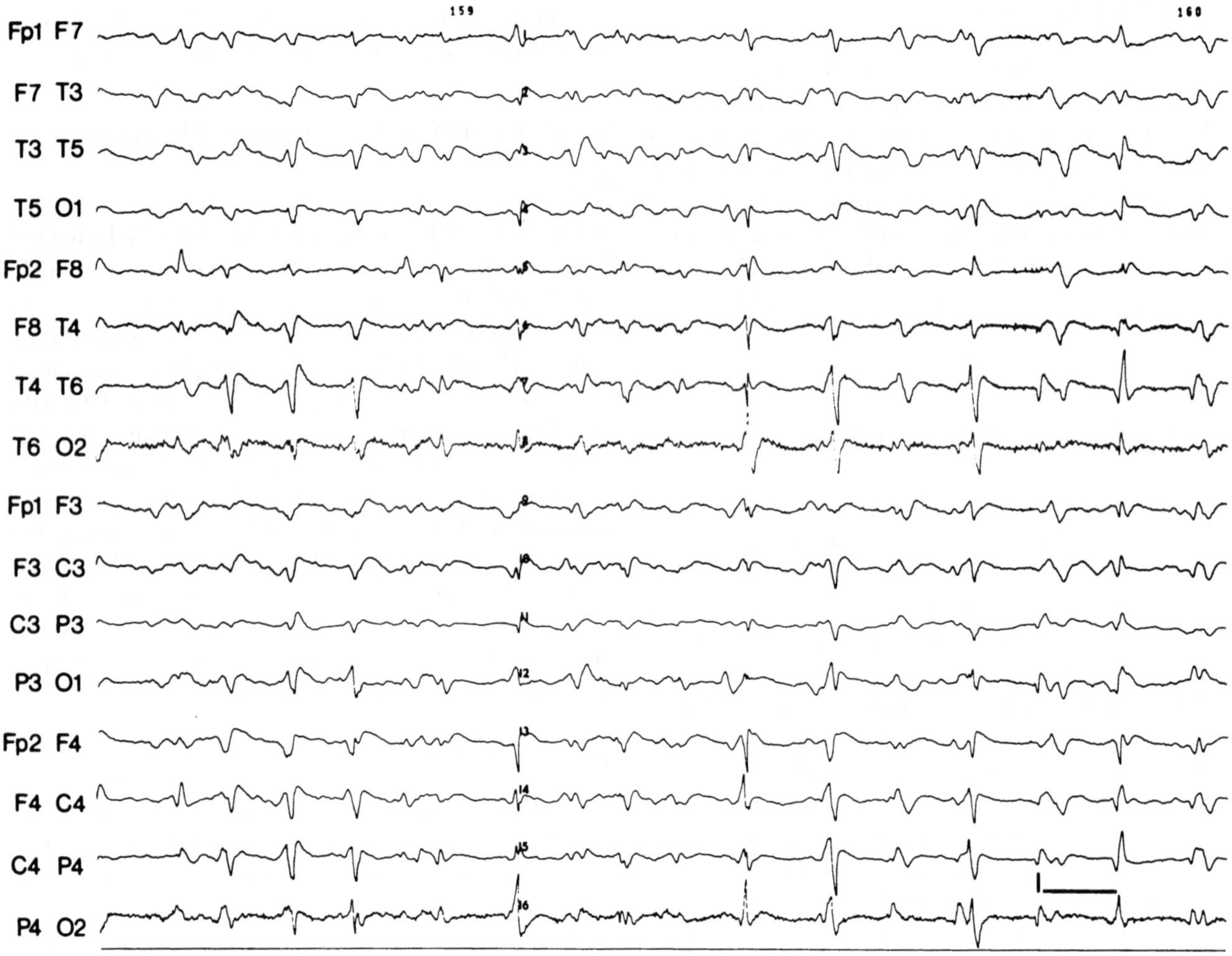

Figure 15.46. Generalized periodic complexes in a patient with Jakob-Creutzfeldt disease. At the time of the recording, the patient was demented and had myoclonic jerks involving the upper extremities. Filters: low frequency = 1 Hz, high frequency = 70 Hz. Calibrations: horizontal = 1 second, vertical = 50 μV.

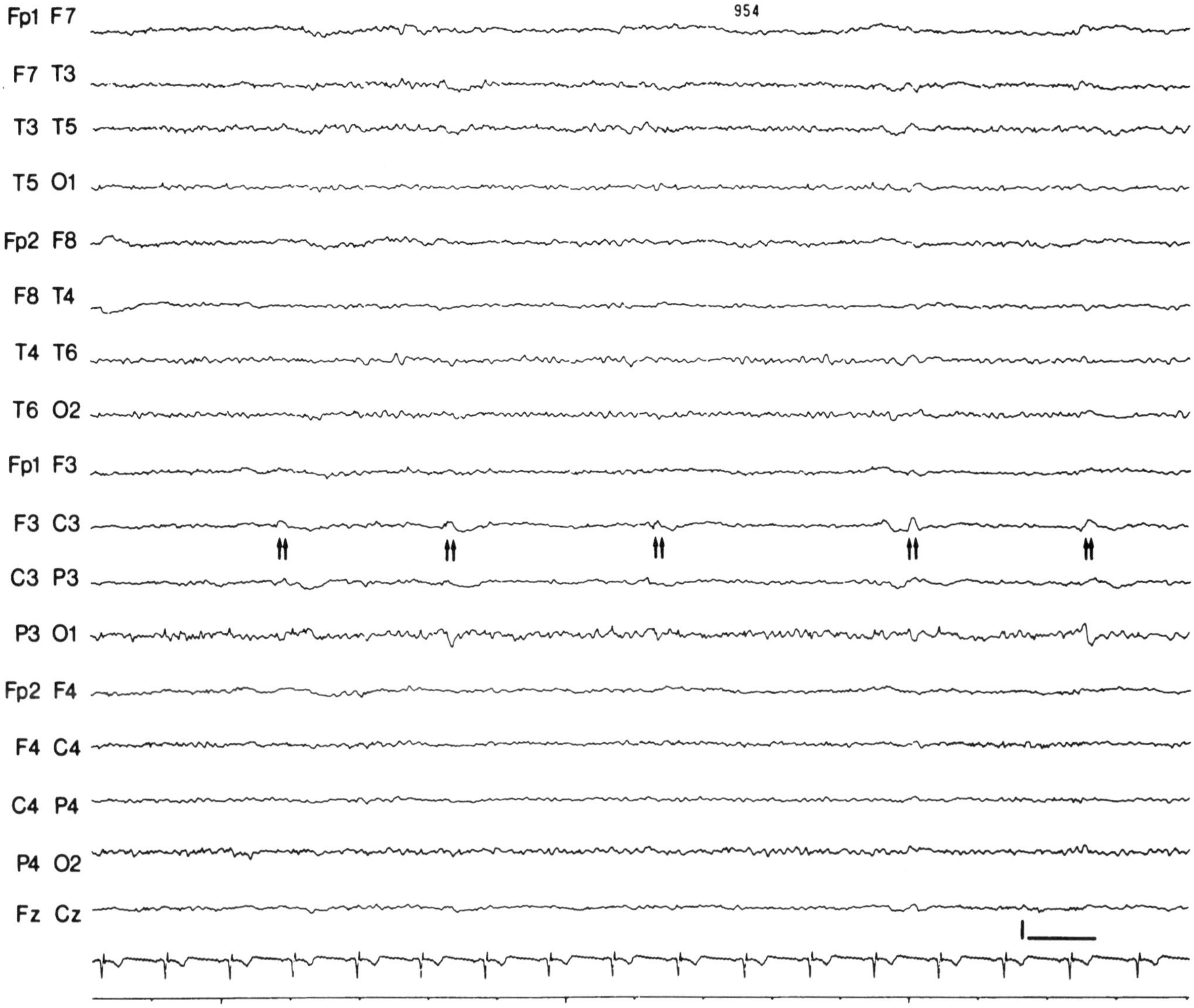

Figure 15.47. Localized periodic complexes (vertical double arrows) in the early phase of herpes simplex encephalitis. The complexes became lateralized and then generalized in the following 2 weeks. Filters: low frequency = 1 Hz, high frequency = 70 Hz. Calibrations: horizontal = 1 second, vertical = 50 µV.

Herpes Simplex Encephalitis

The periodic complexes may be focal, as in Fig. 15.47, or lateralized to start with; but later they become generalized. The usual site is temporal or temperofrontal. The discharges consist of large sharp waves, 100 to 500 µV in amplitude, having a duration of up to 1 second and occurring at variable intervals of 2 to 4 seconds. The background activity may show focal and, later, diffuse slowing. The complexes usually appear between 2 days and 2 weeks after the onset of the illness, and their presence in a patient with the clinical picture of encephalitis strongly suggests HSE.

Other Related Conditions

There are a number of other conditions in which periodic or quasiperidic patterns may be seen. In patients who have suffered from cerebral hypoxia, generalized bisynchronous sharp discharges may occur periodically against the background of a flat EEG. These periodic discharges are often accompanied by myoclonus. Rarely, such discharges may be drug induced. Thus, for example, generalized periodic complexes have been reported in association with phencyclidine intoxication.

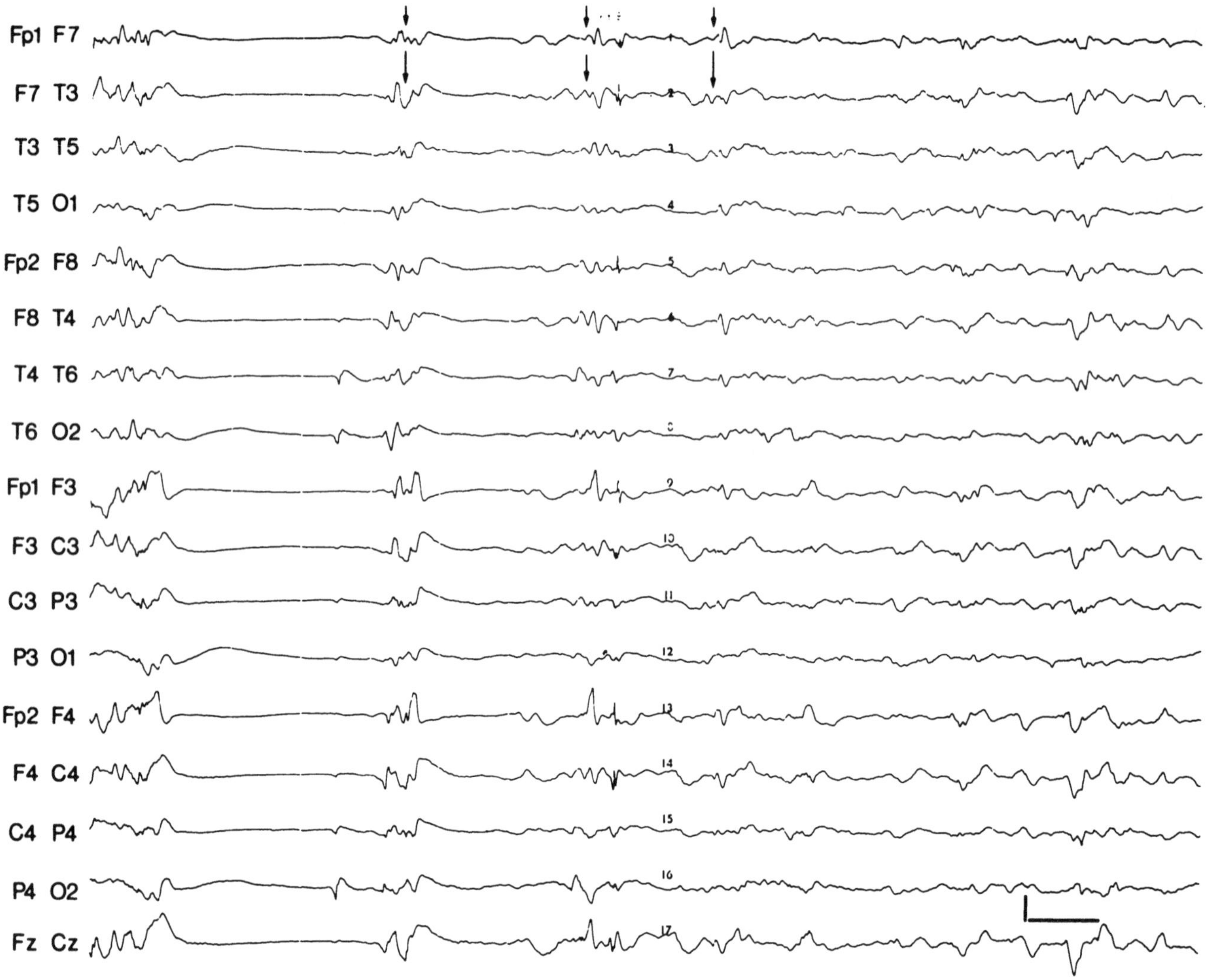

Figure 15.48. Pattern of burst suppression in a 60-year-old patient following cardiopulmonary arrest. The bursts (vertical arrows) contain activity of mixed frequency combined with generalized sharp waves. The patient was unresponsive but free of any obvious convulsions at this time. Filters: low frequency = 1 Hz, high frequency = 70 Hz. Calibrations: horizontal = 1 second, vertical = 150 μV.

Suppression Burst Pattern

This may be considered as a periodic pattern, since it consists of periodic bursts of activity with intervals between in which the background activity is markedly attenuated to less than 10 μV. The pattern is indicative of a severe encephalopathy and may result from cerebral anoxia, head trauma, and severe drug intoxication, as well as deep anesthesia. The bursts consist of mixed frequency (theta and delta) polymorphous activity lasting about 1 second. Figures 15.8, 15.48, and 15.49 are typical of this abnormality. When found in postanoxic tracings, the pattern suggests a grave prognosis.

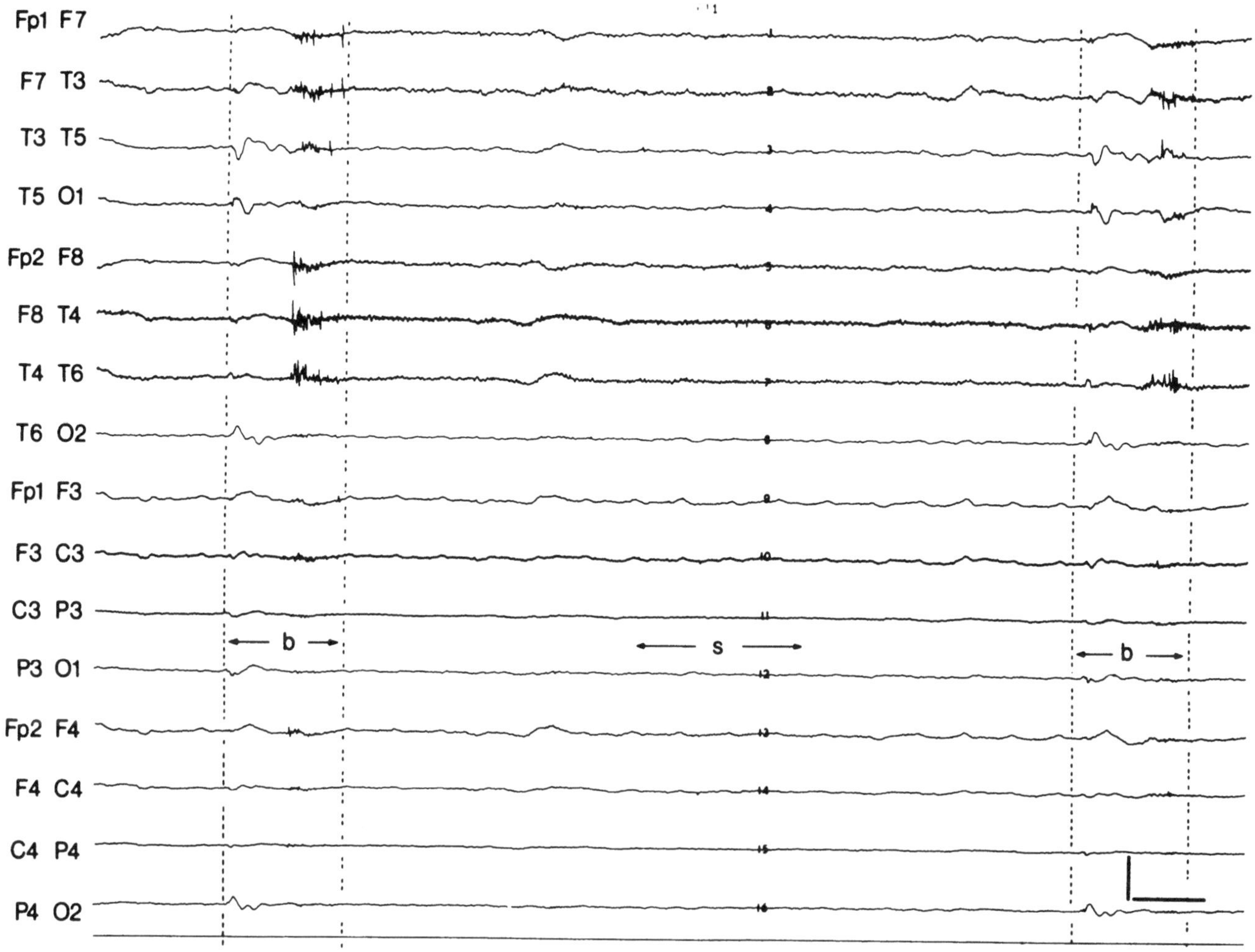

Figure 15.49. Marked attenuation of background activity in a 50-year-old patient following cerebral hypoxia from cardiopulmonary arrest. Area marked "s" shows an epoch of suppression, and areas marked "b" show bursts of low-amplitude slow activity. Muscle artifacts during the bursts correspond to myoclonic jerks. Filters: low frequency = 1 Hz, high frequency = 70 Hz. Calibrations: horizontal = 1 second, vertical = 50 μV.

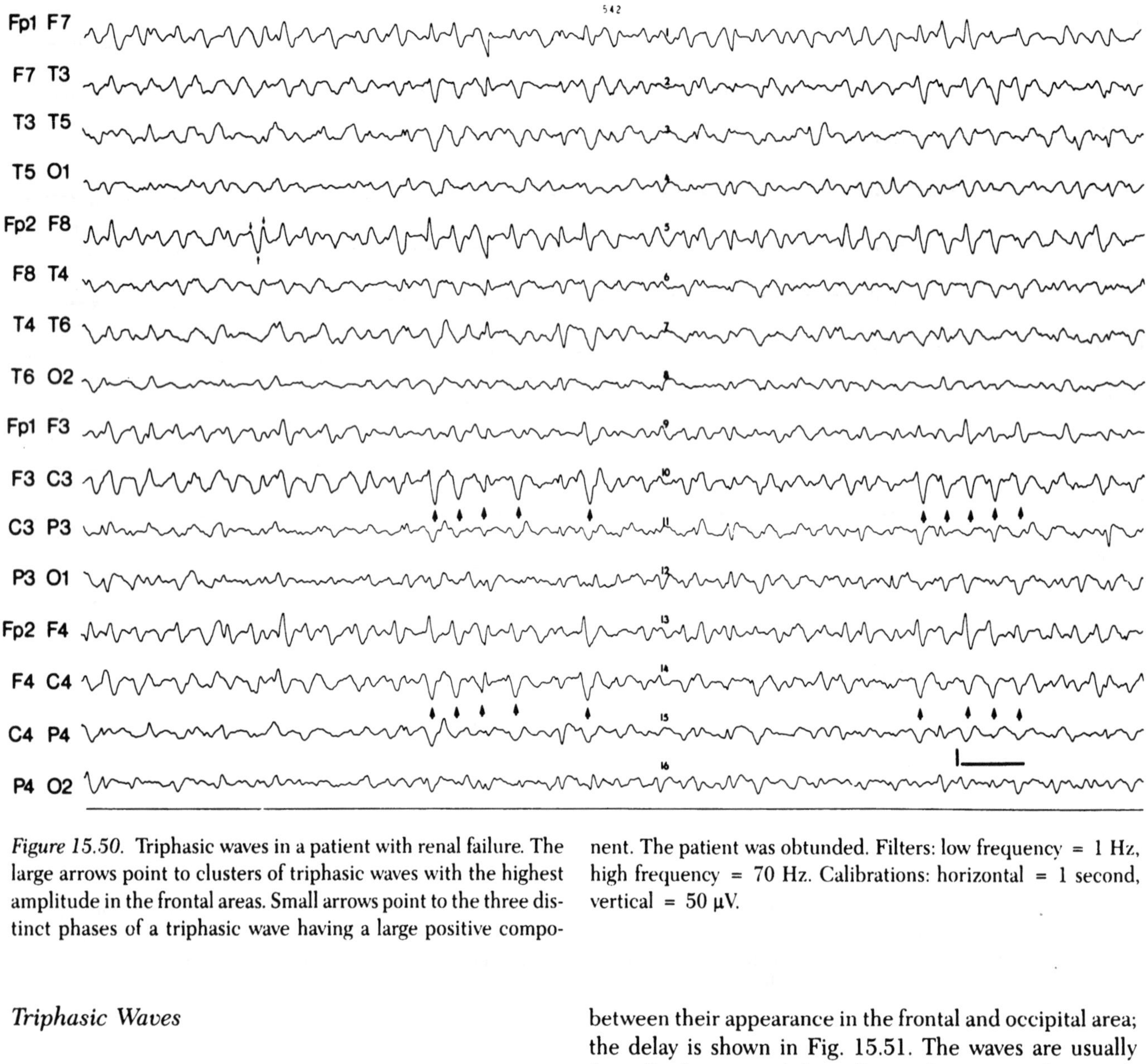

Figure 15.50. Triphasic waves in a patient with renal failure. The large arrows point to clusters of triphasic waves with the highest amplitude in the frontal areas. Small arrows point to the three distinct phases of a triphasic wave having a large positive component. The patient was obtunded. Filters: low frequency = 1 Hz, high frequency = 70 Hz. Calibrations: horizontal = 1 second, vertical = 50 µV.

Triphasic Waves

Triphasic waves may sometimes occur in a periodic fashion (Kuroiwa Y and Celesia G, 1980)and hence are included in this section. Each waveform shows a prominent positive phase (Figs. 15.50, 15.51) preceded and followed by negative phases. Classic triphasic waves show a time delay between their appearance in the frontal and occipital area; the delay is shown in Fig. 15.51. The waves are usually frontally dominant, bilaterally synchronous and generalized. Although originally described in hepatic encephalopathy, this abnormality may occur in other forms of metabolic encephalopathies such as uremia and following cerebral hypoxia. The abnormality accompanies symptoms of impaired consciousness.

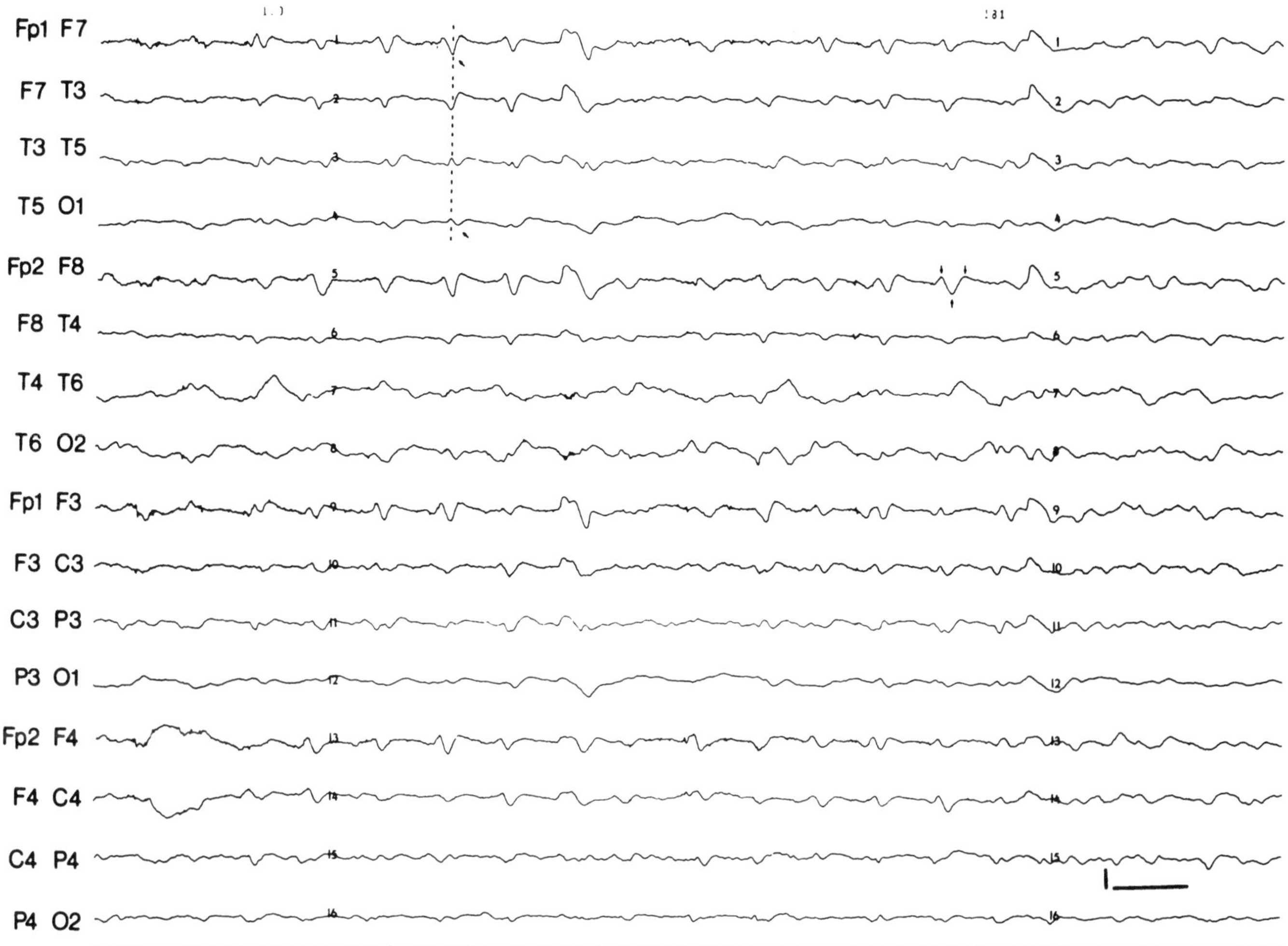

Figure 15.51. Triphasic waves in a patient with hepatic encephalopathy. Vertical arrows point to the three phases of the waveform. The slanted arrows point to the main positive component of a triphasic wave that shows a fronto-occipital delay. Filters: low frequency = 1 Hz, high frequency = 70 Hz. Calibrations: horizontal = 1 second, vertical = 50 μV.

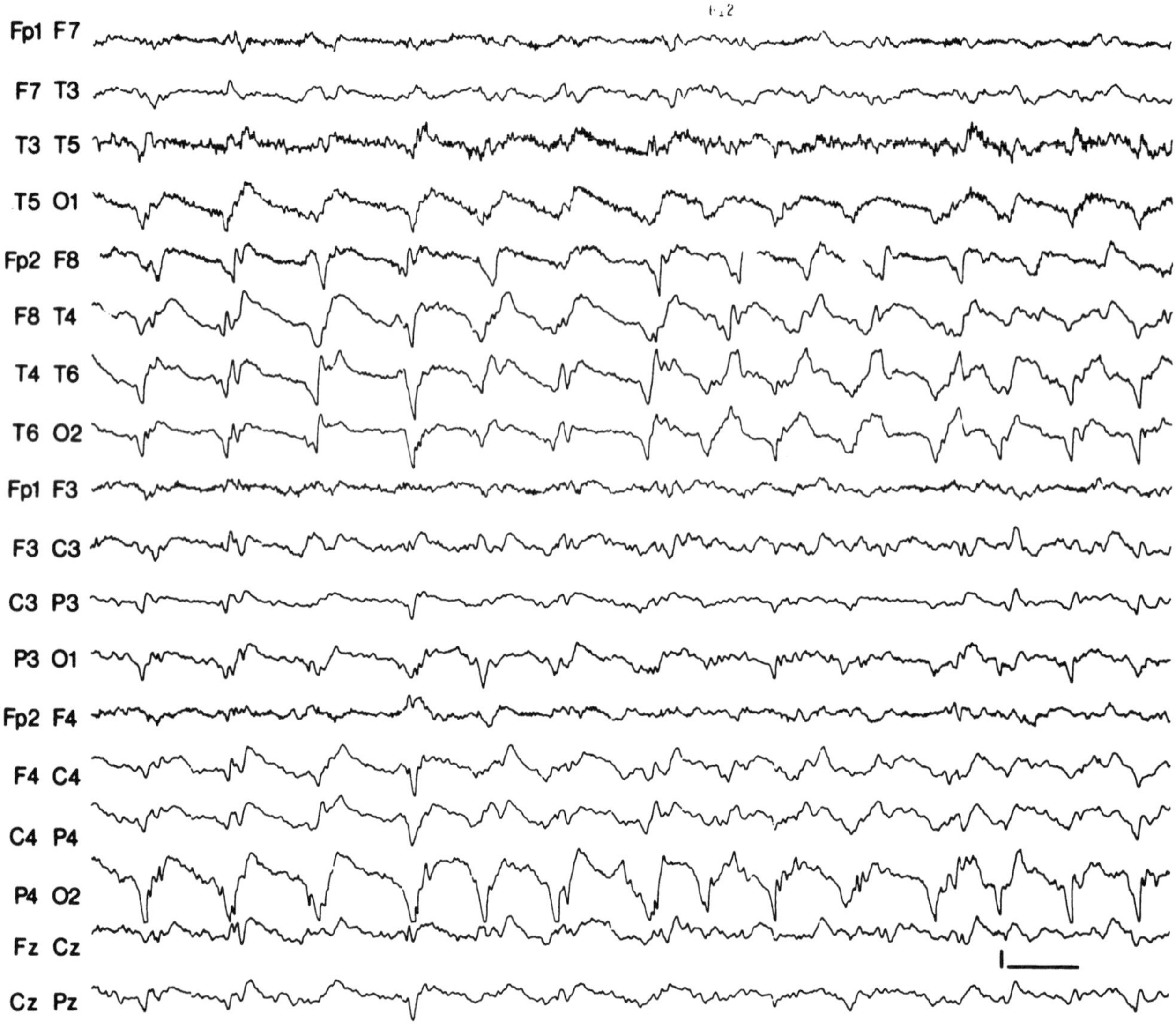

Figure 15.52. Periodic lateralized epileptiform discharges (PLEDS) in a 60-year-old patient with recent onset, massive infarction of the right hemisphere. The patient was comatose and had jerky movements involving the left upper extremity. Filters: low frequency = 1 Hz, high frequency = 70 Hz. Calibrations: horizontal = 1 second, vertical = 50 μV.

Lateralized Periodic Paroxysmal Patterns

Periodic Lateralized Epileptiform Discharges (PLEDS)

The term is used for periodic complexes that are lateralized to one side and repeat at 1- to 2- second intervals. Each complex consists of spikes or sharp waves that are often followed by slow waves (Chatrian GE, Shaw CM, and Leffman H, 1964). They are seen in acute, often destructive lesions involving one hemisphere. The classic examples are infarcts (Fig. 15.52), rapidly growing tumors (Fig. 15.53), and infectious encephalopathies like HSE. Sometimes they may be accompanied by focal seizures that may be difficult to control. The pattern is usually a self-limiting phenomenon that lasts for a few days and is followed by the appearance of PDA in that area.

BIPLEDS

Sometimes PLEDS may occur independently on both sides; in this case, the term BIPLEDS is used. In one study (de la Paz D and Bremer RP, 1981) the most common causes of BIPLEDS were anoxic encephalopathy and CNS infection. It was reported that patients with BIPLEDS were often comatose and had a higher mortality rate than patients with PLEDS.

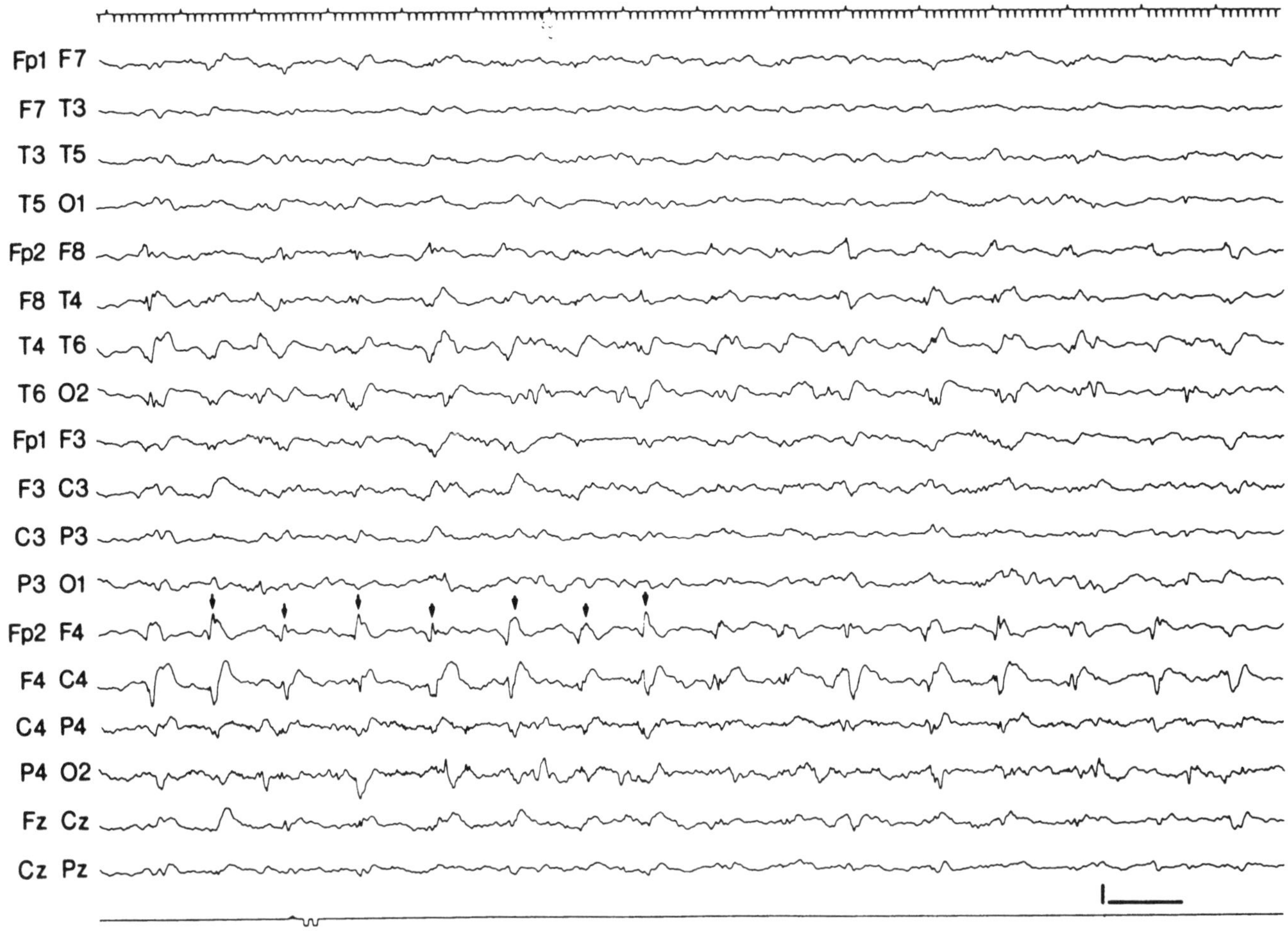

Figure 15.53. Periodic lateralized epileptiform discharges (PLEDS) in a patient with a glioma involving the right frontal area. Vertical arrows point to the periodic discharges. The patient was obtunded and had seizures involving the left side of the face and the left upper extremity. Filters: low frequency = 1 Hz, high frequency = 70 Hz. Calibrations: horizontal = 1 second, vertical = 50 μV.

References

American Electroencephalographic Society Guideline Three: Minimum technical standards for EEG recording in suspected cerebral death. *J Clin Neurophysiol* 1986; 3 (suppl 1): 12–17.

Chatrian GE, Bergamini L, Donday M, et al: A glossary of terms most commonly used by clinical electroencephalographers. *Electroencephalogr Clin Neurophysiol* 1974; 37: 538–548.

Chatrian GE, Shaw CM, Leffman H: The significance of periodic lateralized epileptiform discharges in EEG. *Electroencephalogr Clin Neurophysiol* 1964; 17: 177–193.

de la Paz D, Bremer RP: Bilateral independent lateralized epileptiform discharges—clinical significance. *Arch Neurol* 1981; 38: 713–715.

Kuroiwa Y, Celesia G: Clinical significance of periodic EEG patterns. *Arch Neurol* 1980; 37: 15–20.

Chapter 16
Activation Procedures

When using the EEG as a clinical tool, one should always keep in mind that the EEG recording is simply a random sampling of the person's brain electric activity taken at a particular period of time. Hence, in those neurologic disorders that produce transient abnormalities in the EEG, an EEG may be interpreted as normal unless the time of occurrence of the abnormality coincides with the time of recording. This issue becomes particularly important in investigations of seizure disorders. Contrary to the notions of the uninitiated, a normal EEG in no way "rules out" a genuine seizure disorder, as interictal epileptiform abnormalities may or may not have been present at the time of recording. One of the ways that is employed to mitigate this problem is to increase the probability of occurrence of abnormalities during the recording period. This may be achieved by using various activation procedures that can elicit or enhance certain normal as well as abnormal activity in the EEG. It must be understood, however, that while activation procedures are most valuable in the case of seizure disorders, they may also be useful in the study of many other neurologic disorders.

Hyperventilation

Hyperventilation is perhaps the most widely used activation procedure in EEG laboratories. The procedure, which is simple and relatively safe, consists of three to five minutes of deep breathing. It is, however, difficult to perform in patients who are uncooperative, mentally retarded, or below the age of 4 or 5 years, and it is preferable to avoid in patients with recent stroke or subarachnoid hemorrhage, recent myocardial infarction, chronic obstructive pulmonary disease, and other conditions causing difficulty in breathing. Although hyperventilation has become a common procedure during routine EEG recording, it is of special importance in the case of patients suspected of having seizure disorders, particularly absence seizures.

Procedure

The standard procedure is to have the patient take deep breaths at the rate of about 20 per minute for three to five minutes. The first step is to explain the procedure in detail to the patient. Tell the patient to relax, keep the eyes closed and mouth open, and to breath deeply in and out at a regular pace until told to stop. It is best to show the patient what to do by giving a brief demonstration of deep breathing. Explanations such as "try to completely empty the lungs, and then take in a lung full" may be helpful. Ask the patient to refrain from making any unnecessary movements of the head and upper part of the body, as these can introduce large artifacts into the recording. The patient also should be instructed to report, if possible, any warnings of or actual occurrence of a seizure immediately. Symptoms like tingling of the extremities, muscle spasms, or lightheadedness are known to occur during hyperventilation in some individuals.

During hyperventilation, the technician should carefully observe the patient to detect the occurrence of absence seizures, which may simply consist of a brief (lasting only a few seconds) interruption of hyperventilation with associated staring and unresponsiveness. When this happens, it is important to try to assess the patient's responsiveness by asking questions and/or giving verbal commands promptly. The command given and the patient's response or lack of response should be clearly indicated on the EEG chart. Sometimes the patient may become drowsy halfway through the procedure and may stop hyperventilating. The technician should be on the lookout for such a problem and prevent it from occurring by repeatedly calling the patient by name and encouraging

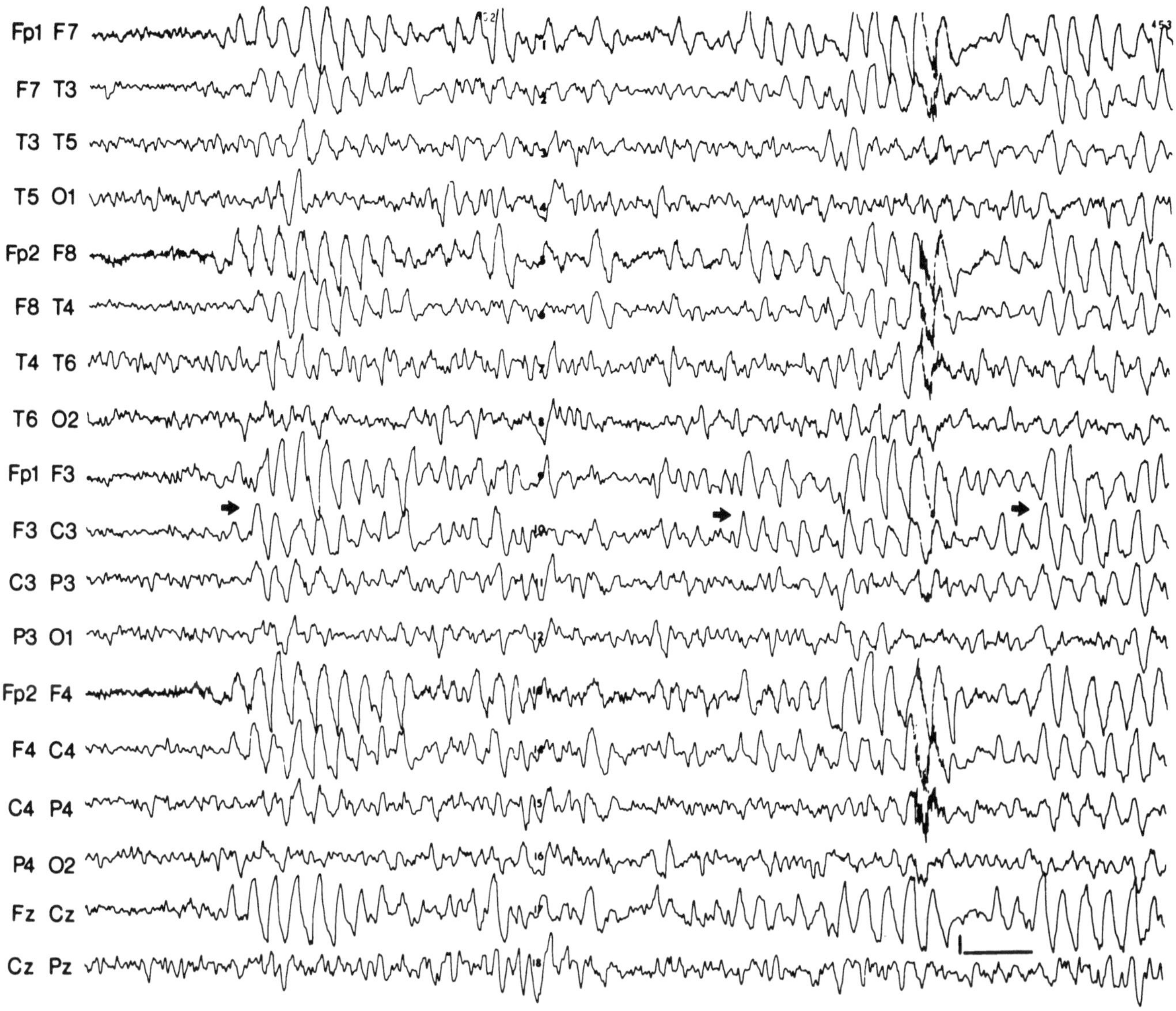

Figure 16.1. Intermittent slow activity during hyperventilation. The high amplitude, 3- to 3.5-Hz rhythmic activity is present on both sides and is symmetrical. Each arrow indicates the onset of a burst of slow activity. This is a normal response to hyperventilation. Filters: low frequency = 1 Hz, high frequency = 70 Hz. Calibrations: horizontal = 1 second, vertical = 50 μV.

him or her to keep up the deep breathing. Also, the technician should carefully observe whether the patient actually stops overbreathing when he or she is told to do so. Sometimes a patient may continue to overbreath for a prolonged period of time even after being told to stop; this can lead to prolonged slowing of the EEG (see below) thus creating problems in interpretation of the tracing.

It is good practice to mark the time elapsed since starting hyperventilation on the EEG chart every 30 seconds so that the electroencephalographer will know how long the patient has been hyperventilating. If the patient is referred for evaluation of absence spells, and if the typical three per second spike and wave discharges (see Chapter 15) occur in the EEG tracing during the first three minutes, there is no need to prolong the hyperventilation. On the other hand, if no abnormalities appear in the first three minutes, then have the patient continue overbreathing for two additional minutes. For such patients, it is a good practice to repeat the hyperventilation twice or even three times if the initial attempt fails to elicit the typical abnormal discharge.

Normal and Abnormal Response

The normal response to hyperventilation consists of the occurrence of symmetrical slow activity on both sides. The absence of any change in the EEG is also normal. Although this slow activity may be diffuse theta activity, a more characteristic finding is the occurrence of intermittent or continuous 3- to 4-Hz high amplitude activity that is frontally or occipitally dominant (Figs. 16.1 and 16.2). If the

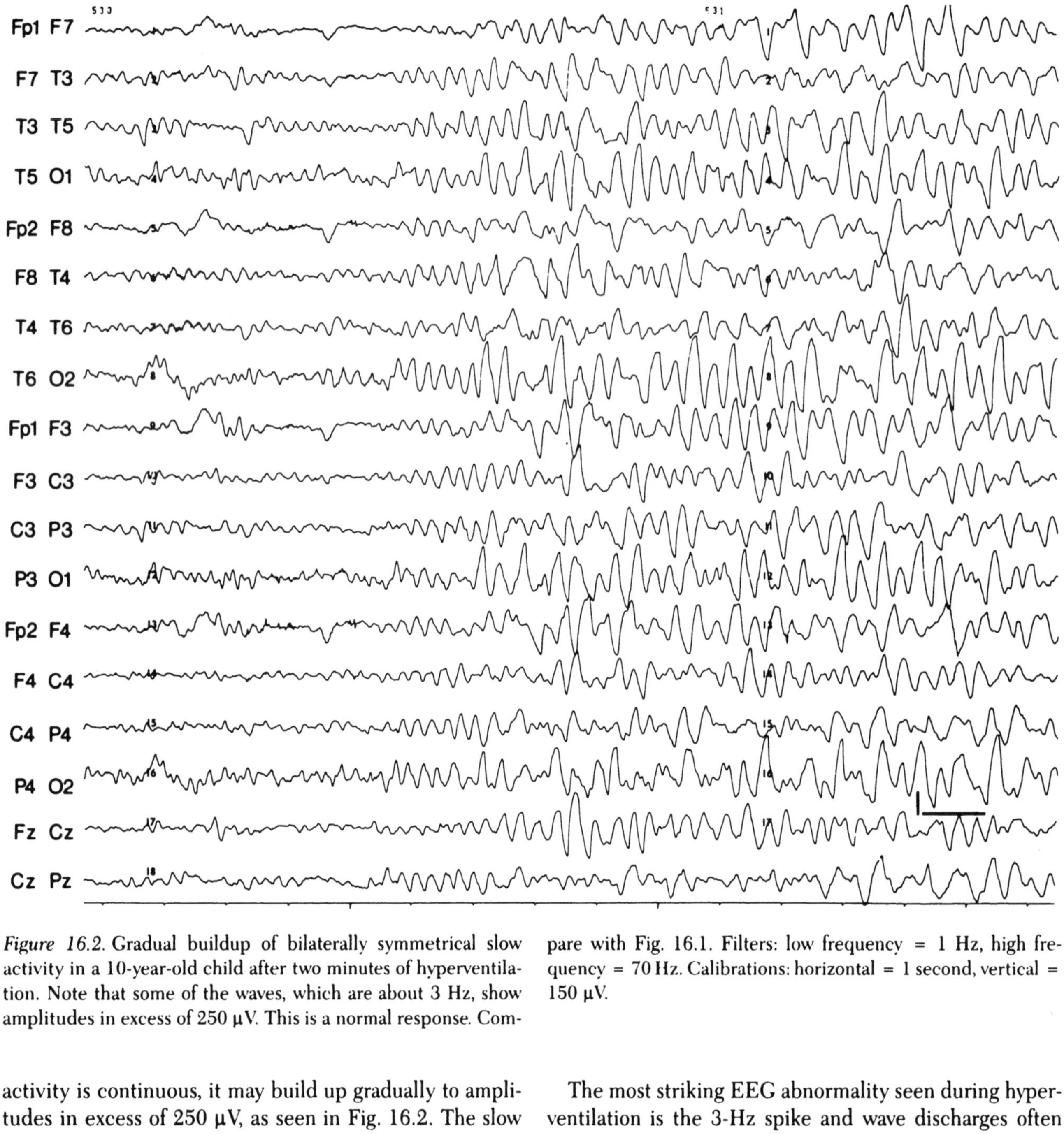

Figure 16.2. Gradual buildup of bilaterally symmetrical slow activity in a 10-year-old child after two minutes of hyperventilation. Note that some of the waves, which are about 3 Hz, show amplitudes in excess of 250 μV. This is a normal response. Compare with Fig. 16.1. Filters: low frequency = 1 Hz, high frequency = 70 Hz. Calibrations: horizontal = 1 second, vertical = 150 μV.

activity is continuous, it may build up gradually to amplitudes in excess of 250 μV, as seen in Fig. 16.2. The slow activity may persist for up to a minute after hyperventilation ceases, and the EEG may not return to its prehyperventilation state for two to three minutes. The amplitude and frequency of the slow activity are of no clinical importance unless there is *consistent* asymmetry between the two hemispheres. The side that shows a slower frequency and/or a lower amplitude is usually considered to be the abnormal side. Figure 16.3 shows an abnormal response to hyperventilation.

The most striking EEG abnormality seen during hyperventilation is the 3-Hz spike and wave discharges often brought on in patients with absence seizures. Figure 16.4 illustrates this. These discharges usually are frontally dominant and may occur in brief epochs, or they may persist for several seconds during which time an episode of unresponsiveness may be documented. Sometimes, other types of epileptiform abnormality, such as generalized spike discharges or even focal spikes, may be brought on by hyperventilation. Hyperventilation may be performed while recording any of the commonly used montages;

Figure 16.4. Generalized 3-Hz spike and wave discharges following three minutes of hyperventilation in a patient suffering from absence seizures. Note that the discharges are frontally dominant and occur in brief epochs. Arrows point to several of the spikes. Filters: low frequency = 1 Hz, high frequency = 70 Hz. Calibrations: horizontal = 1 second, vertical = 50 μV.

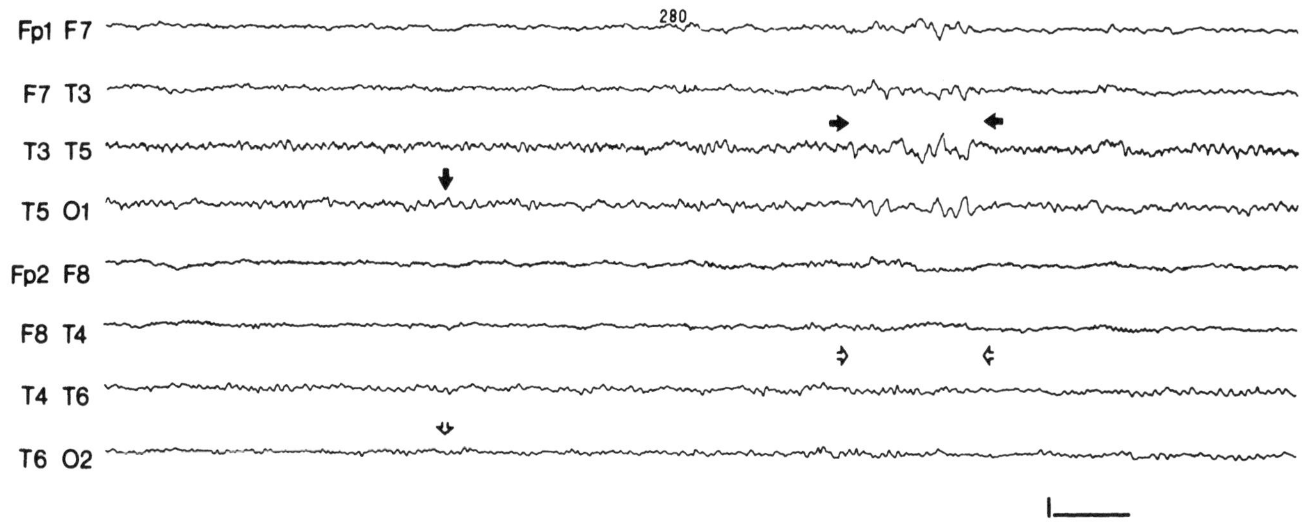

Figure 16.3. Asymmetrical slow activity and sharp wave after 1.5 minutes of hyperventilation. The presence of slow activity in the theta band on the left side (vertical solid arrow) but its absence on the right side (vertical open arrow) is abnormal. Also abnormal are the sharp wave and the brief paroxysm of delta activity on the left between the horizontal solid arrows; these abnormal features are not present on the right side (horizontal open arrows). Filters: low frequency = 1 Hz, high frequency = 70 Hz. Calibrations: horizontal = 1 second, vertical = 50 μV.

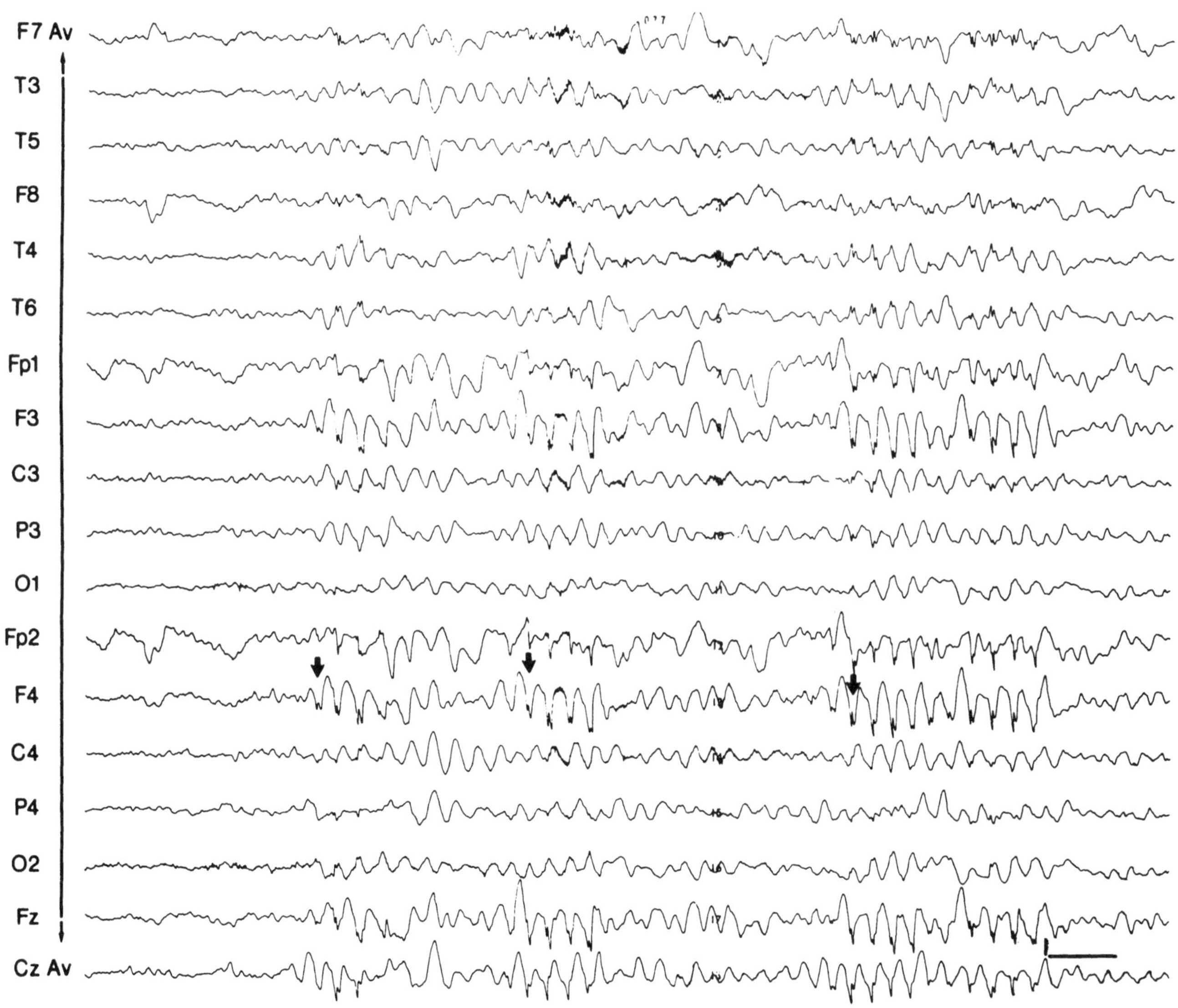

selection of a particular montage is dictated by the suspected abnormality. For example, if absence is suspected it is important to have a montage that includes the frontal, parietal, and occipital electrodes. In the case of complex partial seizures, the montage should include not only the parasagittal but also, and more importantly, the temporal chain of electrodes.

How does hyperventilation bring about such dramatic changes in the EEG? The major biochemical finding during hyperventilation is a drop in carbon dioxide content of the blood (hypocarbia). It is well known that the most important vasodilatory stimulus for the blood vessels of the brain is carbon dioxide. The higher the carbon dioxide content, the greater the vasodilatation. So when there is hypocarbia, the reverse occurs, namely, vasoconstriction. This presumably alters the metabolic rate of the neurons and leads to the slow activity.

The effect of hyperventilation on the EEG is much more marked in children than in adults, with children's EEGs sometimes showing an enormous buildup of slow activity. Blood sugar level also appears to influence the response to hyperventilation. The lower the blood sugar, the more marked the hyperventilation-induced slow activity. When an adult EEG shows marked and prolonged slowing as a result of hyperventilation, one should consider the possibility of hypoglycemia and should repeat the procedure 15 to 30 minutes after giving a drink containing glucose. It should be obvious, in this context, why documentation in the EEG worksheet concerning the time of the preceding meal is important.

Intermittent Photic Stimulation

Visual stimuli are perhaps one of the most effective means of stimulating the brain. The ready availability of user-friendly stroboscopes has resulted in the routine use of intermittent photic stimulation (IPS) as an activation procedure during electroencephalography. The method is most valuable in documenting photosensitivity, which has a high clinical correlation with primary generalized epilepsy.

Apparatus

The device used is called a stroboscope or photic stimulator. It is capable of delivering single or continuous bright flashes of light at frequencies ranging from 1 to 50 flashes per second. One of the commonly used models provides flashes of 10-μs duration with an intensity of 1.5 million foot-candles. In some models the intensity of the flash is adjustable; others are capable of providing colored as well as white flashes (red flashes are said to be more effective in eliciting photoparoxysmal responses). The device also delivers a signal that is synchronous with each flash; this signal is recorded as a stimulus marker on one of the EEG channels. Alternatively, a photoelectric cell may be attached to the patient's forehead and connected to one of the EEG channels.

Technique

The test begins by explaining the procedure to the patient. Tell the patient that he or she will be seeing very bright flashes of light (bright even with the eyes closed) and to keep the eyes closed or open as instructed during the course of the test. The flash lamp is positioned approximately 30 cm in front of the eyes. Start with one or two flashes per second and increase the rate gradually up to 30 flashes per second. Each flash rate is presented for a duration of about 10 seconds, and the eyes are kept closed in the first 5 seconds and open in the next 5 seconds. If a photoparoxysmal response (explained later) is elicited, the IPS should be stopped to avoid precipitating a seizure. If the response occurs only during a brief part of the stimulation, the technician needs to confirm that it is, indeed, a photoparoxysmal response by cautiously repeating the stimulation at the same flash rate. The technician should be on the lookout for evidence of seizures or seizure-like phemonena, and should make appropriate notations in the EEG chart.

Photic Driving

EEG activity that is time-locked to the photic stimuli often is seen over the posterior head regions during IPS. This activity has the same frequency as the photic stimuli, or it may be a harmonic or subharmonic of the flash rate. Normally, the activity is symmetrical on the two sides. Such a response to photic stimulation is called *photic driving*. Photic driving is a physiological response; it may be observed in all age groups. Figure 16.5 is an example of photic driving at different flash rates, and Fig. 16.6 shows photic driving at subharmonics of the flash rate.

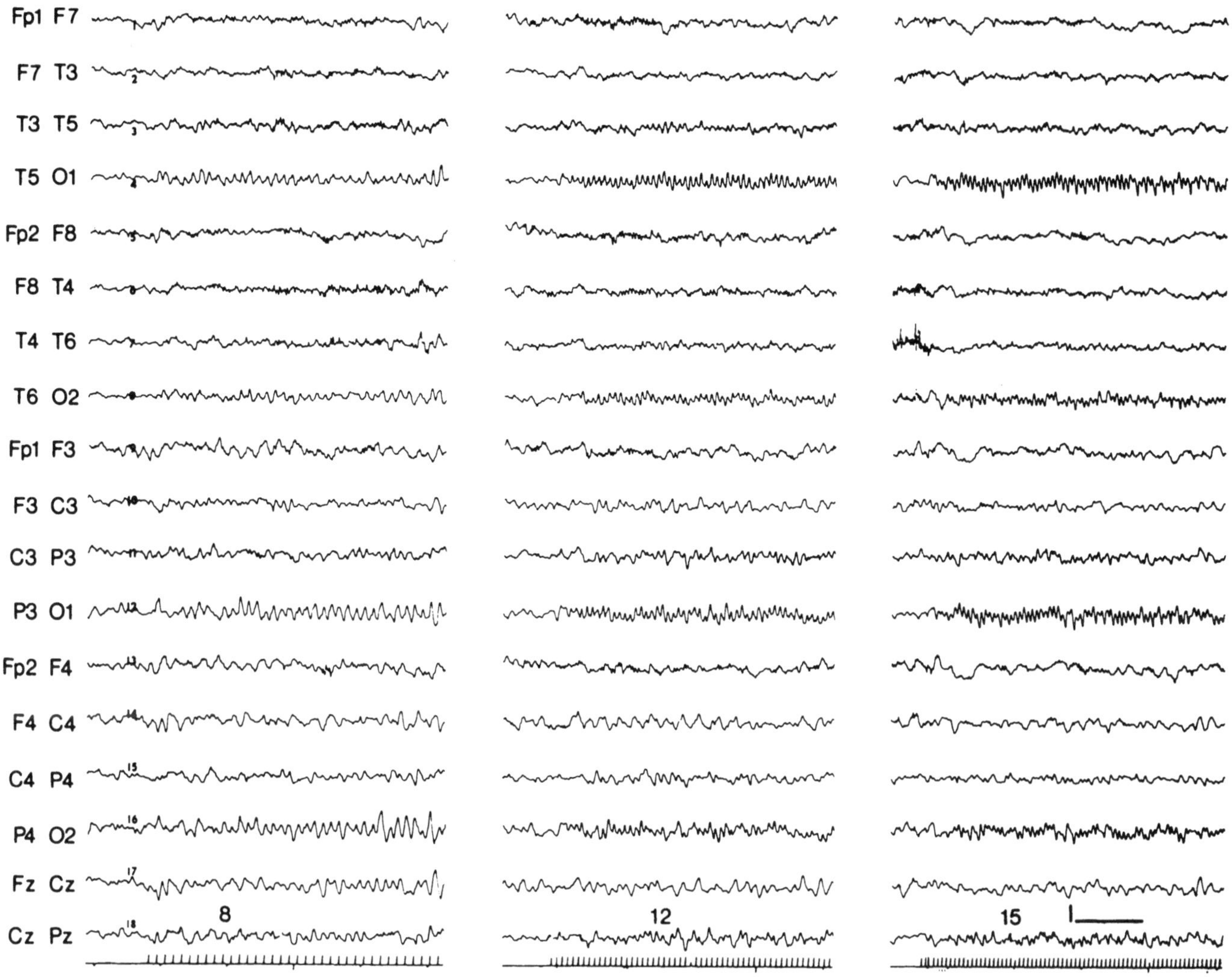

Figure 16.5. Photic driving at 8, 12, and 15 flashes per second. The time-locked activity, which is most prominent in the posterior regions, is a normal response to photic stimulation. Filters: low frequency = 1 Hz, high frequency = 70 Hz. Calibrations: horizontal = 1 second, vertical = 50 μV.

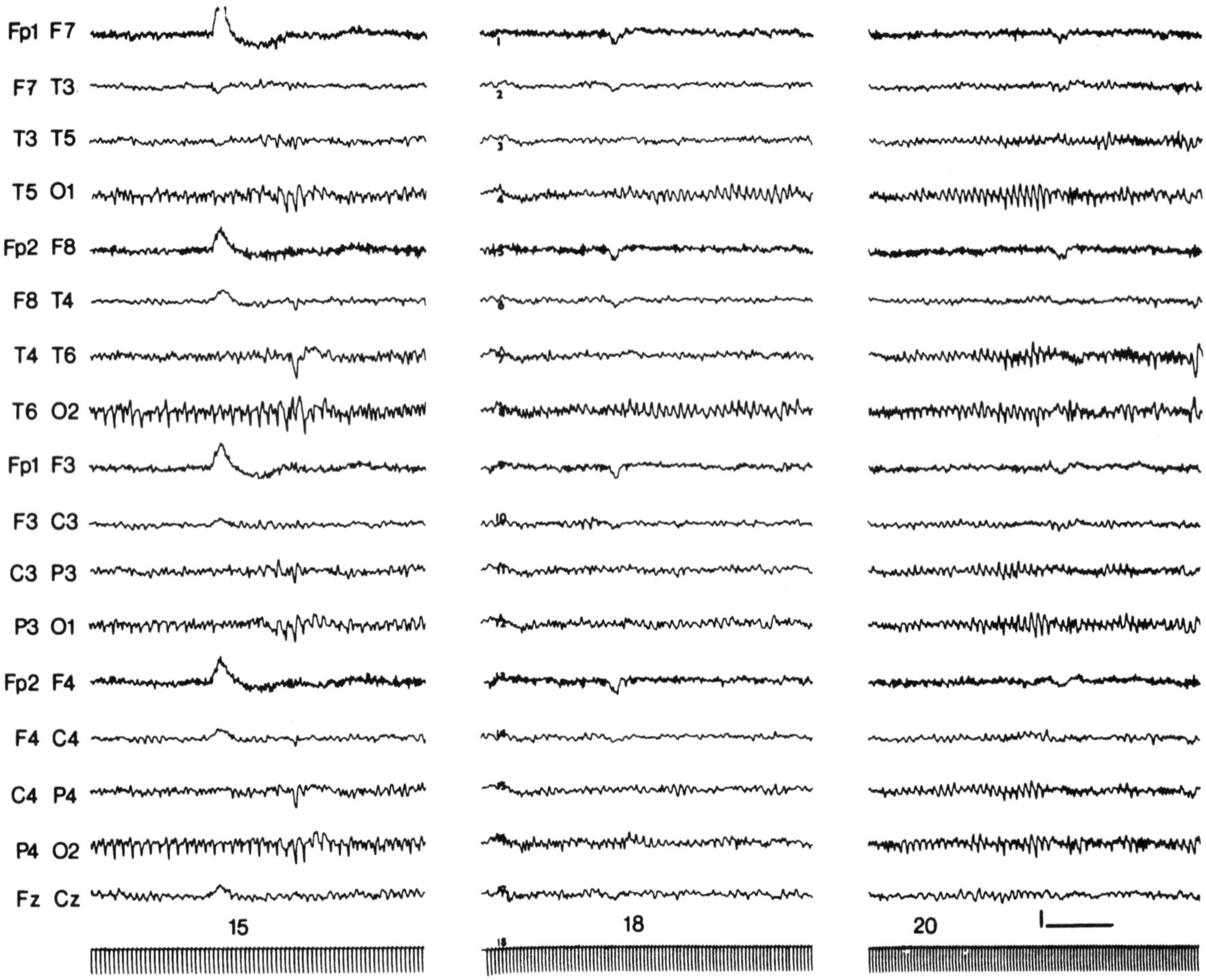

Figure 16.6. Photic driving at 15, 18, and 20 flashes per second, and also at the subharmonics of the flash rates, namely, 7.5, 9, and 10 Hz, respectively. Both kinds of response are normal. Filters: low frequency = 1 Hz, high frequency = 70 Hz. Calibrations: horizontal = 1 second, vertical = 50 μV.

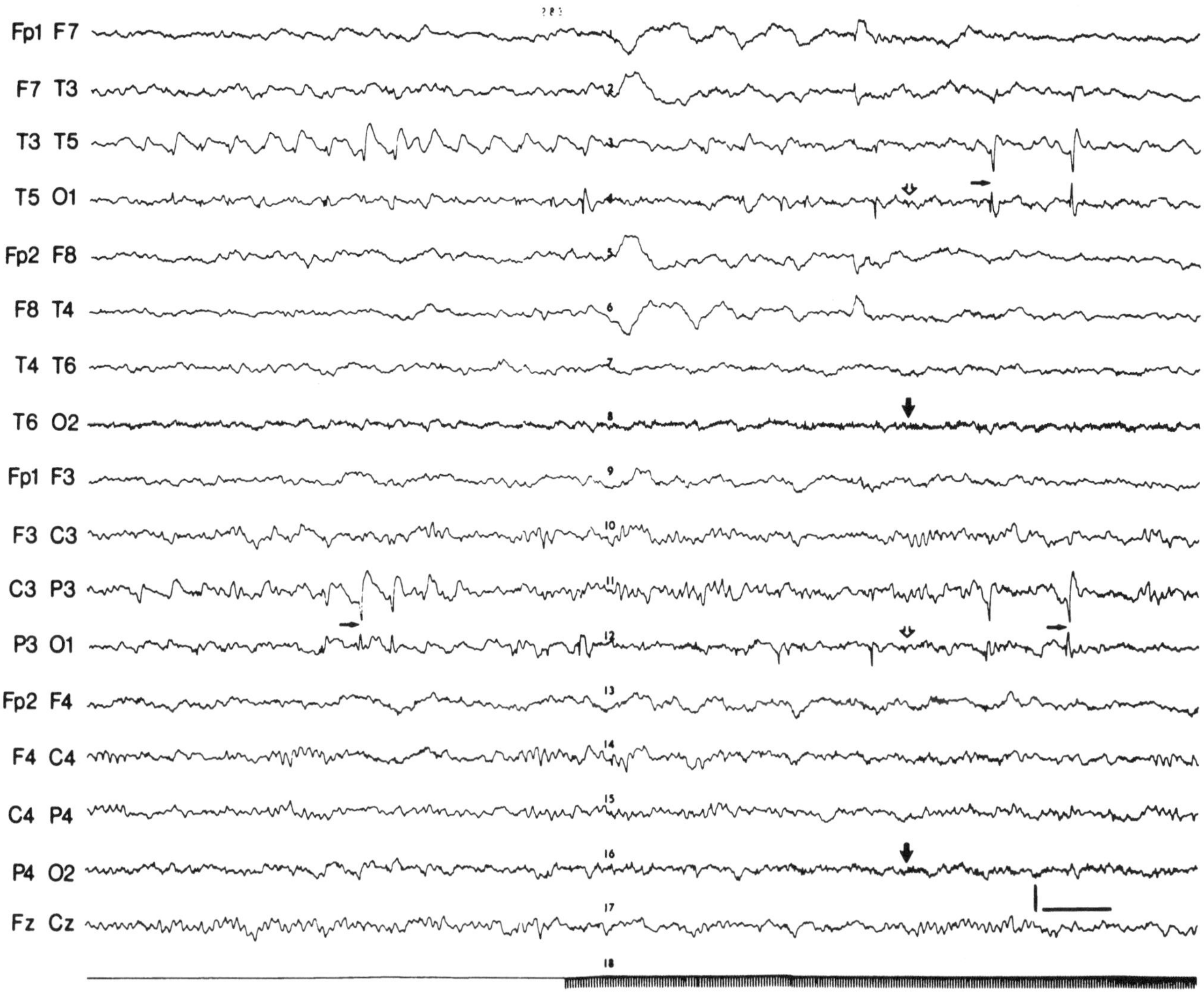

Figure 16.7. Asymmetrical photic driving. Solid vertical arrows indicate low-amplitude driving at 25 flashes per second in the posterior region on the right side; open vertical arrows indicate the relative absence of driving on the left. The left side shows other abnormal features, namely, slowing of the background and spikes in the parietal and posterior temporal regions. Some of the spikes are indicated by the horizontal arrows. Note the mu activity in the central regions on both sides. Filters: low frequency = 1 Hz, high frequency = 70 Hz. Calibrations: horizontal = 1 second, vertical = 100 μV.

Absence of any response to IPS or a bilaterally symmetrical, low-voltage response is not considered abnormal. On the other hand, a marked, consistent asymmetry in amplitude between the two sides is considered abnormal. Similarly, consistent absence of driving to many flash frequencies on one side, with intact driving on the other side, is also abnormal. Thus, in unilateral destructive structural lesions involving the occipital lobe, driving is less pronounced or absent on the side of the lesion (see Fig. 16.7); also, the affected side may show no driving for only certain frequencies of stimulation. Exceptionally, there may be more prominent driving over an irritative lesion in the occipital lobe, such as a meningioma.

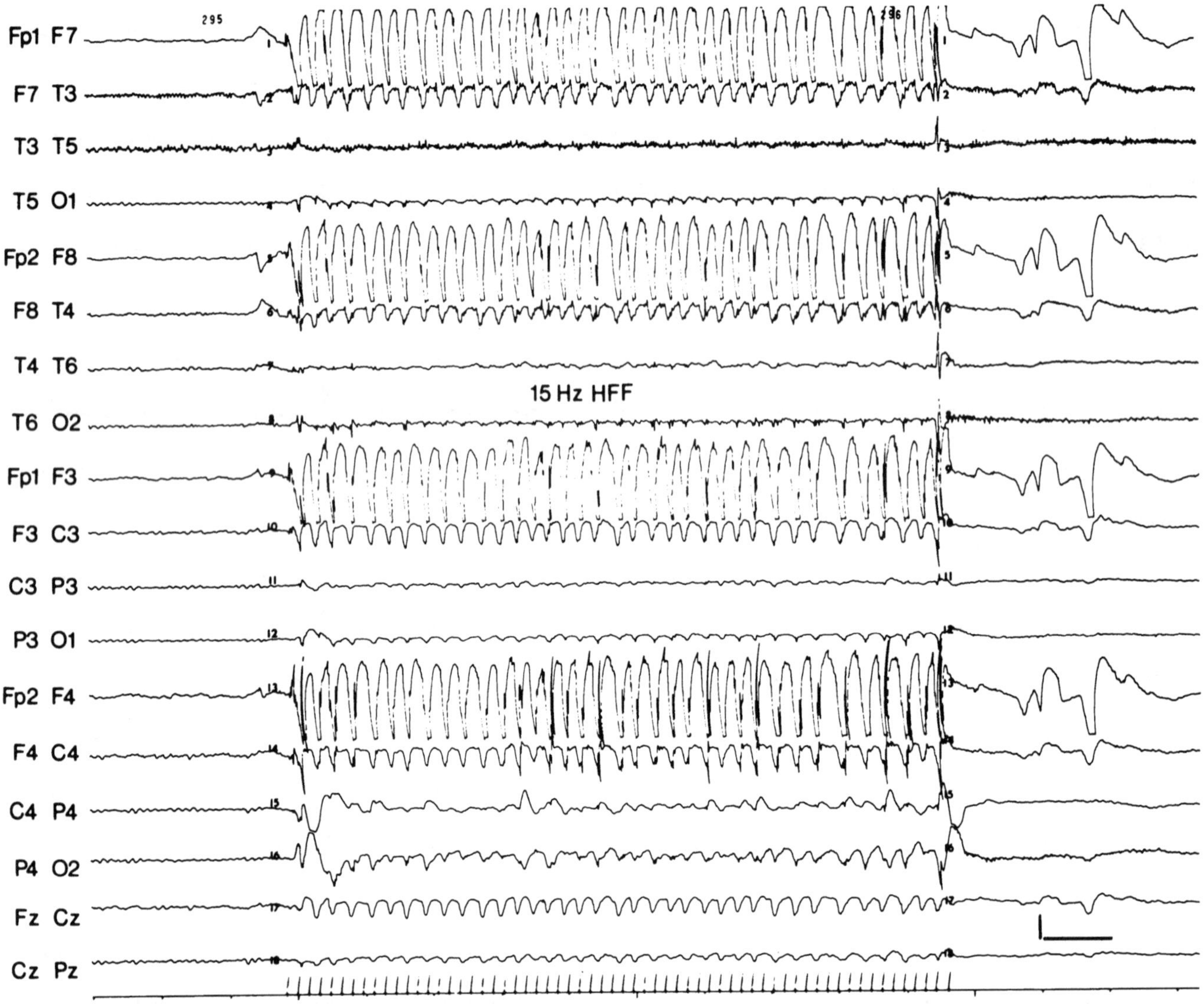

Figure 16.8. Photomyogenic response to photic stimulation at 6 flashes per second. Note that the response is most prominent in the frontal electrodes and stops abruptly when photic stimulation ceases. To accentuate the clonic movements seen in the tracing, the high-frequency response of the EEG machine has been severely reduced. As a result, the large-amplitude muscle spikes that normally accompany photomyogenic responses are greatly attenuated. Filters: low frequency = 1 Hz, high frequency = 15 Hz. Calibrations: horizontal = 1 second, vertical = 100 μV.

Photomyogenic Response

The normal reflex reaction to a flash of light is a blink during which time there is contraction of the orbicularis oculi, the muscles that are situated around the eyes. In some subjects, the response may be accentuated and a number of periorbital and facial muscles may take part. When this happens, the EEG shows prominent muscle spikes that are time-locked to the light flashes and are most conspicuous over the frontal electrodes (Fig. 16.8). This phenomenon is called a photomyogenic response. It is important to distinguish this response from a photoparoxysmal response. Unlike the photoparoxysmal response, the clonic movements of the photomyogenic response stop abruptly as soon as photic stimulation is discontinued (see Table 16.1); it is a nonspecific finding, known to occur in a small per-

Table 16.1. Comparison of Photomyogenic and Photoparoxysmal Responses to Photic Stimulation

	Photomyogenic Response	Photoparoxysmal Response
EEG Findings		
Morphology	Muscle spikes	Spike and wave complexes
Location	Frontopolar and frontal	Generalized
Frequency	Follows flash frequency	Independent of flash frequency
Relationship to onset and offset of the stimulation	Abrupt onset and offset simultaneous with photic stimulation	Often outlasts the duration of photic stimulation

(Continued)

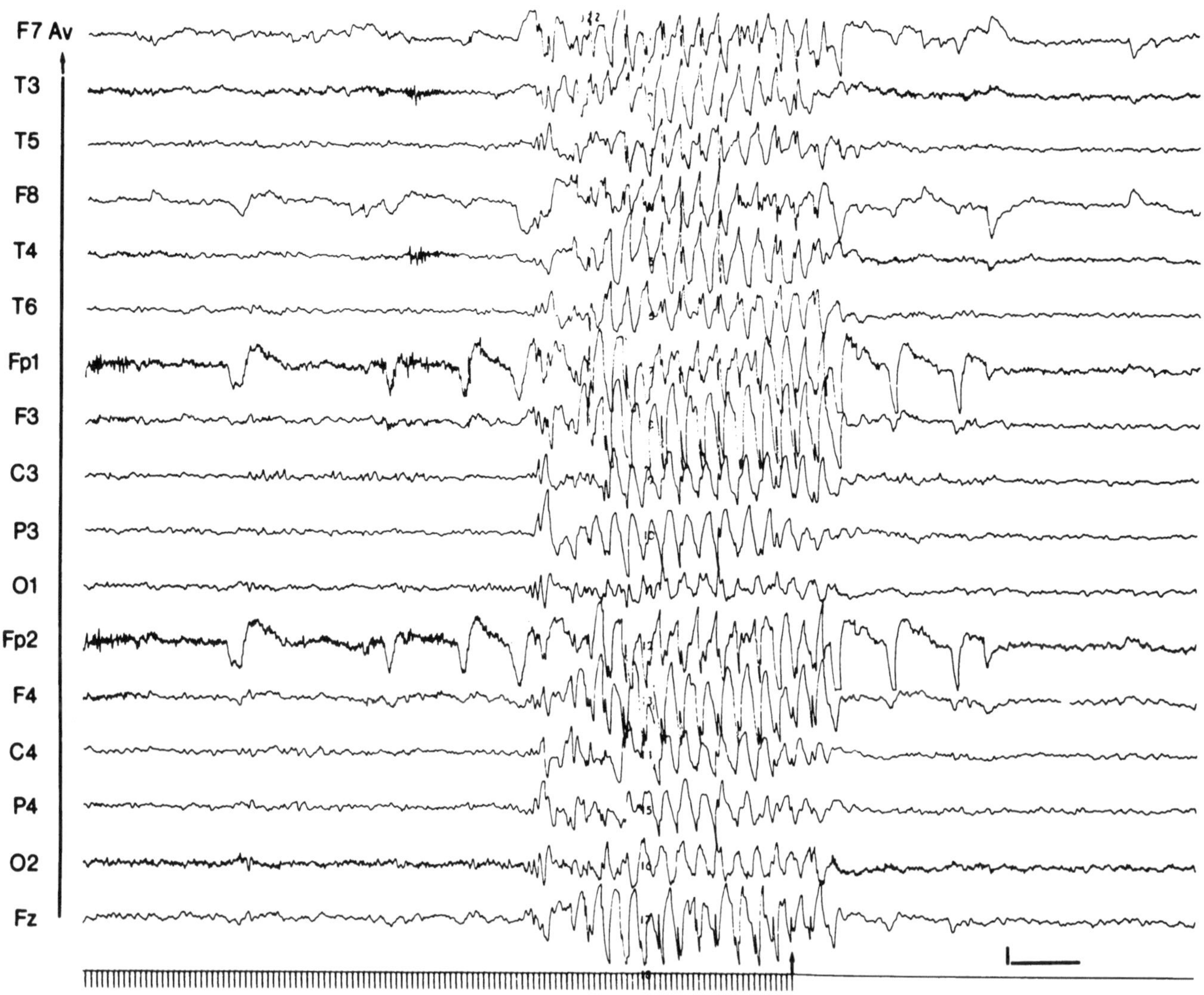

Figure 16.9. Typical photoparoxysmal response to photic stimulation at 12 flashes per second. The generalized, 3- to 4-Hz bilaterally symmetrical and synchronous spike and wave discharges outlast the flashes—which cease after the vertical arrow—nearly by a second. Filters: low frequency = 1 Hz, high frequency = 70 Hz. Calibrations: horizontal = 1 second, vertical = 50 μV.

Table 16.1. Continued

	Photomyogenic Response	Photoparoxysmal Response
Clinical Findings	Contraction of periocular and facial muscles synchronous with flashes	Tonic clonic seizure or myoclonic jerks or absence spell
Significance	None	Suggestive of photosensitivity—high correlation with primary generalized epilepsy

centage of the normal population. No clinical significance is attached to the occurrence of photomyogenic responses.

Photoparoxysmal Response

Also known as a photoconvulsive response, the diagnostic feature of the photoparoxysmal response is the occurrence of spike and wave or multiple spike and wave discharges during photic stimulation. These discharges are bilaterally symmetrical, synchronous, and generalized; they often outlast the stimulus duration (Fig. 16.9). A sustained discharge that persists after the cessation of stimulation is more significant than a brief response (Fig. 16.10). Although generalized, the discharges may be most pronounced in the frontal, central, or occipital areas. Sometimes occipital spikes alone are noted; these are considered to carry little clinical significance. Photoparoxysmal responses occur most often at flash rates of 15 to

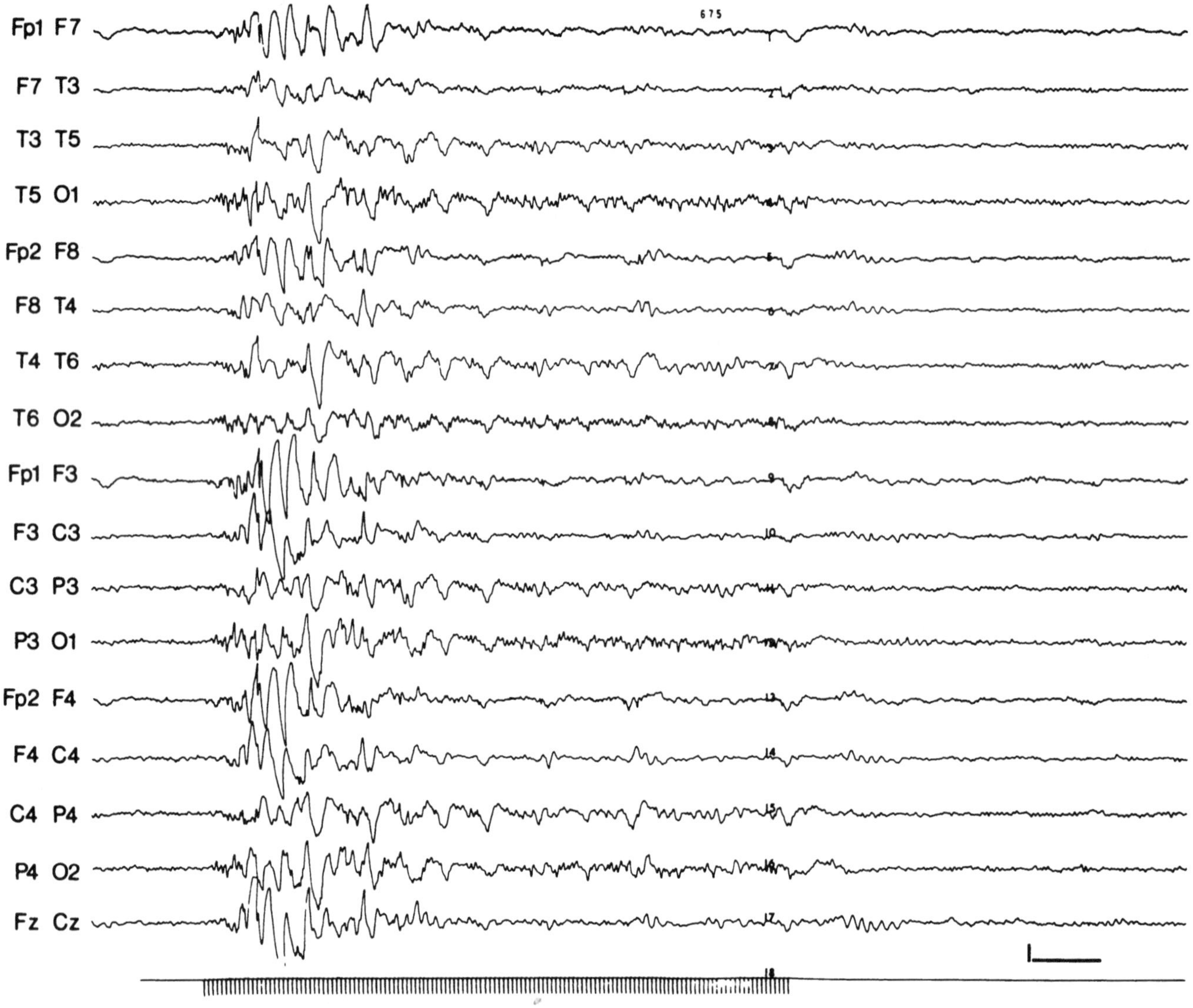

Figure 16.10. Brief photoparoxysmal response to photic stimulation at 15 flashes per second. The generalized discharge, which lasts less than 2 seconds, is most pronounced in the frontal and central areas. Filters: low frequency = 1 Hz, high frequency = 70 Hz. Calibrations: horizontal = 1 second, vertical = 50 μV.

20 per second. Table 16.1 compares these responses with photomyogenic responses.

Photoparoxysmal responses indicate photosensitivity. They are known to occur in a significant proportion of patients having seizure disorders of the primary generalized type, particularly absence seizures. Sometimes a frank seizure is precipitated by photic stimulation. Figure 16.11 shows an absence seizure precipitated by a flash rate of 20 per second. Note that the patient was found to be unresponsive during the spike and wave discharges.

It should be understood that the presence of a photoparoxysmal response does not necessarily indicate that the patient suffers from epilepsy: rarely, photosensitivity may exist unassociated with clinical seizures. A transient photoparoxysmal response may occur in other conditions, e.g., in drug or alcohol withdrawal, and in diffuse encephalopathies including metabolic encephalopathies like uremia. Occurrence of high-amplitude spikes in response to stimulation at three flashes per second has been known to occur in children with the late infantile form of sphingolipidosis.

Photoelectric Electrode Artifacts

It will be apparent that the light flashes produced during photic stimulation subject some of the electrodes placed on the patient's head to some very bright illumination. Occasionally, this light produces a photochemical reaction at the electrode interface that results in the occurrence of artifacts in the EEG tracings. Predictably, these artifacts appear chiefly in the Fp_1–F7, Fp_2–F8, Fp_1–F3, and

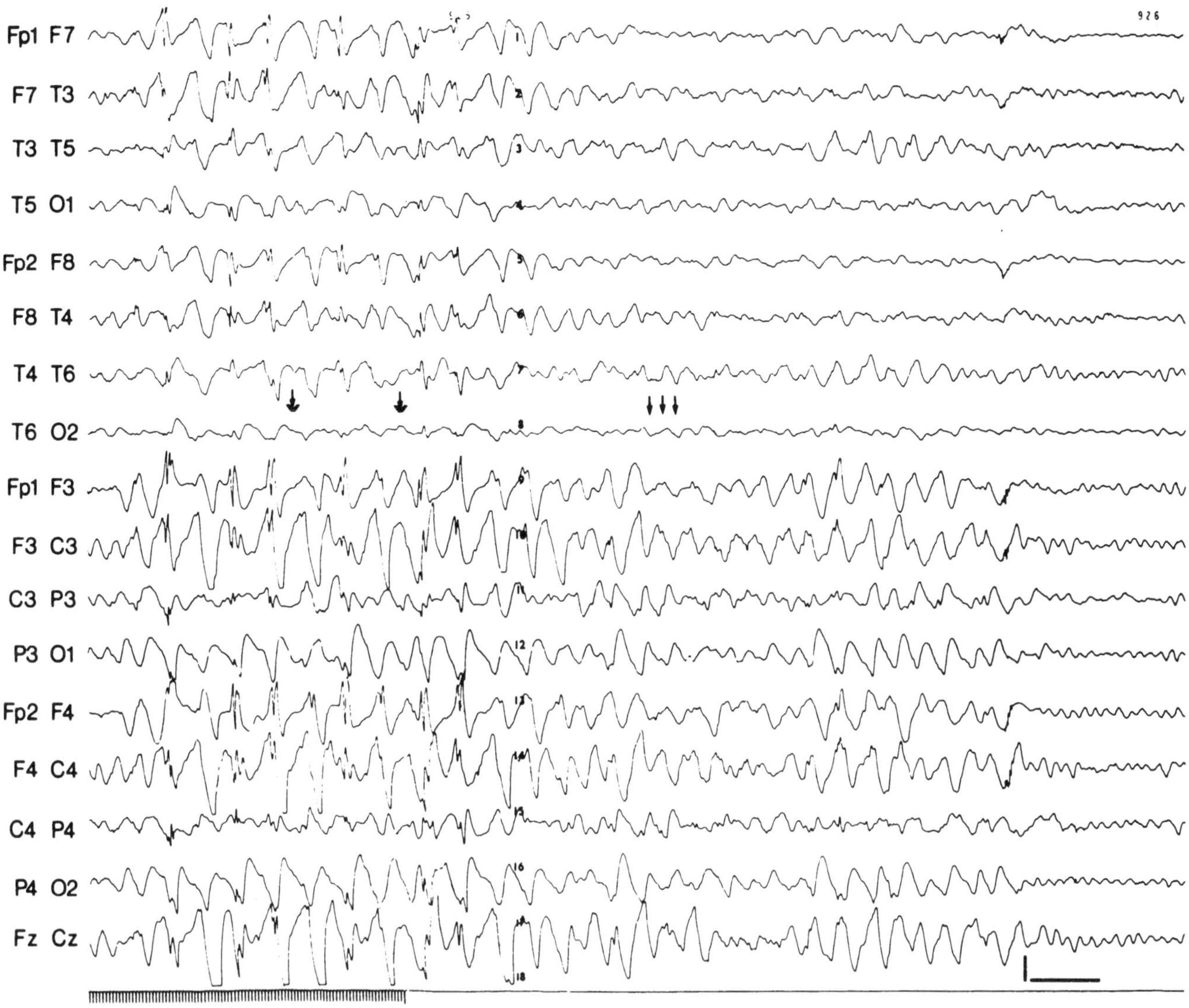

Figure 16.11. Absence seizure produced by photic stimulation at 20 flashes per second. The patient was called by name at the times indicated by the large arrowheads, but did not respond until several seconds later at the three consecutive arrows. Filters: low frequency = 1 Hz, high frequency = 70 Hz. Calibrations: horizontal = 1 second, vertical = 100 μV.

Fp_2–F4 bipolar derivations. The artifacts are seen as spikes or sharp waves occurring in time with the light flashes. They are readily identified as artifacts since they disappear when the electrodes concerned are covered with a piece of opaque cloth or plastic. For the most part, these artifacts can be avoided by ensuring that electrode impedances are below 5K ohms.

Sleep

It is debatable whether sleep, being a natural phenomenon, can be considered an activation procedure. In the past several years, sleep recording has become a routine procedure in many EEG laboratories, especially in patients suspected of having seizure disorders. The impetus derives from a number of studies that have documented the effect of both naturally occurring and hypnotic-induced sleep in bringing out or accentuating epileptiform abnormalities in the EEG. The augmenting effect of sleep is seen both in generalized as well as in focal epilepsies. In a significant proportion of patients with epilepsy, generalized epileptiform discharges may become evident only during sleep. A dramatic increase in the number of spike discharges during drowsiness and light sleep is a characteristic feature of benign Rolandic epilepsy. Also, epileptiform activity may appear for the first time or may be exaggerated during sleep in patients with temporal lobe seizure foci. In general, drowsiness and light sleep (stages I and II) are more effective in bringing out epileptiform abnormalities in the EEG than the deeper stages of sleep.

Sedation

At the time the appointment for the EEG is made, the technician should ascertain whether a sleep recording is required. Since it may not always be possible to obtain natural sleep in the EEG laboratory, the referring physician should authorize the EEG laboratory to administer sedation if a sleep recording is necessary. Orally administered chloral hydrate is the preferred drug because it produces much less beta activity in the EEG (see Chapter 14) than other sedative agents and also because it is rapidly metabolized. Side effects are minimal, although like most other sedative drugs it can also provoke paradoxical excitation, particularly in hyperactive children. The recommended dosage is 25 to 50 mg/kg body weight, with a maximum total dose of 2,000 mg for adults and 1,000 mg for children. Chloral hydrate can be administered in the form of a syrup or a capsule. The technician, however, should make a sincere effort to obtain natural sleep before resorting to administration of the sedative drug. If sedation is to be used, it is important that someone who can take the patient home accompany the patient to the EEG laboratory, as an automobile should not be driven after a sedative has been taken.

Sleep Deprivation

Although there is still some debate as to the value of sleep deprivation as an activation procedure, the majority of electroencephalographers believe it to be a very useful procedure, particularly in patients with seizure disorders. There have been several studies documenting that there is a significant increase in the yield of EEG epileptiform abnormalities following overnight sleep deprivation in epileptic patients. This has been variously attributed to the increased percentage of time that a person is likely to spend asleep during the EEG following overnight sleep deprivation, and also to a specific potentiating effect of sleep deprivation per se on epileptiform activity. The technique followed in sleep deprivation studies varies in different laboratories. Some laboratories require that the patient stay awake only until midnight; others require all-night sleep deprivation prior to taking the EEG.

Pharmacological Activation

A number of pharmacological agents have been used to induce epileptiform activity in patients with seizure disorders, the purpose being to determine whether one is dealing with a primary generalized or a focal onset seizure. The drugs used include certain convulsants such as Metrazole (pentylenetetrazol) and bemegride and the barbiturates. Metrazole by itself or combined with IPS can lead to the appearance of epileptiform abnormalities in the EEG with progression to generalized seizures. Bemegride or β-ethyl-β-methylglutarimide is also a convulsant drug with similar effects. Potentially, these drugs can induce major seizures and, therefore, should be used with greatest caution. There are few occasions for the clinical use of these drugs at the present time. Some centers utilize them in the evaluation of possible candidates for seizure surgery.

Among the barbiturates, thiopental injected intravenously has been used to distinguish focal seizures with secondary bilateral synchrony from primary generalized seizures. This drug may also be useful in determining whether a particular spike focus represents a mirror focus (focus occurring in the contralateral side at a homologous site, driven by the primary focus) or an independent focus in some cases of temporal lobe epilepsy. Brevital (methohexital) is another barbiturate derivative; it is a rapidly acting drug that has been shown to activate focal and generalized epileptiform discharges.

Chapter 17
Average Evoked Potentials

In the decade of the 1980s, many clinical EEG laboratories have added average evoked potential studies to their routine procedures. Indeed, short-latency brain-stem auditory-evoked potentials, as well as short- and intermediate-latency cortical-evoked potentials, lately have proved to be valuable clinical tools for objectively testing afferent functions in patients with neurological and sensory disorders. This being the case, there is need for the EEG technician to become familiar with evoked-potential methods and to add a variety of new skills to his or her repertoire. At the same time, the person reading and interpreting the records will find it necessary to deal with concepts and techniques that are markedly different from those encountered in EEG interpretation.

This chapter is intended to address and, hopefully, to meet some of these needs. It is not intended to be an exhaustive treatment of the topic of average evoked potentials. This would require several chapters or an entire volume, and such texts are already available. Rather, our purpose is only to present a bird's-eye view of the essentials.

Historical Background

It is well known that the sensory modalities are laid out in an orderly fashion throughout the brain right up to the level of the cerebral cortex. In the somatosensory system, for example, peripheral stimulation evokes electrical activity at the cortex that has topographical features—the familiar sensory homunculus. The possible clinical value of recording such evoked electrical activity from the brain to investigate sensory and neurological deficits is obvious and has long been recognized. On the other hand, the routine recording of evoked brain electrical activity by noninvasive procedures using scalp electrodes has only recently been realized, although the basic methods for doing so have long been available. The reason for this is that until recently, there was a lack of adequate technology. Let us review some of the high points that led to this development.

As we have seen in the earlier chapters of this text, the EEG or spontaneous electrical activity of the brain is readily picked up and recorded from electrodes placed on the patient's scalp. Because the activity produced at the cortex by sensory stimulation is also electrical, it is reasonable to expect that this activity could be recorded by means of scalp electrodes as well. But while this is true in theory, serious problems are encountered in practice. When picked up by scalp electrodes, evoked electrical activity appears against a background of spontaneous electrical activity. In other words, what is seen in reality is a mixture of evoked and spontaneous electrical activity. More often than not, the spontaneous activity is of much greater amplitude than the evoked activity.

In technical language, the evoked activity is the "signal" we desire to record and the background activity is "noise." As we just noted, the signal is normally of much lower amplitude than the "noise" so that the proportion of signal in relation to noise—the signal-to-noise ratio—is low. A low signal-to-noise ratio means that although a signal is present, it may go undetected because it is hidden or masked by the noise. To detect an evoked potential it is essential to increase the signal-to-noise ratio; and one can do this either by increasing amplitude of the signal, decreasing amplitude of the noise, or both. Because amplitude of signal is governed by the intensity of stimulation, it is easy to see that the chief way of increasing signal-to-noise ratio is by reducing the amount of noise. One obvious way of doing this is to have the patient keep the eyes open. Under such conditions the alpha rhythm, which for purposes of evoked electrical activity is "noise," will be reduced in amplitude. In the awake, alert individual, this leaves a background consisting mostly of low-amplitude activity in the alpha and beta frequency bands. However, as cortical evoked potentials are commonly less than a few microvolts in amplitude, they still may be hidden in the remaining noise of the background activity. Obviously some other, more powerful method is needed to reduce noise.

It is interesting to note that a method for detecting small, systematic fluctuations among larger, irregular ones was already available in principle during the 18th century. Laplace, the French mathematician whom we mentioned in an earlier chapter, hypothesized that it should be possible to demonstrate a lunar tide in atmospheric pressure that was smaller than the error in reading a barometer by combining and averaging a sufficiently large number of observations. In a similar vein, Sir Francis Galton, the 19th century English scientist, suggested that the facial characteristics common to a number of different individuals could be extracted from the minor details peculiar to each individual by the technique of superimposition. This technique involves optically superimposing a number of similar drawings or photographs of different persons. In doing so, the regular features common to all individuals are emphasized while the irregular, idiosyncratic features wash out and appear only as a diffuse thickening of the composite.[1] The technique was both simple and ingenious and could be applied to enhancing the typical features of a variety of different data. What was lacking at the time, however, was the technology needed for rapidly and accurately combining or superimposing the data.

Method of Superimposition

At the time of World War II, the method of superimposition was put to practical use in the early radar systems. By superimposing many faint blips on a cathode-ray tube, it was possible to detect signals from a target that were otherwise masked by an irregular or noisy background. As early as the late 1940s, G.D. Dawson in England applied this same method to detect and enhance cerebral responses to electrical stimulation of peripheral nerve. Figure 17.1 shows a visual-evoked potential obtained on a cathode-ray tube using the technique of superimposition. For the method to work, all of the separate tracings have to be lined up so that the times when the stimuli occur coincide. Although the method is capable of detecting an evoked potential in a noisy background, it does not permit accurate quantification of the waveform's features. For this purpose, practical methods of signal averaging are needed.

Signal Averaging

The advent of digital computers marked the beginning of a practical, effective, truly quantitative method of enhancing evoked potentials. Although signal averaging is quite complex in the details of the method, it is relatively simple in principle. To begin with, the mixture of electrical activity composed of spontaneously generated voltages and the voltage evoked by stimulation is picked up from scalp electrodes and amplified. This changing pattern of electrical activity with time is divided into segments or epochs of equal duration. The start of each epoch coincides with the presentation of a stimulus, whereas its duration varies, depending on the nature of the evoked potential of interest. In the case of brain stem auditory-evoked potentials (BAEPs), for example, the epoch is only 10 ms long. On the other hand, for cortical evoked potentials the epoch may have a duration of several hundred milliseconds.

The electrical activity contained within each of the epochs is referred to as an *analog voltage.* This voltage is converted to digital form by a process known as analog-to-digital conversion or A-D conversion. The process is akin to feeding the electrical activity contained within the epoch into a voltmeter and then reading and tabulating the voltages shown at consecutive, equal time intervals following the stimulus to the end of the epoch. For example, the voltmeter might be read at the time the stimulus is presented and then at 1 ms, 2 ms, 3 ms, and so on following the stimulus. How frequently readings are taken is known as the *sampling rate.* Just as a curve plotted from many points will show more detail than one plotted from a few, a high sampling rate will better define the way in which the voltage changes within an epoch than a low sampling rate. The string of numbers corresponding to the voltages present at specified times within the epoch is stored in a bank of adding counters having a separate bin for each of the numbers. Thus, for example, if an epoch lasts for 250 ms and the sampling rate of the A-D converter is 1,000 Hz, 251 separate numbers will be entered into a 251-bin counter, one number in each bin.

The process just described repeats itself in the same way for each epoch. If, in our example, the stimulus is presented a total of 100 times, there will be a total of 100 separate 250-ms epochs. When the A-D conversion has been carried out on each, a total of 100 numbers will have been added into each of the 251 bins of the counter. Thereupon, the computer controlling the bank of counters divides the total in each of the bins by 100 to obtain the average. Finally, these 251 mean values are used to plot the average evoked potential. This is accomplished by means of the process of digital-to-analog conversion, which is the reverse of A-D conversion. The average evoked potential is displayed on a cathode-ray tube or a hard copy of it is made

[1] Galton reported that Herbert Spencer, the English philosopher, also had the idea of using the technique of superimposition to obtain a composite photograph of several individuals (Pearson K: *The Life, Letters, and Labors of Francis Galton.* Cambridge, England, The University Press, 1924, vol 2, p 229).

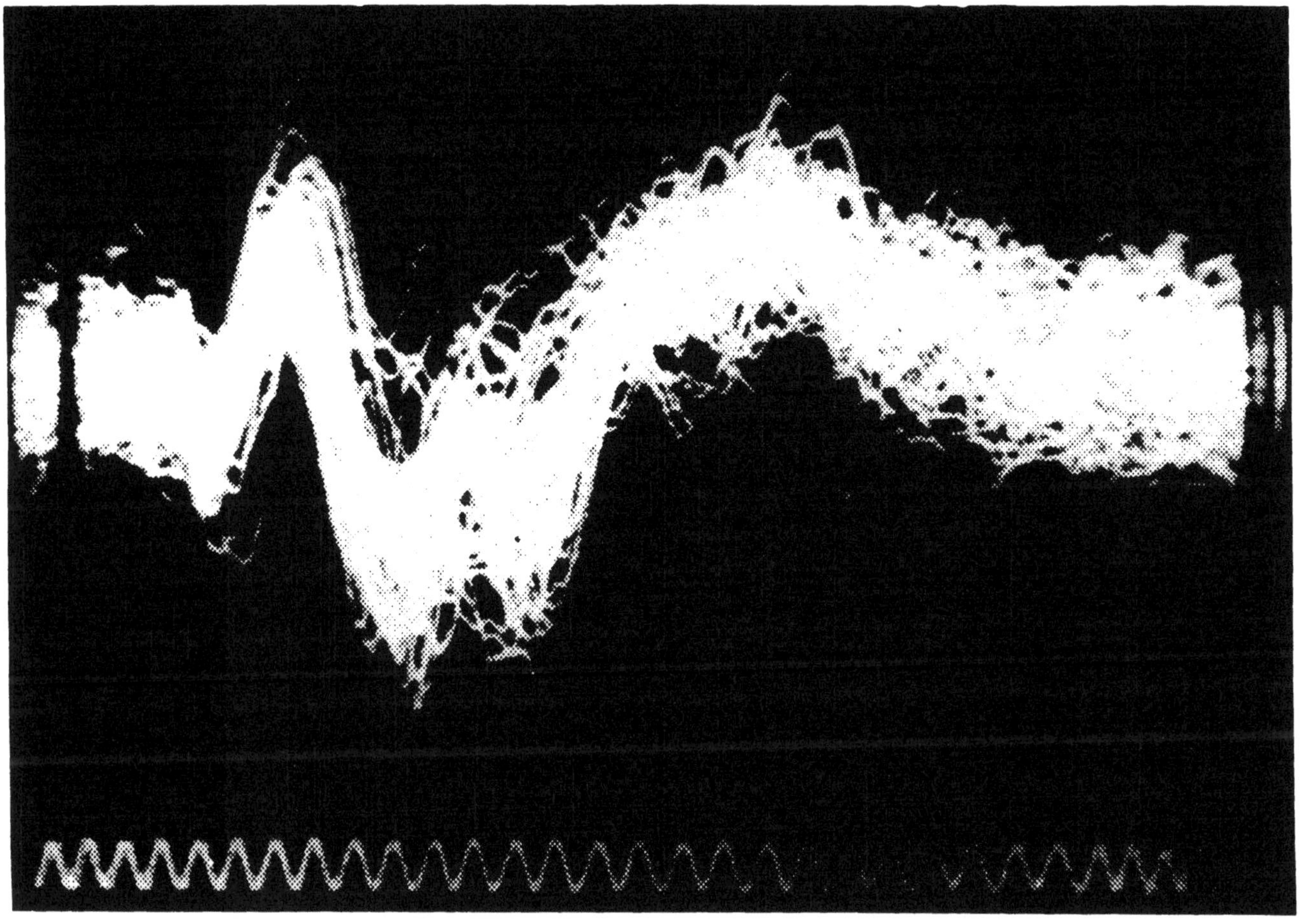

Figure 17.1. Visual-evoked potential obtained on a cathode-ray tube using Dawson's superimposition technique. Calibrations: 5 μV, 10 ms (100-Hz calibration signal). (Reproduced by permission of the author and publisher from Cigánek L: Excitability cycle of the visual cortex in man. *Ann NY Acad Sci* 1964; 112:241–253.)

on an X-Y plotter. The whole process is referred to mathematically as *coherent averaging.*

Coherent Averaging

This term derives from the fact that an evoked potential that may be present in an epoch will be coherent with, or time-locked to, the evoking stimulus. Since the brain's spontaneous electrical activity is essentially random with respect to this stimulus, algebraic summing of the signal containing both evoked and spontaneous activity over a sufficient number of summing cycles causes the spontaneous activity to sum to zero whereas the evoked activity will sum linearly. This happens because, on the average, the polarity of the evoked activity will always be the same at any given point in time relative to the evoking stimulus, whereas the spontaneous activity or noise can be of either polarity and thus will tend to cancel out. Figure 17.2 shows how this process works.

Mathematically speaking, the evoked potentials sum up linearly, and the spontaneous activity sums up as the square root of the number of stimulus repetitions averaged. Thus, after 100 epochs have been summed, the evoked activity or signal may be 100 times larger. By contrast, the spontaneous activity or noise will have increased only by $\sqrt{100}$ or 10 times. It is apparent that the result in this case will be a 10:1 signal-to-noise enhancement. What happens to signal-to-noise enhancement when the number of stimulus repetitions varies is shown in Table 17.1. Note that in order to double the signal-to-noise enhancement, the stimulus repetitions need to be quadrupled. Table 17.1 makes it clear that a very large number of repetitions is necessary in order to achieve high levels of signal enhancement. This means that as the size of the evoked activity decreases relative to the spontaneous activity, a greater degree of enhancement is needed to detect the evoked activity, which, in turn, necessitates a larger number of stimulus repetitions.

As an example, suppose that the amplitude of the evoked potential elicited by some form of stimulation is 5 μV, while the amplitude of the brain's spontaneously generated activity is 50 μV. Now, if there are 200 repetitions of the stimulus, Table 17.1 shows that the signal-to-noise

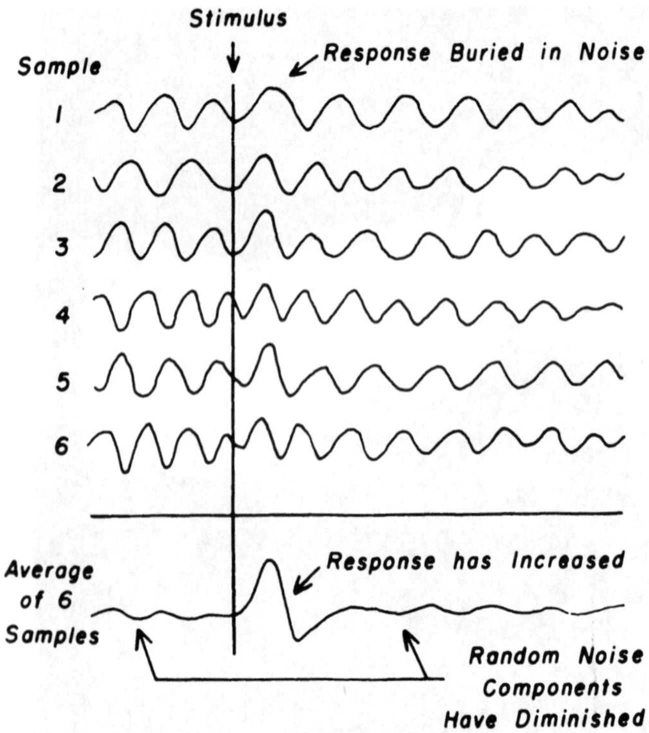

Figure 17.2. Simplified diagram illustrating how coherent averaging enhances a low-level signal. The method depends on the fact that in each sample, the stimulus evokes a response at the same latency. The responses, therefore, are aligned when the six traces are added together. Unwanted signals like the EEG rhythms, on the other hand, are not systematically related to the stimulus and will not be aligned in the six traces. These signals tend to cancel when the traces are added together. When the average is taken, the response becomes enhanced while the other signals present are averaged out. (Taken from Fig. 6, p 457 of Bickford RG: Newer methods of recording and analyzing EEGs, in Klass DW, Daly DD (eds): *Current Practice of Clinical Electroencephalography*. New York, Raven Press, 1979, pp 451–480, by courtesy of author and publisher.)

enhancement is 14.14:1. This is equivalent to reducing the noise by a factor of 14.14, which means that combining and averaging two hundred 5 μV evoked potentials will reduce the 50 μV background activity to 50 × 1/14.14 = 3.54 μV. Although this amount of signal enhancement makes it possible to detect an evoked potential that is otherwise buried in the noise, it is clear that the specific features of the waveform will be marred if not partially obscured by the presence of the residual background activity.

Instrumentation

Signal averaging systems are of two main types. First of all, there is the hard-wired, fixed-program averager. This is a self-contained system that has its own preamplifiers,

Table 17.1. Enhancement of Evoked Activity Achieved by Signal Averaging in Relation to the Number of Stimulus Repetitions

Stimulus Repetitions (SR)	$\sqrt{SR}$	Signal-to-Noise Enhancement
10	3.16	3.16:1
25	5.00	5:1
49	7.00	7:1
81	9.00	9:1
100	10.00	10:1
200	14.14	14.14:1
400	20.00	20:1
800	28.28	28.28:1
1,600	40.00	40:1
2,000	44.72	44.72:1

amplifiers, filters, A-D converter, and microprocessor or built-in microcomputer to perform the summing and averaging operations. The program of the computer is fixed or hard-wired into the instrument so there is no software to deal with. All changes in system parameters—duration of an epoch, number of stimulus repetitions, repetition rate, etc.—are accomplished by operating various switching devices on the machine. This type of averager also has its own self-contained display system, usually a cathode-ray tube. A hard copy of the data may be obtained by use of a peripheral unit like an X-Y plotter.

The other type of signal averager is the integrated, component-type system. This type of system may have its own amplifiers and filters but sometimes makes use of an EEG machine to amplify and filter the raw signal. The amplifier output or IRIG output (see Chapter 3) of the EEG machine is then fed into an A-D converter and the digitized output is connected to a general purpose microcomputer. Appropriate signal-averaging software controls processing of the data. The average evoked potentials are then displayed on a graphics display terminal. Depending upon the software available, a variety of descriptive statistics relevant to the features of the waveforms of the average evoked potentials may be calculated and displayed on the terminal as well. In addition, an assortment of statistical tests using these data may also be carried out at the same time.

It should be evident from the foregoing that the integrated, component-type system is more flexible and more sophisticated than the hard-wired, fixed-program averager. But it is usually more costly, in terms of both time to set up and operate and money to purchase. By contrast, the hard-wired averager is simpler to use, usually cheaper, and frequently more reliable since reloading of programs from disk or tape whenever a program change is needed is eliminated. For these reasons, the hard-wired, fixed-program averager has been the standard in many clinical laboratories. In the interest of simplicity, the material and discussion that follow assume that this type of system is being used. In the case of the newer systems that employ software and microchip control of operations, consult the appropriate instruction manuals for operational details.

Display Systems

The display systems employed have already been mentioned briefly in the previous section. Almost universally, the hard-wired averager will have a self-contained, cathode-ray tube for display of the waveforms of the average evoked potential. Although the appearance of the waveforms varies with the sensory modality stimulated, the location of the recording electrodes, and the duration of the epoch, the waveforms commonly consist of a number of different peaks (positive deflections) and troughs (negative deflections). In general, three kinds of data are derived from the waveforms and used clinically: (1) measures of the latency of the various peaks and troughs from the time of stimulation, (2) measures of the time elapsing between particular peaks and/or particular troughs, and (3) measures of the amplitude of certain peaks and troughs in the waveforms. These data usually are acquired with the help of a clever device referred to as a *cursor* or bug.

The cursor is an intensified portion of the waveform traced on the cathode-ray tube. This bright spot may be shifted from right to left and vice versa by means of a control on the front panel of the instrument whence it appears to ride along on the trace. Associated with the cursor is a readout display that shows the latency of the selected point from the time of stimulation as well as the amplitude from the zero baseline. Thus, by setting the cursor directly on the peak or trough of a wave, the machine will display the latency and amplitude of that wave in digital form on the cathode-ray tube. Some averagers also have a second, independently controlled cursor. When both cursors are used, the display shows the time between two selected points (as the milliseconds between two peaks) and the peak-to-peak amplitude of a particular wave (as the number of microvolts between consecutive peaks and troughs) in the waveform displayed. In actual use, the technician has to jot down the data from the readout display, preferably on a hard copy of the average evoked-potential waveform. Some systems have a data-point printer associated with the cursor, which relieves the technician of this manual operation.

Most signal averagers have some means of providing a hard copy, or permanent record, of the average evoked-potential waveform. In the early days of signal averaging, the user simply photographed the trace on the cathode-ray tube. But this requires a camera, some bulky equipment for mounting it, as well as an optical system for simultaneously viewing and photographing the waveform. At the present time, hard copies are produced either by an X-Y plotter or by a high-speed, computer-controlled printer.

The X-Y plotter performs the function for which it is aptly named. This instrument is a peripheral recording device having a pen that traces out the X and Y coordinates of a voltage (Y) that varies with time (X). The plotting is done on graph paper, which is inserted, a sheet at a time, by the technician before the plot mode of the averager is enabled. If a computer-controlled, high-speed printer is substituted for the X-Y plotter, the production of a hard copy of the average evoked potential becomes completely automated. In this case, graph paper is not used; instead, the printer prints a set of axes and identifying labels directly onto the perforated, blank paper that normally comes with the machine. The waveform of the average evoked potential is printed on it simultaneously. Such display systems usually incorporate a data-point printer associated with the cursor so that latencies or amplitudes of particular waves of interest may be printed on the hard copy.

Practical Clinical Methods

Many of the methods used in conjunction with the recording of clinical EEGs are appropriate for the recording of average evoked potentials. The same kind of recording electrodes and methods of application are employed, and electrode placement follows the 10-20 International System described earlier in the text. As is the case with EEG recording, low-impedance-recording electrodes are essential, and the same precautions regarding the patient's safety need to be observed. But here is where the direct similarity ends.

For routine clinical EEG recordings, a total of 21 electrodes normally are attached to the patient's scalp. In average evoked-potential studies, on the other hand, the full array of electrodes is not needed. Thus, for example, only Cz, A1, and A2 are commonly used in recording BAEPs. Fewer recording electrodes, of course, mean that fewer recording channels are necessary. For this reason, most clinical averagers have no more than two to four channels available for simultaneous recording.

In recording average evoked potentials, the EEG technician encounters some unique problems that are not experienced in routine EEG recording. These problems derive mainly from the fact that whereas in EEG recording the technician receives continuous feedback from the record concerning the state of the patient and the status of the recording electrodes, in evoked potential recording he/she does not. The technician must wait until all the stimuli are presented and the averaging process is completed before knowing with certainty whether or not the technique was adequate and the waveform obtained is acceptable. As one experienced EEG technician has put it: "With EEGs, you always know from moment to moment what's going on; but with average evoked potentials you are working blind." It is true, of course, that many averagers provide a "live" display that allows the technician to continually observe the averaging process and to view the partial results. But it is rarely possible to judge whether a waveform that looks like it may be acceptable after 50 stimulus repetitions will actually *be acceptable* after 500 repetitions. Similarly, a waveform that looks like pure background noise after 50 stimulus repetitions may be transformed into a classic, textbook example of an average evoked potential after 500. This means that the technician must learn to depend more and more on the results of various routine checks—like electrode impedance tests—to assess the adequacy of her/his technique rather than waiting for the actual test data to become available.

The presence of artifacts, whether they be instrumental, environmental, or physiological, can result in formidable problems for the technician recording average evoked potentials. Instrumental artifacts that occur at random with respect to the stimulus will generally average out if their amplitude is not too large and a sufficient number of stimulus repetitions are used. On the other hand, instrumental artifacts that are caused by the stimulus or by stimulus-generating circuits will likely be coherent with the stimulus and are serious sources of contamination. They must be eliminated. Fortunately, their presence is easily documented or ruled out. This is accomplished by going through the usual recording procedure in every detail except that the stimulus is not allowed to reach the patient's sense organ; in other words, a "sham" test is performed. Thus, for example, if the stimulus employed is a flash of light, a tight-fitting cover is placed over the lamp so that the patient cannot see the flash. Under these conditions, the averaged waveform should show no evidence of evoked activity. Any such activity that may be present is an artifact. The elimination of such artifacts is a job for a specialist.

Artifacts from the environment, especially 60-Hz pickup, can be particularly burdensome. This problem comes about because 60-Hz electrical fields from the electric power lines are present almost everywhere. If the stimulus repetition rate happens to be an even multiple of 60 Hz, the latter can become coherent with the stimulus. In other words, the stimulus occurs at the same phase of the 60 Hz activity each time. When this takes place, the 60-Hz activity will sum up as signal rather than as noise, whereupon it may be significantly enhanced. The simple solution of the problem is to ensure that the stimulus repetition rate is *not* an even multiple of 60. It is essential, of course, that the impedance of the recording electrodes be checked. As with EEG recording, impedance should be less than 5K ohms, and the loop area between the wires running to the electrodes must be kept to a minimum.

Physiological artifacts due to eye movements and muscle twitches, as well as artifacts resulting from mechanical movement of the leads, can pose severe problems. Such artifacts may be quite large—sometimes 200 μV or more in amplitude. Occasionally the eye-movement artifacts may be coherent with the stimulus. If there is reason to suspect that this may be the case, an eye electrode has to be attached and the signal from this derivation averaged in the same way as the signal from the scalp electrodes. Should a clearly discernible waveform appear in this average, the possibility exists that the average evoked potential of interest may be contaminated by this eye activity. Aside from employing rather elaborate subtraction procedures that are hardly appropriate in clinical work, the only solution to this problem is to have the patient exert some voluntary restraint over his/her eye movements.

If the physiologic artifacts we have been considering are not coherent with the stimulus, they will average out—assuming, of course, that there are a sufficient number of stimulus repetitions. But this is usually not practically feasible. Consider a simple example: as we learned earlier, 200 stimulus repetitions will reduce a 50-μV background of spontaneously occurring activity to 3.54 μV. But suppose, now, that in one third of the 200 repetitions some large amplitude (200 μV) eye-movement artifacts are present. After the 200 repetitions, the average background activity would be

$$\frac{50 \times 133 + 200 \times 67}{200} \times \frac{1}{14.14} = 7.08\ \mu\text{V}.$$

Note that this is twice the amplitude of the average background activity when the spontaneously occurring activity in each of the 200 repetitions is 50 μV. If the average evoked potential were only 5 μV as in our earlier example, it would be lost in the higher-amplitude noise of the background. To halve the amplitude of the background activity, we would need to increase the stimulus repetitions from 200 to 800.

The foregoing makes it clear that some technique besides the brute force method of increasing the number of stimulus repetitions is required to deal with such artifacts. Most clinical averagers, therefore, contain an artifact rejection feature. Sometimes referred to as an overvoltage detector, the operation of this device is relatively simple. Each epoch of digitized data is first stored in a buffer memory before being entered into the bank of counters. If this signal exceeds preset limits in either the positive or negative direction, this is taken as evidence that an artifact has occurred. The upshot is that the epoch in question is *not* added to the counters and an additional trial is automatically taken to make up for the epoch that was rejected. Adjusting the setting of the artifact rejection "window" represents a compromise. If the window is too narrow, a large percentage of the epochs will be rejected and some good data may be lost in the process. On the other hand, if the window is too wide, epochs containing obvious artifacts will be accepted and the degree of signal enhancement reduced.

As is the case with EEG recording, the best approach to artifact reduction is to eliminate the artifacts at their sources. Whenever possible, the technician should try this approach. Thus, for example, eye movements can frequently be reduced by properly instructing the patient or by the use of eye pads. Similarly, muscle activity can be controlled by repositioning the patient, or by providing some help with relaxation. Finally, artifacts due to mechanical movements of the electrode wires can be minimized by ensuring that the wires are properly supported and are not allowed to dangle, sway, or otherwise move about during the recording.

An aspect unique to the recording of average evoked potentials is the *stimulus repetition rate*. Because a large number of stimuli need to be presented to obtain an average evoked potential, the rate of presentation is important. If the rate is too slow, an untoward amount of time will be consumed in obtaining the necessary data. On the other hand, if a very fast rate is used and one stimulus follows another too closely, the effects of the second stimulus will be contaminated or confounded by the response to the first stimulus. For this reason, the repetition rate varies, depending on the particular evoked-potential study underway. The actual repetition rates used in practice are taken up later in sections dealing with specific average evoked-potential studies.

As we will see in the next section, the first step in the interpretation of an average evoked potential is the identification of the various components of the waveform. In doing this, how can we be sure that a particular wave we pick out is really the component we think it is, and not an artifact or part of the spontaneously generated activity that is noise? The answer is that we cannot be certain, having only the single average evoked potential upon which to base our judgment. But if the study is repeated again and again, and the particular wave is replicated, we can be fairly confident that the wave is indeed signal and not noise. In actual practice, every average evoked-potential study is repeated at least once. If the waveform or waveforms replicate, these data are taken to be representative and are used to define the latency and amplitude parameters of the average evoked potential. Conversely, if the waveforms fail to replicate, the study is repeated a second and sometimes a third time.

Failure to replicate a study—or failure to obtain closely similar waveforms on several studies—is sometimes attributable to restlessness of the patient. Thus, restlessness produces artifacts that contaminate the averaged waveform; this contamination, in turn, obscures identification of the waveform's components. In averagers having an artifact reject feature, restlessness is accompanied by a large number of stimulus repetitions being rejected. As is the case of EEG recording, the technician frequently can reduce restlessness by repositioning, relaxing, or reassuring the patient. If these measures do not help, the failure to replicate might suggest the absence of a normal waveform and represent an abnormality. This is taken up in the following section on interpretation.

General Principles of Interpretation

The useful clinical information in an average evoked potential resides in the latencies and amplitudes of certain waves or components that are typically present in the waveform. Therefore, the first step in interpretation entails the identification of the various components of which the average evoked potential for a particular sensory modality is formed. This may be relatively easy to do in the case of a waveform that is readily replicable and has all of the usual waves present at the expected latencies. But if one or more of the usual components appear to be absent, or if the latency of one or more of the waves is prolonged, the interpreter may be faced with a dilemma. Delayed or absent waves are suggestive of pathology, but they may equally well be due to the presence of artifacts that obfuscate the waveform.

A variety of special techniques have been devised to help identify certain of the waves in an average evoked potential and, hence, to alleviate the dilemma. For example, as we will see in the section dealing with BAEPs, data simultaneously obtained from the contralateral side can aid in identifying waves IV and V of the complex waveform. Similarly, the use of alternating compression and rarefaction clicks as the stimuli can help to distinguish wave I from stimulus artifact and cochlear potentials. However, when these techniques fail, or when special techniques of this kind are not available to help identify certain waves, the person interpreting the data must fall back on replication as the criterion for identifying the waves. The rule here is simple and straightforward. Waveforms that fail to replicate are considered to be unreliable and hence uninterpretable, even though the possibility exists that they might represent some kind of abnormality.

Once the various components of a waveform have been identified, the person interpreting the data is ready for the next step. This involves comparison of the relevant parameters of the patient's recorded waveforms with a set of norms. Normative data for average evoked potentials for the different sensory modalities have been published by a number of laboratories. These data continue to be updated, and so it is essential for the person interpreting average evoked potentials to keep abreast of the latest work in the area . As it is important to be familiar with the published norms, it is likewise important for each laboratory to have its own normative data as well. Norms vary somewhat with sex and vary markedly with age of a patient. For premature infants they show considerable variability as well.

In the following three major sections, we take up specific aspects of technique and interpretation relevant to visual-evoked potentials, BAEPs, and somatosensory-evoked potentials.

Table 17.2. Relationship Between Check Size and Visual Angle at a Distance of 1 m[a]

Number of Checks	Width of One Check (cm)	Visual Angle	
		Degrees	Minutes of Arc
8 × 8	3.2	1.83	110.4
16 × 16	1.6	0.92	55.2
32 × 32	0.8	0.46	27.6
64 × 64	0.4	0.23	13.8

[a] The TV monitor measures 28 × 23 cm.

Visual-Evoked Potential (VEP)

Anatomical Basis of the VEP

Visual-evoked potentials represent electrical activity induced in the visual cortex by light stimuli that reach the macular and perimacular areas of the retina. The macula is rich in cones that are retinal receptors that convert light energy into electrical energy. The electrical impulses are carried through the optic nerve, which consists of axons arising from the ganglion cells of the retina. Axons from the nasal half of the retina (including nasal half of the macula) cross to the opposite side at the optic chiasm. From there, fibers from the temporal half of the ipsilateral retina and from the nasal half of the contralateral retina form the optic tract on each side. The axons forming the optic tract synapse at the lateral geniculate body. Thenceforth, the neuronal cell bodies in the lateral geniculate body give rise to the optic radiation. The optic radiation is located in the subcortical regions of the parietal and temporal lobes. It terminates in the visual cortex, which is situated in the medial part of the occipital lobe, in and around the calcarine fissure and the occipital pole (see Fig. A2.5). The result is that the left visual cortex is connected to fibers that carry signals from the temporal half of the left retina and the nasal half of the right retina and vice versa. This means that objects situated in the right half of the visual field are perceived by the left visual cortex.

It should be obvious that when either eye is stimulated separately (full-field stimulation), signals pass onto both visual cortices because the nasal fibers cross to the opposite side at the optic chiasm. Such stimulation is more useful in evaluating the anterior visual pathways (optic nerves and chiasm) than the retrochiasmal pathways (optic tract, radiation and visual cortices). However, each half of the retina can be stimulated separately (half-field stimulation), thereby enabling one to determine the functional status of the retrochiasmal visual pathways more accurately.

Stimulus Parameters of the VEP

Two types of stimuli are used for recording VEP, namely, pattern and flash. In the case of pattern visual stimuli, the standard technique makes use of a black-white shift or reversal in a checkerboard pattern electronically produced on a TV monitor. The black and white squares are made to reverse without there being any change in the luminance, or total light output of the screen. A number of stimulus parameters are known to influence the VEP. These include the rate of pattern reversal, the size of the checks, and the luminance. There are also patient-related factors that affect the degree of retinal stimulation, namely, visual acuity, visual fixation, and pupillary size.

A pattern-reversal rate of 1 to 2/s has been found to be optimal for recording the *transient* visual-evoked potential, which is used most often in routine clinical practice; a *steady-state* evoked potential is recorded when the stimulus rate exceeds 10 per second. Although the VEP can be obtained with even slower rates of stimulation, there are disadvantages. With a slower rate, the test becomes more time-consuming, and there is a greater chance that the patient may not fixate on the screen. If the stimulus rate is too fast, the response to one stimulus may contaminate the response to the subsequent stimulus.

The visual angle, or the angle that the individual checks subtend at the retina, has a significant effect on the latency of the response and, hence, needs to be standardized. It depends upon the height or width of the checks, as well as the distance between screen and eye. The formula used for the calculation is: Visual angle = 57.3 (W/D), where W is the width of one check and D is the distance between eye and screen, both in centimeters. Table 17.2 gives the relationship between check size and the visual angle at a distance of 1 m.

The brightness setting of the monitor should be kept constant to maintain the same mean luminance of the checkerboard pattern. Similarly, ambient luminance of the room needs to be kept at a constant level. Consistency of these settings is essential for obtaining consistent data. The luminance can be measured using a spot photometer.

The patient is asked to fixate on a dot placed in the center of the TV monitor screen. The purpose of the fixation point is to ensure that the macular and perimacular areas of the retina are stimulated. The visual-evoked potential becomes smaller in amplitude whenever good visual fixation is not achieved.

The size of the pupil determines the amount of light that enters the eye and, hence, can affect the VEP. The patient should not have mydriatics or cycloplegics for at least 12 hours prior to the test. Visual acuity can also influence the latency and amplitude of VEP. The patient should wear corrective glasses, if any, during the test. If there is a marked deficiency in visual acuity, the checks should be made larger so that the patient can see them clearly. The technician should assess the patient's visual acuity—using a Snellen chart—and pupillary size and enter these data in the worksheet before commencing the test.

Flashes from a photic stimulator positioned 30 to 45 cm in front of the subject's eye may also be used as the stimuli. The stimulation rate is usually between 0.5 to 1/s. The flash VEP is useful in patients whose visual acuity is poor or in comatose/anesthetized patients. However, it shows more variability than the pattern-reversal VEP and, hence, is less reliable for clinical use.

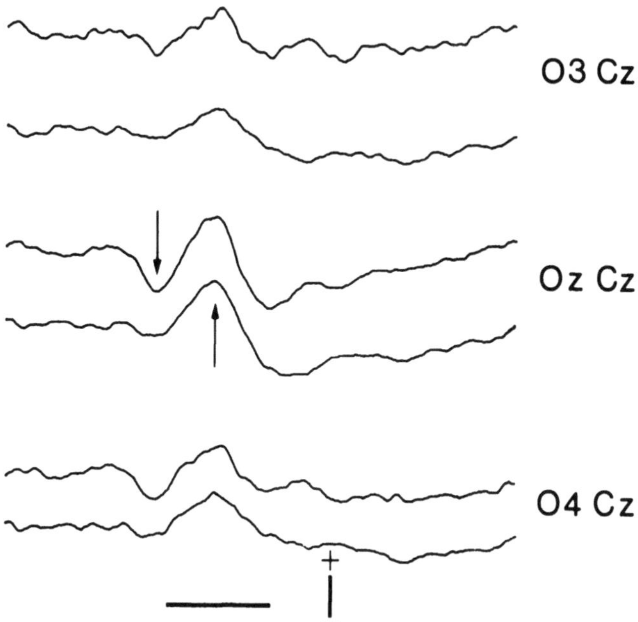

Figure 17.3. Normal VEPs to pattern-reversal, full-field stimulation of the right eye using a checkerboard image of 16 × 16 checks at a distance of 1 m. Stimulation rate at 1.88/s. The bandpass is at 1 to 100 Hz, and 250 trials were averaged. Two separate tests were run, and the paired tracings show good replication. The downgoing arrow points to N75, while the upgoing arrow indicates P100. Calibrations: horizontal = 50 ms, vertical = 5 μV.

Recording Parameters of the VEP

The recommended filter settings are 0.2 to 2 Hz low frequency and 100 to 500 Hz high frequency. Analysis time is 250 ms; but if no response is obtained, the study should be repeated at 500 ms before it is presumed that the VEP is absent. The number of trials averaged can vary from 100 to 200, although a larger number may be needed if the potentials are not well delineated. Recording electrodes are standard EEG electrodes placed over the occipital areas. The American EEG Societey (American Electroencephalographic Society Guidelines, 1986) recommends a midline occipital electrode 5 cm above the inion and right and left occipital electrodes, each 5 cm lateral to the midline electrode. These are referenced to a midfrontal electrode that is located 12 cm above the nasion. The ground electrode may be placed over Fp_z. At least two trials should be done to ensure that the waveforms recorded are replicable.

Normal VEP

The normal VEP often contains three peaks. The initial peak is negative and occurs at a mean latency of 75 ms; it is designated N75. The most prominent and consistent wave is a positive peak (Fig. 17.3), which has a mean latency of 100 ms and is called P100. A subsequent negative peak may be seen at a mean latency of 145 ms (N145). The latency to the peak of the P100 wave is measured along with P100 amplitude, taken from the peak of N75 to the peak of P100. The absolute latency of P100, and any interocular differences in latency and amplitude of P100, are used as criteria for interpretation. Normal values should be established in each laboratory. Latency values exceeding three standard deviations are considered abnormal. An interocular difference in latency of more than 8 to 10 ms is often considered abnormal.

Abnormal VEP

The abnormalities include absence of a VEP, prolonged P100 latency, and an excessive interocular difference in P100 latency.

When the technician discovers that the VEP is absent, a number of steps have to be taken to ensure that a technical problem is not to blame. These include: (1) making sure that the TV monitor is connected to the evoked-potential machine and that the two are synchronized so that a pattern shift occurs each time the averager is triggered; (2) ensuring that the patient is focusing on the checks; and (3) making sure that the electrodes are properly applied, that their impedances are less than 3K ohms, and that the electrode box is connected to the machine. Once these conditions are satisfied, the test should be run again using a 500-ms sweep so that a possible delayed response is not missed. If the VEP is still absent, the finding is definitely abnormal. If monocular full-field stimulation results in an absent response on one side and a normal response on the other, a lesion of the ipsilateral optic nerve is most likely, provided ocular pathology, including retinal lesions, is excluded. Absence of the VEP on monocular stimulation of right and left eyes suggests either bilateral optic nerve or chiasmal lesions, or less commonly, bilateral retrochiasmal lesions.

Prolonged P100 latency characteristically occurs in demyelination of the anterior visual pathways; amplitude attenuation is more typically seen in compressive lesions. If a prolonged latency occurs only on left or right eye stimulation, a lesion of the ipsilateral optic nerve is most likely. Such findings are most characteristically seen in optic neuropathies, particularly the demyelinating type as occurs in multiple sclerosis. On the other hand, if latency is prolonged on stimulation of either side, bilateral optic nerve or retrochiasmal lesions are to be suspected. An excessive interocular difference in latency suggests an optic nerve lesion on the side with the longer latency.

Clinical Correlation of Abnormal VEP

The pattern-reversal VEP is particularly useful in the diagnosis of multiple sclerosis. This is because of the high incidence of optic nerve demyelination in these patients. It has been estimated that 20% to 50% of patients presenting with optic neuropathy may develop multiple sclerosis in the future. The test is particularly useful in confirming a diagnosis of optic neuropathy when the symptoms are atypical. In addition, abnormalities resulting from demyelination of the optic nerve tend to persist for several years even after return of visual functioning to normal. Thus, abnormal VEP is useful in documenting a past optic neuropathy, thereby making it a valuable tool in the diagnosis of multiple sclerosis. In definite cases of multiple sclerosis, abnormalities in VEP have been reported to occur in about 85% of patients. The changes noted in the P100 response include excessive interocular difference in latency, prolonged absolute latency, decreased amplitude, and distorted appearance. The first two have been found to be the most reliable criteria.

The optic nerve and optic chiasm may be compressed by tumors like optic nerve glioma and sellar masses. Decreased amplitude and sometimes prolonged latency of the P100 response may be seen in such cases.

As mentioned earlier, full-field, pattern-reversal VEP is not quite effective in diagnosing retrochiasmal disorders. With half-field stimulation, there is a greater chance of documenting disorders of the optic tract, optic radiation, and occipital cortex. But even in cortical blindness, it has been observed that the VEPs are not consistently lost. There are a number of other clinical indications for VEP, but these are beyond the scope of this text.

Brain-Stem Auditory-Evoked Potential

The short latency BAEP consists of a series of electrical potentials generated in the auditory nerve and the brainstem auditory pathways in response to auditory stimuli. We begin by considering the anatomical basis of these potentials.

Anatomical Basis of the BAEP

It is helpful to have a basic understanding of the mechanics of the ear when attempting to record the BAEP. The external ear—which consists of the auricle or pinna and the external auditory meatus—serves to funnel sound waves onto the tympanic membrane (eardrum), which separates the external ear from the middle ear. The middle ear is a small cavity within the temporal bone. It contains three tiny bones (ossicles) called malleus, incus, and stapes. These bones transmit the vibrations of the tympanic membrane onto the inner ear. The stapes, or innermost ossicle, has a foot plate that sits in the oval window, separating the middle ear from the inner ear.

Unlike the middle ear, the inner ear is a fluid-filled cavity with two components. One component, the cochlea, is concerned with hearing and the other, the vestibular apparatus, is concerned with balance. The cochlea is a spiral-shaped channel that has basal and apical turns. A part of the cochlea called the scala media contains the actual receptors for hearing—the organ of Corti. The major components of the receptors are the hair cells attached to the basilar membrane, which forms one of the walls of the scala media. The movements of the stapes in response to sound waves reaching the eardrum cause movements of selected areas of the basilar membrane, depending on the frequency of the sound. This, in turn, leads to movement of the hair-like processes, which triggers the electrical potentials in the auditory nerve endings. The electric signals are carried through the auditory nerve (axons derived from the spiral ganglion), which serves as the connection between the cochlea and the brain stem.

Entering the brain stem at the pontomedullary junction, the auditory nerve fibers make connections with the ventral and dorsal cochlear nuclei and subsequently with the superior olivary nucleus situated in the pons. The fibers that cross from one superior olivary nucleus to the other form the trapezoid body. Axons arising from the superior olivary nucleus travel dorsally in the lateral lemniscus. Both crossed and uncrossed fibers are present in the lateral lemniscus. The lateral lemniscus makes connections with the inferior colliculus situated in the dorsal aspect of the midbrain. This structure, in turn, sends signals to the medial geniculate body. From there they reach the auditory cortex in the temporal lobe of the brain (Figs. A2.6, A2.7).

The multiple components of the BAEP arise at the different tracts or cell stations comprising the auditory pathway. Wave I is believed to reflect activity in the auditory nerve; waves II and III, activity in the cochlear and superior olivary nuclei of the pons; and waves IV and V, activity in the lateral lemniscus and the inferior colliculi of the midbrain. Thus, I to III interpeak latency reflects conduction between auditory nerve and the pons; III to V interpeak latency reflects conduction between pontine and midbrain components of the brain stem auditory pathways.

Stimulus Parameters of the BAEP

Although the auditory stimulus may be given in the form of clicks, tone pips, or tone bursts, most often broadband clicks are used. A broadband click is one in which a wide range of audio frequencies—from 100 Hz to 8K Hz—is present so that the entire cochlea is stimulated. The clicks are generated by driving a standard audiometric ear speaker with a brief electrical pulse of 100-μs duration. The BAEPs obtained may show variations, depending on a number of stimulus parameters. These include polarity, rate, and intensity of the click.

Stimulus Polarity. Two types of clicks may be produced, one that moves the earphone diaphragm away from the eardrum (rarefaction click), and one that moves it in the opposite direction (condensation or compression click). One may use rarefaction clicks, condensation clicks, or clicks with alternating polarity for the test. Since the response characteristics can vary, depending on click polarity, the type of stimulus used should be specified in the worksheet.

Stimulus Rate. Many of the waveforms are reduced in amplitude at high rates of stimulation. The preferred stimulus rate for BAEP is 8 to 10/s. Most machines on the market are capable of rates ranging from 1 to 70/s.

Stimulus Intensity. There are many ways of defining the stimulus intensity. Most laboratories use two scales: hearing level (HL) and sensation level (SL). To establish the HL value, a number of normal persons are tested to determine the hearing threshold, or the lowest click intensity that can be heard, and the mean value is determined. This is taken as zero dBHL, and may vary from 5 to 30 dB, depending on the characteristics of the stimulator, earphone, and laboratory environment. If we assume zero dBHL to be 20 dB, then a 60-dB click has an actual intensity of 60 minus 20, or 40 dBHL, which is 40 dB above hearing level. Alternatively, the patient's own hearing threshold may be taken as the zero measure. If the hearing threshold is 30 dB and a 60-dB click is used, then the actual intensity is 60 minus 30, or 30 dBSL, which is 30 dB above sensation level. The click intensity used should be recorded in dBHL or dBSL.

During monaural testing, it is important to mask the contralateral ear to avoid recording a crossover response from inadvertent stimulation of the contralateral ear via bone conduction of the stimulus. This is particularly important when a high-intensity click is used on the side with poor hearing, and the contralateral ear happens to be normal. Usually, white noise at 30 dB below the intensity of the stimulating click is used as the masking stimulus.

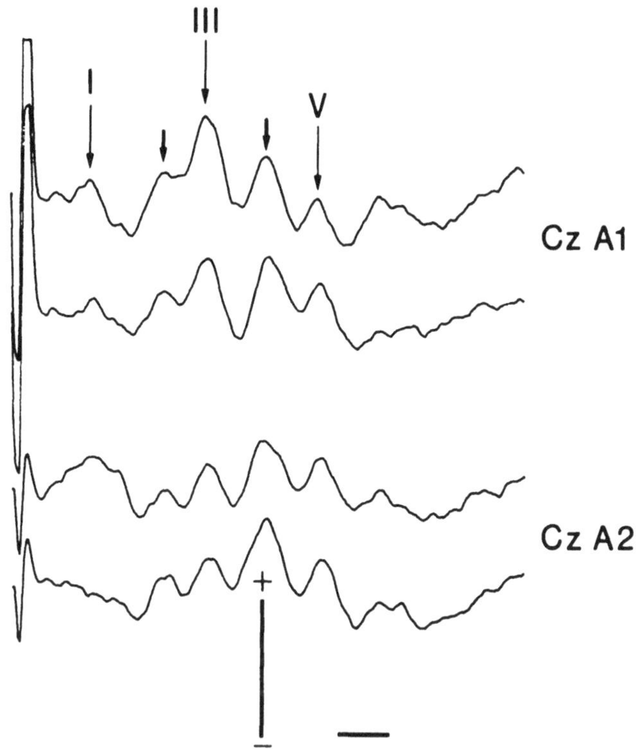

Figure 17.4. Normal brain-stem auditory-evoked potentials to left ear stimulation at 60 dbSL. Rarefaction clicks were used at a stimulation rate of 11.1/s. The bandpass is at 30 to 3,000 Hz, and 2,000 trials were averaged. The two tests run show excellent replication between the paired tracings. Arrows point to waves I to V, which have average latencies of 1.58, 3.00, 3.90, 5.12, and 6.12 ms, respectively, for the Cz-A1 derivation. Calibrations: horizontal = 1 ms, vertical = 0.5 μV.

BAEP Recording Techniques

The recording electrodes are placed on the earlobes (or mastoids) and over Cz, the vertex. The ground electrode may be placed at Fz. With a two-channel system, channel 1 should record between vertex and ipsilateral earlobe, and channel 2 between vertex and contralateral earlobe. The recommended filter settings are 30 to 100 Hz low frequency and 2,500 to 3,000 Hz high frequency. Although an analysis time of 10 ms is often used, it is better to increase it to 15 ms so that any delayed responses will not be missed. The number of trials may vary from 1,000 to 4,000, but usually 2,000 trials are adequate.

Normal BAEP

The normal BAEP typically shows five distinct peaks (Fig. 17.4) in the Cz to ipsilateral ear derivation (positivity at Cz). These are named serially I through V. Wave I is distinguished by its presence in the Cz to ipsilateral ear derivation and absence in the Cz to contralateral ear derivation. Wave V is distinguished by its prominent trough below the baseline. Waves IV and V are sometimes fused, but tend to be more distinct in the Cz to contralateral ear derivation. Wave III is normally equidistant between waves I and V and is less prominent in the Cz to contralateral ear derivation. The measurements taken are the peak latencies of waves I, III, and V and the amplitudes of waves I and V. Interpeak latency values for I to III, III to V, and I to V are calculated. The peak amplitude ratio between V and I is also estimated.

Each laboratory should have its own normative data. Values beyond three standard deviations (SD) of the mean of normal age-matched controls are considered to be definitely abnormal. It should be noted that the latency values may vary with the age and gender of the patient, apart from stimulus parameters. Published normative data (Chiappa KH, Gladstone KJ, and Young RR, 1979) suggest the following upper interpeak latency values (mean + 3 SD) at a click rate of 10/s: I to III, 2.6 ms; III to V, 2.4 ms; and I to V, 4.7 ms for a group aged 15 to 51 years. Values in excess of these are considered to be abnormal. An inter-ear I to V interpeak latency difference of more than 0.5 ms is also considered to be abnormal. The latency values are mildly prolonged in old and very young persons. Females have slightly shorter III to V and I to V interpeak latencies than males. Body temperature also influences latency of the BAEP waves.

It should be obvious that identification of the various waveforms is crucial for proper interpretation of the BAEP. Absolute latency values are of less significance than interpeak latencies. This is because changes in wave I latency that occur from cochlear or other ear disorders can prolong the latencies of the subsequent waves. Without identification of wave I, the interpretation becomes less specific. Sometimes wave I may be obscured by the stimulus artifact or by cochlear microphonics; in such cases, use of alternating clicks may be quite helpful. With alternating clicks, the cochlear potential reverses in polarity and cancels out during averaging so that wave I is easier to detect.

Abnormal BAEP

The BAEP is considered to be abnormal and suggestive of retrocochlear dysfunction when there is (1) complete loss of all waveforms (in the absence of severe middle ear or cochlear disease), (2) absence of waveforms following waves I or III, (3) abnormally prolonged interpeak latencies, or (4) abnormal inter-ear difference in the I to V interpeak latencies. A low V/I amplitude ratio is also considered to be abnormal, especially when accompanied by other abnormalities (American Electroencephalographic Society Guidelines, op. cit).

Clinical Correlation of Abnormal BAEP

Eighth-Nerve Tumor. The BAEP is a very sensitive indicator for tumors that arise from or compress the eighth nerve. In the case of an acoustic neurinoma, wave I may be absent on the side of the lesion or the I to III interpeak latency may be prolonged. The test has been found to be highly sensitive in this regard, with some studies suggesting 90% to 95% sensitivity. With a large cerebellopontine angle tumor compressing the brain stem, the III to V interpeak latency may be prolonged, often on the contralateral side. In the case of intra-axial tumors, such as brain-stem glioma, bilateral prolongation of III to V interpeak latency is the more common finding.

Demyelinating Disease. In a patient with a single episode of neurological deficit such as optic neuropathy or diplopia, it is often difficult to make a diagnosis of multiple sclerosis. Although the magnetic resonance imaging (MRI) scan has become the most useful tool for diagnosing this condition, multiple pathway dysfunction is best documented by a battery of evoked-potential tests. In this context, the BAEP is a very useful technique, particularly to document subclinical lesions. A high incidence of abnormalities has been reported in definite cases of multiple sclerosis. The abnormalities may be in the form of prolonged interpeak latencies and absence or distortion of the waveforms.

Coma. Since the BAEPs are not affected to any significant degree by metabolic derangements or by drugs, BAEP is a good test for detecting structural abnormalities of the brain stem in patients in coma. Thus, if a good wave I is present and all the subsequent waveforms are absent or disorganized, or the interpeak latencies are prolonged, one may conclude that there is some structural abnormality of the brain stem. Of course, if wave I is also absent, such a conclusion cannot be drawn because of the possibility that the clicks may not be stimulating the cochlea and triggering signals in the auditory nerve. Total absence of all waves subsequent to wave I in a patient with suspected brain death may be used as a confirmatory test for the lack of brain stem function. The BAEP also serves to distinguish between metabolic coma and coma resulting from structural lesions of the brain stem.

Apart from the above indications, the BAEP studies are very useful in assessing hearing in pediatric patients who cannot cooperate in standard audiometric testing.

Short-Latency Somatosensory-Evoked Potential (SSEP)

Recording of SSEPs may be viewed as an extension of nerve conduction studies to the central somatosensory pathways. The short-latency SSEPs are electrical responses generated in the sensory pathways normally within the first 50 ms following the stimulus.

Anatomical Basis of the SSEP

Some knowledge of the anatomy of the peripheral nerves and the somatosensory pathways in the spinal cord and above is essential for a clear understanding of the SSEPs. A large number of different types of receptors are present in the skin and other tissues that can be activated by different stimuli. However, since stimuli such as touch and pain are difficult to quantify, electrical stimulation of the nerve fibers is employed in SSEP studies. The most commonly tested nerves are the median and the posterior tibial nerves, but any peripheral nerve such as the ulnar or the common peroneal nerve may also be used. The site for stimulation of the median nerve is in front of the forearm close to the wrist, between the tendons of palmaris longus and flexor carpi radialis. The site for tibial nerve stimulation is at the medial aspect of the ankle, between the medial malleolus and the tendo Achilles. In the case of the common peroneal nerve, the stimulation site is at the neck of the fibula behind the knee. The ulnar nerve is easily stimulated on the medial aspect of the forearm, close to the wrist. Stimulation of these nerves should induce twitching of the muscles supplied by them. In the case of the median nerve, there will be movements of the thumb; for the ulnar nerve, there will be movement of the little finger. In the case of the tibial nerve, movements of the big toe will be observed.

Electrical impulses are carried through the nerve trunks, plexuses, and nerve roots to the spinal cord. For the median and ulnar nerves, the brachial plexus and cervical nerve roots are the pathways, whereas for the tibial and peroneal nerves the corresponding pathways are the lumbosacral plexus and component nerve roots. The signals travel in the dorsal columns of the spinal cord and reach the lower end of the medulla. At this point the fibers synapse, cross to the opposite side, and form the medial lemnisci. The medial lemnisci carry the impulses to the thalamus where these fibers synapse. From the thalamus, the sensory signals reach the sensory cortex through the thalamocortical fibers (see Fig. A2.10).

Electrical potentials generated in response to peripheral nerve stimulation can be recorded percutaneously over the plexuses, the spinal cord, and the sensory cortex.

Stimulus Parameters of the SSEP

As already noted, electrical stimuli are used for inducing SSEPs. The site of stimulation depends on the particular nerve under investigation. The sites for median, ulnar, peroneal, and tibial nerves have already been indicated. The cathode (negative-stimulating electrode) is placed proximal to the anode (positive-stimulating electrode). Stimulus parameters are detailed below.

Stimulus Rate. The number of electrical stimuli delivered per second determines the degree of discomfort experienced by the patient and, to some extent, the response obtained. A stimulus rate of 4 to 7/s has been suggested (American Electroencephalographic Society Guidelines, 1986).

Stimulus Duration. Duration of the electric stimulus, which is in the form of a square-wave pulse, can be varied to obtain an optimum response. With longer durations, there is more discomfort and more artifacts. With very brief durations, the response may not be adequate. Usually 0.1- to 0.2-ms pulses are used.

Stimulus Intensity. This is measured in terms of the amount of current delivered during the stimulus. The current can vary from 1 to 20 mA, depending on the amount needed to elicit a muscle twitch. A stimulus twice the sensory threshold is used as a guideline if an obvious twitch is not observed. As the intensity of the stimulus is increased, more artifacts are likely to occur and more traces will be rejected by the averager, thus prolonging the test.

When using surface stimulating electrodes, reducing the skin resistance decreases the voltage needed to deliver sufficient current. With needle electrodes placed subcutaneously, much smaller currents may be used.

Recording Parameters of the SSEP

Electrode Placement and Montages. For obvious reasons, the electrode placement varies, depending on the nerve that is stimulated. In the case of the upper-extremity nerves, electrodes are placed over the brachial plexus at Erb's point, which is 2 cm above the midpoint of the collar bone, and the cervical spine (C-2, C-5, or C-7 spine). Scalp electrodes are located over the contralateral somatosensory area, i.e., 2 cm behind C3 or C4. These placements are designated C3′ and C4′, respectively. The reference electrode may be placed over Fz. A ground electrode is located on the forearm on the side of stimulation. Channel 1 records between Erb's point on the side of stimulation and Erb's point on the contralateral side. Channel 2 records between the cervical spine electrode and Fz, while channel 3 records between C3′ or C4′ and Fz. Sometimes potentials apparently arising from the medial lemniscus may be recorded by connecting the C3′ or C4′ electrode to the contralateral Erb's point.

For lower-extremity nerves, the recording electrodes are placed over the tibial nerve in the middle of the popliteal fossa, the L-3 spine or spinous process above a line connecting the highest points of the iliac crests of the hip

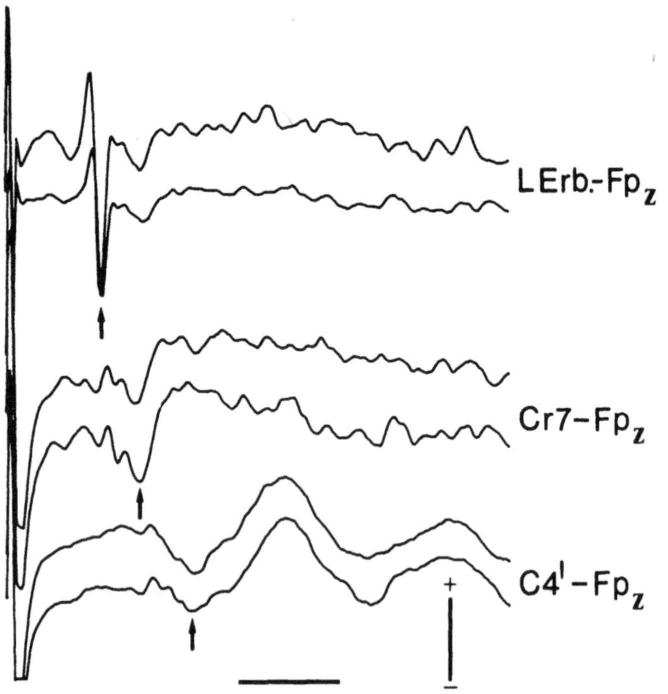

Figure 17.5. Normal somatosensory-evoked potential, left median nerve study. Stimulation rate at 6.1/s. The bandpass is at 20 to 3,000 Hz, and 500 trials were averaged. The two tests run show good replication between the paired tracings. From top to bottom, arrows point to N9, N13, and N19; their average values are 9.60, 13.20, and 19.00 ms, respectively. Calibrations: horizontal = 10 ms, vertical = 2.5 μV.

bones and the T-12 spine. Scalp electrodes are located 2 cm behind the vertex and at Fz. The former is designated Cz'. A ground electrode is placed over the calf. Channel 1 records between the electrode in the popliteal fossa and an electrode placed on the medial surface of the knee. Channel 2 records between the L-3 spine electrode and the T-12 electrode. Channel 3 records between the T-12 electrode and an additional electrode that is placed 4 cm rostrally over the spine. Finally, channel 4 records between Cz' and Fz.

Filter Settings. The recommended filter settings are 5 to 30 Hz low frequency and 2,500 to 4,000 Hz high frequency.

Analysis Time. For upper-extremity studies, a 50-ms epoch is suitable. For lower-extremity studies, 100 ms is more appropriate. This should be increased if no cortical responses are observed so as not to miss a delayed response.

Number of Trials. The number of responses averaged may vary from 500 to 2,000.

Normal SSEP

Median Nerve. Distinct potentials can be recorded over Erb's point, the cervical spine, as well as the scalp. The waves are named after their polarity—N for negative and P for positive—and a suffix, which gives the mean latency in milliseconds. Thus, the negative potential over the brachial plexus is called an N9 response, the one over the cervical spine is called N13, and the first significant negative potential seen over the cortex is designated N19 or N20. The latter is often followed by a positive potential called P22 (Fig. 17.5). The N9 potential is believed to arise from passage of the stimulus through the brachial plexus. The N13 wave is a complex response and may have more than one peak, in which case the component peaks are designated N11 and N13. N11 is believed to be due to the passage of signals through the cervical nerve roots, and N13 to signals passing through the dorsal columns and arriving and synapsing at the lower end of the medulla. The N19 potential signals the arrival of the stimuli at the sensory cortex. P14, a peak representing the passage of signals through the medial lemniscus, may be seen in farfield recording[2] between scalp and contralateral Erb's point.

Tibial Nerve. The negative potential recorded over the popliteal fossa usually has a latency of 6 to 10 ms, depending on limb length and temperature. The lumbar potential is often seen between 14 and 24 ms, again varying with limb length and temperature. The lumbar potential is the benchmark on which further calculations are made and for this reason is crucial for interpretation. The cortical response shows a positive wave at a latency of about 37 ms—the P37 response. This may be followed by a negative deflection at a latency of 45 ms, which is designated the N45 response.

It is important again to point out that each laboratory should have its own normative data. Variation of more than 3 SDs from the mean of the normal group is considered definitely abnormal. Figure 17.6 shows a normal tibial nerve study.

Abnormal SSEPs

In the case of the median nerve SSEP, the interpeak latency between N9 and N13, as well as between N13 and N20, is taken into consideration. Prolongation of the interpeak latency between N9 and N13 suggests a slowing of conduction in the cervical nerve roots and/or the cervical dorsal columns. A prolonged interpeak latency between the N13 and N20 responses would suggest delayed conduction in the pathway between the medulla and the sensory cortex. Absence of cortical potentials in the presence of a normal N13 response suggests a lesion involving the medial lemniscus, the thalamocortical projections, and/or the sensory cortex. Excessive left-right interpeak latency

[2] Farfield recording is a term used to indicate that the source of the recorded potential is a field remote from the recording site.

differences are also used as criteria to detect abnormalities. Since absolute latency values are of much less significance than interpeak latency values, absence of the N9 response makes interpretation very difficult. If none of the potentials are obtained, a troubleshooting procedure should be carried out to make sure that the stimuli are actually being given, that the recording electrodes are properly applied, and that the electrodes are indeed connected to the evoked-potential machine.

In the case of the tibial nerve SSEPs, the interpeak latency between the spinal and the cortical potential is taken as a criterion for abnormality. Prolongation of the central conduction time would suggest a lesion anywhere in the dorsal columns of the spinal cord, the medial lemniscus, or the thalamocortical projections. When central conduction time is found to be prolonged, a median nerve study should also be done to differentiate lesions below the cervical cord from those above the cervical cord. In the case of the latter, both median and tibial nerve SSEPs will be abnormal. It is not uncommon to find absent lumbar potentials and intact cortical potentials. The absolute latency values of the cortical potentials are not very reliable as criteria for abnormality because too many variables are involved in conduction through the peripheral nerve. For this reason, a technician should strive to obtain a lumbar potential. This may involve special effort in getting the patient to relax. Sometimes administration of sedatives becomes necessary.

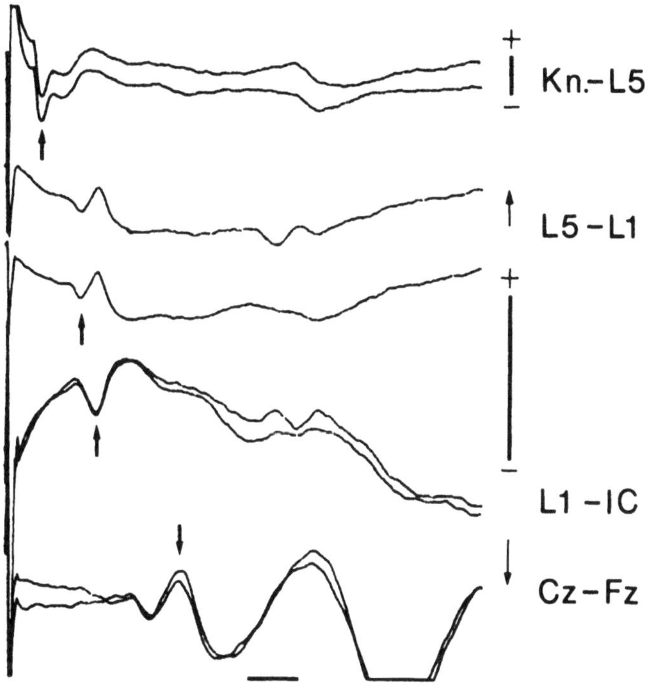

Figure 17.6. Normal somatosensory-evoked potential, left tibial nerve study. Stimulation rate at 6.1/s. The bandpass is at 5 to 1,500 Hz, and 500 trials were averaged. The two tests run show excellent replication between the paired tracings. From top to bottom, the arrows point to potentials at the popliteal fossa, L-5, and L-1, and to the cortical potential designated P37. Their average values are 7.6, 15.6, 19.6, and 37.6 ms, respectively. Calibrations: horizontal = 10 ms, vertical = 2.5 μV.

Clinical Correlates of the SSEP

The major disorders for which SSEPs are of value include multiple sclerosis, spinal cord lesions (traumatic, compressive, and other etiology), and lesions of brachial and lumbosacral plexuses. The SSEP is also useful in cases of head trauma and in the determination of brain death.

The somatosensory pathways are some of the longest myelinated pathways in the body. For this reason, the changc of detecting subclinical demyelinating lesions is high if both upper- and lower-extremity nerves are studied. In patients diagnosed as having definite multiple sclerosis, bilateral upper- and lower-extremity SSEP studies may detect abnormalities in as high as 90% of the cases.

In spinal cord lesions the SSEPs are useful in localizing the lesion and in assessing its extent; they are also useful for prognosis if serial studies are employed. These potentials are monitored in spinal cord trauma and other forms of spinal surgery. Combining tibial nerve and median nerve studies results in better localization of lesions.

Since it is difficult to stimulate the plexus directly, evaluation of SSEPs by stimulation of peripheral nerves, and recording their transit through the nerve roots and spinal cord, may be helpful in documenting lesions of the brachial and lumbosacral plexuses.

In patients with head injury, evoked potentials—both SSEPs and BAEPs—may give prognostic indications. The evaluation of patients with brain death is also made easier by studying SSEPs as they are not significantly altered by drugs and metabolic problems. This is an advantage over the EEG.

References

American Electroencephalographic Society Guidelines in EEG and Evoked Potentials. *J Clin Neurophysiol* 1986; 3 (suppl 1): 54–70.

American Electroencephalographic Society Guidelines in EEG and Evoked Potentials, op. cit, pp 71–79.

Chiappa KH, Gladstone KJ, Young RR: Brain stem auditory evoked responses: Studies of waveform variations in 50 normal human subjects. *Arch Neurol* 1979; 36:81–87.

Chapter 18
Seizure Monitoring and Ambulatory EEGs

Infrequent abnormalities in a person's EEG, particularly those associated with epilepsy, pose severe problems, as they may not be identified in a routine laboratory study. To address this problem, a special technique known as seizure monitoring is increasingly being used at many hospitals and epilepsy centers. In some EEG laboratories, seizure monitoring is used routinely with patients suspected of having seizures. Additionally, some laboratories have begun doing prolonged monitoring on ambulatory patients using a system in which the EEGs are recorded on a portable tape recorder worn by the patient. In view of these recent developments, a discussion of these topics is warranted in a general text dealing with the technique and practice of clinical electroencephalography. We begin with a discussion of seizure monitoring.

Seizure Monitoring

Seizure monitoring provides for simultaneous recording of a patient's EEG and behavior. This is made possible by the use of a split-half TV screen. One-half of the screen shows the patient on camera, while the other half displays his/her continuous EEG tracings. The EEGs appear on the screen of the TV monitor in much the same way as they are seen on the chart paper as it comes out of the EEG machine. In most modern installations, the patient may be seen on the screen either in a closeup or in full figure lying in bed, or both simultaneously. As a result, subtle changes in the patient's behavior may be observed and correlated with events taking place in the EEG. Since both halves of the screen are recorded on magnetic tape, the entire sequence may be played back later on the TV monitor for review and evaluation.

System Design

A variety of different seizure-monitoring systems are commercially available or may be custom designed. Many systems are quite complex and sophisticated; in the case of the custom designs, the ultimate capability possible is limited only by the cost.

A simple, basic system consists of:

1. TV camera with manual zoom feature
2. electronics for displaying the video on one half of the TV monitor screen
3. patient hookup and electronics—sometimes called the signal conditioner—for filtering and amplifying the EEG voltages to IRIG levels
4. electronics for converting the EEG voltages to video signals suitable for display on the other half of the TV monitor; these electronics include circuits for generating moving grid lines on the TV monitor like the grid lines on EEG chart paper
5. digital clock that displays real time on the TV monitor
6. video cassette recorder for recording live video, EEG video, and real time on magnetic tape.

In using this simple basic system, the EEG technologist attaches electrodes and connects the patient in the usual way. Rather than observing the EEGs on the moving chart of the EEG machine, however, the technologist sees them on one half of the TV monitor screen. The patient may be observed either directly or via the TV monitor. The technician watches for electrographic or behavioral evidences of a seizure and notes the times of occurrence from the digital clock on the screen of the TV monitor. These times are then used to determine which portions of the record will be played back by the person interpreting the seizure monitoring test.

Patient Hookup

Three different kinds of hookups are available. The simplest and least expensive as far as equipment costs are concerned makes use of a standard EEG machine. The patient is connected to the electrode board in the routine manner, and the EEG machine filters and amplifies the EEG voltages in the usual way. But the penmotors are not activated and no chart paper is run through the machine. Instead, the IRIG outputs of the machine are connected to the electronics of the seizure-monitoring system.

This type of system has several disadvantages. First, as seizure monitoring can involve many hours of recording, it ties up an expensive EEG machine for long periods of time. Second, the EEG machine is bulky piece of equipment and, together with the basic seizure-monitoring instrument, the total system occupies a good deal of space. Finally, during a seizure, artifacts associated with movement of electrode wires and of the cable running from the electrode board to the EEG console can be quite severe and capable of obliterating the brain electrical activity. The first two aforementioned problems can be eliminated, and the last can be mitigated by the use of a telemetry system, either of the cable or wireless type.

A cable telemetry patient hookup consists of a small box about the size of a bar of soap with jacks for plugging in the EEG leads. This box, which contains the signal conditioner or electronics for amplifying and filtering the EEG voltages, is located on the bed close to the patient. After being amplified, the EEG voltages are combined in a special way and used to modulate a high-frequency carrier signal that is generated within the small box. In other words, the information contained in all of the EEG channels is compressed into a single channel containing a frequency-modulated voltage. This voltage is fed by a thin, flexible cable to the rest of the seizure-monitoring system.

A wireless telemetry system is identical to the cable system except that there is no cable connecting the signal conditioner with the rest of the system. Instead, the signal conditioner contains a radio transmitter and antenna for beaming the high-frequency signal to a nearby receiving antenna located on the seizure-monitoring console. The wireless telemetry patient hookup can be mounted close to the patient's head or even on the patient's body. This means that short lead wires may be used on the electrodes, and these can be joined together into a harness. With such a system, artifacts produced by movements of the patient during a seizure can be greatly reduced.

Operation

Although operating the seizure-monitoring system is not unlike running an ordinary EEG, the EEG technologist quickly discovers that the absence of a hard copy of the tracings is a distinct handicap. It is difficult to appreciate at every moment what is happening on all eight or 16 channels of a 14-in. TV monitor. With no hard copy available, it is impossible to confirm an observation by looking back quickly at, say, the previous 10 seconds of the record. Moreover, if the technician should fail to correctly take note of the time when evidence of a seizure occurred, the entire recording may have to be played back to retrieve the data.

To help with this problem, some seizure-monitoring systems incorporate an alarm button in the equipment. This button is given to the patient with the instruction to press it whenever he/she feels that a seizure may be coming on. The button press causes the time registered on the digital clock to be placed into storage. This time can then be retrieved later by the person interpreting the record and the relevant portion of the record replayed.

As seizure monitoring may go on for a number of hours, electrodes attached to the scalp by means of paste are likely to dry out and require reapplication. Because of this, collodion is frequently used instead of paste to connect the electrodes. The collodion-attached electrodes also have better adhesion properties; this is an advantage because they are less likely to be pulled loose from the scalp by the movements accompanying a seizure.

One serious shortcoming of this simple, basic seizure-monitoring system that has already been alluded to is that no permanent, hard copy of the EEG tracings is available. The EEG tracings, of course, are recorded on videotape cassette. But in this form, they are incompatible with and cannot be played back on the EEG machine. To overcome this limitation, some seizure-monitoring systems provide a means of converting the EEGs back to a signal that can be recorded on the EEG machine. Other systems have an analog tape recorder as well as a videotape machine for recording the EEG voltages directly on magnetic tape. In the case of the latter-type system with an EEG machine used as the patient hookup, the IRIG outputs from the machine are fed into the analog tape recorder at the same time that they are fed into the electronics of the seizure-monitoring system. On the other hand, seizure-monitoring systems like the one described using cable or wireless telemetry do not possess these EEG voltages in a form suitable for recording directly on analog tape. As already mentioned, they are compressed into a single, frequency-modulated video signal. To record them using an analog tape recorder, the video signal needs first to be demodulated and the various EEG voltages reformatted. Once recorded on tape in analog form, the EEG voltages can be played back directly onto the chart paper by connecting the tape recorder to the IRIG inputs of the EEG machine.

Time Synchronization of Data

To synchronize the EEGs recorded on analog tape and played back on the EEG machine with the video signal and the events displayed on the TV monitor, real time must be recorded on both videotape and analog tape. We already noted that a digital clock is part of the video display and that the time shown (to the closest second) is recorded continuously on the videotape. To synchronize the analog data with the video data, a time code is recorded on the analog tape along with the EEGs. The code is in binary form and is referred to as binary-coded decimal or BCD. This code may be written out every 5 or 10 seconds on one of the channels of the EEG machine as the analog tape is played back. The code is learned quickly and is easy to read. In this way, a specific section of the hard-copy tracing can be precisely correlated with the patient's behavior as seen on the TV monitor.

Interpretation

Although the entire recording session consisting of both the patient's EEGs and his/her behavior on camera can be played back on the video display, this is rarely done in practice. Because a seizure-monitoring session may involve many hours of recording, the record is sampled instead. Sampling is based upon the observations made of the patient's behavior by the EEG technician. As mentioned earlier, the technician notes the times when behavioral or electrographic evidences of seizures are observed. Recordings from those intervals are then brought up on the screen for review. In addition, hard copies of the EEGs for these times can be made to assist the electroencephalographer in the interpretation.

If an alarm button is used, the recordings taken at the times it was pressed by the patient are also reviewed. For obvious reasons, data for review here are limited to intervals when the patient is awake. Finally, some samples taken at random from different portions of the recording may be played back.

The foregoing highlights the vitally important role played by the EEG technician in seizure monitoring. A large block of the technician's time, of course, is required. He or she needs to be alert, observant, and quick to report upon the patient's behavior. Barring this, the procedure, which is inherently expensive, can become excessively costly and impractical. Seizure monitoring has other disadvantages besides being expensive. It frequently requires hospitalization and severely restricts the patient's mobility, sometimes for long periods of time. For these reasons, ambulatory EEG recording was developed. In this technique, the patient's EEGs are recorded on a cassette recorder that is carried by the patient.

Ambulatory EEG Monitoring

The idea of using a portable tape recorder to monitor physiological data on mobile patients for prolonged periods of time was first introduced by N.J. Holter in 1961. As is well known, this method has been used successfully in the ECG evaluation of cardiac arrhythmias. It has the advantage of being able to assess cardiac functioning over a wide range of normal activities while the patient is awake as well as during sleep. With the introduction of the Holter monitor, the potential applicability of ambulatory-recording equipment to EEG monitoring became obvious. Ambulatory monitoring affords the patient increased mobility and makes hospitalization unnecessary, thereby freeing the patient from the unfamiliar, artifical environment of the hospital. This is especially important in the case of sleep disorders. In addition, ambulatory monitoring can provide long-term recordings without the continuous supervision of an EEG technician. However, being limited to but a single channel, the early recorders were not immediately applicable to EEG monitoring.

The development of a solid-state, on-head preamplifier chip in the 1970s made three- and four-channel EEG recording feasible. Coupled with this was the introduction of a rapid video/audio playback device that made it possible to play back data on a video display at speeds as fast as 60 times real time. At such fast replay speeds, 24 hours of recording could be reviewed by the electroencephalographer in 24 minutes. Moreover, simultaneous audio reproduction of one data channel was also provided so that certain prominent auditory cues could be used to detect and identify various EEG features and physiological artifacts in the recording. These technological developments resulted in the appearance in 1983 of a cassette system capable of recording eight channels of continuous EEGs. As this is being written, 16-channel systems are becoming commercially available.

System Design

Eight-channel ambulatory EEG systems, or A/EEG systems as they are called, appear to be the current standard. The systems are battery operated and can run continuously for 24 hours. The preamplifier is secured to the scalp along with the electrodes, and the preamplifier output wires are fed to the recorder, a rectangular box weighing about 1.5 lb and worn on a belt or strap by the patient. The eight channels of EEGs are recorded on ⅛- or ¼-in. cassette tape. Technical details need not concern us here; they can be obtained by consulting more advanced, specialized texts.

Playback units of some systems provide a variety of high-technology features. Some of the features available are automatic search to a specified time in the recording; a limited memory that stores portions of the record so that segments occurring before and after the segment being viewed on the monitor can also be viewed without having to rerun the tape; and alphanumeric registry of gain and filter settings. The user should consult the instruction manual that comes with the particular equipment for operational details.

Patient Hookup

Collodion electrodes must be used. They should be applied with the greatest care, as they need to give stable recordings that are free of electrode artifacts for up to 24 hours. Electrode impedances should be 3K ohms or less; impedances should be tested at the beginning and at the end of the recording. It is good practice to keep a record of the impedance of each electrode. Knowing that electrode impedance was excessively high at the end may help to explain the presence of artifacts observed in the EEGs when the tape is reviewed.

Since electrode problems or failures go undetected until after the monitoring session is over, referential recording is not employed, and separate electrodes frequently are used for each input grid instead of bipolar linkages. This provides a greater margin of safety, as only one channel becomes disabled rather than two channels if a single electrode fails. In hooking up the patient, it is essential that the loop area between the electrode wires of a pair be kept to a minimum. Unless this is done, electrical fields present in the patient's surroundings may cause significant electrical currents to be induced in the loops as the result of their being moved through the field when the patient is in motion. Shoes with rubber soles should be avoided, especially when the humidity is very low, to prevent large electrical charges from building up on the patient's body while he/she is walking. If electrical charges produced in this way present a significant source of artifacts, the problem can be alleviated by having the patient wear carbon-soled shoes of the type worn in operating rooms.

Operation

After electrodes have been applied and tested, and a calibration has been run, it is good practice to make a few brief recordings before letting the patient leave the laboratory. A recording while the patient is at rest and recordings obtained while he or she is engaged in various activities that produce artifacts—such as blinking, moving the eyes, chewing, swallowing, coughing, sniffing, and talking—are recommended. These recordings may be helpful in differentiating EEG abnormalities from artifacts when the tape is reviewed. Before letting the patient go, ask him/her to keep a diary while the monitoring is going on. Instruct the patient to record the exact time of day that various suspicious episodes occur and to briefly describe the episodes. Various activities capable of generating rhythmic artifacts such as walking, eating, brushing teeth, talking, scratching, and the like should be entered in the diary as well, along with their times. If the diary is carefully maintained, it can be helpful in identifying artifacts and in selecting samples for review when the tape is interpreted by the electroencephalographer.

When the monitoring session is over and the patient has returned to the laboratory, a brief EEG should be recorded and printed out before the electrodes are removed. This is essential to confirm that the electrodes have, indeed, been functioning satisfactorily. Finally, a postcalibration should be run to verify that the system is operating properly.

Interpretation

We take up here only the essentials of interpretation. A/EEG is a rapidly expanding area with new equipment and techniques under development that are expected to simplify interpretation. The periodical literature should be consulted for details and further information.

With 24 hours of EEG recording obtainable from eight channels, there are clearly more data available than can be assimilated successfully even when using rapid playback. Nevertheless, rapid playback is the only practical approach to this problem. When searching for ictal events, review speeds of 40 to 60 times real time are used in video scanning of the tape. Slower rates of 20 to 40 times real time are employed in the detection of isolated interictal discharges and focal events. Simultaneous monitoring of the audio output of the EEG channels during video scanning has been found to be useful, as seizures, interictal discharges, normal transients, and various artifacts all have characteristic sounds that can aid in their detection. Verification of events discovered in this way is best carried out at video display rates of 30 mm/s. If the playback unit is IRIG compatible, hard copies of selected segments from the taped EEGs can be made by connecting the playback unit to the IRIG inputs of a standard EEG machine.

The differentiation of artifacts from normal activity and from true EEG abnormalities is a major problem in interpreting A/EEGs. Active wakefulness is filled with a superabundance of various artifacts. In contrast to conventional EEGs, artifacts in A/EEGs confront the electroencephalographer without the benefit of an experienced technologist's observational information concerning the patient's behavior. To be sure, the patient's diary can be useful in this regard; but the diary usually lists only major events such as eating a meal, going for a walk, jogging, and the like. This problem is somewhat mitigated by the fact that epileptiform abnormalities in the EEG occur more frequently during sleep—mostly during stage I and II sleep—when artifacts are less numerous than during wakefulness. Nevertheless, as is the case with any newly developed procedure, the ultimate value of the A/EEG technique rests upon its proven clinical usefulness over the long term.

Chapter 19

Clinical Use of Brain Electrical Activity Mapping

The past 10 years have witnessed rapid growth in the use of signal analysis techniques by clinical neurophysiologists. Prominent among these emerging techniques is brain electrical activity mapping (BEAM)[1], defined as the topographic analysis of scalp-recorded EEG or evoked-potential (EP) data (Duffy FH, Burchfiel JL, and Lombroso CT, 1979; Duffy FH, 1986). Technically, topography refers to the analysis of spatial dimensions, whereas topographic mapping is a cartographic science. A better understanding of the current usage of BEAM is to define the technique as incorporating topographic mapping (spatial) and analysis (temporal, statistical) of electrical activity recorded from the scalp.

A good example of topographic mapping is provided by the colorful daily weather maps shown in national newspapers. Regional average temperature gradation is depicted on a national map consisting of a discrete outline and based on discrete points for which measured data are taken. Interpolation between these points using a contouring algorithm provides a basis for drawing isothermal contour lines. The spaces can then be filled in with a color range representing temperature range differences. BEAM is based on very similar cartographic elements and interpolation techniques. Instead of temperature, however, parameters derived from brain electrical activity are mapped. Figure 19.1 (see color insert page 229) provides a standard example of a topographic map based on recorded EEG data.

Why Brain Electrical Activity Mapping?

One may ask, "Why do we need brain electrical mapping? Aren't traditional EEG and EP recording established as the standard practice?" To understand the contribution of topographic analysis, one must know what underlies the use of EEG and EP in clinical diagnosis.

There are basically two elements of EEG. First, the neurologist wishes to distinguish real discontinuities in brain electrical activity, such as epileptic "spikes," from artifactual discontinuities such as electrode "pops" and eye blink. With training and experience, the task of distinguishing these discontinuities becomes relatively easy. However, there is a second element of EEG, the analysis of the background EEG, i.e., the electrical activity not related to discontinuity. This is a more difficult task, as the process consists of a series of complex steps. First, the electroencephalographer must mentally perform a decomposition of the electrical activity recorded at each channel into its spectral content of frequency (delta, theta, alpha, beta). Second, the lengthy EEG record is analyzed for continuity or consistency of electrical activity over time (temporal analysis). Third, the electroencephalographer must create a mental map of the spatial distribution and trajectory of the brain waves. These three kinds of analyses—spectral, temporal, and spatial—must all, in turn, be analyzed in terms of what is normal and what is abnormality. That is, the electroencephalographer must also perform a statistical analysis. This complex process probably represents a major factor in what has been considered the failure of EEG to achieve its full potential. It also points to the potential contribution of modern computer technologies and statistical paradigms.

There are similar problems in the analysis of data based on EPs, i.e., data based on the brain's electrical activity responses to external stimulation. The demonstration of signal averaging in the 1950s (Dawson GD, 1950) engendered an optimism that long-latency EPs would go beyond the diagnostic power of EEG in analyzing brain function, but these long-latency waves proved to be ex-

---

[1] BEAM is a registered trademark of the Nicolet Biomedical Instrument Company.

ceptionally difficult to analyze in terms of latency, locus of origin, spatial extent, and interaction of multiple waves (Callaway E, 1969; Chiappa KH, 1983; Jeffreys DA, and Axford JG, 1972) by unaided visual inspection of the polygraphic record.

The Use of Topographic Analysis in Interpretation of EEG and EP

To assist clinical appraisal of such data, Duffy and colleagues began studies of BEAM (Duffy, Burchfiel, and Lombroso, op cit; Duffy FH, 1982; Duffy FH, Bartels PH, and Burchfiel JL, 1981). It was believed that the major limiting factor in extraction of meaningful data from EEG and EP was the massive amounts of information contained in the ostensibly simple polygraphic tracings. Figure 19.2 (see page 229) provides the paradigm used to resynthesize the spatial and temporal data gathered from EP recordings. Resultant topographic images are displayed on a computer-based video monitor using a colored "gray-scale." To capture temporal change, these images can be displayed sequentially at 4-ms intervals, creating a cartooning effect for observing the spatial range, trajectory, and latency of the EP waveforms.

Topographic Mapping Algorithms and Spatial Analysis

As with the isothermal maps described in the introduction, the paradigm for mapping brain electrical activity must include an algorithm for interpolating values to fill in the scalp areas between the recording electrodes used in performing a topographic analysis. We use a three-point, three-dimensional linear interpolation, creating values for unknown locations based on the three nearest electrode sites. It may be noted that there are numerous approaches to interpolation among current practitioners (e.g., refer to Duffy, op cit, 1986) and that no optimal paradigm has yet been agreed upon. Keeping in mind that all interpolation techniques are approximations, we have adopted the three-point paradigm as providing images that are smooth and biological in appearance when recorded data are real, but are dramatically and reliably discontinuous when electrode artifact occurs. A valuable test of reliability on a given interpolation algorithm is to compare interpolated values with real measured values.

The spectral content of brain electrical activity provides crucial data on the normal and/or abnormal functioning of the various regions of the brain. The electrical signal recorded at a given electrode site can be decomposed into its constituent frequency components: delta (<4 Hz), theta (4 to 8 Hz), alpha (8 to 13 Hz), and beta (>13 Hz). Such decomposition may be compared to listening to an orchestral piece by a symphony orchestra. The overall musical sound may be broken into the contributions of the bass, baritone, tenor, and soprano notes. A means of tracking the separate elements of brain electrical activity is depicted in Fig. 19.3 (see page 230), where a histogram is formed based on the separate frequency components.

However, simple viewing of a series of maps to analyze spectral content does not enhance their utility. In our experience, analysis of spectral data is best accomplished by (1) the creation of a topographic image of spectral content averaged over a user-selected period of time and (2) the creation of a second map for comparison, which provides the coefficient of variation—a statistical measure of standard deviation from the mean spectral values. This comparison readily isolates values in areas that unexpectedly deviate from a relatively uniform visual image and, in doing so, alerts the clinician to an area of potential concern.

As with analysis of EEG data, the use of topographic mapping of EP data provides for spatial analysis directly by viewing each map of the data. Unlike EEG, temporal analysis of EP data is best accomplished by viewing in progression a series of maps ("cartooning"). Our experience has not found spectral analysis of EP data to be clinically productive.

Significance Probability Mapping

Early on in the development of BEAM, it became clear that normal subjects often demonstrated a degree of asymmetry or focality of topographic brain-wave distributions. The question often became whether or not an obvious focality or asymmetry constituted a clinical abnormality or whether the degree of asymmetry could be explained by normal variation. In 1981, in conjunction with Dr. Peter Bartels from the University of Arizona, a technique known as significance probability mapping (SPM) was developed (Duffy, Bartels, and Burchfiel, op cit; Bartels PH, Subach JA, 1976). In this process, a single subject's topographic image can be compared with that of a control or reference data set. This results in a new image in which the original data are replaced by the delineation of individual deviation from the collected data on normal subjects. Essentially, the subject's data are replaced by their Z transform, thus displaying an image of standard deviation from the norm. The SPM process is depicted graphically in Fig. 19.4A (see page 230). This fulfills the final and complicated step in the clinical evaluation of EEG and EP data, namely, the delineation of regional abnormality. Clinically, the technique of SPM has proved singularly valuable in the diagnostic delineation of abnormalities in clinical subjects.

Further, for research applications, one may wish to compare BEAM data for two populations, a research or experimental group and a control group. This may be accomplished via a comparable SPM process using Student's *t* statistic (Fig. 19.4B).

Tips for the Conduct of Successful BEAM Studies

Quantified electroencephalography requires greater care in electrode placement, artifact management, and state control than standard EEG. Whereas electroencephalographers may readily "accommodate" obvious errors or faults of recording, computers do not. Accordingly, all electrode locations must be carefully measured and electrode impedances checked before, during, and at the end of studies to ensure good placement, low impedance, and consistent contact. Impedance asymmetries will produce unusual and asymmetrical changes in spectral analysis of signals from the offending electrodes. Whereas these signals may appear normal on paper, the higher sensitivity of spectral analysis causes artifact to stand out. In the course of a run, if a single electrode "pop" is observed, that run must be discarded, the electrode repaired, and the run repeated. Subjects should generally be studied in the upright position to assist in maintaining alertness. The chairs used must be adjustable so as to minimize truncal muscle tone and thereby reduce muscle artifact. All clinical studies should consist of a minimum of five study conditions. These are eyes open (alert and awake), eyes closed (alert and awake), drowsy to sleep, the flash visual-evoked response (VER), and the click auditory-evoked response (AER).

The somatosensory-evoked response is useful, but primarily for the detection of abnormalities in specific cortical pathways. It is not useful, however, as a screening procedure. The pattern-reversal VER produces activity that is largely occipital and is therefore of limited value in mapping studies. On the other hand, when detailed information is required about the occipital cortex, or when the standard pattern-reversal VER is morphologically complex, multiple-electrode-mapping studies can sometimes be helpful.

In the eyes-open state, subjects should be seated comfortably, and a visual fixation target should be placed in a comfortable position for viewing so as to minimize frontalis muscle tone. During the recording session the subject is instructed to look at the object and not blink, until no longer able to suppress blinking. At that point the subject is instructed to blink as often as desired and, when comfortable again, the run resumes. This procedure is referred to as the "blink holiday" technique. Some subjects prefer to be instructed as to when they should blink and others prefer to demonstrate the need to blink very vividly.

Eye movements and even blinks are surprisingly more difficult to control during eye closure than during the eyes-open state. Upon closing the eyes, a fully awake normal subject may initiate obvious blinking even though the lids are shut. Allowing the patient to "relax" so that the blinking ceases is often tantamount to allowing the subject to fall asleep or at least to become drowsy. The best strategy, here, is to place very lightly applied gauze pads over the closed lids. This gives feedback to the patient; he will sense his eyelashes brushing against the pad. If absolutely necessary, a technologist or the patient can press upon the eyes to prevent eye movement. Care must be taken, of course, not to introduce 50- or 60-Hz interference at this point.

In analyzing the eyes-open and eyes-closed states after digitization on the computer display screen, eye movements (both vertical and horizontal) should be identified according to their appropriate signals in the Fp_1 and Fp_2 electrodes and in the eye-artifact electrodes placed below the eye and lateral to both eyes. In addition to frank blinks and frank horizontal saccades, some subjects have intermittent rhythmic eye movements. By noting phase reversals in the recordings one can clearly demonstrate eye movement. The technologist should be trained to eliminate eye movement artifact but not randomly eliminate all frontal slow activity. The distinction is difficult, but often crucial for mapping studies.

Unfortunately, in our experience even the most experienced observer cannot eliminate all vertical or horizontal eye movement. A crescent of frontal delta activity can be seen on the subsequent maps when small amounts of vertical eye movement remain. In normal subjects with very low background delta, a surprisingly prominent amount of eye-induced delta may be seen.

The flash VER is ordinarily performed with supramaximal stimulus intensity. For the commonly-used Grass photic stimulator, intensity settings of 8 or 16 are used and the strobe is placed within 25 cm of and directly in front of the face. Such high intensities will often induce involuntary blinks. Conversely, placement over the eyes of a thin film of transparent plastic (Saran® Wrap) secured peripherally with transparent tape will often minimize eye blink, while allowing light to pass through. Recording equipment should be adjusted to eliminate movement artifact and random blink on the basis of an "over voltage" or threshold voltage criteria, which may be customized for each patient. On the other hand, some subjects will blink with every flash; in those instances it is best to allow all recordings to pass through. One, of course, must then interpret frontal positivities and negativities on the resulting maps with caution.

The AER should be generated by supra-maximal stimuli via earphones. Commonly, 50-ms tone pips at 92-dB sound pressure level are used. Normal-appearing long-latency AERs will be seen even in the moderately hearing-impaired subjects.

Crucial to good VER and AER recordings is the necessity to maintain the fully alert state. For this reason, the classic EEG must be collected simultaneously with these recordings and the run interrupted whenever drowsiness is detected. Often we find that the evoked-response method is so conducive to drowsiness that it is wise to interrupt a run and actually record drowsiness, taking advantage of the sleep-inducing effects of repetitive stimulation. Both the VER and AER morphologies are extremely sensitive to drowsiness, which may produce alterations of vertex wave activity, either increasing or decreasing amplitude and increasing latency. On subsequent topographic maps, drowsiness appears as abnormal but symmetrical central vertex waves.

It goes without saying that drowsiness must be carefully controlled in the eyes-open and eyes-closed states as well. Our experience indicates that first signs of drowsiness can often be seen in the sagittal midline electrodes with the appearance of some paroxysmal theta. The finding of slow horizontal eye movements constitutes an inadequate criterion, since subjects are almost asleep at this point and they will show marked changes in spectral amplitude in the slow ranges.

The slowing due to encephalopathy can be distinguished from the slowing induced by sleep, based on the topographic distribution. Slowing related to loss of attention, fatigue, or drowsiness is largely maximal in the central vertex region whereas encephalopathic slowing is more uniform and involves more the temporal lobes. Drowsiness also accentuates "time locking" of alpha, which can seriously confound the late portions of an evoked-response recording.

A clinician reading a topographic analysis can detect poor state control by the appearance of excessive central theta during the waking state, abnormal central vertex wave activity, especially in the AER, and large amounts of time-locked alpha in the VER.

The use of control subjects for comparison is absolutely mandatory, since it is impossible to maintain standards in one's own mind. It is, of course, crucial that equipment used for the gathering of control population data be electrically and functionally identical to that used for the study of patients. Subtle differences in amplifier characteristics can produce systematic abnormalities. It is also crucial that the control population be stratified by age since the developmental curve of the EEG rhythms is rather nonlinear and seldom stable for more than a decade. Indeed, norms are needed for every few years of childhood and for every few weeks of early infancy.

It is also imperative that normative data be provided for the artifact electrodes as well. It is very difficult in a topographic map to distinguish real frontal slowing from that induced by eye movement. However, if comparable increases of delta are seen on the eye-movement artifact channel then the probability is high that the slow activity is induced by the eyes. Conversely, if slowing is seen only over the frontal region and not in the eye channels then real slowing of brain origin is inferred. The analogous situation may be seen for increased beta activity over the temporal lobe and increased beta in a temporal artifact channel. Beta-like activity is commonly produced by muscle; it is mandatory, therefore, that no simultaneous changes be seen in the artifact channel if one wishes to declare that increased beta over the temporal region is of brain origin.

In the course of studies it is quite common for small regional abnormalities to be seen that are clearly not artifactual but seem to have no clinical correlation. Before these regions can be considered electrophysiologically aberrant, it must be demonstrated that they reappear when the patient is restudied. Since this happens so frequently, it is now recommended that all studies be repeated a minimum of three times and the results compared. Thus, a small region of increased theta or an EP abnormality in one limited region becomes much more convincing if seen three times in a row than if noted only once.

Finally, it is important to emphasize that the neurologist should not be overreliant on the SPM. This map demonstrates only electrophysiological deviation from the normal data base. Such differences can occur for many reasons besides pathology, e.g., eye blink, muscle artifact, state change, electrode artifact, inappropriate control groups, etc. It is incumbent upon the reader to be sure that there are no trivial explanations for statistically replicable deviations from normal.

The evaluation of drowsiness is extremely difficult for there is no true "standard" state, and it is virtually impossible to prepare a normative data base for the drowsy state. Under those circumstances, relative comparisons to the eyes-closed state are often useful in picking up greater deviations from normal than one might ordinarily expect.

The recommended sequence for evaluation of a topographic study is (1) evaluate the EEG in the normal manner for clinically important information, (2) reevaluate the EEG looking for artifact and drowsiness or state change abnormality, (3) evaluate the spectral data from the EEG, and (4) evaluate the VER and AER data. Spectral data are best understood by developing a hypothesis from the evaluation of the EEG and confirming it by the spectral data. For example, the finding of increased slowing by spectral analysis is much more meaningful if the reader has observed the fully alert EEG state than if the subject was actually allowed to become drowsy. Both VER and AER data are analyzed keeping in mind the state of the subject during data collection. The EEG taken during EP recording is particularly important to evaluate. Changes in vertex wave activity in the evoked response, especially the AER, are somewhat analogous to EEG slow activity. Vertex wave changes may be seen in drowsiness, in response to medications, and from pathological influences. To decide whether a vertex wave abnormality is clinically significant, one must be sure that the subject was not drowsy and that there were no medications that might have affected central nervous system function. Vertex wave change can be particularly valuable in the early detection of encephalopathic change but only if care is taken in interpretation.

It is clear that the conduct of a BEAM study places greater demands on the conduct of the technologist, the patient, and the reader than ordinary EEG studies.

The Application of BEAM to Clinical Practice

The utility of BEAM and topographic analysis in clinical practice lies in its increased sensitivity over traditional EEG and EP analysis and its increased objectivity in establishing the presence or absence of brain-related disease and/or disability. As such, it is an added weapon in the clinician's arsenal of diagnostic tools. We consistently caution the users of topographic techniques that BEAM does not provide a "stand alone" diagnosis, but rather one of several inputs into comprehensive clinical analysis. We feel, however, that it can provide a vital and, in many cases, unique contribution to the diagnostic process.

Following are three case studies that demonstrate the efficacy of the BEAM technique in clinical practice.

Case Study 1: Presenile Dementia in an Adult Male

A 64-year-old successful businessman complained to his internist of failing memory. Neuropsychological evaluation revealed a superior overall performance level with full scale IQ of 148. Tests of memory demonstrated relative reduction but scores were still superior. Despite the patient's complaint, the physician felt that these issues were largely emotional, having to do with the patient's impending retirement. A BEAM study was nonetheless ordered. In this particular analysis, the standard EEG was read as within normal limits. Spectral analysis showed delta entirely within the normal range. Theta, however, was seen to be asymmetrical and, by the SPM process, enhanced in the left parietal region by 3.50 standard deviations. The possibility of a regional organic pathology was raised on the basis of these findings, but a subsequent computed tomography (CT) scan was within normal limits. The patient was seen again 16 months later with complaints of increased difficulty. However, his neurological examination was again normal, including tests of memory. His BEAM study was repeated. In this study, delta activity was now seen to be increased in the left parietal region by 2.49 standard deviations. It had been normal in the previous study. Theta was again abnormal in the left parietal region by 7.65 standard deviations, up from the previously noted 3.50 standard deviations. In this study the EEG was thought to be somewhat slow due to drowsiness. On the other hand, the BEAM analysis was done during the fully alert state. The patient experienced no difficulty remaining alert and appeared normal. Statistical comparison of the first and the second study by the *t*-SPM process showed delta to be increased in the second study by 5.32 *t* units ($P = .00001$), and theta increased in both left and right parietal regions by a maximum of 5.39 *t* units ($P = .00001$) (see Fig. 19.5). The BEAM study was read as consistent with an asymmetrical but bilateral posterior abnormality, left greater than right. Given the patient's complaint of memory difficulty, the possibility of Alzheimer's disease was now singled out among the many possible organic processes that might cause this EEG picture. A repeat CT and magnetic resonance imaging (MRI) remained normal and unchanged, showing no evidence of atrophy or loss of brain substance. Repeat neuropsychological examination showed some decrease in memory performance, but it still remained well above normal, as did the full-scale IQ, which was now at 140.

Approximately 6 months later the patient developed the full-blown symptomology of Alzheimer's disease complete with profound memory loss and loss of other cognitive skills, including judgment and awareness.

This case exemplifies the difficulty in diagnosing Alzheimer's disease early in patients whose overall cognitive abilities are in the very superior range. The patient may be very aware of some intellectual loss, only to have formal testing procedures indicate superior performance. In such cases BEAM studies are particularly useful for detecting unilateral or bilateral evidence of deviation from normal, particularly involving the lower EEG frequency ranges of delta and theta. For this particular patient, BEAM was the only technique that demonstrated clear deviations from normality almost 2 years in advance of the clear and unequivocal presentation of the clinical syndrome.

Case Study 2: Temporal Lobe Epilepsy in a Female Adult

A 29-year-old woman was referred for BEAM evaluation by her clinician whose diagnosis to date was a "chronic ruminative syndrome" not responding to medication. She had previously had a classic EEG read as normal and no seizure discharges were evidenced. The results of her first BEAM, however, indicated abnormal spectral analysis delta readings in the left and right posterior temporal regions of the brain (Fig. 19.6A). These findings were corroborated by the topographic SPM maps of eight EP epochs. The SPM of two of these, the VER and the AER, are provided in Figs. 19.6B and 19.6C.

As such focal findings could not be interpreted as simple depression, the patient was referred for neurological consultation to search for organic pathology. Her subsequent CT scan was normal, as was the classic neurological assessment. However, the patient had an unusual personality profile characterized by obsessive behaviors. She devoutly maintained a detailed diary in which she reported extreme moral concerns, alternating periods of hypo- and hypersexuality, and periods of religious preoccupation. She was intensely verbal, very serious and persistent, and overfocussed in discussing any given topic that gained her attention. These personality characteristics led us to suspect a conditional temporal lobe epilepsy (Bear DM, 1979; Bear DM and Fedio P, 1977).

The patient, however, denied any incidence of seizures in her life. But on further questioning, she acknowledged incidents of "fainting" as a teen-ager, wherein she experienced a strange feeling in her back and stomach that moved into her mouth previous to sensing fear and "fainting." On the basis of the BEAM findings and a suspected history of partial complex seizures, she was placed on carbamazepine.

After 6 months, the woman was reevaluated. The classic EEG report was unchanged. The spectral analysis BEAM was analyzed via SPM and indicated that increased delta was still in evidence but less dramatically so (5.39 standard deviations) than her previous BEAM (Duffy, op cit, 1986). Only one EP abnormality was noted. She was clinically much improved, having ended reliance on family financial support, started her own acting workshop, and discontinued her diary. She cancelled her next 6-month follow-up appointment.

After an additional 6 months, she contacted the laboratory for a return appointment. This time, her EEG was distinctly abnormal, clearly indicating left temporal discharges. Her BEAM data showed increased delta abnormality (6.79 standard deviations), more extensively involving the left hemisphere than before. She had reestablished the extensive and compulsive use of a diary. Upon interview, it was found that she had discontinued her use of carbamazepine, having felt so well some weeks before that she decided medication was no longer necessary. She had recently experienced the first "fainting" spells she had had since those of her teen-age years. She was, of course, restarted on medication, and her well-being was restored.

This clinical example demonstrated the ability of BEAM to suggest an alternative diagnosis leading to successful therapeutic treatment of a behavior dysfunction.

Case Study 3: Sylvian Seizure Syndrome in an 8-Year-Old Boy

This boy was initially referred for minor seizures involving speech difficulty and facial seizure affecting the mouth regions. Nocturnal grand mal seizures were reported and the classic EEG reading located focal seizure discharges emanating from the left frontotemporal, midtemporal region of the brain. This symptomatology led to the diagnosis of Sylvian seizure syndrome, generally a mild, self-limited epileptic syndrome (Lombroso C, 1967). The topographic analysis was also performed and revealed a spike discharge beginning in the left frontal region (Fig. 19.7A), but rapidly reaching maximal value in the left midtemporal region in 19 ms (Fig. 19.7B). These BEAM findings ran counter to the Sylvian seizure syndrome diagnosis as virtually all patients who experience this syndrome will demonstrate central-parietal initiation. The BEAM spectral analysis and VER results revealed nothing unusual, but the AER indicated a large abnormality beginning at 260 ms and lasting 40 ms. During this epoch, the left midtemporal region was excessively positive, while both frontal regions and much of the right hemisphere were excessively negative (Fig. 19.7C).

The simultaneous appearance of both negative and positive deviations from normal on an SPM indicated a "dipole abnormality," which is highly correlated with pathology. A CT scan was recommended for this patient, the results indicating a subtle but definite left thalamic tumor. This finding was then confirmed by MRI. The contribution of BEAM in this instance was the instigation of further radiographic examination leading to early detection of a pathology. This examination probably would not have happened for several more months given the relatively benign course of the condition when initially diagnosed.

A Cautionary Note: Potential Errors in Clinical Usage of Topographic Analysis

We have witnessed the highly successful application of topographic analysis to clinical practice and to neurological research. It cannot, however, be overemphasized that simple inspection of topographic images is insufficient to establish a reliable diagnosis of neurological condition. BEAM is designed to provide more detailed, objective, and efficient analysis of EEG and EP data to assist the clinician in supplementing neurological, clinical, and other technological assessments of a given patient's condition. Following are described a number of examples demonstrating potential errors in topographic analysis if not performed in a thorough and highly trained milieu.

Figure 19.8 provides a topographic SPM image depicting a very abnormal evoked response (Duffy FH, in press). The actual interpretation of this image was nonetheless that of a normal reading. The basis for this interpretation was the simultaneously recorded EEG indicating that the patient in question was extremely drowsy during this phase of the recording and could not be roused to a full waking state by the EEG technician. This instance illustrates how essential it is for the clinician to obtain a background EEG for two reasons: (1) quality control in terms of excision of artifact and avoidance of artifactual recordings, and (2) extraction of useful information intrinsic to the EEG recording itself.

A second example of a common anomaly in topographic analysis is created by the incidence of "time-locked alpha." Topographic images indicated an asymmetrically increased negativity located primarily in the right occipital region for a 40-ms epoch of the late AER that reached 3.83 standard deviations and involved at least four electrodes. The patient had been referred for bipolar depression. The EEG recordings indicated highly asymmetrical alpha, greatest in the right hemisphere. Analysis of the AER waveforms indicated a time-locked alpha component in the right occipital region that was absent on the left. Although the topographic image indicated a significant difference from normal EP reading, the clinician correctly interpreted these data as indications of time-locked alpha, an anomaly that sometimes contaminates late portions of long-latency EPs when the subject is fatigued or drifting off to sleep. The asymmetry recorded by the BEAM image is not an organic abnormality, but an actual asymmetry caused by the drowsiness of the patient, a frequent occurrence with younger subjects. Time-locked alpha is the leading cause of incorrectly detected "abnormalities" in clinical mapping studies.

Another frequent analytic issue that the clinician may have to deal with concerns spectral analysis. The EEG of a patient may contain a spike or a benign variation paroxysm such as "14 and 6." Since it is benign, the technologist may assume that it can be included in a spectral analysis. The clinician must be aware that inclusion of such data would contaminate the analysis, thereby making it impossible to discern legitimate abnormality in the background EEG spectral content.

Yet another analytic issue regards the occurrence of a minor regional abnormality when reading an SPM of a given patient and then not encountering consistent evidence of the abnormality in the rest of the record. The clinician will want to establish the biological significance of this single event. One of the ways to deal with such instances is to repeat the study. One repeats the study three times in an attempt to search for a consistent neurological abnormality or to find evidence that the initial reading was statistical noise or artifact. Figure 19.9 provides an example of a topographic analysis that initially shows abnormality but upon reassessment was normal. With the substantial number of variables in a complete BEAM, one might occasionally experience a variable falling outside the normal range, even though these variables are correlated. This emphasizes the importance of study repetition, examining each state and each EP, and establishing consistent findings.

Topographic Analysis in the Clinical Setting

Today, there are a substantial number of manufacturers and designers of topographic mapping systems throughout the United States and worldwide. This analytic method is gaining increased acceptance as a diagnostic aid in the neurological clinic. However, it is crucial to recognize that the instruments of topographic analysis are not automated to the point where they can be applied by people who are not knowledgeable nor can they be run by technologists without specific training in the equipment. Clinicians still must have legitimate background and knowledge in electroencephalography and EPs. They must be able to recognize artifact and to eliminate artifactual data from the record as well as recognize its occurrence in the topographic images. They must monitor the state of the patient and be able to detect drowsiness. The use of topographic analysis, rather than diminishing the skill requirement of electroencephalographers, actually places greater demands on their skill repertoire. They must know both traditional EEG and EP analysis and the newly emerging agenda of knowledge and skills particular to topographic analysis.

The dividend paid by reliance on the evolving technologies of topographic analysis is increased reliability, sensitivity, and objectivity of EEG and EP analyses. The payment required is increased knowledge levels and care by neurologists and technicians and a full awareness of the strengths and limitations of this technique. It is designed for, and capable of, enhancement of the traditional array of neurological assessment strategies; it is not meant to replace them.

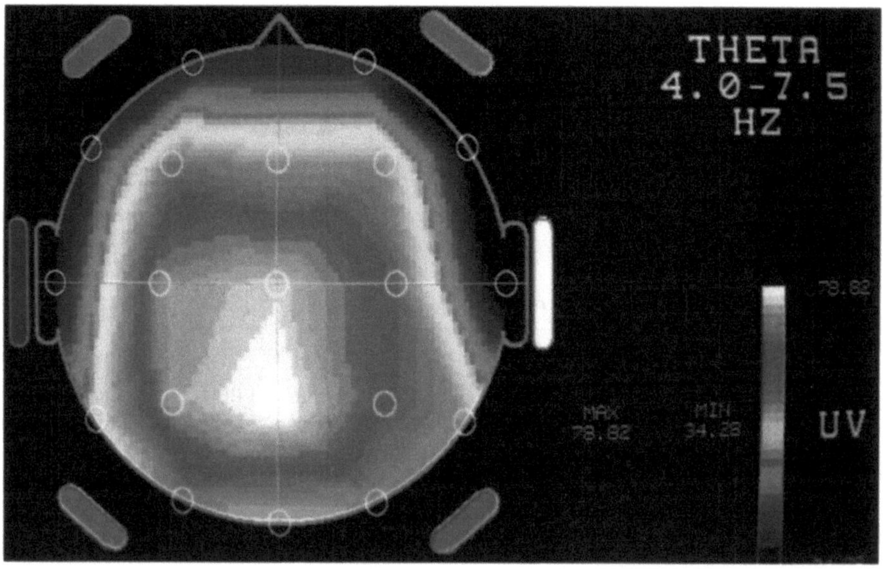

Figure 19.1. Sample topographic map of EEG data recorded during eyes-closed resting state. Image is shown in vertex view, nose toward the top of page, left to reader's left. Rainbow color scale in lower right indicates range of spectral amplitude measured in microvolts. This and subsequent topographic maps shown were obtained using the BEAM system.

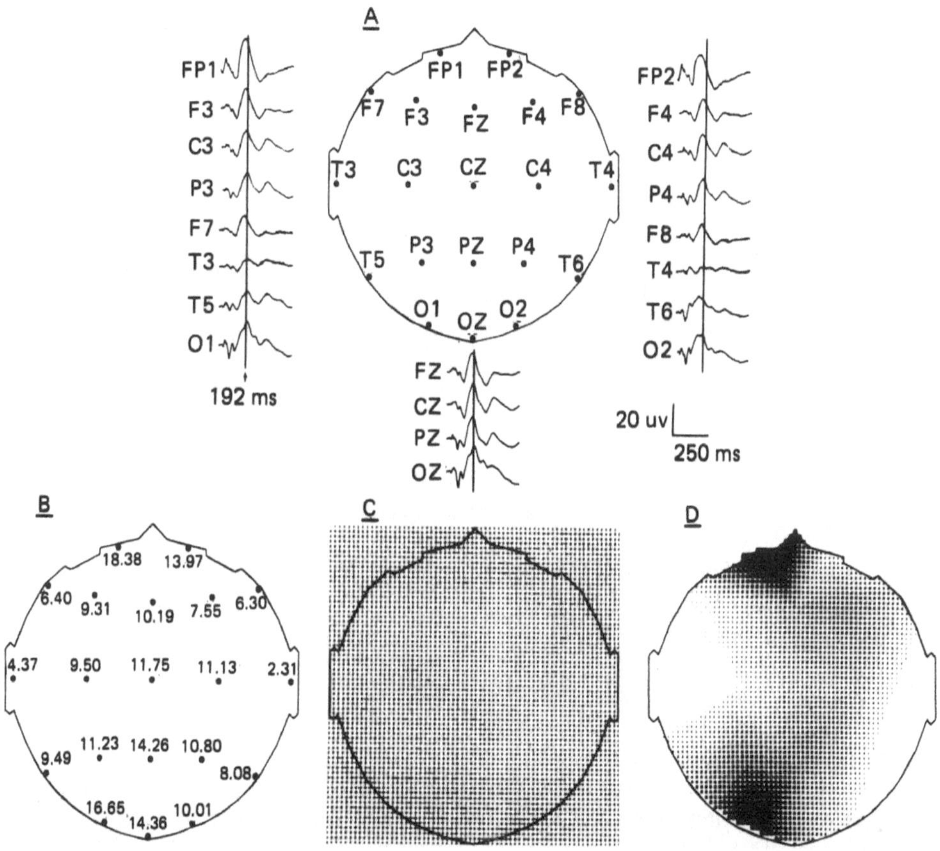

Figure 19.2. Paradigm for the construction of a topographic map for EP data. Mean EPs are formed from each of 20 recording sites. Each EP is divided into 128 4-ms intervals, and the mean voltage value for each interval is calculated. In (A) the individual EPs are shown for the electrode locations indicated on the head diagram. In (B) the mean voltage values at these locations are shown for the interval beginning 192 ms after the stimulus (the vertical line in A indicates this time on the EPs). Next the head region is treated as a 64 × 64 matrix; the resulting 4,096 spatial domains are illustrated in (C). Each domain is assigned a voltage value by linear interpolation from the three nearest known points. Finally, for display, the raw voltage values are fitted to a discrete-level, equal-interval intensity scale as shown in (D). Although a VER is used to illustrate the mapping process, the same procedure is used for mapping other data, including EEG. (From Duffy et al [1979], with permission.)

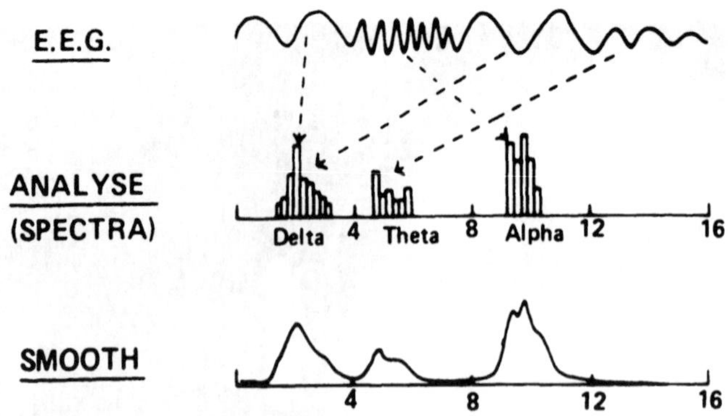

Figure 19.3. A schema illustrating the creation of spectral plots (bottom line segment) from raw EEG tracings (top line segment). Every appearance of delta in the upper tracing is reflected by a cumulative increase in the delta histogram of the lower tracing. This may be repeated for all spectral bands. (From Bickford [1977], with permission.)

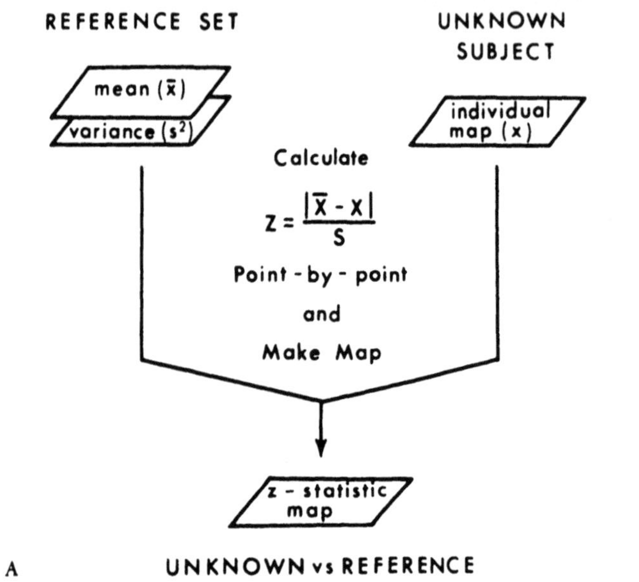

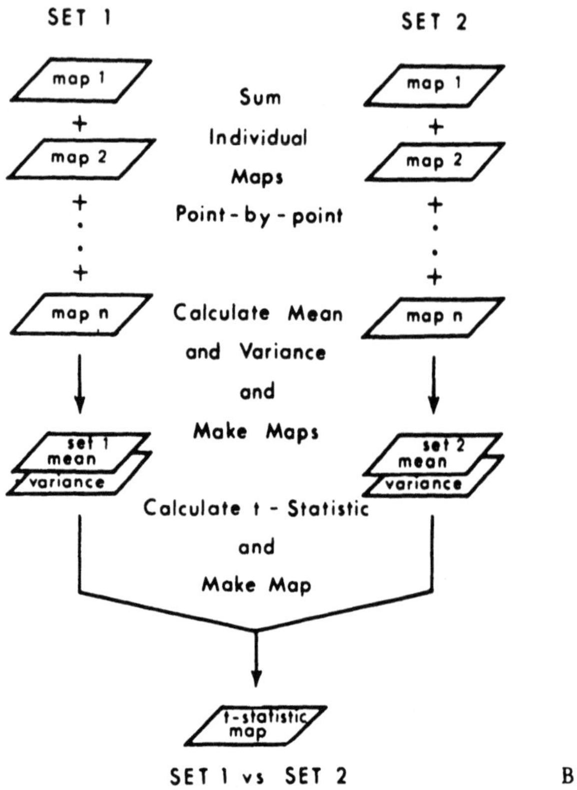

Figure 19.4. Topographic imaging of abnormality: SPM. (A) demonstrates the formation of a Z-statistic SPM. The Z-transform represents the number of standard deviations by which an individual's observations differs from the mean of a reference set. For BEAM, the Z statistic is calculated individually for each of the 20 to 32 scalp electrodes from the data of a single subject and the data of a normal control population. The resulting 20 to 32 Z values are then interpolated according to the procedure shown in Fig. 19.2 to produce the Z-SPM. The result is a display of a subject's deviation from normal in units of standard deviation in such a way that the spatial relations of the original BEAM image are retained. Z-SPMs are ordinarily used in clinical practice to define abnormality in individual subjects. (B) demonstrates the formation of a *t*-statistic SPM. Student's *t* statistic quantifies the separation between two sets of measures, taking into account not only the difference between the mean value of each group but also the variability within each group. Thus the *t* value is lower for the same difference in group means when the variance in either or both groups is larger. For BEAM, the *t*-statistic is calculated individually for each of the 20 to 32 scalp electrodes from the data of the two groups of subjects. The resulting 20 to 32 *t* values are then mapped using the procedure in Fig. 19.2 to produce the *t*-SPM. The result is a topographic display of where the brain electrical activity of one group differs from that of the other. Ordinarily, *t*-SPMs are used in research to delineate the way in which a pathological population differs from a control population. (From Duffy et al [1981], with permission.)

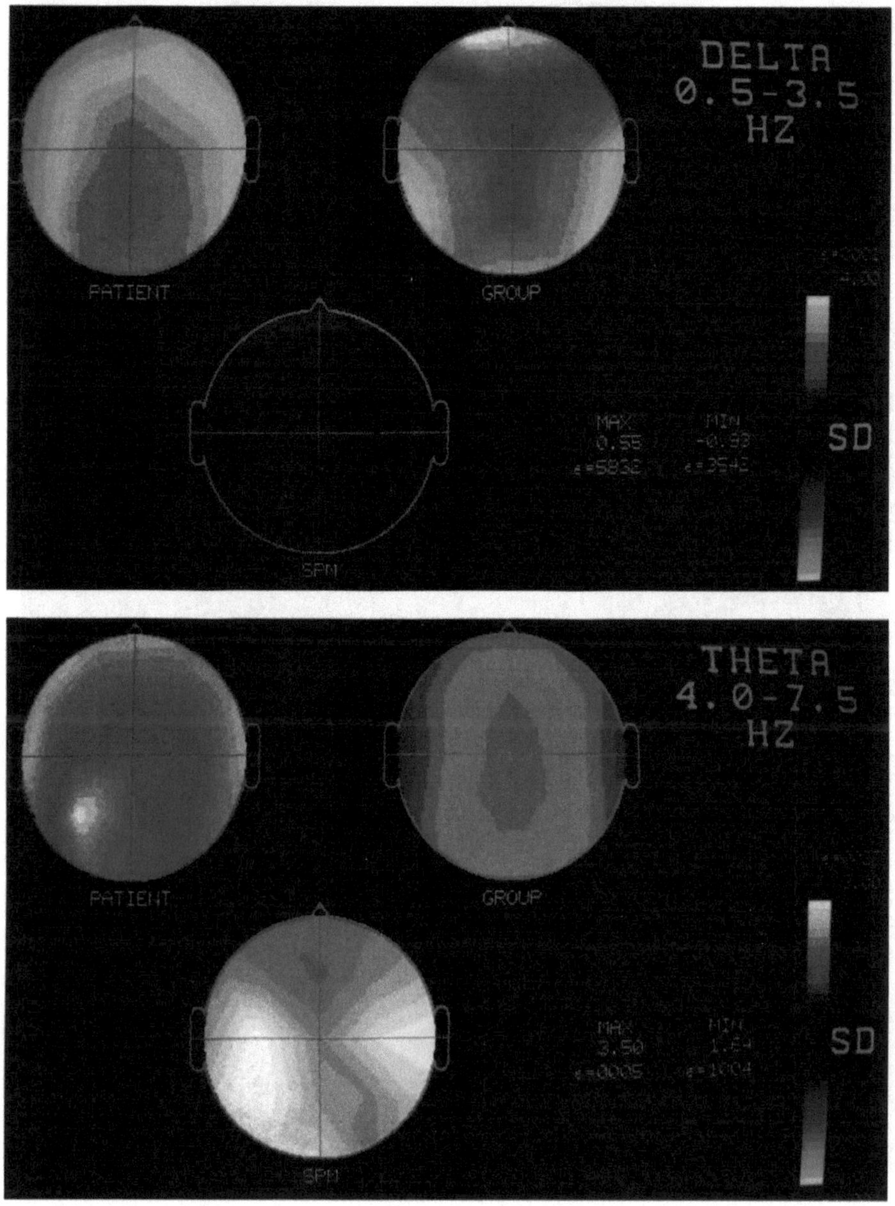

A

B

Figure 19.5. Comparative BEAM topographic images of a man with progressive stages of presenile dementia. Data displayed according to the convention of Fig. 19.1. (A) and (B) provide maps of delta and theta spectra, respectively, based on testing in early 1986. (C) and (D) provide comparable maps in the summer of 1987. Note the progressive increase of delta and theta in the left parietal region. As noted in the text, the patient was clinically diagnosed as basically healthy early on, but was ultimately diagnosed, with input from these images, as having Alzheimer's disease. The apparent difference in the theta frequency maps of the control groups study 1 versus study 2 is the result of scaling convention applied to the second study to enhance the visibility of regional differences demonstrated in the SPM image. The control groups are identical for both studies (*continued*).

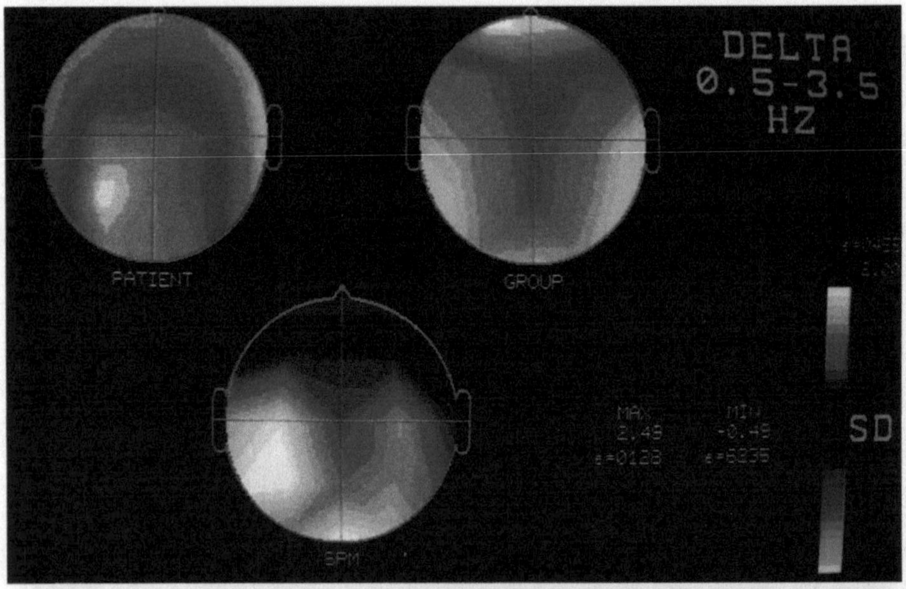

C

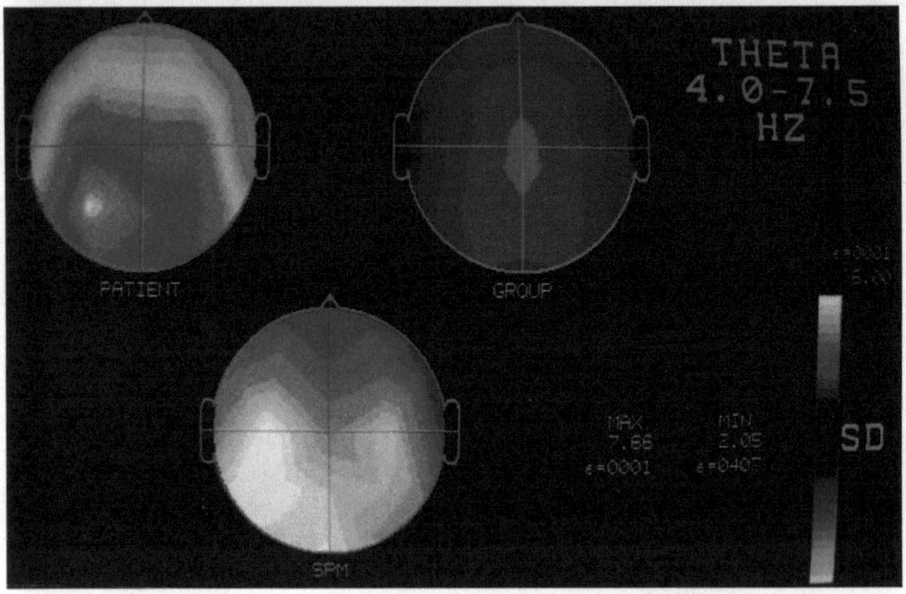

D

Figure 19.6. Left posterior temporal abnormality in a patient with temporal lobe epilepsy personality syndrome. Three Z-statistic SPMs are shown for a 29-year-old woman with the presenting complaint of "chronic ruminative condition" and the final diagnosis of temporal lobe epilepsy personality syndrome. Three sets of three BEAM images are shown. Within each set the patient data are shown to the upper left, the control data to the upper right, and the corresponding Z-SPM below. (A) shows globally augmented 0.5 to 3.5-Hz delta maximal in both posterior temporal regions (left > right) reaching a maximum Z value of 7.41 standard deviations. (B) shows augmented negative activity of the VER from 344 to 380 ms by 3.65 standard deviations in the left posterior temporal/parietal region. (C) shows augmented negative activity of the AER in the left posterior temporal/parietal region from 160 to 196 ms. These three spatially congruent abnormalities lead to the recognition of an electrophysiological abnormality and, eventually, the collection of complex historical data to synthesize the clinically important diagnosis (see text). (From Duffy in *Physiology of the Ear* [in press], with permission.)

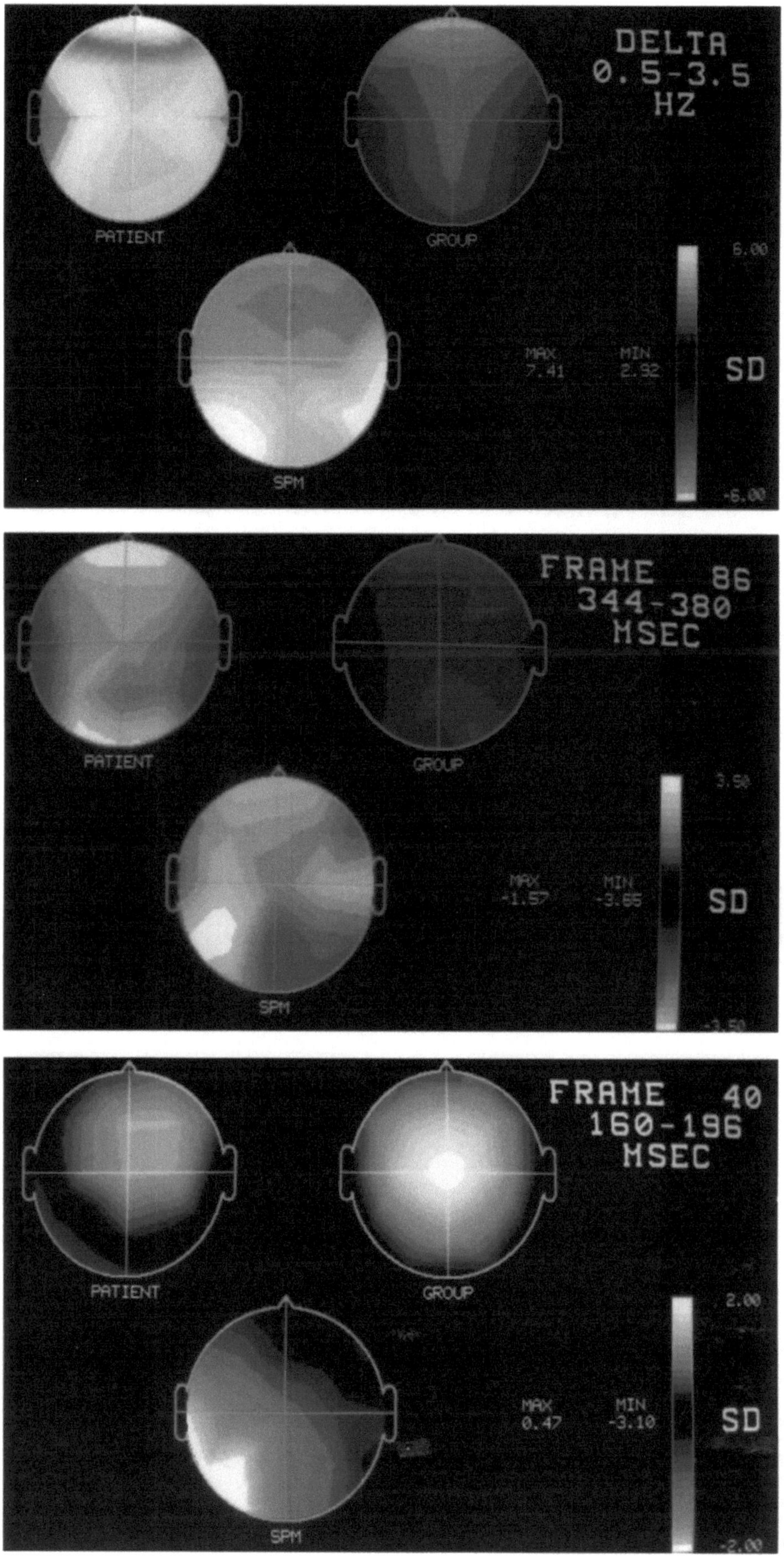
DELTA
0.5-3.5
HZ
PATIENT
GROUP
SPM
6.00
MAX
7.41
MIN
2.92
SD
-6.00
A
FRAME 86
344-380
MSEC
PATIENT
GROUP
SPM
3.50
MAX
-1.57
MIN
-3.65
SD
B
FRAME 40
160-196
MSEC
PATIENT
GROUP
SPM
2.00
MAX
0.47
MIN
-3.10
SD
-2.00
C

Figure 19.6.

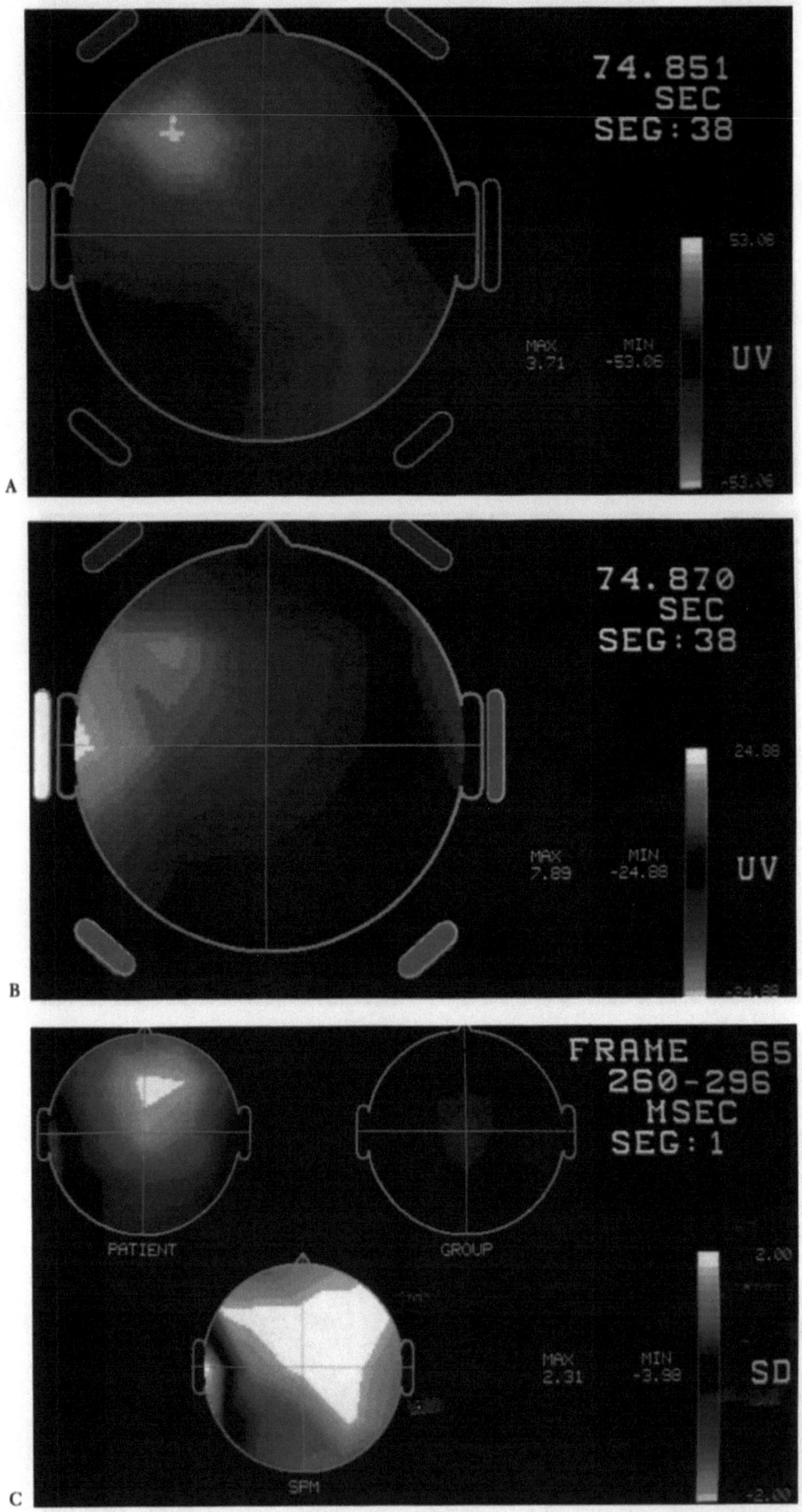
74.851
SEC
SEG: 38
MAX
3.71
MIN
-53.06
UV
A
74.870
SEC
SEG: 38
MAX
7.89
MIN
-24.88
UV
B
FRAME 65
260-296
MSEC
SEG: 1
PATIENT
GROUP
MAX
2.31
MIN
-3.98
SD
SPM
C

Figure 19.7.

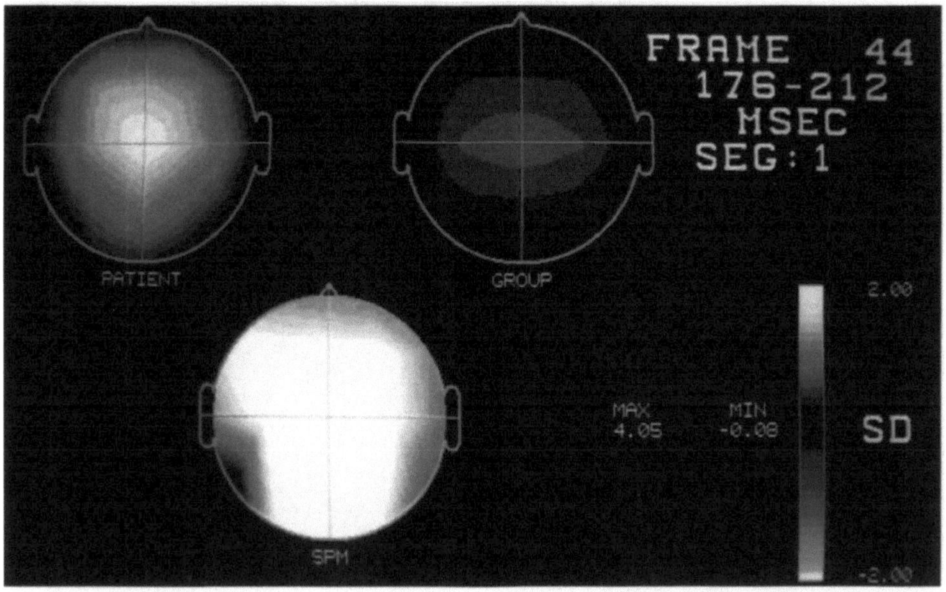

Figure 19.8. Topographic mapping of AER and resulting SPM image showing comparison with normative data. The SPM image would indicate an extreme abnormality. However, traditional EEG data were recorded simultaneously with the EP testing and provided clear evidence of drowsiness in the patient. These results underscore the continued importance of traditional EEG recording in clinical diagnosis.

Figure 19.7. BEAM (eyes open EEG), displayed according to the convention of Fig. 19.1, derived from an 8-year-old boy who presented with clinical signs and symptoms consistent with the Sylvian seizure syndrome, which is generally believed to be a benign, self-limited disorder. In contrast to the usual findings, however, the distribution of activity at the onset of a typical spike was maximal in the left frontal region (A) and 19 ms later moved secondarily to the left midtemporal region (B). Ordinarily, discharges in this syndrome commence in the centroparietal, not frontal, regions. Although spectral and VER data were normal (not shown), the AER demonstrated a very large "dipole" abnormality, with augmented left temporal positivity, by 2.31 standard deviations, and broadly increased frontal and right hemispheric negativity, by 3.98 standard deviations (C). This degree of abnormality is unusual in the Sylvian seizure syndrome. Owing to these unusual BEAM findings, radiographic studies were performed, revealing a left thalamic tumor. Had it not been for the topographic deviations from the normal pattern, prompt radiographic investigations would not have been performed. (From Duffy in *Topographic Brain Mapping of EEG and Evoked Potentials* [in press], with permission.)

A
B
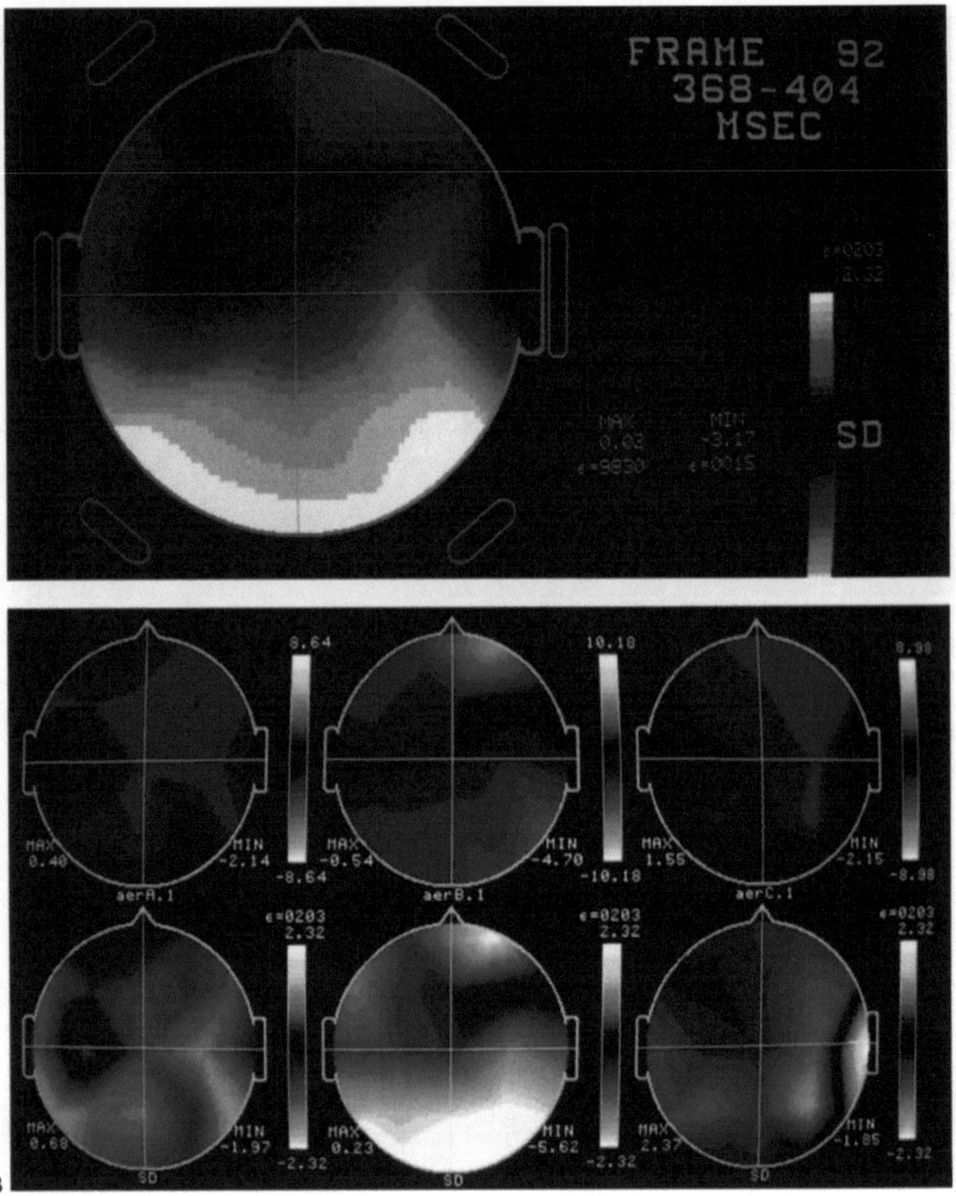

Figure 19.9. Comparable BEAM SPM images based upon testing and retesting of an individual. (A) indicates an AER abnormality in the posterior region, based on comparison with control group data. (B), however, depicts the same individual retested under the same conditions three times. The images in the top row are based on recordings of the patient; those on the bottom are SPMs comparing him with group norms. As can be seen, the abnormality only occurs during one of the three testings, suggesting a transient, probably artifactual abnormality. Supporting neuropsychological and clinical examination resulted in no indication of pathology. These findings stress the importance of replication testings.

References

Bartels PH, Subach JA: Automated interpretation of complex scenes, in Preston K, Onoe M (eds.): *Digital Processing in Biomedical Imagery*. New York, Academic Press, 1976.

Bear DM: Temporal lobe epilepsy—a syndrome of sensory-limbic hyperconnection. *Cortex* 1979; 15:357–384.

Bear DM, Fedio P: Quantitative analysis of interictal behavior in temporal lobe epilepsy. *Arch Neurol* 1977; 34:454–467.

Bickford RG: Computer Techniques in Neonatal EEG, in Werner SS, Stockard JE, and Bickford RG (eds.): *Atlas of Neonatal Electroencephalography*. New York, Raven Press, 1977; 193–200.

Callaway E: Diagnostic uses of the averaged evoked potential, in Donchin E and Lindsley DB (eds.): *Average Evoked Potentials*. Washington, D.C., NASA, 1969; 299–332.

Chiappa KH: *Evoked Potentials in Clinical Medicine*. New York, Raven Press, 1983.

Dawson GD: Cerebral responses to nerve stimulation in man. *Br Med Bull* 1950; 6:326–329.

Duffy FH: Topographic display of evoked potentials: Clinical applications of brain electrical activity mapping (BEAM). *Ann NY Acad Sci* 1982; 388:183–196.

Duffy FH (ed.): *Topographic Mapping of Brain Electrical Activity*. Boston, Butterworths, 1986.

Duffy FH: Topographic mapping of brain electrical activity: clinical applications and issues, in Maurer K (ed.): *Topographic Brain Mapping of EEG and Evoked Potentials*. New York, Springer-Verlag, in press.

Duffy FH: Clinical electroencephalography and topographical brain mapping: technology and practice, in Jahn AF and Santos-Sacchi JR (eds.): *Physiology of the Ear*. New York, Raven Press, in press.

Duffy FH, Bartels PH, Burchfiel JL: Significance probability mapping: An aid in the topographic analysis of brain electrical activity. *Electroencephalogr Clin Neurophysiol* 1981; 51:455–462.

Duffy FH, Burchfiel JL, Lombroso CT: Brain electrical activity mapping (BEAM): A method for extending the clinical utility of EEG and evoked potential data. *Ann Neurol* 1979; 5:309–321.

Jeffreys DA, Axford JG: Source locations of pattern-specific components of human visual evoked potentials. *Exp Brain Res* 1972; 16:1–40.

Lombroso C: Sylvian seizures and mid temporal spike foci in children. *Arch Neurol* 1967; 17:52–57.

Chapter 20

Recommended Standards and Practices for Brain Electrical Activity Mapping and Topographic Analysis

Topographic mapping of neuroelectrical data has experienced a rapid escalation in acceptance and usage over the past several years. Neuroscientists and clinicians are becoming increasingly reliant on the powerful data-interpretation and statistical-analysis capabilities of the various topographic mapping technologies to assist them in clinical diagnosis and research. Unfortunately, the standardization of practice and quality control that should be concomitant with this growth is lacking. There appears to be a substantial range of perceptions by users of these devices as to what BEAM can and should be used to do.

Current topographic mapping practice manifests three very different approaches to clinical application. Some devices, for example, allow the clinician to directly connect the patient to the device, obviating the need for the standard recording of EEGs. These devices produce color-coded maps of spectrally analyzed EEG and, in some cases, single or cartooned topographic images of evoked-potential (EP) data. In these instances, it is not unusual for the user to develop a clinical diagnosis solely on the basis of visual review of the spectral analysis of topographic maps. Alternatively, the more sophisticated devices emphasize the concurrent need to record and analyze EEGs, also providing the clinician with the capability of comparison of their subjects' EEG and EP with normative age-appropriate standard polygraphic data. The clinical assessment then includes the standard EEG, the spectrally analyzed EEG and EP topographic images, and topographic images comparing the patient with a control group, thereby enhancing detection of abnormality. In a third clinical paradigm, the topographic devices are used only as appendage, with raw data and topographic images used primarily for illustrative purposes. These devices have as their basis precalculated multivariate discriminant paradigms used to generate subject assignment into one or more clinical groupings.

It is clear that the practice of neurophysiology in these three settings could vary tremendously. A patient with a given neurological condition might well experience differing diagnosis and treatment, depending on which clinic he or she was referred to. There is currently an unfortunate tendency in some neurology laboratories in the United States to rely solely on the topographic map as the basis of diagnosis. As we have noted in Chapter 19, this approach is limited and potentially harmful. On the other hand, it is also an incorrect perception that all one needs is a "qualified electroencephalographer" as the user responsible for these devices in order to produce consistent and reliable results. In fact, the knowledge required to function in the environment of *quantified* electroencephalography goes substantially beyond the training and experience of a traditionally trained electroencephalographer. The optimal contributions of computerized brain electrical activity mapping devices lie in their reliability, objectivity, flexibility, and speed of analytic function. It is a dangerous misconception to regard them as stand-alone diagnostic devices designed to obviate the need for technically and, in fact, specifically trained professional neurophysiologists.

The intent of this chapter, therefore, is to present a set of recommended minimum standards for the practice of BEAM and quantified electroencephalography. These standards have been cooperatively discussed and designed in concert with developers of several leading topographic mapping devices; as such they do not reflect a prejudice favoring any particular manufacturer's device. Indeed, the recommendations are based on the authors' view of topographic mapping as a part of clinical neurophysiology, including both neurology and psychiatry. The recommendations first address issues of personnel: minimal training and appropriate testing protocol. Then, issues of operating environment and equipment standards will be addressed.

Personnel

EEG Technologists

The technologists are clearly the most critical link in the sound clinical practice of topographic analysis. They stand as the "keystone" in terms of machine operation and patient state. Technicians are in the best position to detect and ameliorate artifact of all types. Moreover, they are in a position wherein it is possible to make costly mistakes that would be very difficult to detect subsequently. The centrality of their role in the topographic process implies thoroughgoing, multifaceted training. Several of the skills are demanded of any qualified EEG technician; BEAM requires additional specialized knowledge.

The key elements in technologists' skill repertoire would be the following:

1. Accurate measurement and proper placement of low-impedance electrodes of a stable sort such as cup (disk) electrodes attached with collodion. Accurate placement is of paramount importance. This is a lengthy process, so speed and dexterity become criteria for judging performance, along with the essential accuracy.
2. Familiarity with operation of the equipment, including all its options. Not only must the technician know the basic operational skills of classic EEG, s/he must acquire the knowledge unique to the topographic mapping devices and their computer-based peripherals.
3. Artifact management skills (equipment). This is of paramount importance. Technologists must possess complete knowledge of sources of artifact. They must know the location of electrodes used to detect and measure artifact and the technique of applying them. They must be capable of recognizing the occurrence of artifact on the polygraph tracings and/or on the computer screen for the purpose of marking artifact-contaminated segments and eliminating them prior to spectral analysis.
4. Artifact management skills (patient/subject management). Technologists must be capable of detecting state changes such as drowsiness; they must be knowledgeable of strategies to control patient state. Similarly, they need to be skilled at recognizing the presence of artifacts such as electrode "pop," eye blink, and muscle contamination, and they must work with patients to reduce or eliminate the occurrence of these artifacts where possible.
5. Artifact management skills (behavioral paradigms). Technicians will be called upon to manage patient performance of specific behavioral paradigms while collecting data (e.g., EPs). They should possess specific awareness of unique potentials for artifact related to given paradigms. They must be capable of modifying behavioral paradigms to minimize such artifact.
6. Recognition of classic EEG features and abnormalities. Finally, it is desirable that technologists have skill in the recognition of epileptiform discharges, normal variant rhythms, drowsiness and sleep, encephalopathy, and any other features commonly occurring in EEG technology. The inclusion, for instance, of undetected segments of drowsiness, or segments of waking state with "14 and 6" discharges, can radically alter the outcome of topographic spectral analysis, causing the appearance of an apparent pathological condition where none actually exists.

In summary, technologists enter the field of BEAM from two backgrounds. The most common background is that of registered EEG technologists. They should come with good skills in measurement and electrode application, and an understanding of EEG rhythms. Many, however, are unskilled in the management of patient state and are prone to allowing subjects to become drowsy. They typically do not have sufficient rigor in artifact detection and removal. Furthermore, they will, of course, need special training in management of topographic mapping equipment. Technologists may also enter the area of BEAM from allied fields such as electrocardiography, radiology, etc. Still others enter the field while deciding whether or not to pursue academic studies. In our experience, intelligent technologists with no background in EEG can be taught the mechanical and technical aspects of the job, including electrode application in approximately 3 months. However, they cannot be expected to have a feel for state control of EEG or for detection or management of unusual EEG rhythms for some time. Such technologists can function well but only with very close supervision by a knowledgeable neurophysiologist or more qualified technologist.

Neurophysiologists, BEAM Readers

Most typically, the BEAM reader is a neurologist with special training in neurophysiology. Alternatively s/he may be a Ph.D. with background in electroencephalography or clinical neurophysiology, or a psychiatrist or pediatrician with particular interest in electroencephalography. Qualification (boards) in EEG is extremely desirable but, in our experience, there are many individuals who have used electroencephalography in their practice extensively but have never bothered to qualify because they have not considered this a major portion of their practice. Such knowledgeable individuals are often quite qualified to enter into the field of quantified EEG and BEAM.

Individuals with a background in electroencephalography, including official qualification, require substantial additional training before they can function adequately in the field of topographic analysis. Given the basic knowledge of electroencephalography, the following are minimal additional requirements of reader/interpreters of topographic images:

1. Complete familiarity with operation of the topographic mapping device as well as its hardware and software, analytic capabilities and limitations, and intended usage.
2. Basic understanding of signal analysis, including such precepts as data-sampling rates, spectral analysis, signal averaging, scaling of spectral plots, differences between microvolt and microvolt squared spectra, and differences between raw spectral and percent (normalized) spectra.
3. Knowledge of normative inferential statistics as applied to topographic analysis. The reader should be familiar generally with univariate and multivariate statistics; and particularly with data Gaussianity; interactive effects of multiple measurements; *t*, Z, and F statistics; and aliasing.
4. Familiarity with artifact detection, reduction, and elimination, not only in raw signal but in statistical outcome.

The reader should understand and be wary of capitalization on chance. The reader must be consistently able to detect change of subject state and its effect on mapped data. As interpreter of topographic maps, s/he should be fully conversant in sources of artifact and the strategies previously discussed to deal with them.

These recommendations generally address the standards to be applied to the training and requisite skills of the professional clinician involved in interpretation and analytic application of topographic mapping systems. Experience has shown that it takes approximately 1 week of work for a bright, qualified electroencephalographer to acquire sufficient skills to begin a practice of BEAM. Obviously, fellowship training of 6 to 12 months is much preferable.

Recommended Standards for Topographic Mapping Environment and Equipment

General Laboratory

Much of the value of BEAM comes from the use of normative data and the individual deviation from the normative data that is indicative of abnormality. Accordingly, standardization is very important. Data in most laboratories are gathered in a controlled environment where light and sound are held at minimal levels. This generally means that within the laboratory a patient must be studied in a room separate from control equipment, telephones, physician/technologist discussions, etc. Nonetheless, it is necessary that there be a means of visualization either through a one-way mirror or through a low-light television system. There must also be a means of verbally communicating with and listening to subjects. The stimulating (EP) equipment and the EEG instrumentation used with the patient must have the same electrical characteristics as the equipment used to gather data from the control population. Any differences in frequency characteristics of the EEG equipment will produce difficulties. All the other requirements of a good EEG laboratory need to be fulfilled such as a relatively noise-free environment that is distant from sources of electrical interference such as certain radiology equipment, elevators, etc.

Equipment Requirements

Regardless of manufacturer, the ideal operating system for display and management of data should be designed by appropriate human engineering methods to allow for simple and rapid data management. Slow operation of equipment can be the most damaging feature of BEAM studies. Minor options that can improve diagnosis should be easily implemented and not require extra time. Even the most conscientious physician, when hurried, will omit something simply because it takes too long to perform. Such systems, therefore, are best built around enhanced microprocessors or minicomputers with hard disks and floating point processors, since these allow faster processing time and generally have better data-handling capabilities. Ideally, interaction should occur through touch screen or mouse as well as keyboard.

There are a substantial number of specific recommendations for equipment performance capabilities:

1. A minimum of the standard 19 scalp electrodes, of the 10-20 International System, must be used. An additional four artifact electrodes are recommended. These should be placed so as to optimally monitor vertical eye movement, horizontal eye movement, muscle and cardiogram artifact. The topographic device should record all artifact channels through to final analysis such that false abnormalities produced by artifact can be detected through the demonstration of statistical differences in the artifact as well as from the scalp electrodes.

2. Frequent monitoring of electrode impedance, preferably on an automatic basis, must be facilitated by the equipment design.

3. There must be means to gather, store, and display EEG data during an entire study. The EEG serves first, to provide ready detection of artifact and, second, as a source of intrinsic clinical information. The EEGs must be recorded both during spectral data collection and during EP recording. All such data should be clinically evaluated. It becomes important to know, for example, if a subject was

drowsy during EP recording or if lambda or "14 and 6" contaminates spectral data. Although automatic artifact detection techniques are useful, no machine system surpasses the trained human eye in the detection and elimination of artifact. Accordingly, the presence of a polygraph (ink writer) or very high-quality video monitor capable of producing near-paper quality EEG tracings is required for every laboratory. Ideally, it should be possible for EEGs gathered at time of analysis to be stored and redisplayed in other montages to assist in interpretation. Finally, the ability to view EP and unprocessed EEG data in a time-sequential manner (cartooning) should be provided.

4. The many analytic techniques and the huge volume of data generated, including individual EEG and EP data, and normative group data on large numbers of parameters, require immense data-handling capabilities. Therefore, very large hard-disk storage is required.

5. The system must provide means to facilitate the detection and elimination of artifact during a recording session, both prior and subsequent to spectral analyses. Visual inspection or automatic artifact filtering of EEG segments to be submitted for spectral or coherence analysis is mandatory as a minimum protective measure. Overvoltage detection must be provided for on-line signal averaging for EP data as a minimal protective measure. The system should allow all analytic processes to be carried out on artifact as well as scalp electrodes.

The system must provide means to inspect the actual EEG segments used for spectral analysis at the time of this analysis. The results of "de-glitching" should be available for later inspection and should be reversible.

6. Topographic imaging devices must be designed to permit calibration of the entire system prior to data collection for DC level and peak-to-peak measures. At a minimum, a sine wave of specified amplitude should be passed through the entire system at a frequency near the main power maximum point of the signals being analyzed (e.g., 10 Hz). Square-wave pulse calibration signals are inadequate for calibration of amplifiers in quantified electroencephalography. Nonetheless, they should be provided as a final check of system performance. There should be means to detect amplifier distortion and nonlinear characteristics. The system should also provide means to facilitate analysis of the sine-wave calibration signal and of the biological calibration signal (biocal), where all channels are presented the very same data. Inspection and assessment of the outcome of this process should be facilitated by statistical printouts and/or maps.

7. Ability to select between topographic mapping in color or in black and white (B&W) should be possible. The user should also have the freedom to modify scaling parameters, i.e., s/he should be able to have a given scale range from zero to maximum, minimum to maximum, and between any two chosen values. S/he should be able to alter visual B&W and color scales as necessary, arraying the colors in a variable scale.

8. Color topographic displays should show positive electrical activity in red and negative in blue, with shading of hue and color determining positive and negative magnitude. Data that range from 0 in only one direction should be scaled with the rainbow colors, with black set at 0, white at maximum and the typical colors of heated metal in between (i.e., from maximum to minimum: red, orange, yellow, green, blue, indigo, violet). Care should be taken that color change promotes a sense of gradient from high to low. Care should also be taken to avoid abrupt changes in color, which potentially create false impression of a region of significance. On the other hand, the use of abrupt color steps to intentionally recognize criterion areas is perfectly legitimate.

9. Basic states available for analysis should include, as a minimum, unprocessed EEG activity, unprocessed EP tracings, results of EEG spectral analysis, and EEG coherence analysis. Wherever topographic images are viewed, the underlying data and/or polygraphic tracings should be readily available. This would apply equally to EEG, EP, and spectral mapping.

10. All systems should provide the user with a normative data base against which to compare individual brain electrical activity data for normalcy. The detailed composition and characteristics of the normative data base must be delineated. The normative group should be stratified by age and sex, and criteria for inclusion or exclusion of subjects should be explained in detail. To facilitate individual versus group comparison, univariate and multivariate statistical capabilities must be available and relatively easily applied.

11. Data derived from the normative population must be addressed for "normality" or Gaussianity if parametric procedures are to be used. Members of the normative subject pool must have been drawn from the normal healthy population, not clinically referred or unknown. However, patients with idiosyncratic findings should not have been eliminated, since this approach tends to excessively narrow the band defined as normal. Caution must be used in the interpretation of data based upon percent spectral values given the lack of operational independence. For example, increased delta as a raw finding may produce concurrent reduction in other frequencies when expressed as a percentage measure. In addition, for example, the presence of excessive muscle activity may artificially reduce the percentage of slow activity present and reduce the sensitivity of a detection procedure.

12. A number of statistical manipulations and analytic capabilities outside of spectral analysis and signal averaging are recommended, and these include coefficient of variation, symmetry functions, the "Hjorth" parameters (Hjorth B, 1986) and coherence. Complete systems should

include the ability to calculate such parameters and provide means to readily add more as they are developed. One should also be able to readily compare these parameters to normative values.

13. The system should incorporate processes to avoid capitalization upon chance or randomly positive results. The most common method is to facilitate the repetition of individual study states to ensure replicability. This is analogous to the overlaying of several EPs in classic neurophysiology.

14. The topographic device should facilitate the gathering of at least one repetition of each data-gathering session with capability of simultaneous comparisons of both unprocessed data (polygraphic tracings) and analyzed data (spectral waveforms and topographic images).

15. There should be means provided for the digitized storage of all data—both processed and unprocessed—on digital tape, hard disk, or optical disk.

16. There should be means to obtain printed copies of all graphics materials such as topographic maps (B&W or color), unprocessed EEG and EP waveforms, and statistical outcomes—numeric or graphic.

With the substantial number of topographic mapping researchers and manufacturers of topographic imaging devices, there are several areas where neurological research scientists hold differing views as to testing protocol. The following two issues are so central to the process of topographic mapping that any enumeration of quality-control standards would be very incomplete without them. They do, however, remain the subject of keen disagreement.

17. The technique for interpolation between electrode recorded data points should be carefully described. Its relative benefits and deficiencies compared with other methods should be addressed. Ideally the user should have the option to employ more than one method for interpolation. Minimally, three-, or four-point linear interpolation should be available.

18. As there exists no unanimity as to the best reference electrode site, it is recommended that data be recorded referenced to a single point, such as linked ears, with provision for unlinked ears, for connecting or calculating another physical reference site, for employing the average reference, and for use of the Laplacian reference (Hjorth, op cit.).

Summary

In summation, we recommend the following primary considerations be intrinsic to the design of any regulatory standards in the clinical use of topographic mapping:

1. Personnel already trained in neurophysiological procedures must receive additional specialized training.
2. Standard EEGs should be collected at the time of data gathering to enhance artifact and state control. The EEGs should be clinically interpreted as well.
3. A normative data base is mandatory for clinical use. Exclusionary and inclusionary criteria should be carefully enumerated. The data base must be stratified by age.
4. Equipment used to evaluate patients should be technically identical or electronically equivalent to that used to gather normative data.
5. At least 19 scalp and four artifact electrodes should be used.
6. As a minimal requirement for topographic maps a vertex view, equal-area display format should be provided.
7. More than one technique for electrode referencing, interpolation, and data normalization should be available.
8. All systems should facilitate the detection and elimination of artifact prior to map formation.
9. Every system must provide means for complete calibration and detection of amplifier distortion.
10. Clinical studies should include repeat testings so as to avoid false conclusions owing to chance findings.
11. Systems providing means for automatic classification must detail appropriate inclusionary and exclusionary criteria to ensure appropriate use.
12. All manufacturers should provide courses of instruction in their instrumentation.

Reference

Hjorth B: Physical aspects of EEG data as a basis for topographic mapping, in Duffy FH (ed.): *Topographic Mapping of Brain Electrical Activity*. Boston, Butterworths, 1986.

Chapter 21

EEG in Clinical Diagnosis and Its Relationship to Other Neurological Tests

Recent years have witnessed an unprecedented development in techniques for diagnosis of neurologic disorders. Notable among them are computerized tomography (CT), magnetic resonance imaging (MRI), neurosonography, positron-emission tomography (PET), brain electrical activity mapping, and averaged evoked potentials. Most of these procedures are imaging techniques in which an image or map of the brain is constructed that reveals structural and, in some cases, functional details. These procedures have virtually revolutionized neurologic diagnosis and management. It may be said without reservation that the practice of neurology and neurosurgery has drastically changed with the advent of the CT scan, the first practical brain-imaging technique to become available to the clinician. It may even be impossible for a current neurology trainee to imagine how one could have practiced neurology or neurosurgery without the help of CT scans and relying purely on clinical judgment and certain "indirect" investigations such as pneumoencephalogram or ventriculogram to arrive at the diagnosis.

In this chapter we discuss the use of electroencephalography in clinical diagnosis and its relative value in the context of some of these recently developed investigative procedures. The aim is to give the physician an insight into the optimal use of electroencephalography in neurological diagnosis. The technologist will also benefit from reading this chapter by gaining some understanding of the various neurodiagnostic procedures. Let us first briefly look at these procedures.

Computerized Tomography

In the routine x-ray examination of the head, only the skull bones are visualized. However, it is well known that there are significant differences in the quantity of x-rays absorbed by the different tissues. In a CT scan of the head, the extent of absorption of the x-rays by various tissues is quantified during small-dose x-ray penetrations of the head from multiple directions. By the use of computer analysis, an entire cross-section of the brain can be mapped out, with clear differentiation of the densities (x-ray absorption coefficients) of the different areas. Such tomographic "slices" can be visualized as images or pictures on a cathode-ray tube in which the shading of each point in an image is proportional to its x-ray absorption coefficient. These images can, in turn, be printed out on an x-ray film. Intravenous injection of iodinated contrast material (intravenous pyelogram dye) may be used to enhance the density of the vascular structures and delineate areas where there is a disruption of the blood-brain barrier.

The CT scan is a relatively non-invasive procedure that gives an image of the brain resembling the tissue slices that the neuropathologist examines. The test is indicated in patients presenting with focal neurological deficits, altered mental status, head trauma, new onset seizures, increased intracranial pressure, suspected mass lesions, and subarachnoid hemorrhage. Although it is highly sensitive and, to some extent, specific in documenting a large variety of intracranial pathology, its limitations include inability to show lesions that are very small or isodense with the brain tissue. Thus, small plaques of multiple sclerosis may not be visible in the CT scan. Even a large infarct, being isodense, may not be visible in the first 24 to 48 hours. Similarly, an isodense subdural hematoma may also be missed. Moreover, it is also important to remember that the CT scan may show abnormalities that are not relevant to the patient's current illness, e.g., old infarcts, agenesis of the corpus callosum, etc.

Thus, although the CT scan is a highly sensitive test, a normal study does not completely exclude intracranial pathology; nor does the presence of a lesion necessarily imply that the particular lesion is responsible for the current illness of the patient. For these reasons the CT scan does not replace other investigations like EEG or arteriography. In many instances these investigations should be viewed as complementary rather than mutually exclusive.

Magnetic Resonance Imaging

Magnetic resonance imaging, which is rapidly becoming the "ultimate test" for neurologic diagnosis, is an excellent method, indeed, for visualizing the structural details of the brain and the spinal cord. Both sagittal and coronal sections can be visualized using this technique. The patient is placed in a powerful magnetic field that tends to make the protons of the tissues align themselves in the orientation of the magnetic field. These protons can be made to resonate and change their axis of alignment by introducing a radio-frequency pulse into the magnetic field; when the radio-frequency pulse is turned off, the protons return to their original position. An image of the tissue is constructed by computer analysis of the radio-frequency energy that is absorbed and then emitted by the protons in the tissue.

It seems likely that the MRI scan will replace the CT scan in the near future as the test of choice in most central nervous system (CNS) disorders as it provides better resolution than the CT scan. Because of its present limited availability and high cost, the MRI scan currently is used primarily for (1) patients suspected of having multiple sclerosis (MS)—it probably is the best means of visualizing the plaques of MS, (2) patients with posterior cranial fossa lesions, as the CT scan tends to show too many artifacts in posterior cranial fossa images, and (3) demonstrating spinal cord lesions. No currently available test is capable of showing spinal cord parenchyma like the MRI scan.

Positron-Emission Tomography

Unlike the CT and MRI scans, which reveal static structural lesions, the PET scan provides insight into the functional or biochemical anatomy of the CNS. Here the brain images are constructed on the basis of the amount of radioactivity emitted from certain chemicals taken up by the brain. Specifically, positron-emitting isotopes of very brief half-life that are produced in a cyclotron are injected into the patient's circulatory system for the purpose of constructing such images. One of the common chemicals used, ^{18}F-fluoro D-oxyglucose (FDG), is handled by the brain cells as glucose, and hence it provides a measure of the cerebral glucose metabolism. One may also study structures that produce or transport a number of neurotransmitters, like dopamine, using suitable radioisotopes. The test is becoming increasingly useful in the study of the metabolic basis of a large number of neurologic disorders including epilepsy.

Neurosonography

Using ultrasound, an image of brain anatomy can be created since the sulci, gray-white junctions, ventricular walls, and blood vessels constitute multiple reflective interfaces. The technique is used to study the structural details of the brain; it is particularly useful in infants as the ultrasound can be introduced through the open fontanelle. It has been quite valuable in the diagnosis and management of intracranial hemorrhage, particularly in the premature infant.

Brain Electrical Activity Mapping

This topic is discussed in detail in Chapter 19: "Clinical Use of Brain Electrical Activity Mapping."

Evoked Potentials

The theory and practice of evoked potential methods are taken up in detail in Chapter 17: "Average Evoked Potentials."

Role of EEG in Relation to Other Neurodiagnostic Tests

The availability of a number of relatively noninvasive diagnostic tests raises the challenge of optimal utilization of these tests in a given clinical setting. To use the tests appropriately, one needs to have a clear understanding of the *sensitivity* and *specificity* of each. The practice of ordering several tests simultaneously without sufficient rationale is not only uneconomical but also often counterproductive. Thus, it is possible to be misled by trivial findings in one or more tests that may be totally irrelevant to the patient's current illness. On the other hand, it must also be pointed out that the tests may not be mutually exclusive; they may even be complementary, depending on the nature of the clinical problem.

With the exception of the PET scan, brain electrical activity mapping, and evoked potentials, the tests mentioned will show abnormalities only if structural or anatomical changes have resulted from the disease process. But there are a number of neurologic disorders in which cerebral dysfunction is known to occur without an obvious structural lesion. The classic examples are those disorders that cause (1) intermittent disturbance of function such as epilepsy and sleep disorders, and (2) persistent disturbance as in the case of diffuse encephalopathies. Presently, the EEG and evoked-potential recordings are the only readily available tests capable of providing information on functional alterations in such disorders. Electroencephalography is totally noninvasive, is of modest cost, and provides exclusive second-to-second information on the functional status of the cerebral cortical neurons over a period of time. These facts are sometimes overlooked and often underemphasized.

Before discussing the specific indications for electroencephalography in relation to other neurodiagnostic tests, it must be stressed that the EEG is a recording of electrical activity originating from large numbers of neurons in the cerebral cortex and that these neurons are known to react in a somewhat stereotyped manner to different types of injuries and insults. Hence, only rarely do we find a situation where the EEG is conclusively and specifically diagnostic of a particular disease. Nevertheless, there are a number of conditions in which certain EEG abnormalities point to the "most likely" diagnosis. In other situations, the EEG and evoked-potential recording are used as complementary investigations to other tests like the CT or MRI scan; in this way, they help to arrive at the final diagnosis.

The conditions in which the EEG provides somewhat specific diagnostic information include seizure disorders, sleep disorders, and certain forms of encephalitis. It is similarly useful in the documentation of brain death. The EEG may provide supportive diagnostic information in conditions such as cerebrovascular disease and head injuries. In addition, it is also quite useful as a prognostic tool in a variety of CNS disorders. Let us take a look at some of these disorders from the viewpoint of the role of electroencephalography in relation to other neurodiagnostic tests.

Seizure Disorders – General Considerations

The EEG is the most important test in the evaluation of seizure disorders, as it provides diagnostic and prognostic information in the majority of patients. Let us briefly review the practical aspects of the role of electroencephalography in the evaluation of seizure disorders.

The two basic questions to address are the sensitivity of the EEG for the diagnosis of seizure disorders and the specificity of individual abnormal EEG patterns relative to the different types of epilepsy. There are a number of investigations that have addressed these questions. The routine EEG study (awake, hyperventilation, and photic stimulation) has been found to show epileptiform abnormalities in 90% of patients with absence. In primary epilepsy manifesting as generalized tonic-clonic seizures, the sensitivity is less and can vary from 20% to 60%, depending on whether the patients have associated features like myoclonic jerks. These data should be adequate to communicate the essential message: a normal EEG does not necessarily rule out a genuine seizure disorder. Interictal abnormalities are more likely to occur if a sleep record is also done, the yield being even greater if the patient has been sleep deprived overnight (see Chapter 16: "Activation Procedures"). Sequential EEG studies also increase the chance of detecting interictal abnormalities.

From the standpoint of specificity, there are a number of EEG patterns that are highly correlated with certain types of specific seizures disorders. Thus, the typical 3-Hz generalized spike and wave discharge present on a normal background rhythm has a high correlation with absence; however, it should be pointed out that about one third of asymptomatic siblings of patients with absence seizures may also show this EEG abnormality. In other words, the presence of this EEG pattern does not necessarily indicate that the patient suffers from absence and needs treatment: it is important to make sure that there is adequate clinical correlation. In this context, it is appropriate to point out that one is treating the patient and not the EEG abnormality.

Certain patterns like hypsarrhythmia and slow spike and wave discharges show a high correlation with infantile spasms and the Lennox-Gastaut syndrome, respectively. On the other hand, there are certain patterns like 6-Hz phantom spike and wave, 14- and 6-Hz positive spikes, and small sharp spikes (SSS) that mimic epileptiform activity but have no significant correlation with seizure disorders (see Pseudoepileptiform Patterns, Chapter 15).

The distinction of seizures from seizure-like phenomena such as syncope, transient ischemic attacks (TIA), cataplexy and conversion reaction may, in some instances, ultimately depend on EEG studies. Sometimes, long-term EEG recording with simultaneous video monitoring (see Chapter 18) may be necessary in such cases.

There are a number of other questions that may uniquely be answered by EEG evaluation of patients with seizure disorders. Determining the probability of recurrence of seizures after a patient has had his first seizure is a common problem. The EEG is particularly useful in this regard. In one study involving children, it was found that the risk for recurrence of seizures is twice as great if the EEG showed an epileptiform abnormality as compared with a normal EEG. The distinction between primary generalized seizures and focal seizures with secondary synchrony may be impossible without an EEG recording, especially in patients who present with seizures that occur primarily during sleep. The type of abnormality noted also helps in categorizing the seizure and in choosing the appropriate anticonvulsants. Selecting suitable candidates for surgery from among patients with intractable epilepsy also depends on adequate EEG studies. Decisions regarding discontinuing anticonvulsants following a prolonged seizure-free period are often based on whether the EEG has remained normal or abnormal; the chances of recurrence tend to be higher if the EEG has continued to be abnormal. Yet another situation where EEG recording is invaluable is in the diagnosis of nonconvulsive status epilepticus (see below).

Some important aspects relating to the EEG evaluation of various specific seizure disorders are discussed in the following sections.

Febrile Seizures

The interictal EEG is normal in patients with febrile seizures except in the immediate postictal period. Slow activity, focal or generalized, may persist for up to 1 week following a seizure and should not be considered significant for predicting future seizures. However, in a child apparently presenting with febrile seizures, the occurrence of a persistent, definite focal epileptiform abnormality in the EEG shifts the diagnosis from benign febrile seizures to seizures probably resulting from an underlying structural disorder. In such a patient, one may need further tests such as the CT scan. Thus, the EEG plays a significant role in the investigation and management of a child with febrile seizures.

Infantile Spasms

Hypsarrhythmia is the characteristic EEG pattern; it consists of a high-amplitude, disorganized and chaotic background rhythm with shifting multifocal spikes and episodes of electro-decrement, which may accompany the seizures (Fig. 15.33). The clinical presentation consists of minor motor seizures manifesting as flexor or extensor spasms, usually associated with arrest of psychomotor development. This is an age-related syndrome and has an onset usually in the first year of life. In the proper clinical setting, the pattern of hypsarrhythmia is diagnostic of infantile spasm (West's syndrome), which may be idiopathic or secondary to disorders like cerebral anoxia, tuberose sclerosis, phenylketonuria, etc. Although electroencephalography is the most important diagnostic test in infantile spasms, other investigations like CT scan and neurometabolic screening are complementary procedures for arriving at an etiological diagnosis.

Lennox-Gastaut Syndrome

The diagnosis of this condition is again based on the clinical presentation coupled with a typical EEG abnormality. The disorder occurs in early childhood; it presents with a multiple variety of seizures (myoclonic, tonic, tonic-clonic, absence, etc.) accompanied by mental subnormality. The typical EEG pattern is one of generalized sharp- and slow-wave complexes (Fig. 15.30) that usually occur at a frequency of less than 2.5 Hz. This slow spike and wave activity, or petit mal variant as it is termed, differs from that in absence which typically shows 3-Hz spike and wave activity. Unlike the tracing in absence, the background activity in Lennox-Gastaut syndrome tends to be slow. During seizures, the initially high-amplitude sharp and slow-wave complexes may be followed by paroxysmal activity that consists of low-voltage fast activity in the alpha or beta frequency band; at this time, there may be transient suppression of the interictal sharp and slow-wave activity.

Primary Generalized Epilepsy

There are two major types of primary generalized epilepsy, namely, absence (petit mal) and tonic-clonic seizures (grand mal). Interictal patterns are extremely important in the diagnosis of these disorders, as ictal patterns may not occur during routine EEG recording.

Absence Seizures

The most common clinical presentation is in the form of staring spells. These are best described as lapses of consciousness that last for a few seconds. Characteristically, the onset is in childhood, and the seizures seldom persist beyond adolescence. The typical EEG pattern of absence is high-amplitude, generalized 3-Hz spike and wave complexes occurring bisynchronously. The rate may vary from 2½ to 4 Hz; the highest amplitude is in the frontal area, although rarely it may occur in the occipital area instead. The discharges may last from one to several seconds. Although discharges of very brief duration (< 3 seconds) are considered interictal and prolonged discharges ictal, the distinction is rather nebulous and depends on the feasibility of detecting changes in level of consciousness within very short intervals of time. Prolonged discharges lasting 12 seconds or more often lead to automatisms and confusional states.

A characteristic feature of absence seizures is the ease with which they can be brought on by hyperventilation (Fig. 16.4). As mentioned earlier in Chapter 16, hyperventilation should be continued for five minutes in a suspected case of absence if no discharges occur in the first three minutes. During sleep, the discharges tend to become less well organized. In most patients, high-amplitude generalized 3-Hz delta paroxysms are seen; but by themselves these are not diagnostic of absence seizures. The background activity tends to be normal in children with absence, and there is no postictal slowing. When the typical EEG pattern is noted in a child with a clinical history suggestive of absence, there is little indication for further tests like the CT scan. However, if there is a consistent asymmetry of the discharges or the onset occurs later in life, one may have to rule out an underlying lesion by other tests like the CT or MRI scan.

Generalized Tonic-Clonic Seizures (Grand Mal)

The ictal pattern is quite typical in generalized tonic-clonic seizures, but the interictal pattern may take different forms. During a seizure, there initially is a repetitive discharge of spikes, or fast rhythmic activity, which may be in the range of 10 to 20 Hz. Following this, there is a progressive slowing of the discharge rate and an increase in the amplitude of the activity. During these periods, the patient goes through the tonic phase. Soon the clonic phase sets in, during which time generalized spikes coinciding with the clonic jerks are noted. These are bilaterally synchronous and symmetrical, although muscle artifacts usually make identification difficult. The spikes are often followed by slow waves, which correspond to the periods of relaxation noted in between the clonic jerks. The clonic movements gradually become less frequent and then stop abruptly; thereupon, there is a marked suppression of EEG activity for varying periods of time. Following this period of attenuation, the EEG is very slow (postictal slowing) after which it gradually returns back to the preictal pattern.

Interictal abnormalities may consist of 2 to 4 Hz or faster, bilaterally synchronous, spike and wave complexes or spikes in generalized distribution (Fig. 15.29). Sometimes polyspikes (multispikes) or polyspike and wave complexes are noted. The terms generalized atypical, irregular spike and wave, and fast spike and wave, have all been used to describe these discharges in order to distinguish them from the 3-Hz spike and wave discharges that are characteristic of absence.

A child with grand mal seizures who has a positive family history and an EEG pattern of generalized epileptiform activity may not need other tests like the CT scan. However, adult onset of the seizures or suspicion of secondary synchrony in the EEG (Fig. 15.41) should alert the physician to the possibility of an underlying structural lesion. Grand mal seizures with a definite aura should also suggest that one is probably dealing with a focal-onset seizure with secondary generalization. Under such circumstances, CT or MRI scan should be the next step.

Partial (Focal) Epilepsy

Unlike generalized epilepsy, patients with partial epilepsy show clinical and/or EEG evidence of focal onset of the seizures. The initial symptoms will depend upon the site of origin of the discharge. Thus, for example, a seizure focus in the motor cortex may cause clonic movements of the contralateral upper or lower extremity, whereas an occipital focus will cause visual sensations like flashes of light. These are simple partial seizures. Further components of the seizure are dictated by the mode of spread of the epileptic discharge. Transcortical spread leads to Jacksonian march,[1] which sequentially involves areas of the body represented in the motor and sensory cortex (see Appendix 2). If the discharges spread rapidly and involve both cerebral hemispheres through secondary bilateral synchrony, the patient develops a generalized tonic-clonic seizure. When the discharge originates in certain areas like the temporal or orbitofrontal cortex, there is associated confusion and automatisms; the term complex partial epilepsy is used in this context.

Partial seizures should alert the physician to the possibility of an underlying focal structural lesion. However, there are forms of familial partial epilepsy in children where no structural lesions are found; such cases are designated benign partial epilepsy of childhood. The role of EEG testing in the partial epilepsies is discussed below.

Rolandic Epilepsy

Known by several eponyms, such as benign epilepsy of childhood with Rolandic spikes (BECRs), Sylvian epilepsy, and centromidtemporal epilepsy, this is an electroclinical syndrome in which the diagnosis depends upon a characteristic EEG pattern. The epileptiform activity, which is markedly accentuated in drowsiness and stage II sleep, consists of spikes arising from the central and midtemporal (C3, T3 and C4, T4) or central and parietal (C3, P3, and C4, P4) areas unilaterally or bilaterally. When appearing in both hemispheres, the spikes switch sides from time to time (Figs. 15.19 and 15.20). This is considered to be a common seizure disorder, accounting for about 25% of childhood epilepsies. The most common age of onset is between 5 and 8 years. Rolandic seizures seldom persist beyond 14 to 15 years of age.

The clinical manifestations include parietal or sometimes generalized seizures, usually nocturnal. Partial seizures manifest as clonic movements of facial and oropharyngeal muscles (leading to guttural sounds), tongue, and upper extremities. Speech arrest and salivation are common. There is usually a positive family history. Although the EEG pattern in conjunction with the clinical picture is specific or diagnostic, it must be pointed out that Rolandic spikes may occur without an accompanying seizure disorder; hence, the presence of spikes alone is not an indication for treatment. A CT scan may not be neces-

[1] Named after Hughlings Jackson, the 19th century neurologist and father of epileptology who first described the disorder.

sary in a child with the typical type of seizure, a positive family history, and the characteristic bilaterally shifting centromidtemporal spikes on a normal background. Nevertheless, any unusual features—e.g., presence of focal neurologic deficits, focal slowing in the EEG, or delayed age of onset of the seizure—should alert one to the possibility of an underlying disorder such as an arteriovenous malformation, a tumor, or a scar. In such cases, further neurologic testing should be undertaken.

A similar syndrome with occipital or parieto-occipital spikes or spike and wave complexes has also been described. The clinical seizures are characterized by a visual aura followed by hemiclonic, tonic-clonic, or complex partial seizures. However, occipital spikes that are unassociated with epilepsy may occur in children with early-onset visual problems.

Other Forms of Partial Epilepsy

Except in conditions like Rolandic epilepsy, patients with clinical focal seizures need tests like the CT scan because the EEG does not reveal the specific cause of the seizure. In newly occurring seizures, especially those starting after the age of 20 years, a CT scan should be done even when the clinical presentation is one of a generalized seizure. This is necessary because a focal onset may not be obvious clinically. The additional neurologic testing becomes even more important if there is a history of aura, focal neurologic deficit, or an EEG focus.

It must be remembered that the EEG is not very sensitive in the case of simple partial seizures and may even be normal in focal motor seizures. Here the clinical findings are more important in diagnosing the type of seizure and ordering of further investigations.

In complex partial seizures, special recording techniques like sphenoidal leads may be necessary. Moreover, as there is a higher incidence of spikes in a sleep EEG, it is important to obtain a sleep recording. Anterior and mesiotemporal spikes are considered more significant in the diagnosis of complex partial epilepsy than midtemporal spikes. Spikes in frontopolar, orbitofrontal, and tempero-occipital areas may also be associated with complex partial seizures. In most patients with complex partial epilepsy, the CT or MRI scan may be necessary to rule out underlying structural lesions in the temporal or frontal lobes.

Nonconvulsive Status Epilepticus (NCSE)

The EEG is valuable in confirming a diagnosis of NCSE. Sudden onset of confusion, inappropriate behavior, memory problems, decreased responsiveness, and automatisms should arouse the suspicion of NCSE. The EEG pattern may be one of continuous, generalized 2.5- to 3-Hz spike and wave complexes, which would suggest absence status (Fig. 15.34). Repetitive focal or lateralized activity consisting of spikes, sharp waves, spike and wave complexes, rhythmic slow waves, or fast activity characterize partial complex status. Without the EEG, the distinction between the two conditions cannot be made.

Sleep Disorders

Recording of the sleep EEG together with other physiologic parameters (polysomnography) like respiratory movements, air flow, eye movements, muscle activity, and blood oxygen concentration is essential for confirming the diagnosis of sleep apnea syndrome and categorizing into obstructive, central, or mixed apnea. For the diagnosis of narcolepsy, the multiple sleep latency test is perhaps the most useful investigation. In this test, one tries to document excessive sleepiness as well as rapid eye movement (REM)-onset sleep in order to confirm the diagnosis of narcolepsy. For details about technology and information concerning interpretation of such sleep recordings, the reader is referred to various available monographs that deal with the subject.

The Comatose Patient

Performing an EEG study in comatose patients is a challenge to the EEG technologist as the procedure is usually done at the bedside in the hostile environment of the intensive care unit (ICU). An equally formidable challenge faces the electroencephalographer who interprets these tracings, as numerous artifacts are all too common in such tracings.

To obtain an acceptable recording from the comatose patient, there are a number of steps the technologist must follow. One of the first is to determine the patient's level of consciousness. Find out whether the patient responds when you call his or her name aloud or when a verbal command is given. If there is no response, this should be clearly noted in the worksheet. It is important to repeat some of these commands while taking the EEG and to enter them, as well as any response, directly on the tracing. In an unresponsive patient the reactivity of the EEG to passive opening of the eyes as well as to painful stimuli like pinching or squeezing the calf muscle tendon should also be noted. A reactive tracing will show a decrease in the amplitude of the background activity; sometimes higher amplitude slow activity may appear. Reactivity suggests a more favorable prognosis.

Another aspect of recording in the ICU is the presence of various artifacts not usually seen in the EEG laboratory. Intravenous pumps, respirators, and other devices commonly found in the ICU produce artifacts that have certain distinctive characteristics. The technician needs to become familiar with these various artifacts so that their sources can be identified and the artifacts eliminated if possible. Monitoring of respiration, ECG, eye movements, and bodily activity—which can be picked up by an electrode placed on the forearm—can be quite useful in this regard.

The EEG recording should be considered as complementary to the CT scan in the evaluation of the comatose patient. If the CT scan is normal, the differential diagnosis can be further narrowed down by EEG evaluation; however, the EEG abnormalities seen in a comatose patient are often nonspecific. There is a progressive slowing of the background activity with decreasing levels of consciousness, and a comcomitant increase in the *amount* of activity in the theta or delta frequency bands. The delta activity may be intermittent and rhythmic in the case of diffuse encephalopathies like metabolic disorders; it may be diffuse and polymorphic as in certain forms of encephalitis; or it may be focal or lateralized and polymorphic in cases of supratentorial lesions. The focal and diffuse encephalopathies are discussed in detail in later sections of this chapter.

The EEG does not give adequate clues concerning the specific cause of coma, although it is easy to make a distinction between psychogenic stupor, diffuse encephalopathy, and focal encephalopathy. Nevertheless, there are certain EEG findings that may aid in identifying the possible cause of the stupor or coma. These are considered in what follows.

The presence of typical triphasic waves (Figs. 15.50 and 15.51), although not pathognomonic, should suggest the possibility of hepatic encephalopathy. A combination of slow background activity, rhythmic delta waves, and a photoparoxysmal response to photic stimulation is often seen in uremic encephalopathy. Presence of generalized beta activity on a diffusely slow background in a comatose patient is most likely to be the result of drug intoxication. Diffuse slowing, burst suppression, or electrocerebral silence may follow cerebral hypoxia; myoclonic jerks accompanying periodic epileptiform activity on a markedly attenuated background is another common pattern (Fig. 15.49). Yet another EEG pattern that occurs following hypoxia is alpha coma. In alpha coma, activity in the alpha frequency band that is unresponsive to passive eye opening is seen in diffuse distribution (Fig. 15.3), often with greater amplitude in the anterior regions. This pattern is also known to occur following bilateral brain-stem lesions that involve the pontine tegmentum. Such lesions usually are due to infarction or hemorrhage. Thus, a patient who appears comatose and has an alpha rhythm in the EEG may have alpha coma resulting either from cerebral hypoxia or brain-stem lesions or from drug intoxication; or he/she may have psychogenic coma or locked-in syndrome. In the latter two conditions, the EEG shows an alpha rhythm having a normal distribution and normal reactivity; hence, these patients are not really comatose.

Another condition that is diagnosed on the basis of the EEG pattern is NCSE-causing coma. Additionally, the differential diagnosis of a patient with prolonged confusional state requires that a number of conditions be considered. These include postictal state, transient global amnesia, drug intoxication, electrolyte imbalance, and psychiatric disorders. The EEG shows focal discharges in complex partial status as against 3-Hz generalized spike and wave activity in absence status. In psychiatric disorders, the EEG tends to be normal, whereas in electrolyte imbalance the pattern is one of diffuse slowing.

Electrocerebral Silence

Absence of electrocerebral activity above 2 μV is considered to be indicative of electrocerebral silence (ECS). The American EEG Society has stipulated the technical standards to be followed when such recordings are done. The technician should consult the American EEG Society guidelines for details (American EEG Society Guidelines Three, 1986). Electrocerebral silence may result from brain death, hypothermia, or drug intoxication. The last two are reversible and should be excluded before suggesting brain death as the cause of ECS.

Diffuse Encephalopathies

Diffuse encephalopathies may result from a number of etiological factors ranging from infections to dementia. The clinical presentation may vary from seizures to coma. Except in a few circumstances, the EEG abnormalities are nonspecific and do not point to the cause. Thus, in encephalitis resulting from a number of different viruses, the finding may simply be diffuse slowing. Similar findings are noted in metabolic encephalopathies as well. In many of these conditions when the EEG is grossly abnormal, the CT scan may not reveal any specific change. Although the EEG findings are generally nonspecific, there are a few instances where they may give important clues to the underlying disorder. For example, the presence of triphasic waves suggests a metabolic encephalopathy like hepatic encephalopathy—although similar EEG findings have been reported in uremic and hypoxic encephalopathies as well as in encephalopathy following head trauma. In the following sections, we discuss the metabolic and infectious encephalopathies.

Metabolic Encephalopathies

These disorders lead to progressive obtundation and eventually coma. The EEG pattern shows parallel changes consisting of diffuse slowing of background rhythms resulting in theta or delta activity. Intermittent bursts of rhythmic high-amplitude, often frontally dominant delta activity are a common feature. This is referred to by the acronym FIRDA, or frontal intermittent rhythmic delta activity (Fig. 15.11). Sometimes the amplitude may be posteriorly dominant, in which case the acronym OIRDA—which stands for occipital intermittent rhythmic delta activity—is used (Fig. 15.12). The various metabolic encephalopathies are taken up in turn.

Hepatic Encephalopathy

As mentioned earlier, about one third of these patients may show typical triphasic waves (Figs. 15.50 and 15.51), although the finding is in no way pathognomonic. Reye's syndrome, which is a form of hepatic encephalopathy occurring in children, seldom shows triphasic waves, although all the nonspecific findings of diffuse encephalopathy may be present.

Hypoglycemia and Hyperglycemia

During hypoglycemia there is an accentuated EEG response to hyperventilation and progressive slowing of background activity. Later, intermittent rhythmic delta activity (IRDA) appears. The condition is reversible, with the EEG becoming normal on restoration of normal blood sugar. Hypoglycemia may also enhance preexisting epileptiform activity.

In hyperglycemia that accompanies the diabetic nonketotic, hyperosmolar state, nonspecific slowing of the background rhythms is present. Apart from this, focal seizure activity is not uncommon. However, a focal structural lesion is seldom demonstrable.

Renal Disorders

Uremic encephalopathy is characterized by slowing of the background rhythm, FIRDA, occasional triphasic waves, and photoparoxysmal or photomyogenic response to photic stimulation. The slow activity gradually disappears following dialysis. Some patients show a temporary worsening of the EEG pattern and a decline in mental status (dialysis dysequilibrium syndrome) during dialysis. In some patients who have undergone dialysis for several years, dialysis dementia (progressive dialysis encephalopathy) may set in. Clinically, the patient becomes progressively demented and devlops focal abnormalities in the EEG pattern. Epileptiform abnormality in the form of sharp- and slow-wave complexes may be seen. As the EEG abnormalities may precede the onset of the clinical syndrome, they may be of predictive value.

Infectious Encephalopathies

In the acute phase the EEG shows diffuse slowing, often with epileptiform abnormality. There are certain EEG findings that may provide clues to the underlying cause. The types of encephalitides that show more specific changes include conditions like herpes simplex encephalitis (HSE) and subacute sclerosing pan encephalitis (SSPE).

In HSE, focal slowing and epileptiform discharges are noted over the temporal or frontotemporal areas early in the course of the disease (Fig. 15.47), followed by periodic complexes either lateralized to one side or generalized. Indeed, if HSE is suspected in a given patient, an emergency EEG study is indicated so that specific antiviral agents may be started promptly if the EEG shows the typical features.

Another condition in which EEG abnormalities may be diagnostic is SSPE. In this condition, generalized periodic complexes are noted at specific intervals, often associated with myoclonus. The complexes are of high voltage (100 to 1,000 μV) and recur at intervals of 4 to 14 seconds. An infectious encephalopathy that leads to dementia is the slow viral infection known as Jakob-Creutzfeldt disease. As in the case of SSPE, periodic complexes are seen that recur at about 1-second intervals (Fig. 15.46). In these diseases, the EEG is much more useful than the CT scan for an early diagnosis.

Dementia

In patients suspected of having dementia, the EEG is often quite informative. A decrease in the frequency of the alpha rhythm, although nonspecific, is a consistent finding in different forms of dementia—including Alzheimer's disease—early in the course of the disease. In making the distinction between pseudodementia (often due to depression) and true dementia, definite slowing of the background rhythm of the awake EEG is a very helpful finding. Of course, one needs to remember that a patient with pseudodementia may be taking one or more antidepressant medications that can potentially cause fast activity and/or some degree of slowing of the background rhythms. In a patient with dementia, the presence of generalized periodic complexes strongly suggests the possibility of Jakob-Creutzfeldt disease. A very low-amplitude EEG with background slowing suggests advanced dementia and has been classically described in Huntington's disease. In patients with dementia, the CT scan is an essential complementary test. Apart from ruling out conditions like chronic subdural hematoma and intracranial tumors, the CT scan documents the amount of cortical atrophy and can detect specific findings like the atrophic caudate nucleus seen in cases of Huntington's disease. The CT scan is also essential for the diagnosis of normal pressure hydrocephalis, a treatable form of dementia.

Focal Encephalopathies

It is perhaps in the field of focal cerebral lesions that the approach of the neurologist has undergone such drastic changes with the advent of CT and MRI scans. Until recently, patients with focal neurologic deficits had EEGs taken prior to having invasive procedures such as angiography or ventriculography carried out. With the universal availability of CT scanners, this pattern has changed. In this context, it may be stated at the outset that there are some situations where the EEG provides valuable information, and others where it is worthless. Take, for example, a suspected case of intracranial tumor. Irrespective of whether the EEG findings are normal or abnormal, it is the CT scan—especially when both routine and contrasted scans are obtained—or the MRI scan that will determine the diagnosis as well as the management. However, this may not always be true. In a patient with a space-occupying lesion and rapidly progressive obtundation, the latter might be the result of accompanying subclinical seizures; without an EEG this cannot be established nor can appropriate treatment be instituted. It should also be noted that there are certain situations in which the EEG may show abnormalities while the CT scan may be spuriously normal, e.g., isodense subdural hematoma and early infarct.

In focal lesions the type of EEG abnormality depends on the site and size of the lesion. When the focal lesion is in the subcortical white matter, thereby disrupting the thalamocortical connections, one can expect to see dysrhythmic and polymorphic delta activity (PDA). The area over which such activity is seen will depend on how extensively the thalamocortical connections are severed. If there are multiple areas of slow activity, the site of the lesion is likely to be underneath the area that shows the slowest and most irregular activity, irrespective of its amplitude. If the lesion involves both the white and the gray matter, one also can expect PDA; but the amplitude is likely to be smaller in the area directly over the lesion than in the surrounding areas. When the lesion is superficially placed and only involves the cortex, the EEG abnormalities may be less striking. The tracing may show only a decrease in amplitude of the background activity—particularly the beta activity—on the side of the lesion, or it may even be normal. Remember that beta attenuation may occur not only with cortical involvement but also in conditions like subdural hematoma, epidural fluid collection, or even scalp edema.

Electroencephalography is useful as a screening tool in the evaluation of patients with chronic and recurrent headaches. In such patients, presence of focal slowing should suggest the possibility of an intracranial tumor, although similar findings may also occur transiently in classical migraine. The EEG tends to be normal in pseudotumor cerebri. Similarly, the presence of focal PDA in a patient being evaluated for a seizure disorder should alert the physician to the possibility of an intracranial tumor. Electroencephalography is of little value in distinguishing between different types of intracranial tumors. While it can correctly lateralize supratentorial tumors, accurate localization is not always possible. However, this is of little concern in the present era of CT and MRI scans.

If the EEG shows focal PDA but the CT scan is normal, one needs to consider the possibility of a recent nonhemorrhagic infarct, trauma, recent episode of migraine, or postictal slowing. Another situation where EEG testing tends to be extremely valuable is in patients with HSE, where focal slowing and/or epileptiform activity may precede the appearance of CT abnormalities.

Cerebrovascular Disorders

In a patient presenting with an acute neurologic deficit like hemiplegia, the CT scan is the most important test to distinguish between a hemorrhagic and a nonhemorrhagic lesion so that appropriate management can be instituted. In those patients with normal CT findings, electroencephalography is useful in assessing the degree and extent of damage. With a large infarct involving subcortical and cortical areas, the abnormalities are striking and consist of PDA, lack of fast (beta and/or alpha) activity, suppression of sleep spindles, and sometimes periodic lateralized epileptiform discharges or PLEDS (Fig. 15.52). In the case of small cortical lesions, the EEG abnormalities may be minimal. In deep lesions at the capsular and upper brainstem level (for example, lacunar infarcts), they also may be minimal. Transient ischemic attacks cause transient EEG changes, mainly focal or lateralized slowing, but persistence of slowing should alert the physician to the possibility of impending infarction.

Subarachnoid hemorrhage may or may not cause significant EEG changes. Often nonspecific slowing occurs, depending on the patient's level of consciousness. Presence of focal abnormalities may be helpful in locating the site of the aneurysm, but at best this is only a supporting piece of evidence to be used in conjunction with angiographic findings.

In arteriovenous malformations, focal or lateralized slowing and/or focal epileptiform abnormality may occur. In Moyamoya disease, a condition associated with childhood hemiplegia that has subsequent potential for intracranial hemorrhage, excessive and prolonged slowing in response to hyperventilation may occur.

Head Trauma

It may be stated at the outset that EEG testing has only a secondary role in the diagnosis and management of head trauma. Imaging techniques such as the CT scan are the most important investigations for this purpose. This is particularly true in the case of severe head injury when life-threatening complications like intracranial hemorrhage may occur. When considering the role of electroencephalography in head trauma, a number of questions are often asked. How can EEG recording help in delineating the extent and severity of brain damage? Can the EEG serve as a prognostic index? Is EEG recording useful in predicting the occurrence of posttraumatic epilepsy?

Diffuse slowing of the EEG is a nonspecific finding that may be seen after concussion. Localized slowing, especially polymorphic delta, suggests a contusion, even in the absence of clinical or CT abnormalities. The gradual change of the EEG pattern back to normal but with symptoms persisting may be helpful in separating neurologic sequelae from psychological problems. Early EEG patterns following head trauma have shown no consistent predictive value for posttraumatic epilepsy. In recent investigations, multimodality-evoked-potential studies have been shown to be better predictors of the extent of brain injury and sequelae than EEG studies. For details about the indications for evoked-potential studies, refer to Chapter 17.

Reference

American EEG Society Guidelines Three: Minimum technical standards for EEG recording in suspected cerebral death. *J Clin Neurophysiol* 1986; 3 (suppl 1):12–17.

Appendix 1
Glossary of Major Terms Used in the Text

A_1: Denotes an electrode location on the left ear lobe.

A_2: Denotes an electrode location on the right ear lobe.

Absence: Refers to an epileptic attack characterized by a brief lapse of consciousness in which there is a sudden, momentary pause in conversation or movement.

AC: Alternating current.

Action potential: A propagated change in which the membrane potential of a neuron suddenly becomes positive for a brief period of time. This change is known as depolarization. Neurons communicate with each other through generation of action potentials.

Activation: (1) Any procedure designed to enhance or elicit normal or abnormal EEG activity, especially paroxysmal activity. Examples are hyperventilation, photic stimulation, sleep, and injection of convulsant drugs. (2) An EEG tracing consisting of low-voltage activity that results from the attenuation of EEG rhythms by physiological or other stimuli.

Active ear: The case in which a focus of activity is located in the temporal lobe adjacent to the ear.

A/EEG: Acronym for ambulatory cassette EEG.

Alpha activity: See alpha rhythm.

Alpha band: Frequency band of 8 to 13 Hz denoted by the Greek letter α.

Alpha rhythm: Rhythmic activity at 8 to 13 Hz occurring during wakefulness over posterior regions of the head, generally of higher amplitude over the occipital areas. Amplitude is variable but is mostly below 50 µV in the adult. It is best seen with the eyes closed and under conditions of physical relaxation and relative mental inactivity. The alpha rhythm is attenuated by attention, especially visual attention, and by mental effort.

Alpha variant rhythm: Applies to certain characteristic EEG rhythms that are recorded most prominently over the posterior regions of the head and differ in frequency but resemble the alpha rhythm in reactivity. The frequency of the variant rhythm may be either an even fraction or multiple of the person's alpha rhythm.

Alpha wave: Wave with duration of 1/8 to 1/13 (0.125–0.077) seconds.

Alternating current: An electric current that undergoes periodic reversals in the direction of its flow. The current supplied by power lines in the United States is almost universally of this type and has a frequency of 60 Hz; in Europe the frequency is 50 Hz. Abbreviated AC.

Ammeter: An instrument used to measure the amount of current flowing in an electric circuit.

Ampere: The unit of measurement of electric current that expresses its rate of flow. In electroencephalography, current flow is more conveniently expressed in milliamperes (thousandths of an ampere) or in microamperes (millionths of an ampere).

Amplification: The process of increasing the strength of an electrical signal; it is expressed in a ratio called the amplification factor, or gain.

Amplification factor: See gain.

Amplifier: A combination of electrical and electronic components designed to increase the voltage, current, or power of an electrical signal.

Amplitude: Voltage of EEG waves. Generally expressed in microvolts (µV) and measured peak-to-peak. The amplitude of EEG waves recorded from the surface of the head is influenced by a number of extracerebral factors. These include the impedances of the meninges, cerebrospinal fluid, skull, scalp, and electrodes.

Anode: The positive electrode of a battery. In a vacuum tube the anode is the element to which electrons are attracted from the cathode. This element is referred to as the plate.

Artifact: In EEG, any recorded signal that does not originate in the brain. The EEG artifacts may be of physiological origin (e.g., muscle activity), instrumental origin (e.g., defects in the recording amplifiers), environmental origin (e.g., 60-Hz voltage from the power line), or may originate from the recording electrodes.

Asymmetry: Unequal amplitude and/or form and frequency of EEG tracings over homologous areas on opposite sides of the head, or unequal development of EEG waves about the baseline.

Asymptote plot: The straight line approximation of a frequency-response curve.

Asynchrony: The nonsimultaneous occurrence of EEG activities over regions on the same or opposite sides of the head.

Attenuation: Reduction in amplitude of EEG activity. This may occur transiently in response to stimulation or result from pathological conditions; cf blocking. May also be used to denote the reduction in sensitivity of an EEG channel, as when the pen deflection is decreased by adjustment of sensitivity or filter controls.

Av: Abbreviation for average potential reference.

Average potential reference: The average of the potentials of all or many EEG electrodes, used as a reference. Synonyms: Average reference, Goldman-Offner reference.

Average reference: Same as average potential reference.

Averaged common reference: Same as average potential reference.

Axon: The single, long nerve fiber that conducts impulses from a nerve cell body to the synaptic terminals.

Background activity: Any EEG activity representing the setting in which a given normal or abnormal pattern appears and from which such pattern is distinguished. Not used as a synonym for any individual rhythm such as the alpha rhythm.

BAEP: Acronym for brain-stem auditory-evoked potential.

Balanced amplifier: An amplifier consisting essentially of two identical single-ended amplifiers that are operated as a pair in opposite phases.

Band: Portion of the EEG frequency spectrum; cf alpha, beta, delta, and theta bands.

Bandwidth, EEG channel: Range of frequencies that an EEG channel is capable of recording. Determined by the frequency response of the writer unit and the settings of the frequency filters. The EEG channel bandwidths specified by different manufacturers are not standardized. Thus, in one instrument a bandwidth of 0.5 to 50 Hz may indicate that 0.5- and 50-Hz signals are attenuated 30% (3 dB), whereas in another instrument they may be attenuated only 20%.

Base (or base element): An electrode of a transistor that corresponds roughly to the grid of a triode vacuum tube in that it serves to control electron flow.

Baseline: In an EEG tracing, the line obtained when an identical voltage is applied to the two input terminals of an EEG amplifier, or when the instrument is in the calibrate position but no calibration signal is applied.

BCD: Acronym for binary-coded decimal.

BEAM®: Acronym for brain electrical activity mapping.

Beta band: Frequency band over 13 Hz. Denoted by the Greek letter β. Normally, this defines frequencies up to and including 35 Hz.

Beta rhythm: In general, any EEG rhythm over 13 Hz. Characteristically it denotes a rhythm from 13 to 35 Hz recorded over the frontocentral regions of the head during wakefulness. Amplitude of the frontocentral beta rhythm is variable but is mostly below 30 μV. Other beta rhythms may also be present; these are most prominent in other locations or are diffuse.

Beta wave: Wave with a duration of less than 0.077 second and usually forming a part of a beta rhythm.

Binary-coded decimal: A system of indicating real time on an EEG chart in which the numbers corresponding to hours, minutes, and seconds are represented by a pattern of deflections on a marker channel.

Bio-cal (bio-calibration): Procedure in which one pair of EEG electrodes is connected to all channels of the EEG machine.

Bioelectricity: Electrical activity of living tissue. In the 18th and 19th centuries was referred to as animal electricity; cf biopotential.

Biopotential: A difference of potential (voltage) of biological origin.

Bipolar, bipolar derivation: A method of recording that uses two scalp electrodes connected to an EEG channel. Both leads pick up significant activity from the underlying brain.

Bipolar montage: Multiple bipolar derivations with no electrode being common to all derivations. In most instances, bipolar derivations are linked so that adjacent derivations along the same array have one electrode in common.

Blocking: In an EEG tracing, the apparent, temporary obliteration of EEG rhythms in response to stimulation; cf attenuation. In an amplifier, a condition of temporary unresponsiveness of the amplifier that is caused by a major overload. Amplifier blocking is manifested by extreme, flat-topped pen excursion(s) sometimes lasting up to a few seconds.

Buffer amplifier: An amplifier, generally with a voltage gain of 1, having a high-input impedance and a low-output impedance. Buffer amplifiers are used to isolate the input signal from the loading effects of an immediately following circuit.

C: Abbreviation for a capacitor in electrical circuits.

C/sec, c/s, cps: Abbreviations for cycles per second. Equivalent to Hz, which is the preferred form.

Calibration: (1) Procedure for testing and recording the responses of EEG channels to voltages of known magnitude applied to the inputs of the machine, and (2) refers to the procedure of testing the accuracy of the paper speed by means of a time marker.

Cancellation: In EEG, the effect produced when the input signals on the two grids in any channel are similar with respect to frequency, amplitude, polarity, waveform, or any combination. When the signals at the two inputs have similar characteristics, the output will be decreased; when they are identical, the output becomes equal to zero.

Capacitor: A device for storing electric charge. The simplest type consists of two conductors separated by an insulator. Used in the filters of an EEG machine to attenuate both high- and low-frequency waves. Direct currents are blocked by a capacitor; alternating currents can pass across a capacitor. Also called condenser.

Cathode: The negative pole or electrode of a battery. In a vacuum tube, the cathode is the element that emits electrons.

Channel: The complete system for the detection, amplification, and display of potential differences between a pair of electrodes. The tracings on the chart coming off an EEG machine are referred to as channels.

Charge, electric: The quantity of electricity held by an object. When the object has more electrons than normal, it has a negative charge; when it has fewer electrons than normal, it has a positive charge. The term also refers to energy stored in a battery as when one says that a battery is charged.

Circuit: Any complete path over which electric current can flow. Also refers to the drawing that shows the way in which the components that comprise an electrical or electronic device are hooked together.

Coherent averaging: The process whereby signal averaging is carried out by a computer. See signal averaging.

Coil: A number of turns of wire, sometimes wound on an iron core. A coil is one of the basic elements of a penmotor or writer unit.

Collector: The element of a transistor that collects charges. This corresponds roughly to the plate of a triode vacuum tube, since it is usually part of the output circuit.

Common-mode rejection (CMR): A characteristic of differential amplifiers whereby they provide markedly reduced amplification of common-mode signals, compared with differential signals. Expressed in quantitative terms as common-mode rejection ratio (CMRR), or the ratio of amplifications of differential and common-mode signals. For example:

$$\frac{\text{amplification, differential}}{\text{amplification, common mode}} = \frac{20{,}000}{1} = 20{,}000{:}1$$

Common-mode signal: Common component of the two signals applied to the two respective input terminals of a differential EEG amplifier. In EEG recording, external interference like 60-Hz pickup from the power line occurs as a common-mode signal.

Complex: A sequence of two or more waves that occur together and repeat at fairly consistent intervals.

Condenser: See capacitor.

Conductor: Any material (solid, liquid, or gas) that offers little opposition to current flow. Metals and salt solutions are examples of conductors. Silver and copper are among the best conductors.

Coronal: Refers to a crown or circle about the head; a coronal montage consists of electrodes placed in transverse arrays across the head from left to right.

Convulsion: An involuntary paroxysm of muscular contraction. It may be either tonic (without relaxation) or clonic (showing alternate contractions of opposing groups of muscles). It may be generalized or focal (localized to a specific part of the body).

Current: An electrical current is the directional movement of free electrons in a conductor under the force provided by a difference of potential. A biological electrical current is the movement of ions.

Cursor: A mark or indicator, such as an intensified dot or pip on the screen of a cathode-ray tube. The cursor can be positioned to measure amplitude and/or time of a waveform as in an evoked potential; or it can select a particular item or event from the screen. Sometimes called a "bug."

Cycle: The complete sequence of potential changes undergone by a wave before the sequence is repeated.

Cycles per second: Unit of frequency. Abbreviated as c/sec, c/s, or cps. The synonym Hertz (Hz) is preferred.

DC: Abbreviation for direct current.

Deflection sensitivity: See sensitivity.

Delta band: Frequency band under 4 Hz denoted by the Greek letter Δ.

Delta rhythm: Rhythmic activity at frequencies less than 4 Hz.

Delta wave: Wave with duration greater than ¼ second.

Dendrite: One of the many branching processes of the nerve cell that serve as the points of input of signals to the neuron.

Depolarize or depolarization: Process whereby the negative resting membrane potential (negative on the inside compared with the outside) is made positive by the influx of positive charges inside the cell membrane.

Derivation: The process of recording from a pair of electrodes on an EEG channel; or, the EEG record obtained by this process.

Differential amplifier: An amplifier whose output is proportional to the voltage difference between the signals applied to its two input terminals. The input stages of EEG machines employ differential amplifiers.

Differential signal: The difference between two signals applied to the respective two input terminals of a differential EEG amplifier.

Diffuse: Used to describe activity that occurs over large areas of one or both sides of the head.

Diphasic wave: Wave consisting of two phases, one negative and the other positive.

Dipole: In EEG, a theoretical concept in which an EEG pattern is assumed to originate from small areas of brain tissue. Each area is thought to contain equal but opposite electrical poles or charges that are separated in space and that change their positions or magnitudes.

Direct-coupled amplifier: An amplifier in which the successive stages are connected together (coupled) by devices that are not frequency-dependent; i.e., condensers are not used.

Direct current (DC): A flow of electrons in only one direction. This type of current does not alternate or change directional flow as with AC.

Disk (disc) electrode: Metal disk that is attached to the scalp with collodion, paste, or wax. Disks are frequently cup-shaped to provide space for the electrolyte.

Double banana: Popular name for the 16-channel longitudinal bipolar montage consisting of left and right temporal and parasagittal chains.

Driving: A condition in which waves in the EEG occur at the same or harmonically related frequency as rhythmically presented stimuli in a phase-locked relationship.

E: Abbreviation sometimes used for voltage. Stands for electromotive force.

Earth connection: Cf ground connection.

Electrical double layer: Two layers, one consisting of positively charged ions and an adjacent layer consisting of negatively charged ions that form when a metal electrode comes in contact with an electrolyte.

Electrocardiogram: Electrical activity generated by the heart. Abbreviated ECG or EKG.

Electrode board: A device with multiple receptacles (jacks) into which the plug ends of electrodes are inserted and from which a multiconductor cable connects the signals to the console of the EEG machine. Also known as an electrode box or jack box.

Electrode box: See electrode board.

Electrode, EEG electrode: A conducting device that makes physical and electrical contact with the subject. An electrode provides a direct connection between the subject and one terminal of the amplifier input. Sometimes referred to as a lead.

Electrode impedance: The opposition offered by an electrode and the scalp to the flow of an AC current. It is measured between pairs of electrodes or, in some electroencephalographs, between each individual electrode and all the other electrodes connected in parallel. Expressed in ohms (generally kilohms, KΩ).

Electrode paste: A blend of materials used to decrease the electrical resistance between skin and electrodes and sometimes, as in the case of disk electrodes, to hold the electrode in place.

Electrode potential: The voltage that develops between metal and an electrolyte when a metal electrode is placed in contact with the electrolyte. Also known as a half-cell potential.

Electrode resistance: Opposition offered by an EEG electrode and the scalp to the flow of a direct current. Resistance is measured between pairs of electrodes or, in some EEGs, between each individual electrode and all the other electrodes connected in parallel. Expressed in ohms (generally kilohms, KΩ).

Electrode selectors: The switching system on an EEG machine that permits the operator to connect various pairs of electrodes to different channels of the machine.

Electron: A negatively charged particle; the basic element of electricity.

Emitter: An electrode of a transistor corresponding roughly to the cathode in a triode vacuum tube in that it emits electrons.

Epilepsy: Cf seizure.

Epoch: A particular interval of time in an EEG tracing, the duration of which is arbitrarily selected.

EPSP: Acronym for excitatory postsynaptic potential.

Equipotential contour: Certain limited region in a volume conductor surrounding a dipole where the voltages referred to a distant reference electrode are all the same. Equipotential contours demarcate the electrical field that surrounds the dipole.

Equipotential zone: Refers to a region on the head in which two or more adjacent electrodes record the same potential at a particular instant in time.

Error of the arc: In curvilinear tracings, the error due to the circumstance that the deflection of the pen describes an arc of a circle rather than a straight line perpendicular to the baseline.

Ethmoidal electrode: Special electrode of flexible, insulated silver wire with a bulbous tip. Introduced into the nose under topical anesthesia so that its tip lies in contact with the cribriform plate of the ethmoid bone, it is useful in recording activity from the orbitofrontal cortex.

Farad: A unit of electrical capacitance. In EEG work, the commonly used unit is microfarad or millionth of a farad, abbreviated μF or MFD.

Filter: A device that allows preselected frequencies or wave components to pass without change but attenuates all other frequencies. On the EEG machine, the low-frequency filters attenuate the amplitude of the low-frequency components of the signal but pass the higher frequencies unchanged; the high-frequency filters do the exact opposite. The 60-Hz notch filter affects primarily those frequencies close to 60 Hz but allows others to pass unchanged.

FIRDA: Acronym for frontal intermittent rhythmic delta activity.

Focal: Refers to activity that is restricted to one region, as, for example, a focal seizure or focal slowing. See focus.

Focus: In general, the center of a limited region displaying the electrical activity generated by a dipole; in EEG, the limited region on the scalp displaying a particular kind of activity, whether normal or abnormal.

Frequency: The number of complete cycles of repetitive waves or complexes in 1 second. Measured in Hertz (Hz), a unit that is preferred to its equivalent, cycles per second.

Frequency response: In general, the response of an electrical circuit to an alternating applied voltage of constant amplitude that is allowed to vary in frequency. In EEG, the term represents the range over which the EEG machine records waves or components in a fairly linear fashion without significant attenuation or distortion; the lower limit of response is set by the low-linear frequency switch, the upper limit by the high-linear frequency control.

Frequency-response curve: A graph of the amplifier output and/or pen deflection in response to an input signal of constant amplitude at all relevant frequencies. A separate curve is derived for each combination of low- and high-frequency filter settings, and these curves may be shown individually or in groups.

Frequency spectrum: The range of frequencies composing the EEG. The spectrum is divided into four bands termed alpha, beta, delta, and theta; cf alpha, beta, delta, theta bands.

F wave: A sharp transient that occurs during sleep with maximal amplitude appearing in the frontal region at the midline.

G1: Abbreviation for grid 1 of the differential amplifiers of an EEG machine.

G2: Abbreviation for grid 2 of the differential amplifiers of an EEG machine.

Gain: The ratio of the voltage obtained at the amplifier output to the voltage applied to the input of an EEG channel. Example:

$$\text{Gain} = \frac{\text{output voltage}}{\text{input voltage}} = \frac{10 \text{ volts}}{10 \text{ microvolts}} = 1{,}000{,}000$$

Often expressed in decibels (dB), a logarithmic unit. Examples: a voltage gain of 10 = 20 dB; of 1,000 = 60 dB; of 1,000,000 = 120 dB. Cf sensitivity.

Gain control switch: The switch that controls the deflection sensitivity of the channels on the EEG machine.

Galvanometer: A device for converting variations in electrical energy to mechanical deflections. In EEG machines, usually a coil of wire (with an attached stylus or pen) that is suspended in a magnetic field and rotates in response to the electrical signals from the output of the amplifier. Commonly called a writer, penmotor, or oscillograph.

Generalized: Used to describe activity that occurs over all regions of the head.

Goldman equation: Equation for estimating the membrane equilibrium potential when more than one ion is involved.

Goldman-Offner reference: Synonym for average reference.

Grand mal: Refers to an epileptic seizure that is characterized by a sudden loss of consciousness, tonic convulsion, followed by clonic spasms. Following the convulsion, the patient is confused and frequently sleeps (postictal state).

Grid: The control element in an amplifier vacuum tube. (In a transistor the corresponding element is the base.)

Grid 1: Input terminal 1 of a differential amplifier. Abbreviated G1.

Grid 2: Input terminal 2 of a differential amplifier. Abbreviated G2.

Ground: The earth; the arbitrarily designated point of zero potential from which potential differences are measured. See also ground connection.

Ground connection: Conducting path that runs between the earth and the EEG machine. Also, the electrode wire that connects the subject to the ground connection of the EEG machine.

Ground loop: The condition that occurs when a patient or a piece of equipment has more than one connection to ground. To ensure the patient's safety and to minimize the amount of externally generated interference in the EEG recording, ground loops should be avoided.

Gyrus: One of the prominent ridges or rounded elevations on the surface of the hemispheres of the brain.

Half-cell potential: See electrode potential.

Hertz: Unit of frequency. Abbreviation: Hz. Preferred to the synonym cycles per second.

High-frequency filter: A circuit that reduces or attenuates the sensitivity of the EEG channel to relatively high frequencies. For each position of the high-frequency filter control, the attenuation is expressed as a percent reduction in pen deflection at a given, stated frequency, relative to frequencies unaffected by the filter.

High-frequency response: Sensitivity of an EEG channel to relatively high frequencies. Determined by the high-frequency response of the amplifier-writer combination and the high-frequency filter used. Expressed as the percent reduction in pen deflection at certain, stated high frequencies, relative to other frequencies in the midfrequency band of the channel.

High-pass filter: Synonym for low-frequency filter.

Hyperpolarize or hyperpolarization: Process whereby the resting membrane potential, which is negative on the inside with respect to the outside, is made more negative by the influx of negative charges inside the cell membrane.

Hyperventilation: Deep respiration performed at a slow, regular rate for a period of several minutes. Used as an activation procedure.

Hypnagogic hypersynchrony: Paroxysmal, high-amplitude rhythmic slow activity that occurs during drowsiness, mainly in infants and young children.

Hypnopompic hypersynchrony: Paroxysmal, high-amplitude rhythmic slow activity that occurs upon arousal from sleep, mainly in infants and young children.

Hypsarrhythmia: A chaotic EEG pattern consisting of high-amplitude, arrhythmic slow waves interspersed with multifocal spike discharges. The activity is without consistent synchrony between left and right sides, or between different areas on the same side.

Hz: Abbreviation for Hertz.

I: Abbreviation for current.

Ictal: The adjective that refers to seizure (ictus).

Impedance: The total opposition to the flow of current in an AC circuit. Includes direct resistance (from wires plus any resistors) and resistance offered by other components (as capacitors).

Impedance meter: An instrument used to measure impedance; cf electrode impedance.

Induction: The production of an electrical field around, or current in, any conductor; due to a change with respect to current in a conductor or magnetic source around the conductor. In EEG recording, 60 Hz can be *induced* in the electrode wires by the magnetic field produced by current flowing in the power lines.

Inion: The prominent protrusion or bump at the back of the head.

In-phase: Coincidence in time of different voltages as, for example, peaks of the same polarity. Commonly, the simultaneous appearance of components being deflected in the same direction in two or more channels of the EEG machine.

In-phase discrimination: Characteristic of a differential amplifier that permits in-phase signals to be discriminated against—or, in preferred terms, common-mode signals to be rejected.

In-phase signals: Waves with no phase (or time) difference between them.

Input: The signal fed into an EEG amplifier. Also, the input terminal or terminals of an amplifier.

Input circuit: The total circuit composed of EEG electrodes and intervening tissues, the electrode lead wires, jack box input cable and electrode selectors. Also used to denote the first part of the electrical circuit of the amplifier.

Input impedance: Impedance that exists between the two input connections of an EEG amplifier. Measured in ohms (generally megohms, MΩ).

Input terminal 1: One of the input terminals of the differential amplifier of an EEG machine. When this terminal is negative relative to the other input terminal, the pen deflects upward. Synonyms: "Grid 1," G1, or black lead.

Input terminal 2: One of the input terminals of the differential amplifier of an EEG machine. When this terminal is negative relative to the other input terminal, the pen deflects downward. Synonyms: "Grid 2," G2, or white lead.

Instrumental phase reversal: Simultaneous pen deflections in opposite directions in two bipolar derivations. This reversal is purely instrumental in nature and is due to the same signal being simultaneously applied to input terminal 2 of one differential amplifier and to input terminal 1 of the other amplifier.

Insulator: Any material (solid, liquid, or gas) offering very high resistance to the flow of any electric current. Commonly used to support or separate conductors so as to prevent undesired flow of current between them or to other objects.

Interface: When used as a noun, the boundary or junction between two objects, media, or bodies. The *metal-electrolyte interface* is the boundary between the metal of an EEG electrode and the electrolyte between it and the surface of the scalp.

Intermittent activity: Activity that appears only from time to time. The term discontinuous is also used.

Ion: An atom or group of atoms carrying an electric charge and forming one of the elements of an electrolyte.

IPSP: Acronym for inhibitory postsynaptic potential.

IRDA: Acronym for intermittent rhythmic delta activity.

IRIG level: Inter-Range Instrumentation Group. Convention that specifies a standard interface and recording level for instrumentation tape recorders (1 volt RMS or 2.828 volts peak-to-peak).

Isoground: A specially designed ground connection, used for safety purposes, that limits any current flowing in the grounding electrode to a safe level in the event of equipment failure.

Isopotential line(s): See equipotential contour.

Jack: A terminal that is a spring-type receptacle used in electronic apparatus for temporary connections. The electrode board or electrode box of the EEG machine uses terminals of this type.

Jack box: See electrode board.

K complex: A biphasic slow-wave transient of variable appearance occurring during sleep with maximal amplitude at the vertex. May occur in association with a sleep spindle.

Lambda wave: A sharp transient that occurs over the occipital regions of the head while awake, with the eyes open, and the subject is engaged in visual exploration. In the main, the waves are positive relative to other regions.

Lateralized: Used to describe activity that is present only on one side of the head.

Lead: Strictly speaking, the wire connecting an electrode to the electrode board of the EEG machine. Loosely, a synonym for electrode.

Leakage current: Inherent and undesirable currents that flow through conductive paths, such as between the AC power line and the chassis of an EEG machine. This current normally flows through a third wire that connects the EEG machine to ground.

Linear: In EEG refers to the recording of signals on the chart in such a manner that the pen deflections are directly proportional to the magnitudes of the voltages at the input. This means that if 50 μV gives a pen deflection of 7 mm, then 100 μV should equal 14 mm, and so on.

Localized: See focal.

Low-frequency filter: A circuit that reduces the sensitivity of the EEG channel to relatively low frequencies. For each position of the low-frequency filter control, this attenuation is expressed as a percent reduction in pen deflection at a given, stated frequency, relative to frequencies unaffected by the filter. Positions of the low-frequency filter control may also be designated by the corresponding time constant in seconds.

Low-frequency response: Sensitivity of an EEG channel to relatively low frequencies. Determined by the low-frequency response of the amplifier and by the low-frequency filter (time constant) used. Expressed as the percent reduction in pen deflection at certain, stated low frequencies, relative to other frequencies in the midfrequency band of the channel.

Low-pass filter: Synonym for high-frequency filter.

Master electrode selector: See montage switch.

Membrane potential: Generally, the voltage that exists between the two sides of a permeable membrane. In the case of a neuron, the potential difference present between the inside and outside of the cell.

Microvolt: One-millionth of a volt; the standard voltage unit in EEG; symbol, micro V or μV.

Millivolt: One-thousandth of a volt; symbol, mV or mv.

Monophasic wave: Wave having only a positive or negative phase.

Monorhythmic: Term sometimes used when a particular EEG pattern shows rhythmic components of a single frequency.

Montage: The particular arrangement by which a number of derivations are displayed simultaneously in an EEG record. Also known as a run.

Montage switch: A device that permits the EEG technician to change from one montage to another simply by changing the position of a single switch. Also called a master electrode selector.

Mu rhythm: EEG rhythm at 7 to 11 Hz; composed of arch-shaped waves (points being negative polarity) occurring over the central or centroparietal regions of the head during wakefulness. Amplitude varies but is mostly below 50 μV. Blocked or attenuated most clearly by contralateral movement, thought of movement, readiness to move, or tactile stimulation.

Nasion: The indentation where the nose joins the forehead.

Nasopharyngeal electrode: Special rod electrode introduced through the nose so that its tip is in contact with the roof of the nasopharynx. Useful in picking up activity from the uncus, hippocampus, and orbitofrontal cortex. Abbreviated NPG.

Negative deflection: By clinical EEG convention, the upward excursion of the pen when the voltage of the electrode connected to input terminal 1 (grid 1) is negative with respect to the voltage of the electrode connected to input terminal 2 (grid 2) of that channel.

Nernst potential: The membrane equilibrium potential for an ion, which is the electrical force required to balance the ionic movements across a membrane that are caused by diffusion.

Noise: Random variations in the output of an amplifier that are seen on the chart as fluctuations of the baseline in the absence of any input signal. Noise is usually due to random movement of electrons in the amplifier components. In some cases it originates from a poor electrode-scalp junction.

Notch filter: A filter that selectively attenuates a very narrow frequency band and, in so doing, produces a sharp notch in the frequency-response curve of an EEG channel. A 60-Hz notch filter is used in some EEG recordings to provide attenuation of 60-Hz interference under extremely unfavorable recording conditions.

NPG electrode: Acronym for nasopharyngeal electrode.

Ohm: The unit of measurement of electrical resistance. Designated by the Greek capital letter omega (Ω). Also, the unit of measurement of impedance.

Ohm's law: The relationship between the *steady* voltage, current, and resistance of any circuit. The law states that the current (I) is directly proportional to the voltage (E) and inversely proportional to the resistance (R) in the circuit. Expressed symbolically as $I = E/R$, $R = E/I$, or $E = IR$.

Ohmmeter: An instrument used to measure resistance; cf electrode resistance.

OIRDA: Acronym for occipital intermittent rhythmic delta activity.

Out-of-phase: Lack of coincidence in time of different voltages as, for example, when waves recorded simultaneously on two or more channels show completely opposite deflections. When the voltages are exactly opposite, they are referred to as being 180 degrees out-of-phase.

Out-of-phase signals: Two waves of opposite phases.

Paper speed: The rate at which the EEG chart paper moves through the EEG machine. Expressed in millimeters per second (mm/s) or centimeters per second (cm/s).

Parameter: A characteristic element or feature of a phenomenon, object, or system. Resistance and capacitance are parameters of electrical circuits.

Paroxysmal activity: Denotes activity that is distinguishable from the background activity and that occurs with sudden onset and offset.

Pattern: Any characteristic EEG activity seen in a tracing.

PDA: Acronym for polymorphic delta activity.

Peak: Point of maximum amplitude of a wave.

Peak-to-peak: A way of measuring and expressing the amplitude of sinusoidal waves. Amplitude peak-to-peak corresponds to the voltage measured between consecutive peaks and troughs of the waves. Compare with RMS voltage.

Penmotor: The electromechanical device that converts the output signal of the amplifier to a mechanical movement that deflects the writing stylus. Synonyms: galvanometer, galvo, writer unit.

Period: Duration of a complete cycle of a wave or complex that occurs in a sequence of regularly repeated waves or complexes. Period is the reciprocal of frequency.

Periodic: Term applied to EEG waves or complexes that recur at approximately regular intervals.

Petit mal: See Absence.

Phase: The time or polarity relationship between a point on a wave displayed in one derivation and the identical point on the same wave recorded simultaneously in another derivation.

Phase reversal: The condition existing when similar waves recorded on two or more channels are 180° out-of-phase, that is, when the waves reach peak values of opposite polarity at the same instant. See instrumental phase reversal and true phase reversal.

Photic driving: The rhythmic activity elicited over the posterior regions of the brain by repetitive photic stimulation at frequencies of about 5 to 30 Hz. This activity is time-locked to the stimulus; frequency is identical to or harmonically related to the stimulus frequency.

Photic stimulation: In clinical EEG, an activation procedure in which a series of brief, brilliant light flashes from a stroboscope provide visual stimulation. Intensity and flash frequency may be varied. The purpose is to evoke latent paroxysmal activity.

Photic stimulator: Device for delivering intermittent flashes of light.

Photomyogenic response: A response to intermittent photic stimulation in which brief, repetitive muscle spikes and movement artifacts associated with eyelid flutter are seen in tracings from anterior regions of the head. The response ceases abruptly as soon as the stimulation is discontinued.

Photoparoxysmal response: A response to intermittent photic stimulation in which generalized, bilaterally symmetrical, synchronous spike and wave or multiple spike and wave discharges are seen in the EEG.

Pin plug: The $\frac{1}{16}$-in. diameter metal plug on the end of an electrode wire. Pin plugs are frequently gold-plated; they fit into the jacks found on the electrode board of an EEG machine.

Polarization: The accumulation of electrical charges on the surface of an electrode resulting from chemical or other changes in the electrode or its medium.

Polymorphic: Irregular, as electrical activity that assumes various forms. Sometimes used to describe irregularly shaped waves appearing in the delta and theta bands.

Polyrhythmic: Term sometimes used when a particular EEG pattern shows rhythmic components of several different frequencies.

Polysomnography: The simultaneous recording of multiple physiological parameters during sleep, such as EEG, ECG, eye movements, muscle tonus, respiration, etc.

Positive occipital sharp transients of sleep: Sharp transients that occur during sleep; they are maximal over the occipital regions and are positive relative to other areas. Abbreviated POSTS.

Posterior slow waves of youth: Slow transients recorded in the waking EEG of normal children from the posterior regions.

POSTS: Acronym for positive occipital sharp transients of sleep.

Postsynaptic potential: See synaptic potential.

Potentiometer: A continuously variable resistance. When suitably connected to a source of voltage, the potentiometer becomes a voltage divider.

Power amplifier: An amplifier in which current as well as voltage is increased in order to augment the input signal.

Power supply: A device that converts alternating current from the power line to direct current of various voltages for use by the amplifiers and other electrical and electronic devices in EEG machines. When circuitry is included to maintain voltages at constant levels that are unaffected by changes in line voltage, the device is referred to as a "regulated" power supply.

Preauricular point: The point in front of the ear that is situated just above the tragus, or triangular piece of cartilage that covers the opening of the external ear canal. Used as a landmark in the 10-20 System of electrode placement.

Proton: A positively charged particle; protons reside within the nucleus of an atom and are electrically balanced by the negative charges of the electrons that surround the nucleus.

R: Abbreviation for resistance in electrical circuits.

Reactivity: Refers to alterations in the amplitude and waveform of activity in response to stimulation.

Record: The end product of the EEG recording process.

Recording: (1) The process of obtaining an EEG record. Synonym: tracing. (2) The end product of the EEG recording process. Synonymous with record.

Reference electrode: In general, any electrode against which the potential variations of another electrode are measured. A suitable reference electrode is an electrode that is customarily connected to input terminal 2 of an EEG amplifier and that is so placed as to minimize the likelihood of picking up the same EEG activity recorded by an electrode connected to input terminal 1 of the same amplifier.

Referential derivation: Recording from a pair of electrodes, one of which is any EEG electrode (usually connected to input terminal 1) and the other, a reference electrode (usually connected to input terminal 2 of an EEG amplifier).

Referential montage: A montage employing referential derivations, or recordings simultaneously taken from different electrodes in comparison with a common reference electrode.

Reformatting (of montages): The process whereby the pattern of activity observed in one montage may be used to derive or infer the pattern of activity that would be seen in another.

Rejection ratio: The gain of an amplifier for out-of-phase signals, as for example the EEG, divided by the gain for in-phase signals such as 60 Hz artifact. See common-mode rejection.

REM: Acronym for rapid eye movement. Used to describe one of the stages of sleep.

Resistance: The property of a substance that limits the flow of electricity through it; or the opposition a conductor displays to the passage of an electric current. The unit of measurement is the ohm.

Resistor: A device that has electrical resistance and is used in an electrical circuit. Abbreviated by the capital letter R.

Resting membrane potential: The voltage between the inside and outside of a neuron when at rest. The inside is negative compared with the outside.

Rhythm: EEG activity consisting of waves of approximately the same period.

RMP: Acronym for resting membrane potential.

RMS: Abbreviation for root-mean-square, which is a way of measuring and expressing the amplitude of sinusoidal waves; 1 volt RMS = 2.828 volts peak-to-peak; cf peak-to-peak.

Rolandic seizure: Refers to seizures arising from the Rolandic area of the brain. They manifest as clonic movements of facial and oropharyngeal muscles, tongue, and upper extremities; speech arrest and salivation may also occur.

Run: A recording in which a particular group of derivations is displayed simultaneously. Preferred term in montage.

Salt bridge: The situation that occurs when the areas to which two scalp electrodes are attached have such an excess of electrolyte spread over them that they overlap each other. When this happens, the electrolyte creates a short-circuit between the two electrode sites. Under such conditions, the recording from this derivation is meaningless.

Scalp electrode: Electrode held against, attached to, or inserted into the scalp.

Seizure: A disorder of the brain manifested by transient episodes of motor, sensory, or psychic dysfunction that occur with or without loss of consciousness or convulsive movements. Epilepsy is a disorder in which there is a tendency for recurrent seizures.

Semiconductor: A class of solids whose electrical conductivity is between that of a conductor and that of an insulator. Commonly, the elements germanium and silicon. Semiconductors are used in transistors; cf solid state.

Sensitivity: Ratio of input voltage to output pen deflection in an EEG channel. Sensitivity is measured in microvolts per millimeter (μV/mm). Example:

$$\text{Sensitivity} = \frac{\text{input voltage}}{\text{output pen deflection}}$$

$$= \frac{50\ \mu\text{V}}{10\ \text{mm}} = 5\ \mu\text{V/mm}$$

Sharp wave: A transient having a pointed peak that is clearly distinguishable from the background activity. Sharp waves may have a duration of 70 to 200 ms. Generally, the main component is negative relative to the surrounding areas. Compare with spike.

Shielding: The covering of an apparatus, cables, or rooms with metal or metal-screen that is connected to ground; used to diminish pickup of external interference, such as 60 Hz artifact and radio-frequency signals.

Short circuit: The condition that exists when any two points on an electrical circuit are connected together.

Signal: Any electrical wave or activity that enters the input of an amplifier. Usually used to describe the "wanted" EEG activity as contrasted to the "unwanted" concomitant activity, which is noise.

Signal averaging: In general, the technique of enhancing the signal-to-noise ratio of signals that are coherent in time. Used in recording evoked potentials.

Signal-to-noise ratio: A ratio, the numerator of which is the voltage of a wanted signal and the denominator is the unwanted signal or noise.

Significance Probability Mapping (SPM): A statistical method of comparing the topographical image of a single individual with that of a reference group, or the topographical images of two groups. The method produces a new image that shows the probabilities of any differences between the original images occurring by chance.

Sine wave: A fundamental form of wave that represents periodic oscillation in which the amplitude at each point is proportional to the mathematical sine function.

Single-ended amplifier: An amplifier that operates on signals that are asymmetric with respect to ground; hence, an amplifier having only a single input terminal. Compare with differential amplifier.

Sinusoidal: Term applies to EEG waves that resemble sine waves.

Sleep spindle: A burst of rhythmic 11- to 15-Hz activity that occurs during sleep, frequently in widespread distribution. Amplitudes, which are mostly below 50 μV, are commonly highest in the central regions.

Sodium-potassium pump: The process or mechanism whereby sodium ions are actively expelled from within a cell and potassium ions are taken in. In so doing, the ionic concentrations are maintained, and a constant resting membrane potential is assured.

Solid state: Electronic components that transfer or control electron flow within solid materials, such as transistors, crystal diodes, and integrated circuits.

Step function: An instantaneous change in voltage. A step function is commonly used in the calibration of an EEG machine.

Spatial summation: The summation of several postsynaptic potentials produced simultaneously or nearly simultaneously at different sites on a postsynaptic membrane.

Sphenoidal electrode: Special electrode of thin, flexible, insulated platinum wire. Inserted under local anesthesia so that its tip lies in close proximity to the foramen ovale, it is useful in recording activity from the basal and mesial temporal cortex.

Spike: A transient having a pointed peak that is clearly distinguishable from the background activity. Spikes have a duration of from 20 to less than 70 ms. Generally, the main component is negative relative to the surrounding areas. Compare with sharp wave.

Spindle: A group of rhythmic waves in which there is a gradual increase and then a decrease in amplitude.

SPM: Acronym for significance probability mapping.

SSEP: Acronym for somatosensory-evoked potential.

Stroboscope: A device that delivers high-intensity flashes of light of extremely short duration. The flash rate is variable and can be controlled by the user. In EEG work, the device is called a photic stimulator.

Sulcus: Groove or cleft in the cerebral hemispheres. The deepest sulci are called fissures.

Summation: The effect produced when the input signals on terminal 1 and terminal 2 of a channel are different with respect to frequency, amplitude, polarity, waveform, or any combination thereof. An increase in amplitude, change of frequency, or distortion of polarity and waveform can occur.

Suppression: Denotes that little or no electrocerebral activity is discernible in a tracing or tracings.

Symmetry: In general, refers to the occurrence of approximately equal amplitude, frequency, and form of EEG activities over homologous areas on opposite sides of the head.

Synapse: The interface or functional junction between two neurons. The synapse is the point at which a nerve impulse is transmitted from one neuron to another.

Synaptic potential or postsynaptic potential: The electrical change in the postsynaptic membrane produced by the action of neurotransmitter at the synapse. Synaptic potentials are either excitatory (EPSPs) or inhibitory (IPSPs).

Synchrony: Refers to the simultaneous appearance of morphologically identical waveforms in areas on the same side or opposite sides of the head.

TC: Abbreviation for time constant.

Temporal summation: The summation of successive potential changes at a single site on the postsynaptic membrane such that one postsynaptic potential is superimposed onto another.

Ten-percent system: An extension of the 10–20 System in which the number of available derivations is expanded so that more than 60 electrodes may be placed on the scalp.

Ten-twenty (10–20) International System: See ten-twenty System.

Ten-twenty (10–20) System: System of standardized scalp electrode placement recommended by the International Federation of Societies for Electroencephalography and Clinical Neurophysiology. According to this system, electrode placements are determined by measuring the head from external landmarks and basing electrode locations upon 10% or 20% of these measurements.

Theta band: Frequency band from 4 to less than 8 Hz. Denoted by the Greek letter θ.

Theta rhythm: Rhythm with a frequency of 4 to under 8 Hz.

Theta wave: Wave with duration of 0.25 to more than 0.125 second.

Time axis: In the EEG the time axis is the chart paper speed. The vertical axis is the voltage.

Time constant, EEG channel: The time (in seconds) required for the pen to fall to 37% of the deflection initially produced when a DC potential difference (a step input) is applied to the input terminals of the amplifier. The value (TC) is determined by the values of resistance and capacitance in the circuit and is mathematically related to the frequency response. Filters to control the low-frequency response are sometimes marked with both parameters. Time constant in this context is a decay time constant. Amplifiers also have a rise time constant, related to high-frequency response.

Tracing: See recording.

Transducer: A device that converts energy from one form to another. Examples of transducers include microphones, earphones, loudspeakers, photo cells, strain gauges, and galvanometers.

Transient: An isolated wave that stands out from the background.

Transient response: The response of an electrical circuit to an instantaneous step change in applied voltage. It refers to the behavior of the circuit during the interval of time that a change is applied to it, and the circuit is still adjusting to the change.

Transistor: A solid-state device made from semiconductor materials, such as germanium or silicon, which can act as electrical insulators or conductors, depending on the electrical charges placed upon them.

Triphasic wave: Wave consisting of three distinct components or phases.

True phase reversal: Simultaneous pen deflections in opposite directions in two adjacent referential derivations having a common reference electrode. Occurs when the axis of a dipole that is the source of the electrical activity is *not* perpendicular to the scalp.

V or v: Abbreviation for voltage.

Vacuum tube: An evacuated tube containing two or more electrodes between which conduction of electricity through the partial vacuum may take place. Also, a general term for all electronic tubes.

Vertex sharp transient: Synonym for vertex wave.

Vertex wave: A sharp transient that occurs during sleep with maximal amplitude appearing at the vertex. Usually of high amplitude, it can attain magnitudes of 250 µV; it is negative with respect to other areas.

Volt: The unit of measurement of voltage, electromotive force, or potential difference.

Voltage: The potential (energy) present when different electrical charges are separated. This potential represents the force or pressure that can cause the movement of free electrons between the two points when a complete circuit is present. Also termed potential difference.

Voltage divider: An electrical circuit that, in its simplest form, consists of two resistors R_1 and R_2 connected in series with a voltage. This voltage divides itself across the two resistors in proportion to their values. A voltage divider may contain any number of resistors or it may use a potentiometer, in which case the voltage may be divided in an infinite number of ways.

Volume conductor: An electrical conductor that occupies three-dimensional space so that current flow can take many pathways.

V wave: Abbreviation for vertex wave.

Wave: In EEG, any change of voltage between any pair of electrodes used in EEG recording. May arise in the brain (EEG wave) or outside it (extracerebral potential).

Waveform: The shape or morphology of an EEG wave.

Writer: Device for direct write-out of the output of an EEG channel. Most writers use ink delivered by a pen. In some instruments the ink is sprayed as a jet stream. Some writers use carbon paper; others heat sensitive paper instead of ink.

Z: Abbreviation for the impedance of an electrical circuit. Measured in ohms. See impedance.

Zygomatic electrode: Special electrode placement situated over the zygomatic arch; used for picking up activity from the tip of the temporal lobe.

Appendix 2
Neuroanatomy for EEG Technologists

An understanding of the basic anatomy of the brain is an essential requirement for an EEG technologist. Without such knowledge, the technician will have little insight into the appropriate anatomical placement of the electrodes or the significance of what he or she is recording. Some knowledge of the anatomy of the spinal cord and peripheral nerves has also become crucial, since in recent years many EEG technologists are called upon to perform evoked potential recording as well. As explained in Chapter 17, evoked potential recording is a technique for troubleshooting the various sensory pathways and is based simply upon sound anatomic and physiologic concepts. In this appendix, the technologist will find descriptions of essential aspects of neuroanatomy as it is applicable to EEG and evoked potential recording and interpretation. This material is meant only as an introduction; readers interested in more detailed information should consult one of the many excellent neuroanatomy textbooks that are currently available.

The nervous system has two major subdivisions: the central nervous system, which consists of the brain and spinal cord, and the peripheral nervous system, which comprises the cranial and peripheral nerves. We will take up the anatomy of the brain first.

Topographical Relationship Between Skull and Brain

It is important to examine the topographic relationship between the various parts of the skull and the underlying areas of the brain in order to gain some perspective into the most likely source of the electrical activity that is recorded by individual scalp electrodes. The skull may be looked upon as a bony box with a space inside (the cranial cavity) that accommodates the brain. The cranial cavity is divided into two compartments by a horizontal dural partition (the dura is the thick membrane that covers the brain) called the tentorium cerebelli. The portion above the tentorium, known as the supratentorial compartment, contains the cerebrum. This compartment is subdivided into left and right spaces—each of which contains the corresponding cerebral hemisphere—by a vertical dural partition called the falx cerebri. The space below the tentorium is the posterior cranial fossa, which accommodates the cerebellum and the brain stem.

The skull bones are named after the area of the brain that they overlie. Thus, the frontal bone is over the frontal lobe, the parietal bone over the parietal lobe, the temporal bone over the temporal lobe, and so on. Figure A2.1 shows the relationship between the skull bones and the underlying areas of the brain along with the location of nasion and inion; Fig. A2.2 shows the cranial compartments formed by the dural partitions. The brain lies directly above the base of the skull, which has a number of holes (foramina) through which the cranial nerves exit and the carotid arteries enter the cranium. The largest of these foramina, the foramen magnum, accommodates the lowest part of the brain stem, which continues downward and exits as the spinal cord.

The Cerebral Hemispheres

As seen in Fig. A2.1, the brain can be divided into four major, anatomically distinct areas. These are the cerebral hemispheres, which are connected to each other by the corpus callosum; the diencephalon; the brain stem, which consists of midbrain, pons, and medulla oblongata; and the cerebellum. Each cerebral hemisphere has within its center a cavity filled with cerebrospinal fluid (CSF) called the lateral ventricle.

If one looks at a cross-section of the cerebrum, two distinct areas can be seen throughout, namely, an outer layer

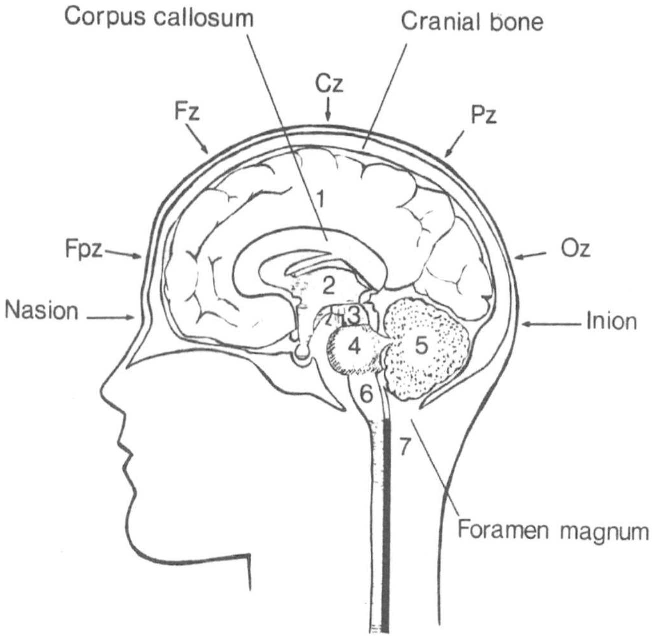

Figure A2.1. Median section of the brain showing the divisions of the central nervous system: 1. cerebral hemisphere; 2. diencephalon; 3. midbrain; 4. pons; 5. cerebellum; 6. medulla oblongata; 7. spinal cord. Also shown are the locations of the frontal pole (Fp_z), frontal (Fz), central (Cz), parietal (Pz), and occipital (Oz) regions along the midline.

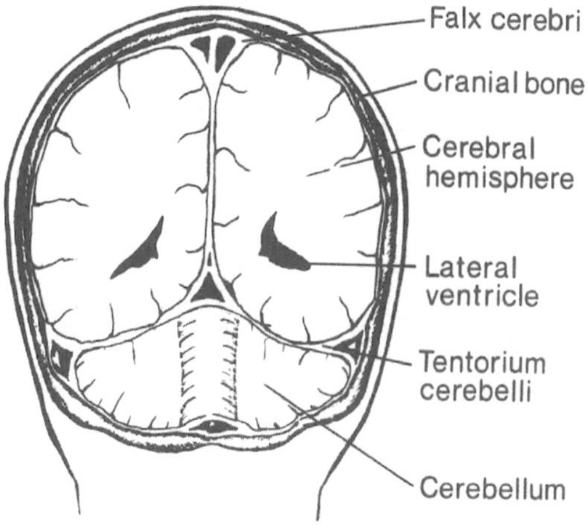

Figure A2.2. Cross-section of the cerebral hemispheres through Pz showing the cranial compartments formed by the dural partitions.

of gray matter and an inner area of white matter. The gray matter, which forms a mantle covering the entire surface of the cerebrum, constitutes the cerebral cortex. It is about 5 to 6 mm thick and is made up of billions of neuronal cell bodies. The cerebral cortex is the most important structure from the point of view of the electroencephalographer since the scalp-recorded EEG is generated by the activity of the neurons of the cerebral cortex (see Chapter 10). The white matter, on the other hand, consists mostly of axons originating from the neurons of the cerebral cortex as well as axons of neurons located elsewhere in the central nervous system. The whitish appearance is attributed to the myelin sheath that surrounds the axons.

The Cerebral Cortex

The gray matter of the cerebral cortex is thrown into folds leading to the presence of ridges (gyri) and fissures (sulci), which constitute a distinct macroscopic feature of the cerebral cortex. By this means, the total surface area of the cortex is greatly increased. It is important to be familiar with some of the easily identifiable gyri and sulci. The most obvious fissure when viewed from the top or front is the midline superior longitudinal fissure that divides the cerebrum into the left and right hemispheres (see Fig. A2.3). At the bottom of this fissure is the corpus callosum (Fig. A2.1), a structure made up of a large number of axons that connect the two cerebral hemispheres together. When viewed from the side, the cerebral hemisphere shows a very prominent fissure that seems to cause a cleavage starting at the inferior surface and extending upward and backward (Fig. A2.4). This is called the Sylvian or lateral fissure and is responsible for the reversed C-shape of the left cerebral hemisphere when it is viewed from the side.

Another important sulcus is the central sulcus, which can be traced from a point close to the vertex to the Sylvian fissure, coursing obliquely downward and forward over the

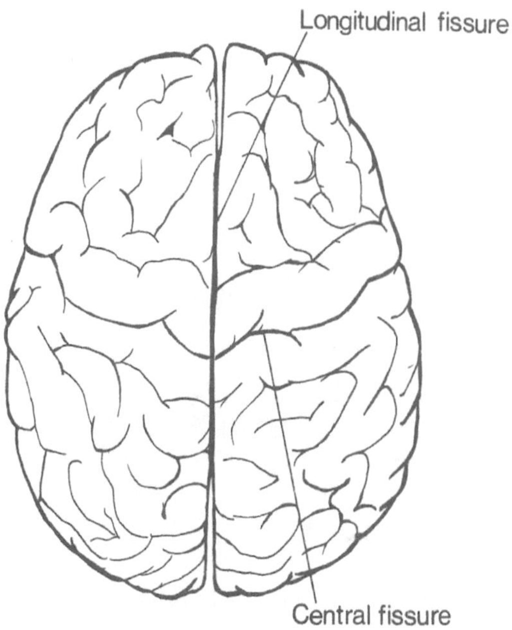

Figure A2.3. Top view of the cerebral hemispheres.

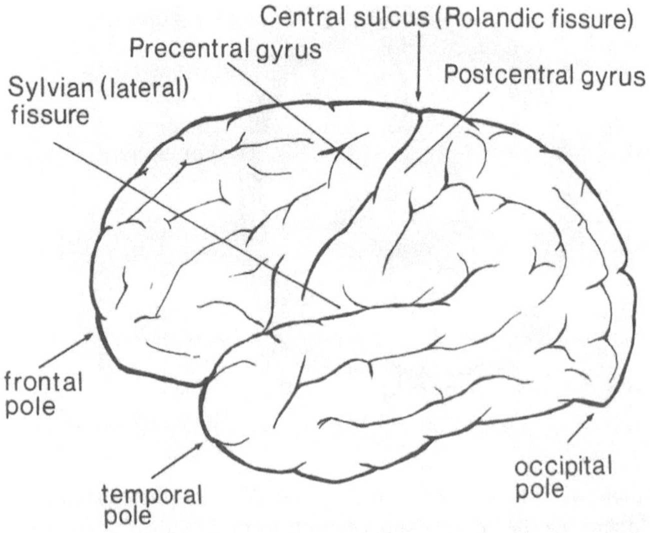

Figure A2.4. Lateral view of the left cerebral hemisphere showing major sulci and gyri.

superolateral surface. The central sulcus is also called the Rolandic fissure after Luigi Rolando, the Italian scientist, who first described it in 1825. This fissure separates the motor area in front from the sensory area behind. The Rolandic fissure is a very important landmark; the letter "C" used in the 10-20 System of electrode placement refers to the central fissure. It may be pointed out that in the placement of electrodes for recording somatosensory-evoked potentials (SSEPs), the Cz′ (Cz prime, 2 cm directly behind Cz) electrode is behind the central fissure and close to the sensory area for the lower extremities. Correspondingly, the C3′ and C4′ electrodes are over the left and right somatosensory areas for the upper extremities. Another important fissure to be noted is the calcarine sulcus, which extends from the medial surface of the occipital lobe toward the occipital pole, as shown in the medial surface view of the right cerebral hemisphere in Fig. A2.5.

It will be apparent from a perusal of Fig. A2.6 that the division of each cerebral hemisphere into different lobes depends on many of the anatomic landmarks noted above. Thus, the frontal lobe is the part that lies in front of the central fissure and above the Sylvian fissure. The part below the Sylvian fissure constitutes the temporal lobe. The occipital lobe is the part that lies behind an imaginary line joining the parieto-occipital fissure and the preoccipital notch. Finally, the area between the central fissure and this line above the Sylvian fissure constitutes the parietal lobe.

Table A2.1 gives a summary of the functional anatomy of the cerebral cortex; it should be studied in conjunction with Fig. A2.5. As mentioned earlier, it is important for the EEG technologist to know the localization of various functions in the cerebral cortex. This is especially important if he or she is to obtain a clear understanding of the techniques of recording evoked potentials. Thus, for example, Fig. A2.5 shows why Cz′ is chosen as the recording site for tibial nerve SSEP and C3′ and C4′ for median nerve SSEP.

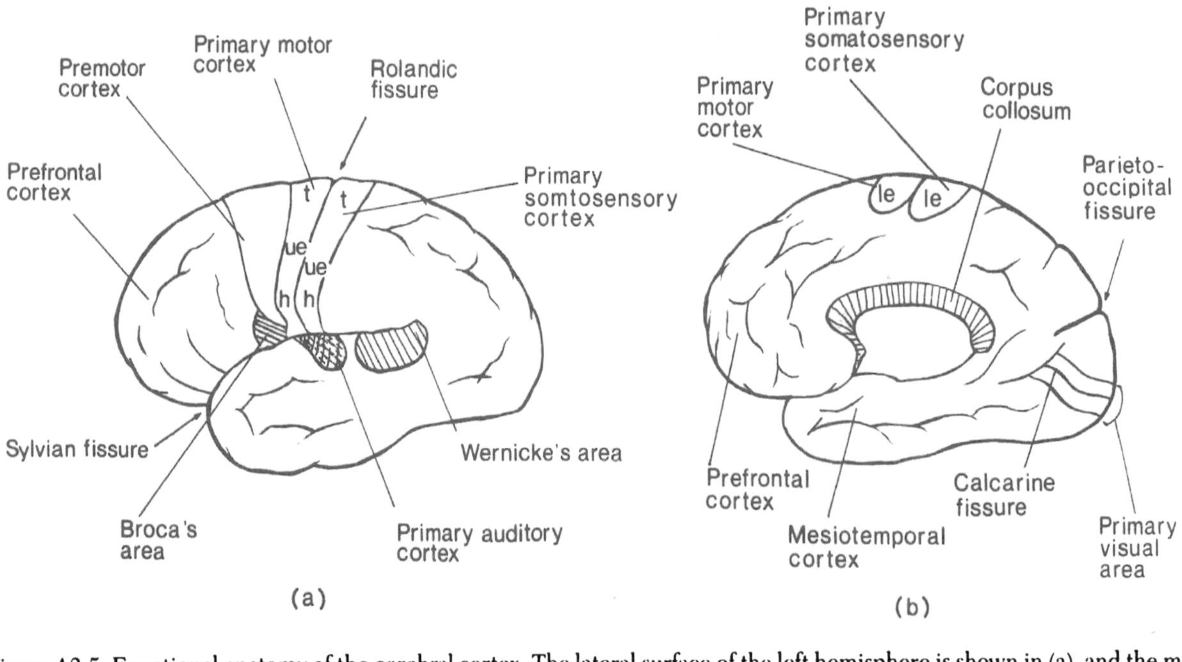

Figure A2.5. Functional anatomy of the cerebral cortex. The lateral surface of the left hemisphere is shown in (a), and the medial surface of the right hemisphere is shown in (b). Abbreviations are as follows: ue, upper extremity; le, lower extremity; t, trunk; h, head.

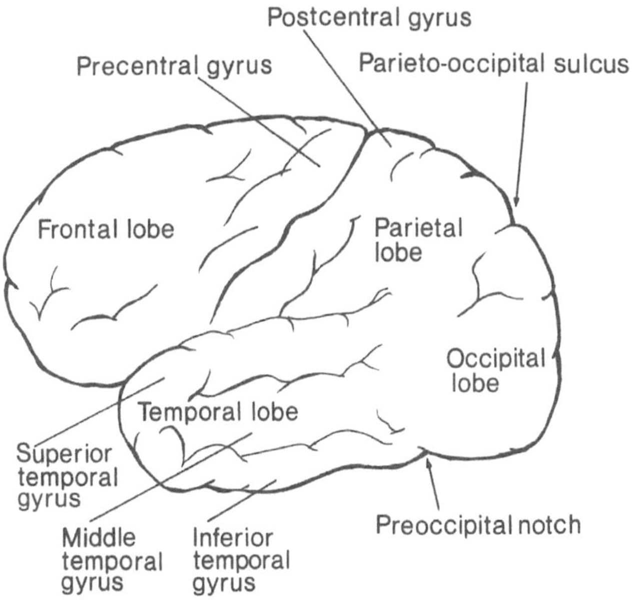

Figure A2.6. Lobes and important gyri of the cerebral hemisphere.

Table A2.1. Functional Anatomy of the Cerebral Cortex[a]

Area	Location	Function
Broca's	Anterior to precentral gyrus close to Sylvian fissure (posterior part of frontal lobe)	Motor speech
Somatosensory	Postcentral gyrus (anterior part of parietal lobe)	Interpretation of sensory information from opposite side of body
Wernicke's	Temporal lobe (upper and posterior region, at the junction of parietal, occipital, and temporal lobes)	Interpretation of speech
Visual	Occipital lobe	Interpretation of visual sensations
Auditory	Temporal lobe (upper part)	Interpretation of auditory sensations
Olfactory	Temporal lobe (medial part)	Interpretation of smell
Short-term memory	Temporal lobe (inferior part)	Temporary memory for visual and auditory events
Prefrontal	Frontal lobe (anterior part)	Elaboration of thought, behavioral control, inhibition

[a] Refer to Fig. A2.5 for the anatomic locations.

It is also important to know where the primary areas for vision and hearing are located so that the appropriate electrode placement can be used for the particular sensory modality that is being stimulated.

Although a large portion of the cerebral cortex is visible on the superolateral surface, it must be realized that a significant proportion of this structure lies hidden in the inferior as well as the medial aspect of the cerebrum. This will be clear from a perusal of Fig. A2.7, which shows the inferior surface of the cerebrum. Portions of the temporal lobe cortex that lie medially and inferiorly are particularly noteworthy because of their known propensity for seizure genesis. The usual array of 10-20 System electrodes often fails to record activity from these hidden areas of the temporal lobe; hence, there is need for specially placed electrodes such as the sphenoidal and nasopharyngeal electrodes. These electrodes are discussed in Chapter 11. Another significant observation is that the medial portion of the occipital lobe around the calcarine sulcus represents the primary visual area (Fig. A2.5b). It is interesting to note that with this location of the primary visual cortex, the electrical activity from the right visual area is better seen over O1 than O2, and vice versa.

Microscopic Anatomy of the Cerebral Cortex

The cerebral cortex needs to be discussed in more detail since it is thought to be the generator of the EEG wave forms. With the exception of the hippocampal region of the temporal lobe, one can identify six distinct layers in the cortex. The six layers from surface to depth are shown diagrammatically in Fig. A2.8. It will be noted that layer I contains dense arborizations of dendrites and axons (the term neuropil is used for this). There is general consensus

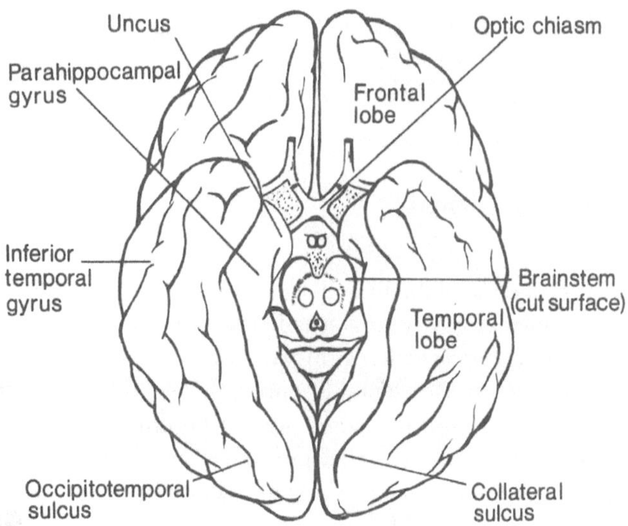

Figure A2.7. Sulci and gyri in the inferior surface of the cerebrum.

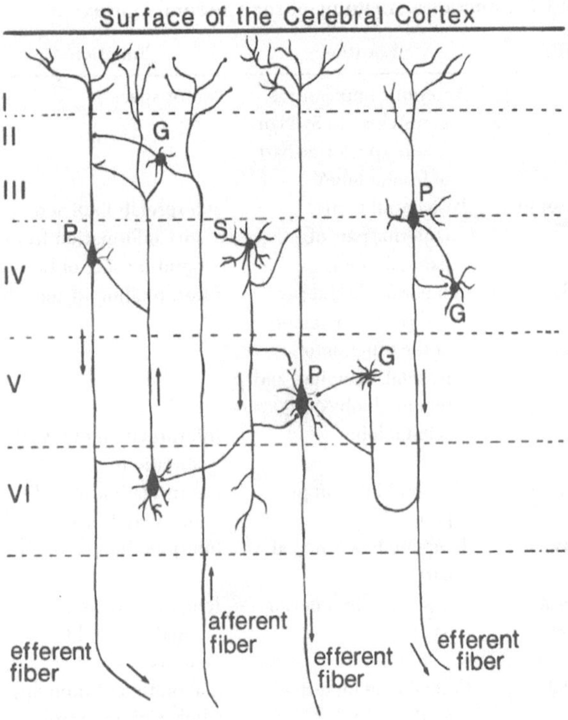

Figure A2.8. Simplified microscopic anatomy of the cerebral cortex. P = pyramidal cells; G = granule cells; S = stellate cells; I = molecular layer; II = external granular layer; III = external pyramidal layer; IV = internal granular layer; V = internal pyramidal layer; VI = multiform layer.

among scientists that inhibitory and excitatory postsynaptic potentials originating in this layer are the agents responsible for the production of the waveforms recorded in the EEG (see Chapter 10). Certainly, the vertical orientation of the pyramidal neurons, with their apical dendrites arborizing in layer I and their cell bodies located in the deeper layers, is ideally suited for the creation of vertical dipoles with their resultant field currents. These, in turn, could generate the EEG waveforms.

Diencephalon

If a cross-section of the brain is examined, one sees a number of large masses of gray matter in the depths of the brain surrounding the ventricles. Figure A2.9 shows some of these structures. Most notable among them is the mass of gray matter situated on either side of the third ventricle called the thalamus. The thalamus, which is a major division of the diencephalon, contains a number of distinct collections of neurons (nuclei). It has a close topographic relationship to different parts of the cerebral cortex; so much so, that the cortex may actually be considered as an

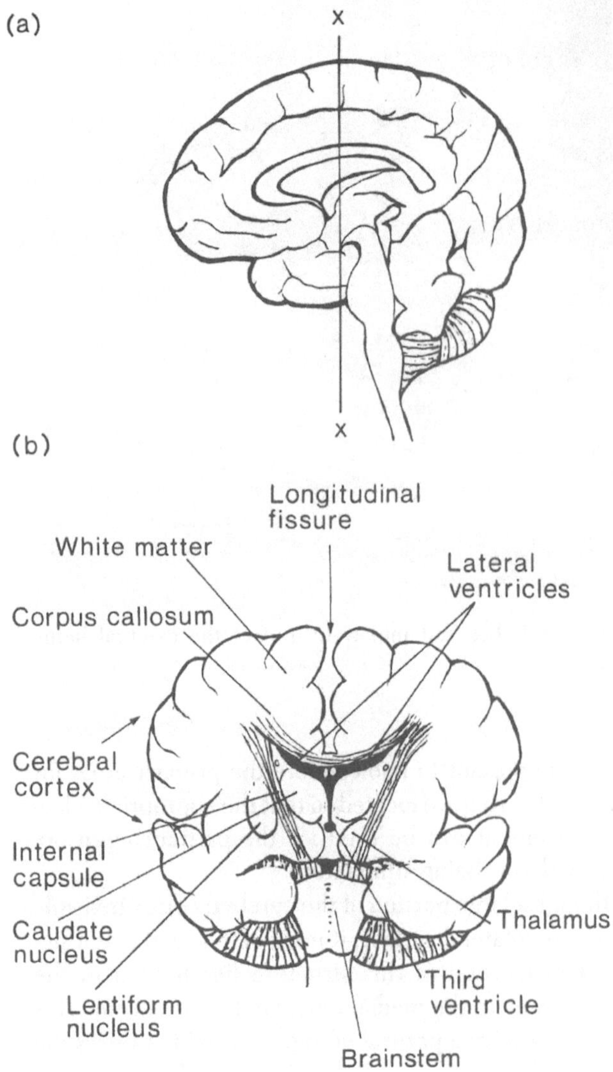

Figure A2.9. Section through the brain as in (a) showing in (b) the thalamus and structures of the basal ganglia.

outgrowth or projection of the thalamic system. The thalamus receives sensory input from all parts of the body and relays information on to specific portions of the cerebral cortex. Although stimulation of the specific sensory nuclei of the thalamus leads to activation of discrete regions of the cerebral cortex, stimulation of the nonspecific or generalized thalamic nuclei activates much larger areas of the cerebral cortex. The thalamus is one of the most important structures from the point of view of the generation of EEGs. As noted in Chapter 10, thalamic neurons are considered to be crucial in the production of the rhythmicity that is characteristic of EEG waveforms.

The other major part of the diencephalon is the hypothalamus. This structure is situated on the floor of the third ventricle and is important in the control of many of the autonomic functions of the body such as the heart rate, temperature, etc. As seen in Fig. A2.9, there are also other collections of gray matter surrounding the thalamus. These are called the basal ganglia and consist of the caudate nucleus, which is adjacent to the lateral ventricle, and lentiform nucleus, which is lateral to the thalamus and is separated from the caudate nucleus and thalamus by the internal capsule, a large bundle of fibers from the overlying cortex. The basal ganglia play an important role in certain aspects of motor function.

The Brain Stem and Cerebellum

The brain stem consists of the midbrain, pons, and medulla oblongata, all structures being contained in the posterior cranial fossa. The cranial nerves (other than the olfactory and optic nerves) arise from the brain stem. The brain stem is an important structure in the coordination of eye and head movements. In addition, it contains the ascending reticular activating system (ARAS), which consists of a large number of neuronal collections and interconnecting fiber systems important in the maintenance of alertness. The ARAS can cause synchronization or desynchronization of the EEG through its connections with the thalamus. The brain stem is also involved in the mechanisms that control and regulate sleep. Thus, lesions of the raphe nuclei of the lower pons and medulla are known to abolish sleep. Similarly, lesions of the locus ceruleus, a collection of neurons situated at the junction of pons and midbrain, can reduce the amount of rapid eye movement (REM) sleep. Finally, the pons and medulla contain important centers for the control of respiration and vasomotor function.

The cerebellum is a structure important in motor control. As mentioned earlier, it is situated below the tentorium, in the posterior cranial fossa (Fig. A2.2). The cerebellum has close connections with the brain stem, the cerebral cortex, and the thalamus. Details concerning this structure need not concern us here.

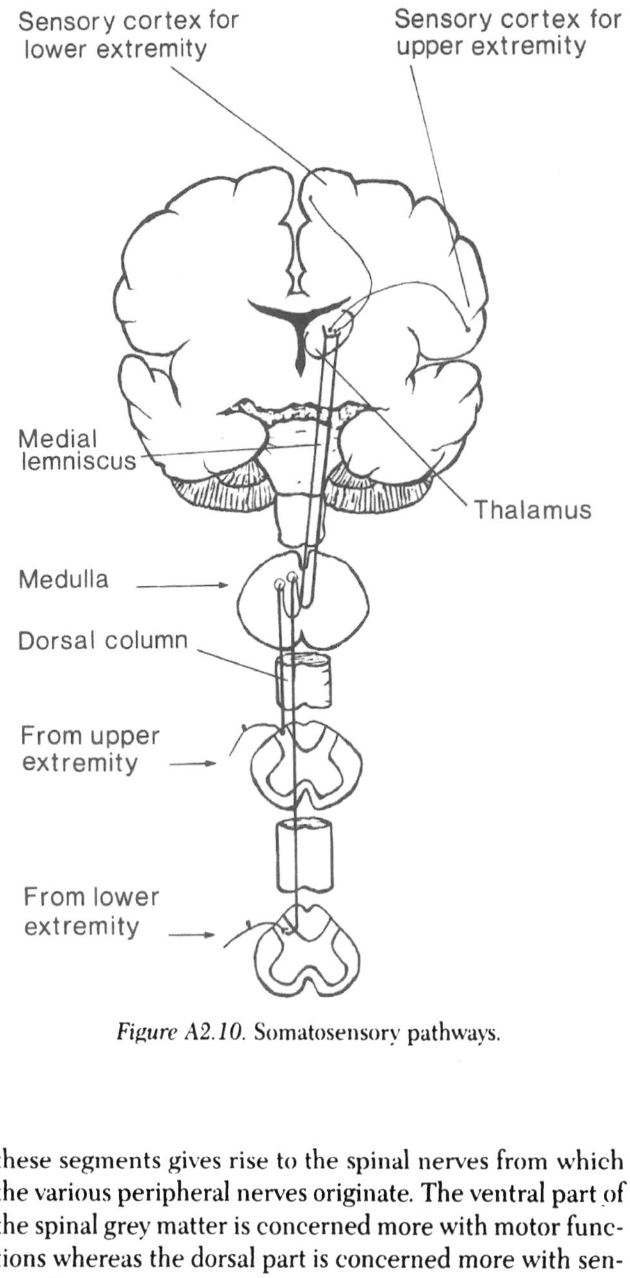

Figure A2.10. Somatosensory pathways.

Spinal Cord

The spinal cord is located in the spinal canal of the vertebral column and extends from the level of the foramen magnum to L-1, the first lumbar vertebra. The spinal cord has a central canal with gray matter surrounding it and which contains the cell bodies of the neurons. Superficial to the grey matter is the white matter. White matter is made up of columns of axons and dendrites traveling to and from the brain. The spinal cord consists of several segments (cervical, thoracic, lumbar, and sacral). Each of these segments gives rise to the spinal nerves from which the various peripheral nerves originate. The ventral part of the spinal grey matter is concerned more with motor functions whereas the dorsal part is concerned more with sensory functions.

The white matter is made up of distinct bundles of nerve fibers called tracts, columns, or fasciculi. The tracts that are the most important to be familiar with are the ascending dorsal columns (sensory) and the descending corticospinal tracts (motor). In the study of SSEPs, one attempts to record the passage of signals evoked by electrical stimulation at the periphery through the somatosensory pathways, of which the dorsal columns are perhaps the most important. Fig. A2.10 is a schematic of these pathways. The corticospinal tracts, which connect the motor cortex to the spinal motor neurons, are important for initiation of voluntary movements. These pathways can be studied by the recently developed technique for the recording of motor-evoked potentials.

Peripheral Nervous System

The peripheral nerves of the extremities arise from plexuses, e.g., the brachial plexus in the case of the upper extremity and the lumbosacral plexus in the case of the lower extremity. Some knowledge of the anatomic position of peripheral nerves as well as the plexuses is important for the recording of SSEPs. Thus, the *median nerve* can easily be stimulated over the wrist between the tendons of palmaris longus and flexor carpi radialis, the *ulnar nerve* on the medial aspect of the wrist near the tendon of flexor capri ulnaris, and the *tibial nerve* below and behind the medial malleolus (refer to Chapter 17).

The foregoing has been a brief excursion into human neuroanatomy. For the EEG technologists who would like to pursue the topic in greater detail, there are many excellent texts available. Consult the director of your EEG laboratory for some suggested titles.

Appendix 3
Grounding Checks

The importance of properly grounding the EEG machine or any other electrical equipment used in the EEG laboratory is taken up in detail in the chapter on electrical safety. By connecting the three-prong plug on the end of the EEG machine's power cord to an appropriate electrical outlet on the wall or floor of the laboratory, the EEG technician should be properly grounding the machine. But in doing so, he or she is assuming that there are no faults in the electrical outlet and that there is electrical continuity between the plug, cable, and chassis of the machine. To assure that the EEG machine is indeed grounded, the technician should carry out a few simple continuity tests.

Using an ordinary ohmmeter set to its lowest scale setting,[1] connect the end of one of the probes to the metal chassis of the EEG machine and the end of the other probe to some metal in the room (e.g., a cold-water pipe) that ought to be connected to earth. The pointer of the ohmmeter should deflect to the right and then settle down to a reading somewhere under 10 ohms. This indicates that the machine is indeed grounded and that the building's grounding system is intact. If the meter does not deflect to the right, or shows a reading significantly higher than 10 ohms, try connecting the second probe to another ground connection—for example, the ground wire on another wall socket in the room or in an adjacent room. If, now, a proper ohmmeter reading is obtained, the cold-water pipe in the room may not be making proper contact with the building's earth connection. Since this could point to a possibly dangerous condition, consult the building's electrician. On the other hand, if the meter still fails to give the proper reading, you need to test the ground system within the EEG machine.

To carry out this test, unplug the power cord of the EEG machine from the electrical outlet and check for continuity between the ground prong on the plug and chassis of the machine. This is done by connecting the ohmmeter between the ground prong (the longest of the three) and any metal point on the chassis. The ohmmeter should show a reading close to zero ohms. If it does not, contact the manufacturer of the machine for help; for safety's sake, do not use the machine. On the other hand, if this test shows that there is continuity in the EEG machine's ground system, the fault may be in the electrical outlet you are using or in the building's ground system. These are serious and potentially dangerous conditions, and you should consult the building's electrician for help at once.

[1] All ohmmeters need to have their zero settings adjusted before using them. After selecting the scale that you would like to use, touch the two probes of the meter together. As is the case whenever using an ohmmeter, make sure that the probes are touching surfaces where the metal is clean and free of corrosion. With the probes connected together in this manner, turn the zero-adjust knob on the meter until the pointer reads exactly zero ohms. If this adjustment cannot be satisfied, the battery inside the meter will need to be replaced. Consult the instruction manual for help.

Appendix 4

Measurement of Chassis Leakage Current

The way in which leakage currents affect the safety of patient and technician during EEG recording is discussed in Chapter 8. One important source of such leakage currents is the chassis leakage current. We already noted in the chapter on electrical safety that the maximum allowable chassis leakage current with the ground wire of the EEG machine's power cord interrupted is 100 μA.

Measurements of chassis leakage current are relatively easy to make if an accurate AC ammeter capable of measuring currents in the 0.025 to 0.25-mA root-mean-square (RMS) range is available. Proceed as follows: disconnect the power cord of the EEG machine from the electrical outlet and reconnect it to the electrical outlet via a three-way to two-way converter.[1] Because doing this leaves the machine ungrounded, be careful not to touch the chassis of the machine at the same time that your body is touching ground. Also, never use an extension cord of any kind to make this hookup as this will generally increase the amount of the chassis leakage current measured. Next, connect one probe of the meter to the ground socket of the electrical wall outlet and the other probe to the metal chassis of the EEG machine. The meter should read no more than 100 μA. If larger readings are observed, contact the manufacturer immediately.

[1] This converter is commonly referred to as a "cheater." A cheater should *never* be used with the EEG machine for any purpose except to measure chassis leakage currents. It should be removed immediately after the measurement is made.

Appendix 5
The 10-20 International System of Electrode Placement

The essential features of the 10-20 System have already been taken up in Chapter 11: "Recording Systems." The following is a step-by-step "how to" procedure for the EEG technician who is starting her/his training. This procedure should be carefully followed and thoroughly mastered, as the importance of accurately placed electrodes cannot be overemphasized. Electrodes that are not symmetrically placed over homologous areas on opposite sides of the head can result in left-right asymmetries in the EEG tracings that may lead to an erroneous interpretation of the record.

As explained in Chapter 11, electrodes in the 10-20 System are defined by two characters—a capital letter and a number. The letter refers to the particular region of the cerebral cortex, and the number indicates whether the electrode is on the right or left side. Odd numbers represent the left side of the head and even numbers represent the right side. For electrodes located along the midline, the second character is a lower-case z. Thus, for example, P3 represents the left parietal region; O2, the right occipital region; A1, the left auricle or left earlobe, and Cz, the central region at the midline. The frontal pole electrodes are an exception and are defined by two letters and a number. Thus, the frontal pole electrode on the left is designated Fp_1.

The location of the 19 scalp electrodes of the 10-20 International System is based upon four fixed landmarks on the head. These landmarks are the *nasion*, or indentation where the nose joins the forehead; the *inion*, a prominent protrusion or bump at the back of the head located by running your finger from the back of the neck toward the top of the head; and the *preauricular points*. As the name suggests, the preauricular point is in front of the ear. It is located just above the tragus, or triangular piece of cartilage that covers the opening of the external ear canal. Figure A5.1 shows the location of these landmarks.

To correctly locate the positions of the 19 scalp electrodes, follow the instructions contained in the succeeding six steps. The tools you need include a narrow measuring tape scaled in centimeters and millimeters, a pair of straight calipers, and wax pencils (e.g., china marker pencils) in two different colors. One color is used for the initial marking of the locations; the second color is used only if corrections or modifications to the original markings become necessary. The second color avoids the confusion that results from having more than one mark at a particular site.

Step 1. In this step, you establish the positions that the z electrodes assume along the midline. Using the measuring tape, measure the distance from nasion to inion. To the nearest millimeter, compute 10% and 20% of the total nasion-to-inion distance; using these values, mark off the locations of the frontal pole, frontal, central, parietal, and occipital regions on the scalp with the wax pencil, as shown in Fig. A5.2a. Draw a vertical mark in the exact middle of the patient's forehead through the frontal pole loca-

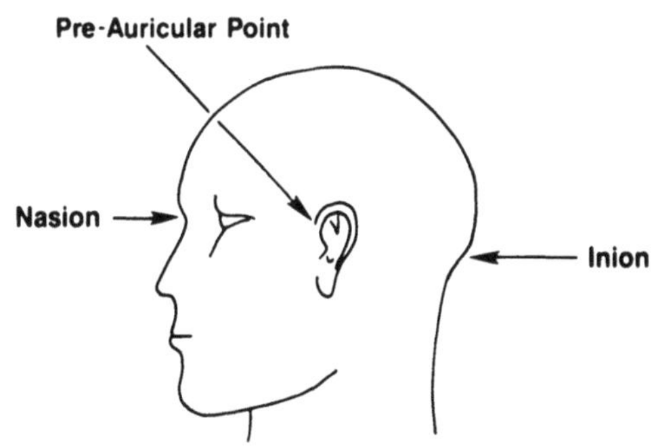

Figure A5.1. Landmarks of the 10-20 International System of Electrode Placement.

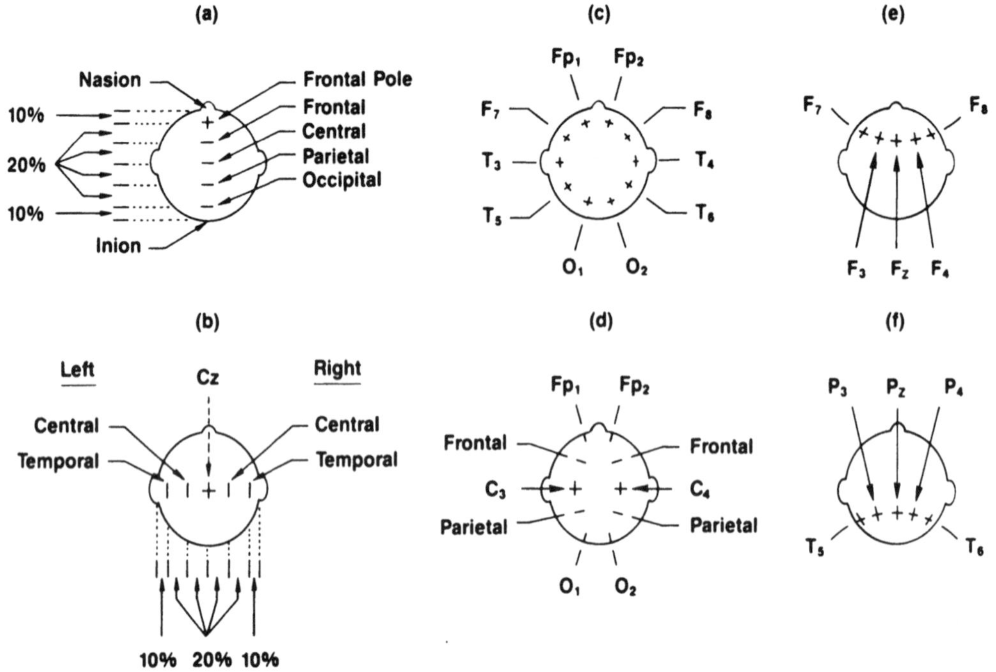

Figure A5.2. The six steps (a through f) used to locate the positions of the 19 scalp electrodes in the 10-20 International System. The two other electrodes, A1 and A2, are located on the left and right earlobes. In all cases, the view is from the top of the head.

tion. This establishes a primary reference point, Fp_z, that will be used in step 3.

Step 2. This step establishes the positions of the electrodes on the coronal line that runs across the head from ear to ear. Measure this distance from the preauricular point on the left to the preauricular point on the right, and compute 10% and 20% of the total distance to the nearest millimeter. Mark off the locations of the central and temporal regions on the right and left sides of the scalp and the midline central position, as shown in Fig. A5.2b. Together with the mark locating the central region in step 1, the midline central mark defines Cz, the vertex electrode.

Step 3. In this step, you establish the positions of the electrodes that are placed along the circumference of the head. Using the measuring tape, measure and record the head circumference, making sure that the tape lines up with the frontal pole, and temporal and occipital marks already traced on the scalp. Set the straight calipers to 5% of the total head circumference and, using them as a guide, mark off 5% of this distance to the left and right of Fp_z, the primary reference point established in Step 1. Moving backward from the 5% point on the left side, mark off 10% of the head circumference each for the anterior temporal, midtemporal, posttemporal, and occipital positions. Repeat these steps going backward from the 5% point on the right side. As a check on the accuracy of the measurements and the precision of the markings, verify that the distance between the occipital positions on the left and right sides is equal to 10% of the head circumference measurement. Finally, again place the tape measure along the circumference of the head as previously instructed and draw *horizontal* marks through the 5% and three of the four 10% points on each side (the midtemporal regions already have horizontal marks, made in step 2).

Step 3 marks the locations of 10 electrodes: Fp_1, F7, T3, T5, and O1 on the left, and Fp_2, F8, T4, T6, and O2 on the right. These are shown diagrammatically in Fig. A5.2c. It is apparent from the diagram that each electrode is distant from its immediately adjacent, neighboring electrodes by 10% of the head circumference measurement.

Step 4. This step establishes the locations of the electrodes along the left and right parasagittal rows. Begin by measuring the distance from Fp_1 to O1 with the tape measure; in doing so, be sure that the tape lines up with the central mark already placed on the scalp in step 2. Now locate the frontal region by a mark that is 25% of the way backward from Fp_1. Continuing backward, place similar marks to define the central and parietal regions. These marks are placed so that the frontal to central and central to parietal distances are also 25% of the distance from Fp_1 to O1. Repeat these measurements and markings on the right side of the head. Figure A5.2d diagrams this step. Note that in completing step 4 you establish the positions of the left and right central electrodes, C3 and C4.

Step 5. This step fixes the locations of three electrodes in the frontal coronal row, namely, F3, Fz, and F4. Starting from F7, measure the distance across the top of the head to F8; make sure that the tape measure lines up with the left, midline, and right frontal marks already present on the scalp. Next, measure and mark off 25% of the way to the right of F7. Continuing to the right, measure and mark off 50% of the way and 75% of the way from F7. Figure A5.2e diagrams this step. Note that the 25%, 50%, and 75% marks, together with marks already placed in steps 1 and 4, establish the locations of the F3, Fz, and F4 electrodes.

Step 6. This step fixes the locations of three electrodes in the parietal coronal row, namely, P3, Pz, and P4. Beginning from T5, measure the distance across the top of the head to T6, making sure that the tape measure lines up with the left, midline, and right parietal marks already present on the scalp. Next, measure and mark off 25% of the way to the right of T5. Continuing on to the right, measure and mark off 50% of the way and 75% of the way from T5. Figure A5.2f diagrams this step. Note that the 25%, 50%, and 75% marks, together with marks already placed in steps 1 and 4, establish the locations of the P3, Pz, and P4 electrodes.

The procedure is completed by attaching an electrode to each earlobe. These electrodes are designated A1 on the left and A2 on the right. Together with the others, they make up the full complement of 21 electrodes of the 10-20 International System. Finally, a ground electrode is added; this usually is attached to the middle of the patients's forehead, between Fp_1 and Fp_2.

Appendix 6

A Glossary of Common Artifacts in the EEG

Many of the different artifacts that are commonly encountered in electroencephalography have been discussed in various chapters throughout the text. Electrode artifacts and physiological artifacts, which are probably the most common of all artifacts, were discussed in Chapters 7, 11, and 13, respectively. Along with these discussions, ways of eliminating or reducing the artifacts were considered.

The following illustrations are presented as an aid to the

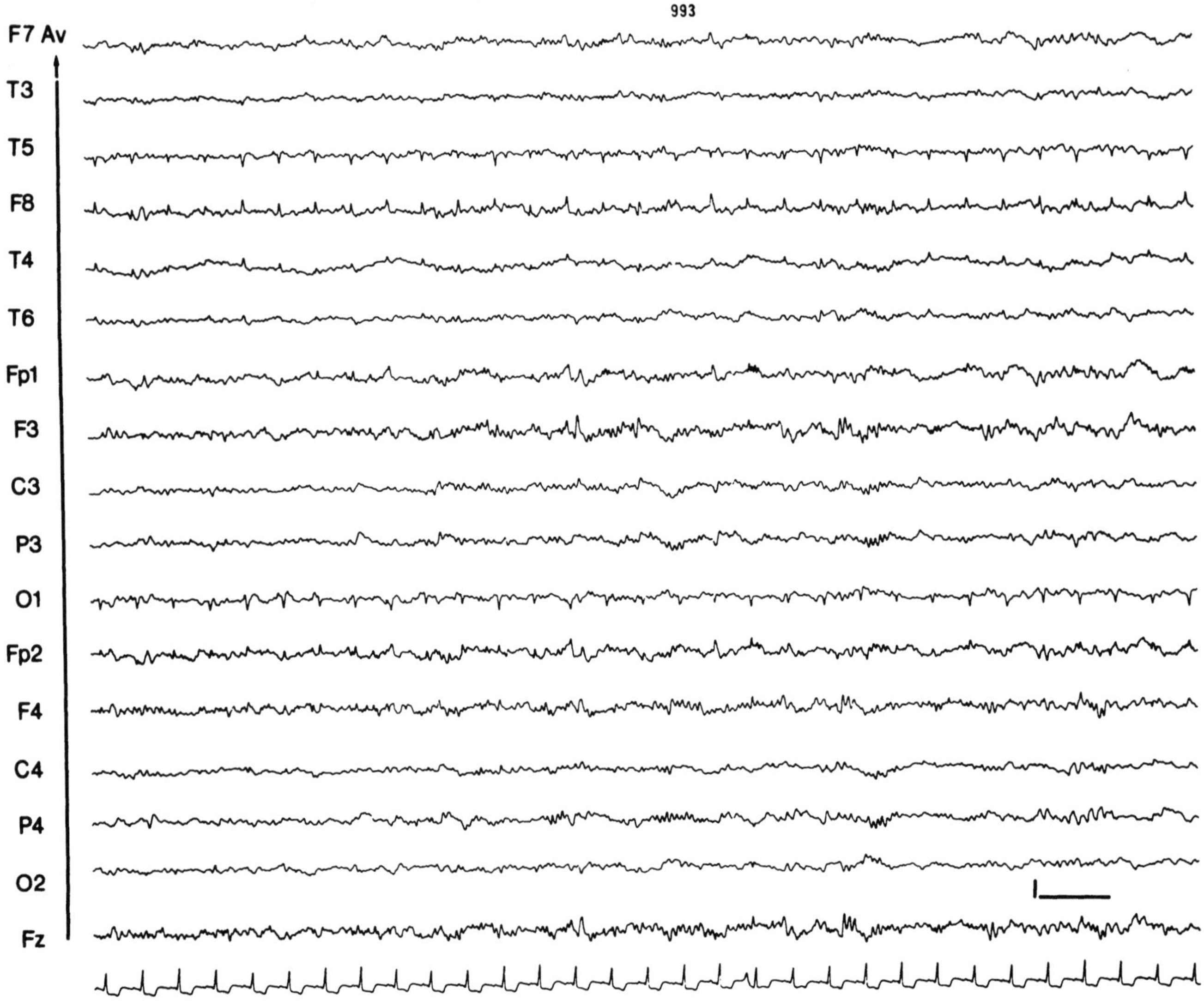

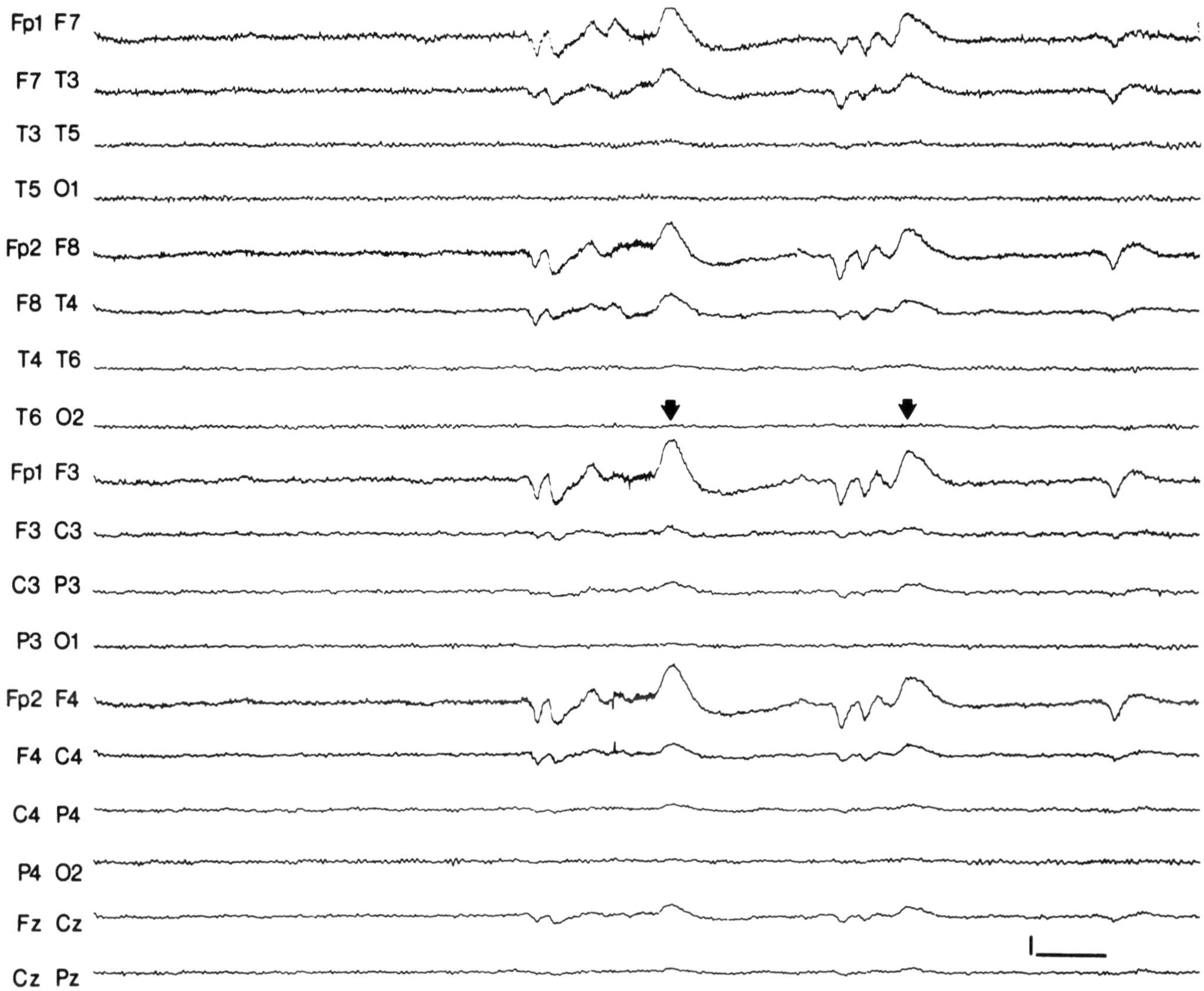

Figure A6.2. Eye-movement artifact. These artifacts may have amplitudes greater than 100 μV, as is the case with those marked by the arrows. They are readily identified by their anterior-posterior distribution; amplitude is highest in the derivations containing the frontal pole electrodes and diminishes as the electrodes move backward. At times, the artifacts can be quite rhythmic and may be mistaken for FIRDA. Any doubts are resolved by attaching an electrode to the cheek below the eye and recording between this and the ipsilateral ear lobe. This derivation will show waves that are out of phase with the deflections in question if they are indeed eye-movement artifacts. Filters: low frequency = 1 Hz, high frequency = 70 Hz. Calibrations: horizontal = 1 second, vertical = 50 μV.

recognition of such artifacts. They are offered with only brief comments in the legends. If the reader has studied the text, he/she will already be familiar with these artifacts. Some of them are present in the numerous illustrations that comprise the main body of the text and are pointed out along the way. Although the illustrations given herewith represent typical examples, the same kind of artifacts sometimes may assume a variety of different forms. Only through experience can the EEG technologist, neurology resident, and electroencephalographer gain the knowledge needed to quickly and correctly identify all of them.

This glossary includes illustrations of the following artifacts: ECG, eye movement, eye lid (blink), yawn, swallowing, respiration, muscle spike, electrode "pop," sweating, nasopharyngeal electrode, and ventilator.

Figure A6.1. An ECG artifact. The artifact is readily identified by its repetitive, periodic character. Any lingering doubts concerning the origin are settled by attaching an electrode to the person's chest and recording the ECG directly. The bottom channel shows the ECG as recorded between the chest electrode and A2. Note that the artifacts are exactly in phase with the voltages from the heart. The ECG artifacts are more common when using a referential montage. Filters: low frequency = 1 Hz, high frequency = 70 Hz. Calibrations: horizontal = 1 second, vertical = 50 μV.

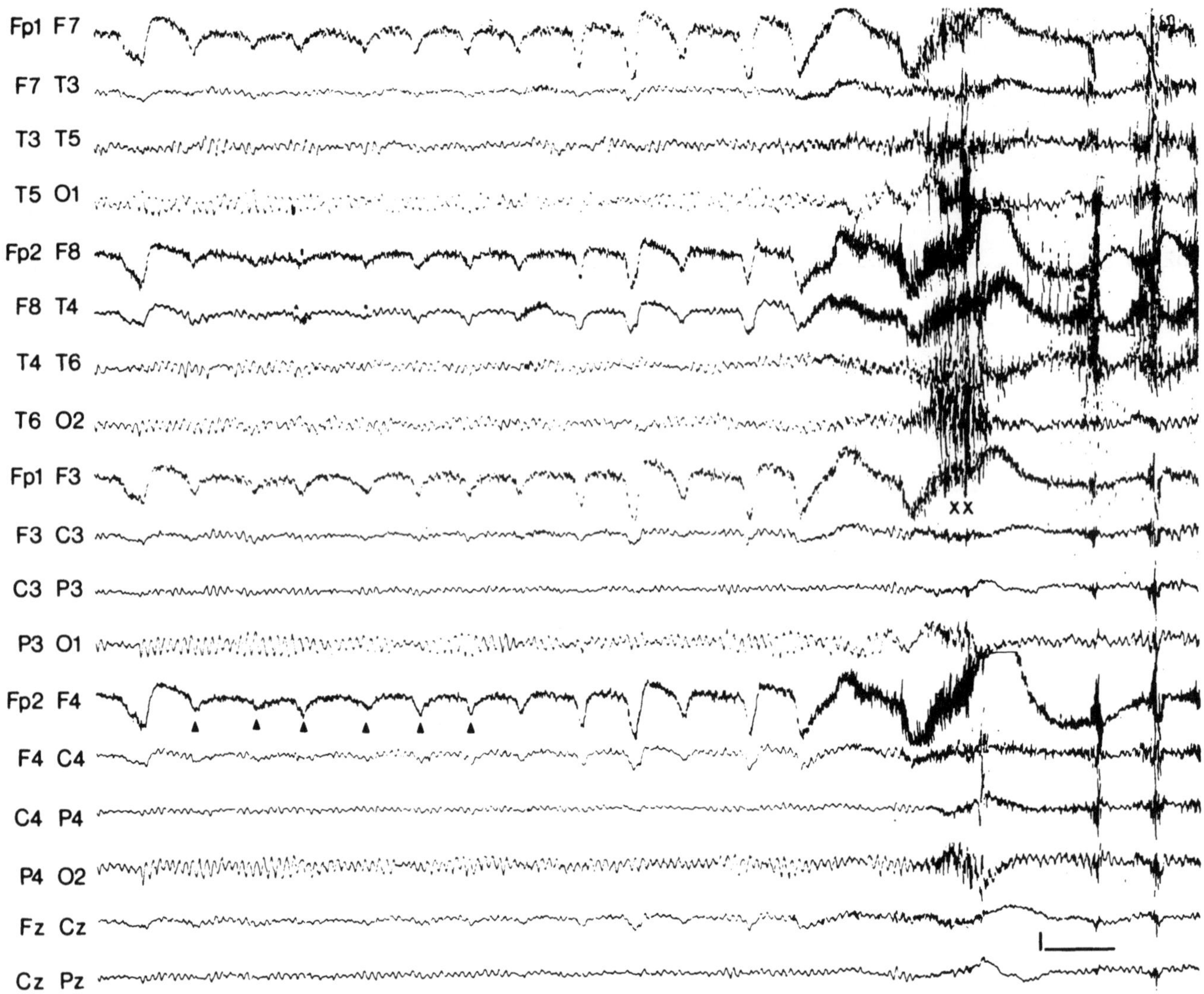

Figure A6.3. Eye-lid (blink) and yawn artifacts. The blink artifacts (triangles) are recognized by the sharp contours of the waves, their prominent appearance in the anterior derivations, and their often repetitive nature. Placing pads over the eyes may help, but sometimes doing so makes the problem even worse. The yawn is marked "XX." Aside from consisting of large-amplitude muscle spikes, it has no specific distinguishing features. Filters: low frequency = 1 Hz, high frequency = 70 Hz. Calibrations : horizontal = 1 second, vertical = 50 μV.

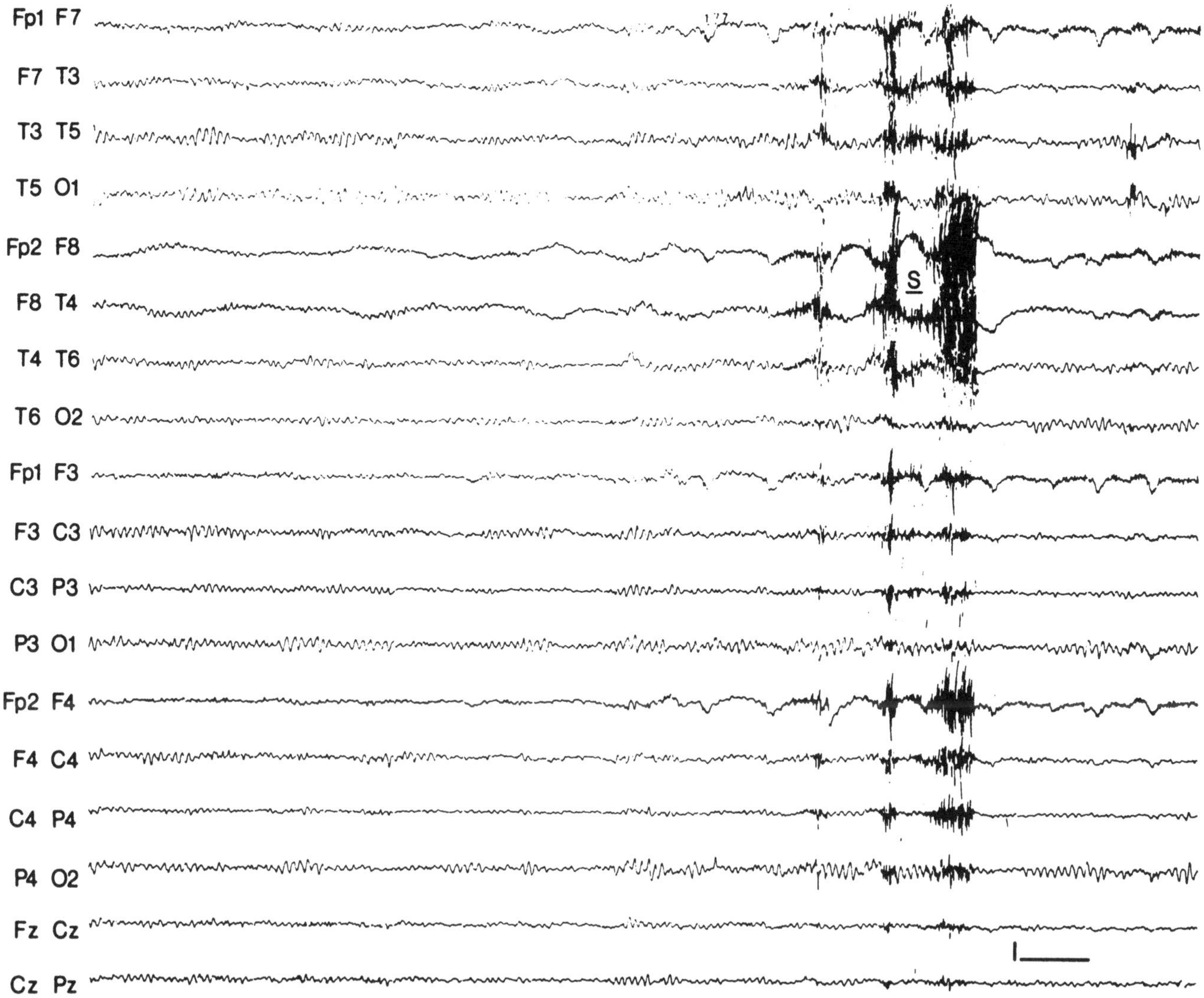

Figure A6.4. Swallowing artifact. Occurring at "S," this artifact is distinguished by its rhythmic character. Filters: low frequency = 1 Hz, high frequency = 70 Hz. Calibrations: horizontal = 1 second, vertical = 50 μV.

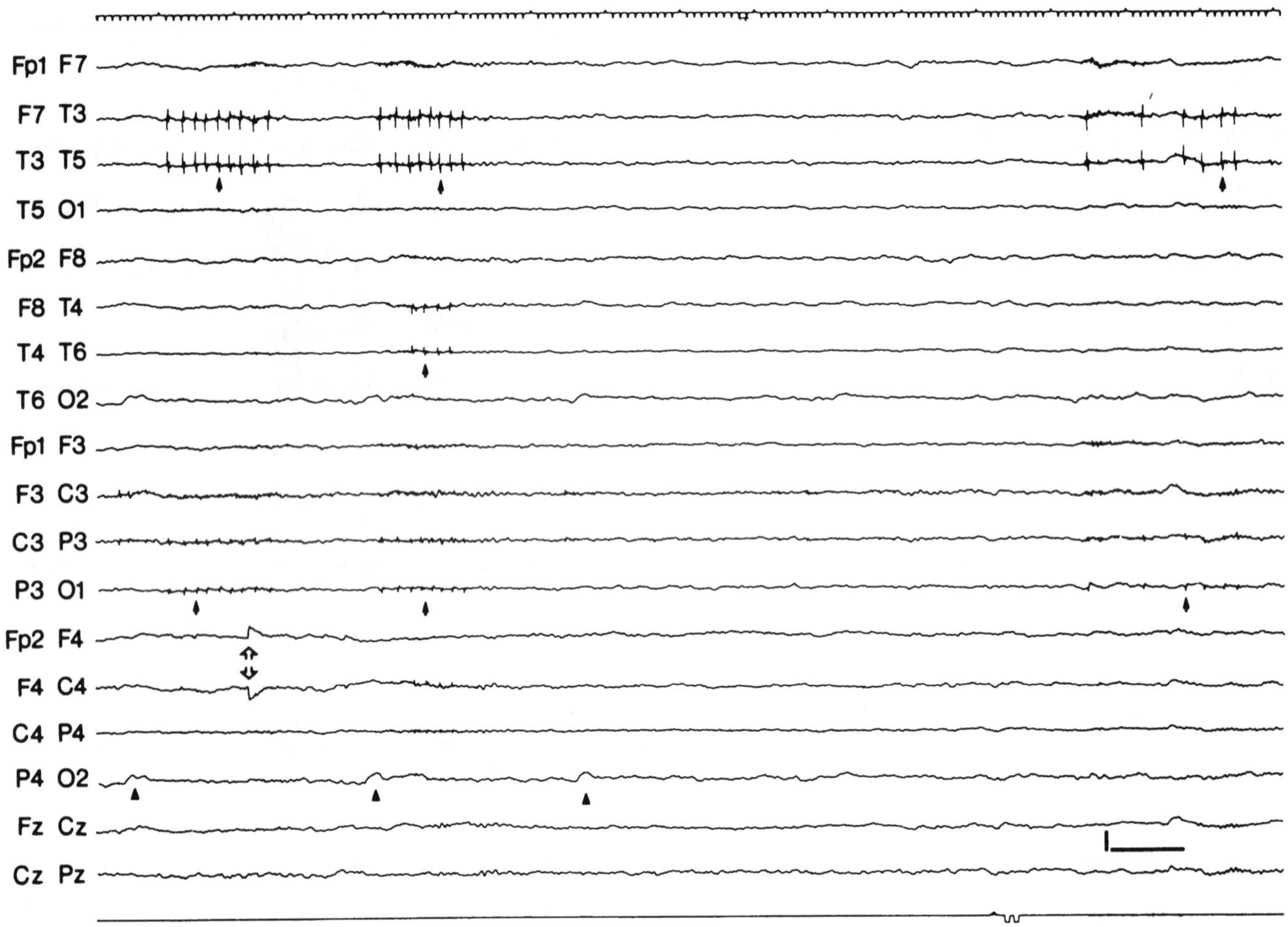

Figure A6.5. Muscle-spike artifacts (solid arrows), respiration artifacts (triangles), and an electrode "pop." Muscle spikes commonly occur in clusters and are recognized by their sharp contours at the standard 70 Hz, high-frequency filter setting (see also Fig. 14.10). Respiration artifacts are best confirmed simply by observing that the rise and fall of the patient's chest coincide with the artifacts. Electrode "pops" are always mirror-image deflections in adjacent channels of a bipolar montage. The "pop" originates in the electrode that is common to the two channels (F4 in the tracing). Filters: low frequency = 1 Hz, high frequency = 70 Hz. Calibrations: horizontal = 1 second, vertical = 50 μV.

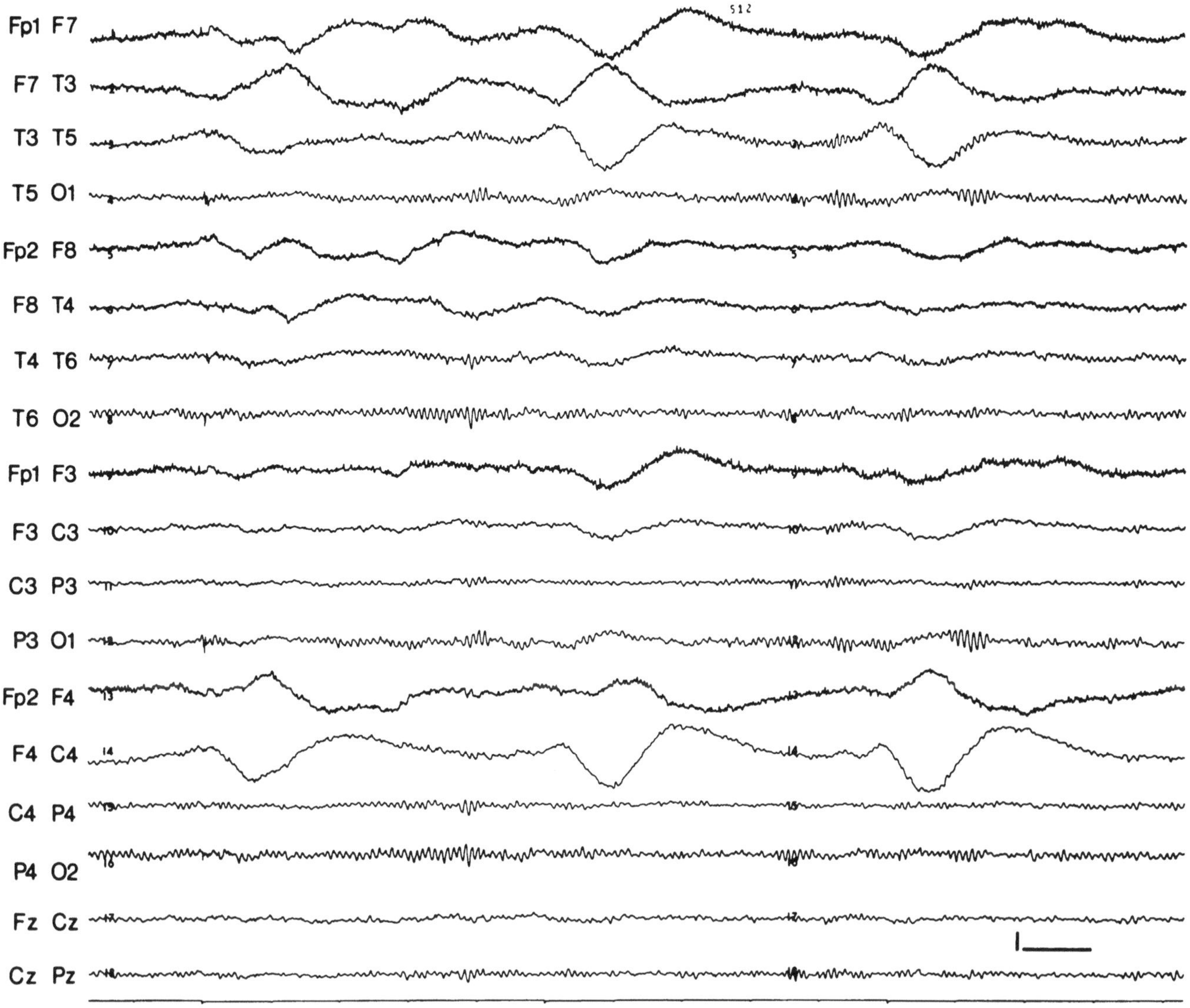

Figure A6.6. Sweating artifact. These artifacts are recognized by the slow, pendulous character of the waves. As in the present tracing, they usually occur in the frontal and temporal regions. Sweating artifacts may be reduced or sometimes even eliminated by switching the low-frequency filter setting from 1 to 5 Hz; but of course, this will reduce low-frequency cortical activity as well. Filters: low frequency = 1 Hz, high frequency = 70 Hz. Calibrations: horizontal = 1 second, vertical = 50 μV.

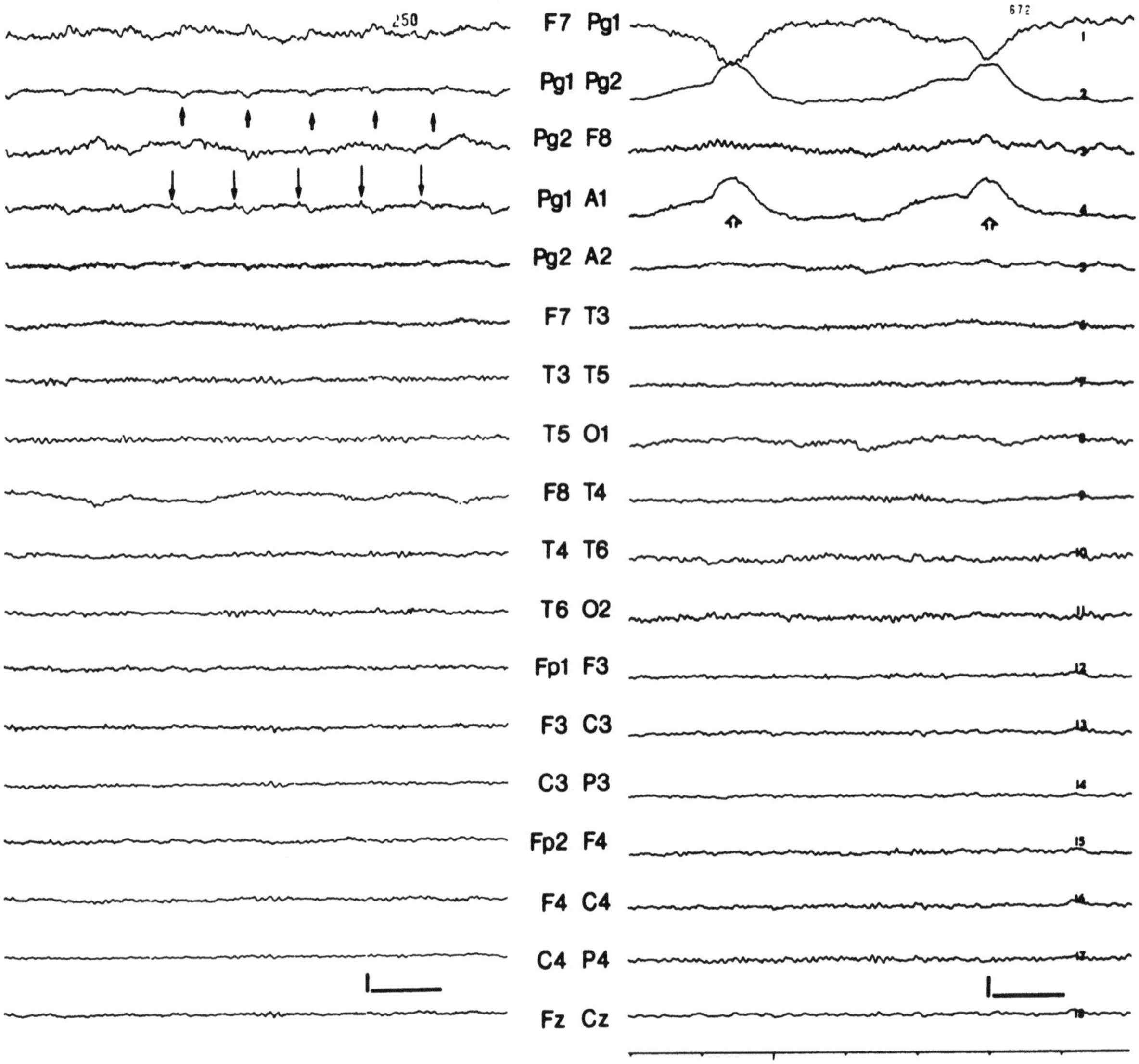

Figure A6.7. Pulse artifacts (left) and respiration artifacts (right) in nasopharyngeal electrodes. Such artifacts are common and difficult to get rid of. Filters: low frequency = 1 Hz, high frequency = 70 Hz. Calibrations: horizontal = 1 second, vertical = 50 µV.

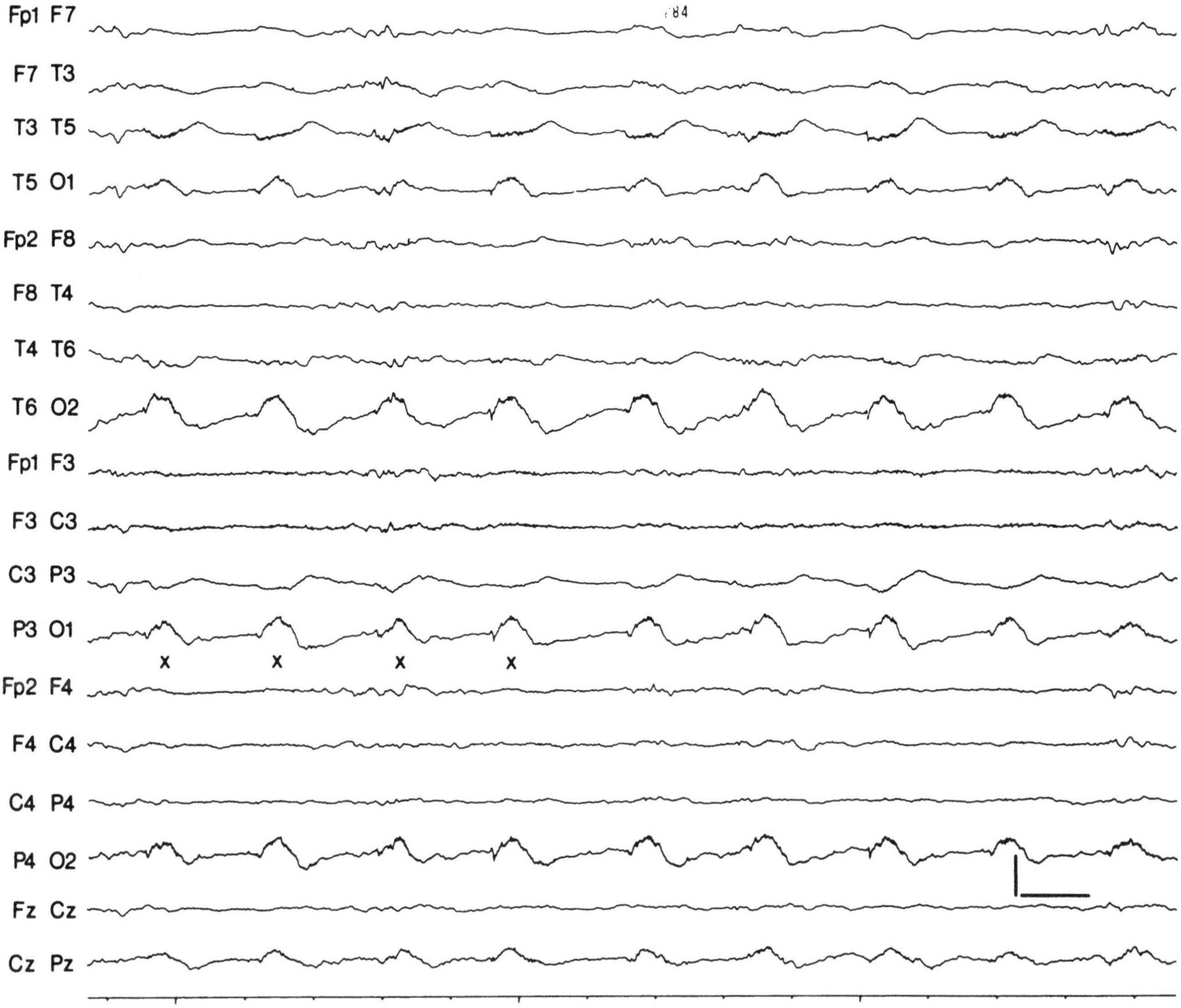

Figure A6.8. Ventilator artifact. These artifacts, which are marked by "X's," are among the artifacts most commonly seen in EEGs taken in the intensive care unit. Filters: low frequency = 1 Hz, high frequency = 70 Hz. Calibrations: horizontal = 1 second, vertical = 50 μV.

Appendix 7
EEG Recording in Patients with Infectious Diseases

Although unlikely, a potential exists for the transmission of infectious diseases in the EEG laboratory. Transmission of disease may occur from a patient to the technician and also from one patient to another. As in any other laboratory or workplace, airborne or contact transmission of bacterial and viral infections are possible. Inadequate cleaning of the laboratory, the equipment, or the electrodes; poor ventilation systems; or inappropriate handling of infected patients by technicians may all contribute to the spread of infection. For this reason, careful attention should be paid to general principles of hygiene and infection control.

The three major diseases where risk for infections exists are Jakob-Creutzfeldt's disease (JCD), viral hepatitis B, and acquired immune deficiency syndrome (AIDS). Viral hepatitis B and AIDS are known to be spread by contaminated blood or serum. In the case of JCD, such possibility also exists; but the primary mode of transmission has been through infected brain tissue. In this appendix we outline some of the more important procedures to follow to avoid transmission of these diseases in the EEG laboratory. For a detailed treatment of the topic, refer to the Report of the Committee on Infectious Disease of the American EEG Society (Report of the Committee on Infectious Diseases, 1986), and the Special Report on Infection Control for Patients with AIDS (Special Report on Infection Control for Patients with AIDS, 1983).

Precautions with Known Cases of JCD, Hepatitis, or AIDS

1. The technician should wear gloves, mask, and apron.
2. Needle electrodes should not be used.
3. Avoid using nasopharyngeal electrodes. If they must be used, disinfect them as described below.
4. Avoid abrading the skin. If electrode impedances cannot be brought down to sufficiently low values, the blunt tip of a needle may be used. After using the needle, make sure that it is bagged in a special container, autoclaved, and discarded.
5. After removing the electrodes from the patient, they should be placed in a disinfectant solution before the electrode paste is cleaned off. Soaking in Dakin's solution[1] for several hours has been recommended (Report of the Committee on Infectious Diseases, 1986). Thereupon, the electrodes are cleaned of electrode paste using tap water (the technician should wear gloves while doing this), bagged in a container, and autoclaved. Steam autoclaving at 132°C for one hour inactivates the JCD virus (Committee on Health Care Issues, ANA, 1986), as well as the hepatitis B virus and AIDS virus (Report of the Committee on Infectious Diseases, 1986).
6. Alternatively, a solely chemical method of disinfection may be employed. For this purpose, use of a 5% solution of sodium hypochlorite (full strength household bleach) has been suggested (Brown P, Gibbs Jr. CJ, Amyx HL et al, 1982; Bond WW, Peterson NJ, Favero MS, 1977; Gajdusek DC, Gibbs CJ, Ashor DM et al, 1977). Unfortunately, such a high concentration of sodium hypochlorite is corrosive to metal electrodes and reduces their useful life. As noted below, such high concentrations are not needed to inactivate the AIDS virus.

General Precautions Relevant to AIDS

The precautions outlined in the previous section are appropriate when a patient is known to have AIDS. More

[1] Dakin's solution is a dilute solution of household bleach with sodium bicarbonate added. Consult *The Merck Index* for formulation.

often, however, there is risk from patients in whom the diagnosis has not yet been made. This risk is greatest in places where AIDS is highly prevalent. Recently, the Centers for Disease Control declared that blood and other body fluids from all patients should be considered infective (Centers for Disease Control, 1987). With this in mind, a prudent policy for the EEG technician to follow is to wear gloves during electrode application and removal, and electrode cleaning. Care should be taken to avoid breaking the skin at the electrode sites by too vigorous scrubbing. After the electrodes are removed, they should be soaked in a disinfectant. Soaking in a 2% solution of glutaraldehyde for 30 to 60 minutes has been recommended (Report of the Committee on Infectious Diseases, 1986). A 10% solution of ordinary household bleach is used in some laboratories. But this would seem to be an excessively high concentration, as a recent study reported that the AIDS virus is inactivated by treating for 10 minutes with a 0.1% solution of household bleach at room temperature (Martin LS, McDougal JS and Loskoski SL, 1985). Electrodes should be rinsed in tap water to remove the disinfectant before they are used on a patient.

References

Bond WW, Peterson NJ, Favero MS: Viral hepatitis B: aspects of environmental control. *Health Lab Sci* 1977;14:235–252.

Brown P, Gibbs, Jr, CJ, Amyx HL, et al: Chemical disinfection of Creutzfeldt-Jakob disease virus. *N Engl J Med* 1982;306: 1279–1282.

Centers for Disease Control. Recommendations for prevention of HIV transmission in health care settings. MMWR 1987;36/ 2S:8S–18S.

Committee on Health Care Issues, ANA. Precautions in handling tissues, fluids, and other contaminated materials from patients with documented or suspected CJD. *Ann Neurol* 1986;10: 75–77.

Gajdusek DC, Gibbs CJ, Asher DM, et al: Precautions in medical care of, and in handling materials from patients with transmissible virus dementia (Creutzfeldt-Jakob disease). *N Engl J Med* 1977;297:1253–1258.

Martin LS, McDougal JS, Loskoski SL: Disinfection and inactivation of human T lymphotropic virus Type III/lymphadenopathy-associated virus. *J Infect Dis* 1985;152:400–403.

Report of the Committee on Infectious Diseases. *J Clin Neurophysiol* 1986;3(suppl 1):38–42.

Special Report on Infection Control for Patients with AIDS. *N Engl J Med* 1983;309:740–744.

Index